Edited by
Arturo M. Baró and Ronald G. Reifenberger

Atomic Force Microscopy in Liquid

Related Titles

Bowker, M., Davies, P. R. (eds.)

Scanning Tunneling Microscopy in Surface Science

2010

ISBN: 978-3-527-31982-4

Haugstad, G.

Understanding Atomic Force Microscopy

Basic Modes for Advanced Applications

2012

ISBN: 978-0-470-63882-8

García, R.

Amplitude Modulation Atomic Force Microscopy

2010

ISBN: 978-3-527-40834-4

Fuchs, H. (ed.)

Nanotechnology

Volume 6: Nanoprobes

2009

ISBN: 978-3-527-31733-2

van Tendeloo, G., van Dyck, D., Pennycook, S. J. (eds.)

Handbook of Nanoscopy

2012

ISBN: 978-3-527-31706-6

Edited by Arturo M. Baró and Ronald G. Reifenberger

Atomic Force Microscopy in Liquid

Biological Applications

WILEY-VCH Verlag GmbH & Co. KGaA

The Editors

Prof. Arturo M. Baró
Instituto de Ciencia de Materials
de Madrid (CSIC)
Sor Inés de la Cruz
Madrid 28049
Spain

Prof. Ronald G. Reifenberger
Purdue University
Department of Physics
525, Northwestern Avenue
West Lafayette, IN 47907-2036
USA

Library of Congress Card No.: applied for

British Library Cataloguing-in-Publication Data
A catalogue record for this book is available from the British Library.

Bibliographic information published by the Deutsche Nationalbibliothek
The Deutsche Nationalbibliothek lists this publication in the Deutsche Nationalbibliografie; detailed bibliographic data are available on the Internet at <http://dnb.d-nb.de>.

Cover Design Grafik-Design Schulz, Fußgönheim
Typesetting Laserwords Private Limited, Chennai, India
Printing and Binding Markono Print Media Pte Ltd, Singapore

Printed in Singapore
Printed on acid-free paper

Print ISBN: 978-3-527-32758-4
ePDF ISBN: 978-3-527-64983-9
ePub ISBN: 978-3-527-64982-2
mobi ISBN: 978-3-527-64981-5
oBook ISBN: 978-3-527-64980-8

Contents

Preface

The atomic force microscope (AFM) is a member of the broad family of scanning probe microscopes and arguably has become the most widely used scanning probe instrument in the world. The high image resolution coupled with a new class of spectroscopic tools has enabled AFMs to perform real-time dynamic studies as well as controlled nanomanipulation, resulting in significant breakthroughs in many different realms of science and engineering.

The book *Atomic Force Microscopy in Liquid: Biological Applications* broadly focuses on phenomena relevant to AFM studies at a solid–liquid interface, with an emphasis on biological applications ranging from small biomolecules to living cells. As far as we know, there are no books closely related to this one. The ability of an AFM to study samples in a liquid environment provides a significant advantage when compared to other microscopies such as SEM and TEM. This unique capability allows measurements of native biological samples in aqueous environments under physiologically relevant conditions. The weakness of van der Waals interactions and the absence of capillary forces in liquid drastically reduce the tip–sample interaction, resulting in little damage to soft biological samples. The highly local character of AFM that directly results from probe proximity to the sample coupled with tip sharpness not only allows high-resolution images but also permits laterally resolved spectroscopic measurements capable of reaching the level of a single molecule, as, for example, in single molecule force spectroscopy applications.

This book provides a thorough description of AFM operation in liquid environments and will serve as a useful reference for all AFM groups. It is organized into two sections in an effort to be especially useful for new researchers who desire to start bio-related studies. The first part of the book is focused on the study of features unique to AFM including instrumentation, force spectroscopic analysis, general imaging and spectroscopy considerations, single molecule force spectroscopy, operational modes, electrostatic forces in liquids containing ions, high-speed imaging, nanomanipulation, and lithography. Historically, there have been a number of AFM studies on biological systems in liquid by contact-mode AFM. Recently, dynamic force spectroscopy (DFS) experiments have appeared that utilize noncontact imaging, further reducing sample damage. Therefore, DFS is discussed in two chapters, one connected with experimental work and a second that deals with the theory of dynamic AFM. We also include a chapter on the combination of AFM with more traditional optical techniques such as fluorescence. A growing trend will be the simultaneous utilization of one or more auxiliary techniques with AFM to exploit the many advantages when complementary techniques are combined.

The second part of the book deals with applications of AFM to the study of biological materials ranging from the smallest biomolecules (phospholipids, proteins, DNA, RNA, and protein complexes) to subcellular structures (e.g., membranes), and finally culminating with studies of living microbial cells. Single cell force spectroscopy and the manipulation of biological material with an AFM are also included. The goal is to feature recent advances that emphasize *in vivo* experiments.

This book is timely and up-to-date. It is aimed at a mixed audience that includes starting graduate students, young researchers, and established scientists. Physicists need to learn how to handle/prepare biological samples; biologists need to understand the important issues related to imaging of complex samples in liquid. We hope the book is useful, especially for those who enjoy breaking ground in a new and interdisciplinary field.

We would like to thank all the distinguished scientists and their coauthors for their timely and well-referenced contributions. Grateful acknowledgments are offered to the Wiley-VCH editorial staff, in particular Lesley Belfit, Project Editor, and Publisher Dr. Gudrun Walter.

UAM, Madrid — *Arturo M. Baró*
Purdue University — *Ronald G. Reifenberger*

List of Contributors

Rehana Afrin
Tokyo Institute of Technology
Innovation Laboratory
4259 Nagatsuta
Yokohama 226-8501
Japan

Toshio Ando
Kanazawa University
Department of Physics
Kakuma-machi
Kanazawa 920-1192
Japan

Arturo M. Baró
Instituto de Ciencia de Materiales
de Madrid (CSIC)
Sor Inós de la Cruz
28049 Madrid
Spain

Mariano Carrión-Vázquez
Instituto Cajal
Consejo Superior de
Investigaciones
Científicas, IMDEA
Nanociencia and Centro de
Investigación Biomédica en Red
sobre Enfermedades
Neurodegenerativas (CIBERNED)
Avenida Doctor Arce 37
28002 Madrid
Spain

Edward D. de Asis Jr.
Santa Clara University
Departments of Electrical
Engineering and Bioengineering
School of Engineering
500 El camin Real
Santa Clara
CA 95053
USA

Yves F. Dufrêne
Université catholique de Louvain
Institute of Condensed Matter
and Nanosciences
Croix du Sud 2/18
1348 Louvain-la-Neuve
Belgium

Clemens M. Franz
Karlsruhe Institute of Technology
DFG-Center for Functional
Nanostructures
Wolfgang-Gaede-Str. 1a
76131 Karlsruhe
Germany

Takeshi Fukuma
Kanazawa University
Frontier Science Organization
Kakuma-machi
Kanazawa 920-1192
Japan

Albert Galera-Prat
Instituto Cajal
Consejo Superior de
Investigaciones Científicas,
IMDEA Nanociencia and Centro
de Investigación Biomédica en
Red sobre Enfermedades
Neurodegenerativas (CIBERNED)
Avenida Doctor Arce 37
28002 Madrid
Spain

Marina Inés Giannotti
CIBER de Bioingeniería
Biomateriales y Nanomedicina
(CIBER-BBN)
Campus Rìo Ebro, Edificio I+D,
Poeta Mariano Esquillor s/n,
50018 Zaragoza
Spain

and

University of Barcelona (UB)
Physical Chemistry Department
1-3 Martí i Franquès
08028 Barcelona
Spain

and

Institute for Bioengineering of
Catalonia (IBEC)
15-21 Baldiri I Reixac
08028 Barcelona
Spain

Julio Gomez-Herrero
Universidad Autónoma de
Madrid (UAM)
Departamento de Física de la
Materia Condensada
28049 Madrid
Spain

Àngel Gómez-Sicilia
Instituto Cajal
Consejo Superior de
Investigaciones Científicas,
IMDEA Nanociencia and Centro
de Investigación Biomédica en
Red sobre Enfermedades
Neurodegenerativas (CIBERNED)
Avenida Doctor Arce 37
28002 Madrid
Spain

Jürgen J. Heinisch
Fachbereich Biologie/Chemie
Universität Osnabrück
AG Genetik
Barbarastr. 11
49076 Osnabrück
Germany

Rodolfo Hermans
University College London
London Centre for
Nanotechnology
17-19 Gordon Street
London WC1H 0AH
UK

Rubén Hervás
Instituto Cajal
Consejo Superior de
Investigaciones Científicas,
IMDEA Nanociencia and Centro
de Investigación Biomédica en
Red sobre Enfermedades
Neurodegenerativas (CIBERNED)
Avenida Doctor Arce 37
28002 Madrid
Spain

Michael J. Higgins
University of Wollongong
AIIM Facility
ARC Centre of Excellence for
Electromaterials Science
Intelligent Polymer Research
Institute
New South Wares 2522
Australia

Atsushi Ikai
Tokyo Institute of Technology
Innovation Laboratory
4259 Nagatsuta
Yokohama 226-8501
Japan

Daniel Kiracofe
Purdue University
School of Mechanical
Engineering and the Birck
Nanotechnology Center
1205 W. State Street
West Lafayette
IN 47906
USA

Noriyuki Kodera
Department of Physics
Kanazawa University
Kakuma-machi
Kanazawa 920-1192
Japan

Joseph Leung
NASA Ames Research Center
Moffett Field
CA 94035-1000
USA

Yuri L. Lyubchenko
Department of Pharmaceutical
Sciences
University of Nebraska
Medical Center
4350 Dewey Avenue
Ohama, NE 68198
USA

Shin-ichi Machida
Innovation Laboratory
Tokyo Institute of Technology
4259 Nagatsuta
Yokohama 226-8501
Japan

Gerald A. Meininger
Dalton Cardiovascular
Research Center
Department of Medical
Pharmacology and Physiology
University of Missouri-Columbia
134 Research Park Drive
Columbia
MO 65211
USA

Kenith Meissner
Texas A&M University
Department of Biomedical
Engineering
College Station
TX 77843
USA

John Melcher
Purdue University
School of Mechanical
Engineering and the Birck
Nanotechnology Center
1205 W. State Street
West Lafayette
IN 47906
USA

Fernando Moreno-Herrero
Departamento de Estructura de Macromoleculas
Centro Nacional de Biotecnología
Consejo Superior de Investigaciones Científicas (CSIC)
Darwin 3
28049 Madrid
Spain

Takahiro Watanabe-Nakayama
Tokyo Institute of Technology
Innovation Laboratory
4259 Nagatsuta
Yokohama 226-8501
Japan

Cattien V. Nguyen
Eloret Corporation
NASA Ames Research Center
M/S 229-1
Moffett Field
CA 94035-1000
USA

Arvind Raman
Purdue University
School of Mechanical Engineering and the Birck Nanotechnology Center
1205 W. State Street
West Lafayette
IN 47906
USA

Lorena Redondo-Morata
Institute for Bioengineering of Catalonia (IBEC)
15-21 Baldiri I Reixac
08028 Barcelona
Spain

and

University of Barcelona (UB)
Physical Chemistry Department
1-3 Martí i Franquès
08028 Barcelona
Spain

and

CIBER de Bioingeniería Biomateriales y Nanomedicina (CIBER-BBN)
Campus Rìo Ebro, Edificio I+D
Poeta Mariano Esquillor s/n
50018 Zaragoza
Spain

Fausto Sanz
Institute for Bioengineering of Catalonia (IBEC)
15-21 Baldiri I Reixac
08028 Barcelona
Spain

and

CIBER de Bioingeniería Biomateriales y Nanomedicina (CIBER-BBN)
50018 Zaragoza
Spain

and

University of Barcelona (UB)
Physical Chemistry Department
1-3 Martí i Franquès
08028 Barcelona
Spain

Mikihiro Shibata
Kanazawa University
Department of Physics
Kakuma-machi
Kanazawa 920-1192
Japan

Zhe Sun
University of Missouri-Columbia
Dalton Cardiovascular Research Center
134 Research Park Drive
Columbia
MO 65211
USA

Anna Taubenberger
Queensland University of Technology
Institute of Health and Biomedical Innovation
60 Musk Avenue
Kelvin Grove
QLD 4059
Australia

Andreea Trache
Texas A&M University
Department of Systems Biology & Translational Medicine
Texas A&M Health Science Center
and Department of Biomedical Engineering
336 Reynolds Medical Bldg
College Station
TX 77843
USA

Takayuki Uchihashi
Kanazawa University
Department of Physics
Kakuma-machi
Kanazawa 920-1192
Japan

Daisuke Yamamoto
Kanazawa University
Department of Physics
Kakuma-machi
Kanazawa 920-1192
Japan

Hayato Yamashita
Kanazawa University
Department of Physics
Kakuma-machi
Kanazawa 920-1192
Japan

Part I
General Atomic Force Microscopy

Atomic Force Microscopy in Liquid: Biological Applications, First Edition.
Edited by Arturo M. Baró and Ronald G. Reifenberger.

1
AFM: Basic Concepts

Fernando Moreno-Herrero and Julio Gomez-Herrero

1.1
Atomic Force Microscope: Principles

A conceptually new family of microscopes emerged after the invention of the scanning tunneling microscope (STM) by Binnig and Rohrer in 1982 [1]. This family of instruments called scanning probe microscopes (SPMs) is based on the strong distance-dependent interaction between a sharp probe or tip and a sample. The atomic force microscope therefore uses the force existing between the probe and the sample to build an image of an object [2, 3]. AFMs can operate in almost any environment including aqueous solution, and that opened myriad uses in biology [4, 5]. When thinking about how an AFM works, all notions of conventional microscope design need to be disregarded, since there are no lenses through which the operator looks at the sample. In AFM, images are obtained by sensing with the probe rather than by seeing.

The central part of an AFM is therefore the tip that literally feels the sample. A nanometer-sharp AFM tip made by microfabricating technology is grown at the free end of a flexible cantilever that is used as the transductor of the interaction between the tip and sample. The reflection of a laser beam focused at the back side of the cantilever is frequently used by most AFMs to amplify and measure the movement of the cantilever, although other detection methods may also be used (Section 1.4). The reflected beam is directed to a photodiode that provides a voltage depending on the position of the laser beam. For imaging, the tip is scanned over the sample, or as in some designs, it is the sample that moves with respect to the fixed tip, which is only allowed to move in the vertical direction. In both cases, the fine movements of the tip and sample are provided by piezoelectric materials that can move with subnanometer precision. At each position, the cantilever deflection is measured, from which a topography map can be constructed. This scanning technique in which the tip is brought into mechanical contact with the sample surface is known as *contact mode*, and it was first described by Binnig and coworkers [2]. Both the tip and scanner are the key features in any AFM setup.

Atomic Force Microscopy in Liquid: Biological Applications, First Edition.
Edited by Arturo M. Baró and Ronald G. Reifenberger.

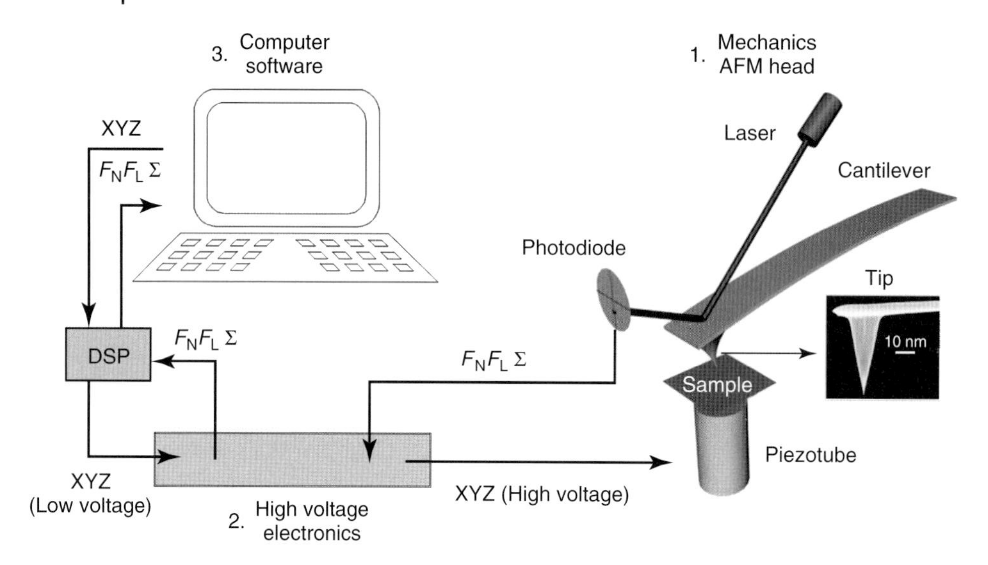

Figure 1.1 Components of a standard atomic force microscope. 1. The AFM head and the piezoelectric stage. The cantilever and its detection system as well as the sample movement are the main parts of this component. From the photodiode, the signals related to the normal and lateral forces (F_N and F_L) as well as the total intensity of light (Σ) are obtained and transferred to the high voltage electronics. 2. The high voltage electronics. This component amplifies voltages from the digital signal processor to perform the movement of the piezotube (XYZ voltages). It also collects signals from the photodiode (F_N, F_L, and Σ) and transfers them to the DSP. 3. The computer, DSP, and software that controls the AFM setup.

One of the most common models of AFM is schematically depicted in Figure 1.1. In this model, the sample is scanned over the tip, but the opposite is also possible. The latter are the so-called stand-alone AFMs that are commonly used combined with an inverted optical microscope to image biological samples in liquid. In either case, a standard AFM setup consists of three main components. (i) *the AFM head and base stage*: The AFM head contains the tip holder, the laser, and the photodiode. It also includes positioning mechanisms for focusing the laser beam on the back side of the cantilever and photodiode and small electronics for processing the signals coming from the photodiode. From this, the vertical (F_N) and lateral (F_L) deflections of the laser beam and its total intensity (Σ) are obtained. The AFM head is placed over a base stage that holds the piezoelectric scanner that moves the sample, and also a coarse, micrometer-ranged approaching mechanism, usually based on step motors.[1] (ii) *The high voltage (HV) electronics*: It amplifies the signals coming from the digital signal processor (DSP) (XYZ low voltage) to drive the piezoelectric scanner with voltages of about 100 V (XYZ HV). The electronics also transfers the analog voltage signals from the photodiode (F_N, F_L, Σ)

1) On a stand-alone AFM, the tip is scanned over the sample and the approaching mechanisms (step motors) are placed on the AFM head rather than on the mechanical stage.

to the DSP. The HV electronics must be able to amplify small signals from the computer (of some volts) to hundreds of volts needed to move the piezoelectric tube over micrometer distances. It is therefore essential that this amplification does not introduce electrical noises that may affect the resolution of the AFM. (iii) *The DSP, the computer, and the software*: The DSP performs all the signal processing and calculations involved in the real-time operation of the AFM. The DSP is mainly located in a board plugged in the computer. It contains the chips to perform the translation from digital to analog signals (digital to analog converters (DACs)), which are further managed by the HV electronics. Analog signals from the HV amplifier are converted to digital signals also at the DSP board using analog to digital converters (ADCs). Finally, all computer-based systems need software to run the setup. Nearly all AFMs in the market come with purpose-made acquisition software. Raw images can later be processed with any of the many imaging-processing freeware available in the Internet.

Operating the AFM in liquid conditions requires modifications of some parts to prevent wetting of electrical components such as piezoelectric ceramics. For instance, the sample holder must be large enough to accommodate the sample and the buffer in which it is immersed. Some authors simply use a small droplet of some tens of microliters, which covers the sample; others use a small container filled with several milliliters of buffer. The first approach has the advantage of a smaller mass (droplet) added to the piezo, but experiments suffer from evaporation, which results in a change in the concentration of solvents. On the other hand, using the container approach, concentrations are kept roughly constant, but a considerable mass must be moved by the piezo scanner, which reduces its resonance frequency and therefore the range of imaging speeds. The tip holder, also known as *liquid cell* (Section 1.4.2.1), must be designed to prevent contact between the liquid and the small piezo that drives movement of the cantilever. Finally, in some AFMs, the piezo tube is protected and covered to prevent wetting in case of liquid spill (Section 1.2.1).

1.2 Piezoelectric Scanners

Piezoelectric ceramic transducers are used to accurately position the tip and sample in AFMs. The direct piezoelectric effect consists of the generation of a potential difference across the opposite faces of certain nonconductive crystals as a result of the application of stress. The reverse piezoelectric effect is also possible because of a change in dimensions of the crystal as a consequence of the application of a potential difference between two faces of the piezoelectric material. This method is used to position the tip and sample with respect to each other with subnanometer precision since piezoelectric ceramic transducers are highly sensitive, stable, and reliable. Regardless, if the tip is moved over the sample or vice versa, the scan in AFM is performed using piezoelectric transducers.

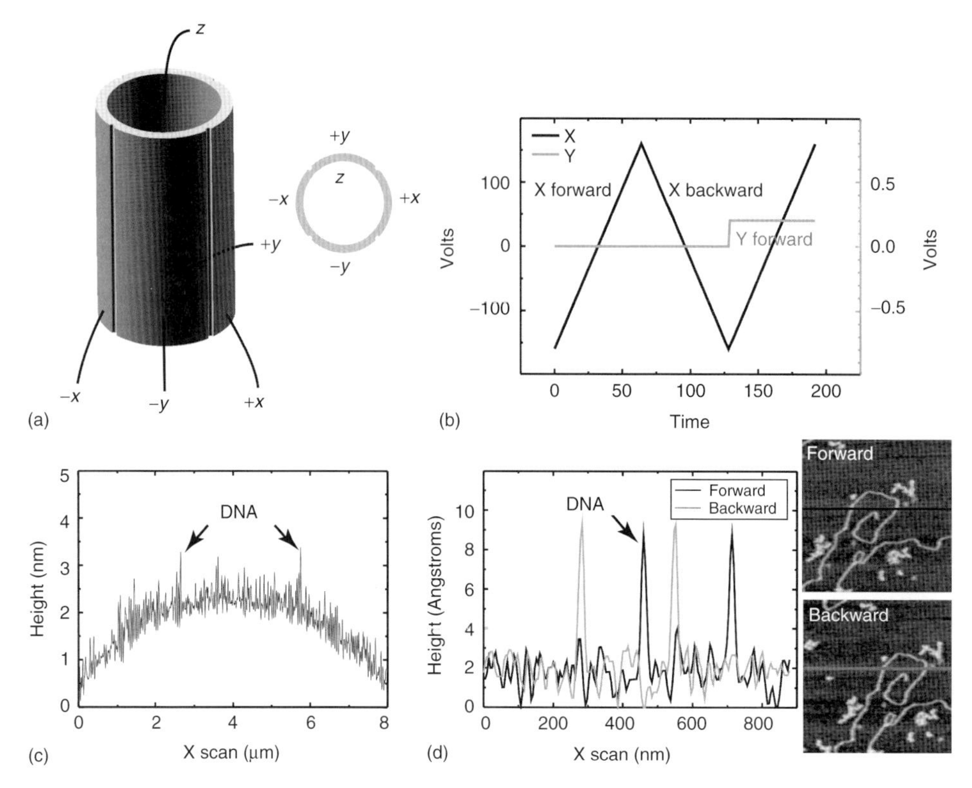

Figure 1.2 Piezoelectric scanners. (a) Piezotube architecture based on four sectors. Voltages are applied between opposite sides, and as a consequence, movement of the piezo is generated. (b) Sequence of voltages applied to X (fast scan) and Y (slow scan) to generate an image scan. Each step in y-axis is associated with the change of an imaging line. (c,d) Two typical problems of piezotube scanners. (c) The plane where the sample is situated describes an arc rather than a straight line. (d) Clear effect of piezo hysteresis when imaging DNA molecules.

Many AFMs use piezoelectric tube scanners such as the one shown in Figure 1.2a. They consist of a thin-walled hard piezoelectric ceramic, which is polarized radially [6]. The external face of the tube is divided into four longitudinal segments of equal size and electrodes are welded to the internal and external faces of the tube. To achieve extension or contraction, a bias voltage is applied between the inner and all the outer electrodes. The scan movement is performed by applying a bias voltage to one of the segments of the outer wall. To amplify this bending effect by a factor of two, a voltage with opposite sign is applied to the opposite segment. With a correct synchronization of applied voltages ($\pm x$ and $\pm y$), a sequential scan can be generated (Figure 1.2b). Typically, tube scanners of $10 \times 10\,\mu$m range have sensitivities of ~40 nm/V. This means that voltages up to ±125 V must be generated with submilivolts precision to achieve a basal noise of ~0.01 nm. This operation is performed by the HV electronics.

Tube scanners have some drawbacks. For instance, the plane in which the sample is located describes an arc rather than a straight line. This effect is more pronounced when large areas are scanned and the height of the objects to be imaged is small compared to the scan area. For instance, Figure 1.2c shows a profile of a flat surface with adsorbed DNA molecules. The effect of the arc trajectory described by the piezo is clearly visible, and the detection of small molecules such as DNA (height < 1 nm) is challenging. This effect can easily be corrected by subtracting a polynomial function to each scan line or to the overall surface. Piezo tubes have relatively low resonance frequencies, of the order of kilohertz, which limits the scan speed. Recently, some manufacturers have employed small stacks of piezoelectric ceramics to increase the resonance frequency of these devices and therefore the imaging speed. However, stacked piezos have a quite limited scan range.

A different approach used to move the sample is based on stick-slip motion. These positioners rely on the controllable use of the inertia of a sliding block. In brief, a sliding block slips along a guided rod, which is otherwise clamped in frictional engagement. A net step is obtained by first accelerating very rapidly the guiding rod over a short period (typically microseconds) so that the inertia of the sliding block overcomes the friction. The sliding block disengages from the accelerated rod and remains nearly nondisplaced. Then, the guiding rod moves back to its initial position slowly enough so that the sliding block sticks to it and thus makes a net step. Periodic repetition of this sequence leads to a step-by-step motion of the sliding block in one direction. The movement of the guiding rod is performed by a piezoelectric ceramic, which can pull or push as required. Stick-slip positioners have long travel ranges of several millimeters, but their performance is dependent on the mass to be moved, which can be significant in liquid imaging. These devices also have the limitation of a relatively large step (few nanometers) and a low resonance frequency (much lower than that of stacked piezos). Hence the main use of these devices is as nanopositioners rather than as fast scanners.

Piezoelectric scanners are inherently nonlinear, and this nonlinearity becomes quite significant at large scans. Typical piezos suffer from hysteresis in the forward and backward traces. This effect can be clearly seen in the forward and backward profiles shown in Figure 1.2d. Piezo scanners are also subjected to creep after changing the polarity (direction) of the scan or just after setting the voltage to zero. This is due to some sort of relaxation, which occurs under constant stress. It has an effect on the images distorting the dimensions of the objects to be imaged. In general, this problem can be solved by repeating the scan, allowing the piezo to relax. To minimize unwanted motions in the piezoactuator, some AFMs incorporate a combination of piezoactuators and metal springs. These devices have flexure-guided stages, acting as springs and restricted to move only in one direction. A piezoactuator pulls against the spring, and therefore a forward and backward movement of the flexure guide can be achieved by changing the voltage in the piezoactuator. This combination effectively decouples the unwanted motions in the piezoactuator and produces a pure linear translation while keeping high resonance frequencies at relatively high loads ($\sim$2 kHz for a 100 g load). Finally, many AFMs have capacitive sensors incorporated in their piezos that allow for measuring the position independent of the applied voltage. With this feature, a closed-loop circuit can be designed, being able to cancel any hysteresis, creep, or nonlinearity by applying additional correction voltages.

1.2.1 Piezoelectric Scanners for Imaging in Liquids

In many AFMs, the piezoelectric used to image in air is the same as that used to image in liquids, but some precautions must be taken. The main concern is related to the electrical isolation of the piezo to avoid any shortcut due to wetting. HVs (hundreds of volts) are applied to the piezo, and if any water gets into it, the expensive piezo tube will almost certainly be destroyed. Therefore, some caution must be taken when imaging in liquids to avoid any spill of water into the piezo. In most AFMs, some silicone or rubber is added to prevent any liquid from getting into the piezo.

For imaging in liquids, it is often recommended to move the tip relative to the sample instead of keeping the tip fixed and move the sample. In the latter scenario, volumes of milliliters should be moved at kilohertz frequencies, affecting the mechanical stability of the piezo. This is equivalent to considering an effective mass in Eq. (1.3). That will lower the piezo resonant frequency and will slow down the imaging speed of the AFM. Instead, when moving the tip relative to the sample in liquids, the added effective mass is small because only the tip and parts of the tip holder are immersed in the buffer container. When imaging in environments with large viscosity such as liquids, it is important to keep the mass of moving objects as low as possible. This is also important for oscillating the tip; a need in dynamic modes. Some users oscillate the complete tip holder, exciting many mechanical vibrations in the buffer container, which hide the genuine mechanical resonance of the cantilever (Section 1.5.3.1).

1.3 Tips and Cantilevers

In contact mode, to be able to feel the surface with atom resolution, the stiffness of an AFM cantilever should be much smaller than the spring constant that maintains the atoms confined together on the surface. This bonding force constant in a crystalline lattice is of the order of $1\ \mathrm{N\ m^{-1}}$ [7], meaning that to use the AFM in contact mode (Section 1.5.1), the spring constant of the cantilevers (k) should be much smaller than $1\ \mathrm{N\ m^{-1}}$. To achieve this value of k, a beamlike cantilever made of silicon or silicon nitride should have micrometer dimensions if one considers the formula for the spring constant of a cantilever

$$k = \frac{Et^3 w}{4l^3} \tag{1.1}$$

where E is the Young's modulus of the material (i.e., for silicon nitride $E = 1.5 \times 10^{11}$ N m^{-2}) and t, w, and l are the thickness, width, and length of the cantilever, respectively. For example, a silicon nitride cantilever of dimensions $(t, w, l) = (0.3, 10, 100)$ μm will yield a value of k of 0.01 N m^{-1}, well below the spring constant of the atoms in a crystalline lattice. In principle, one could think that it may be advantageous for imaging soft materials such as proteins to fabricate cantilevers with an arbitrarily low constant. However, there is a fundamental limitation for lowering k. A cantilever in thermodynamic equilibrium with a thermal bath at temperature T has a thermal energy k_BT, k_B being the Boltzmann's constant. Considering the cantilever as a system with just one degree of freedom (it can move only up and down), the thermal energy increases the elastic energy stored in the cantilever as

$$\frac{1}{2}kA^2 = \frac{1}{2}k_B T;\ A = \sqrt{\frac{k_B T}{k}} \tag{1.2}$$

where A is the oscillation amplitude of the cantilever. For a cantilever with $k = 0.1$ N m^{-1}, the thermal noise amplitude is $A \sim 0.2$ nm, a value similar to the atomic corrugation of a surface.

$$f = \frac{1}{2\pi}\sqrt{\frac{k}{m}} = \frac{1}{4\pi}\frac{t}{l^2}\sqrt{\frac{E}{\rho}} \tag{1.3}$$

Equation (1.3) combines the resonance frequency of a harmonic oscillator with the stiffness values of a cantilever. It should be noted that Eq. (1.3) is only exact for a point mass particle, but it is generally considered a good approximation for a continuum mass such as a cantilever. A cantilever of the abovementioned dimensions will have a mass of ~1 ng (considering a density of 3.44 g cm^{-3} and not taking into account the mass of the tip) and a resonant frequency of approximately few kilohertz, well above the mechanical resonances of the building, for instance. The resonant frequency of the cantilever has important implications related to the imaging speed of the AFM. Let us assume a surface with a corrugation that can be approximated to a sinusoidal function with a wavelength of 2 nm. This means that the cantilever should move up and down at a frequency of ~5 kHz when imaging a surface of size 10 μm × 10 μm at a speed of 1 line per second (a typical value for AFM). In other words, a wavelength translates to a time period when the cantilever scans the surface at a given speed. In order to respond to such a corrugation, the resonance frequency of the cantilever should be well above the frequency associated with the corrugation. Therefore, it turns out that the use of cantilevers with high resonant frequencies and low stiffness is highly advantageous for fast imaging of soft materials.

The offer provided by several manufacturers of different tips, cantilevers, and materials is very extensive. This large offer has opened the AFM field to multiple applications, such as noncontact AFM (NC-AFM) and dynamic modes, conductive AFM, electrostatic AFM, and so on, and has also given the possibility to image in different environments. The criterion to choose the most appropriate tip and cantilever depends on the application. In general, soft cantilevers are used in contact mode to avoid damage to the sample. On the other hand, stiff cantilevers

Table 1.1 Standard properties of cantilever used for each imaging mode and for liquids and air.[a]

Imaging mode (environment)	K (N m^{-1})	f (kHz) (in air)	Notes
Contact mode (*air* and *liquids*)	<1	10–30	Major requirement is to use soft cantilevers V-shaped cantilevers are preferred to minimize lateral bending Rectangular cantilever are used to measure friction
Jumping/pulsed mode (*air*)	1.5–3	25–70	Requires relatively large K values to overcome capillary and adhesion forces
Jumping/pulsed mode (*liquids*)	<0.1	20	Absence of capillary and low adhesion forces enable use of softer cantilevers than in air
Dynamic modes (*air*)	15–60	130–350	Stiff lever required to give a high Q factor and to overcome capillary adhesion between tip–surface if working in air Moderate amplitude oscillation (>5 nm)
Dynamic modes (*liquids*)	<0.1	30–50	Rectangular or V shape Requires a Q-value over 1 (long tips ~10 μm) Small amplitude oscillation (<5 nm)

[a]Data extracted from manufacturers: Nanosensors and Olympus. *www.nanosensors.com and http://probe.olympus-global.com/en/.*

are preferred for imaging in dynamic mode to overcome capillary forces. Ideally, cantilevers of high resonant frequency and low spring constant are preferred, but this is only possible by reducing the cantilever dimensions, which turns out to be complicated. However, recently, some manufacturers have presented new small cantilevers that are meant to meet the expectations of demanding users. Table 1.1 gives an overview of cantilever properties and their uses in the different imaging modes (see Section 1.5 for reference to imaging modes).

1.3.1
Cantilever Calibration

For some quantitative AFM applications, for instance, force–distance spectroscopy, the spring constant of the cantilever must be precisely determined since the quoted value provided by the manufacturer is only an approximation based on the dimensions of the cantilevers. In some cases, the spring constant of the cantilever can be over 20% off from the quoted value.

Equation (1.3) can be used to calculate the cantilever spring constant if its mass is known. However, the AFM cantilever beam is not a simple point mass added at the end of a spring but has its weight distributed along its length, so Eq. (1.3) is usually modified by considering an effective mass, m_0. In any case, measuring the effective mass of a cantilever is rather complicated, and Cleveland and coworkers [8] proposed a method based on measuring the changes in resonant frequency, f, as small masses, m^*, were added to the cantilever (Eq. (1.4)).

$$\omega^2 = \frac{k}{m^* + m_0} \quad \text{and} \quad f = \frac{\omega}{2\pi} \tag{1.4}$$

A plot of added mass, m^*, versus ω^{-2} has a slope equal to k, and an intercept equal to the effective cantilever mass m_0. By carefully performing the measurements described in [8], Cleveland and coworkers derived the following equation, which allows calculation of k with reasonable accuracy by just measuring the unloaded resonant frequency of the cantilever, assuming that one has accurate information on the length and the width of the cantilever.

$$k = 2w(\pi l f)^3 \sqrt{\frac{\rho^3}{E}} \tag{1.5}$$

where l is the length of the cantilever, w its width, ρ is the density of the material, E the elastic modulus or Young's modulus, and f the measured resonant frequency.

A more accurate method was proposed in 1995 by Sader *et al.* [9], and it was improved and applied to rectangular cantilevers in 1999 [10]. This method required to know the width and length of the cantilever, the experimentally measured resonant frequency and quality factor of the cantilever, and the density and viscosity of the fluid (properties of air: density $\rho_{air} = 1.18\ \text{kg m}^{-3}$ and viscosity $\eta_{air} = 1.86 \times 10^{-5}\ \text{kg m}^{-1}\ \text{s}^{-1}$). Sader's method was extended to enable simultaneous calibration of the torsional spring constant of rectangular cantilevers in 2004 [11]. The advantages of these methods are that the thickness, density, and resonant frequency in vacuum of the cantilever are not needed. In addition, they are rapid to use, simple to implement experimentally, noninvasive, and nondestructive (*http://www.ampc.ms.unimelb.edu.au/afm/calibration.html*).

1.3.2
Tips and Cantilevers for Imaging in Liquids

In solution, the charge of an object (i.e., of the tip and cantilever) is normally screened by mobile ions in the surrounding electrolyte. Coions (ions with the same sign of charge) are repelled from the surroundings of the object. Counterions (ions with opposite charge) are electrostatically attracted to the object, but this attraction diminishes their entropy. Their spatial distribution is a compromise between these two opposite tendencies. The resulting arrangement of screening charges around the object is known as the *electric double layer*, and its structure has a major impact on interactions between charged objects in solution [12]. Therefore, in contrast to air imaging, where capillary forces play a central role in the tip–surface interaction, in liquid imaging, electrostatic interactions are the most relevant ones. This means that resolution, which is closely related to tip–sample distance, depends on the arrangement and screening of charges in solution, and therefore, high resolution in buffer can only be achieved if contact takes place between the tip and sample. The fact that imaging in buffer at high resolution requires contact is a major issue when imaging soft samples because soft cantilevers are needed. Originally, soft cantilevers could only be used in contact mode because of their low resonant frequency. Imaging in contact mode implies the presence of shear and lateral forces that could damage soft samples or drag objects along the surface. This problem was partially overcome with the use of pulsed modes such as jumping mode, pulsed force mode, or force volume mode that minimized lateral forces. Recently, some cantilever manufacturers have developed cantilevers designed for imaging soft samples in buffer with low spring constants and reasonably high resonant frequencies, thus allowing the use of dynamic modes. This has increased the range of measurements and has minimized shear and lateral forces. However, fine tuning of electrolyte concentrations is still required to minimize the distance of electrostatic interactions between the tip and sample and to achieve high resolution.

Figure 1.3 exemplifies the relevance of proper tuning of the amount of monovalent and divalent ions when operating in liquids. It shows experimental data of force (cantilever deflection) as a function of piezo extension obtained in different liquids using a soft cantilever of 0.08 N m^{-1} and freshly cleaved mica as a surface. The deflection or force is zero until an attractive (negative) force pulls the cantilever toward the mica. Once the cantilever is in contact with the surface, the deflection follows the piezo movement. When using water (black curve), this attractive force is

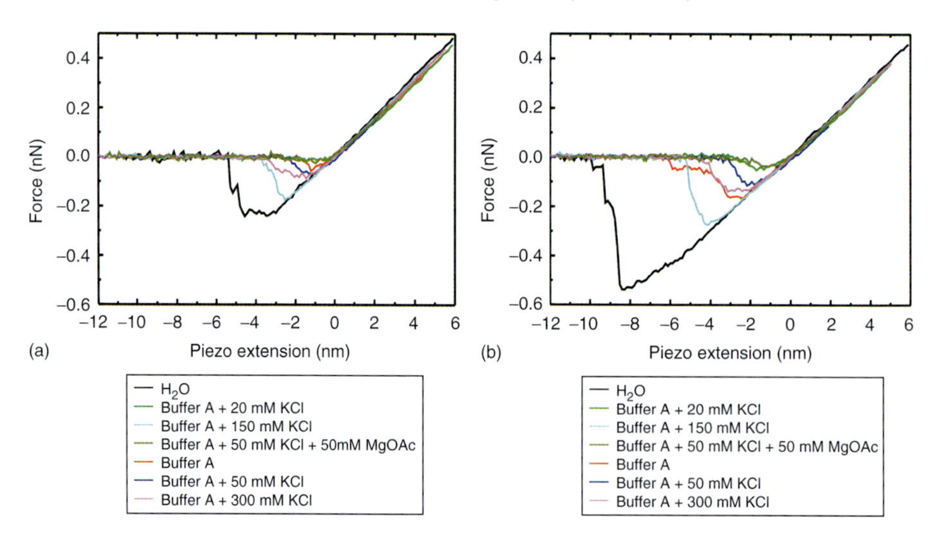

Figure 1.3 Approaching (a) and withdrawing (b) force-extension curves for a mica surface immersed in different buffers. The cantilever used had a force constant of 0.08 Nm^{-1}, and Buffer A is 25 mM tris-acetate pH (7.5) and 2 mM MgOAc supplemented with KCl or extra MgOAc when stated.

over 0.5 nN, a rather large force when imaging biological materials. Data obtained using 25 mM tris-acetate (pH 7.5), 50 mM KCl, 50 mM MgOAc (dark yellow curve) shows an effective absence of adhesion and attractive force.

1.3.3 Cantilever Dynamics in Liquids

According to Eq. (1.1), the spring constant k only depends on the material properties of the cantilever and its geometrical dimensions. This means that k is independent of the surrounding environment in which the cantilever may be immersed. However, viscosity of the surrounding media does affect the mechanical response of the cantilever, and this is important when operating in liquids. Equation (1.3) already indicates that moving the cantilever from air to a more dense fluidlike liquid will have an immediate effect on its resonant frequency, which is reduced because it has to move an extra mass [13]. In addition, the liquid viscosity is much higher than the viscosity of air. The movement of a cantilever driven by an external oscillatory force $F(t) = F_0\cos(\omega t)$ can be described as a forced harmonic oscillator with damping [14].

$$m\ddot{z} + k\dot{z} + \gamma z = F_0\cos(\omega t) \tag{1.6}$$

$$\gamma = \frac{m\omega_0}{Q} \tag{1.7}$$

$$\omega_r = \omega_0\sqrt{\left(1 - \frac{1}{2Q^2}\right)} \tag{1.8}$$

Here, z describes the vertical movement of the cantilever, m is its mass (note that an effective mass m^* can substitute m in all these equations), ω_0 is its free (vacuum) resonant frequency and ω_r its resonance frequency in a fluid, Q is the quality factor, γ the damping coefficient, and F_0 the amplitude of the oscillatory force at a time t.

The solution of Eq. (1.6) has a transient term and a steady term [15].

$$z = B\exp(-\alpha t)\cos(\omega_r t + \beta) + A\cos(\omega t - \varphi) \tag{1.9}$$

The transient term is reduced by a factor of $1/e$, after a time $1/\alpha = 2Q/\omega_0$. From then, the motion of the tip is dominated by the steady term. The steady term is a harmonic function with a phase lag with respect to the external excitation force. Amplitude and phase lag can be calculated with the following equations:

$$A(\omega) = \frac{\frac{F_0}{m}}{\sqrt{\left(\omega_0^2 - \omega^2\right)^2 + \left(\frac{\omega\omega_0}{Q}\right)^2}} \tag{1.10}$$

$$\tan(\varphi) = \frac{\frac{\omega\omega_0}{Q}}{\omega_0^2 - \omega^2} \tag{1.11}$$

Examples of amplitudes and phase lags are shown in Figure 1.4 for different values of Q.

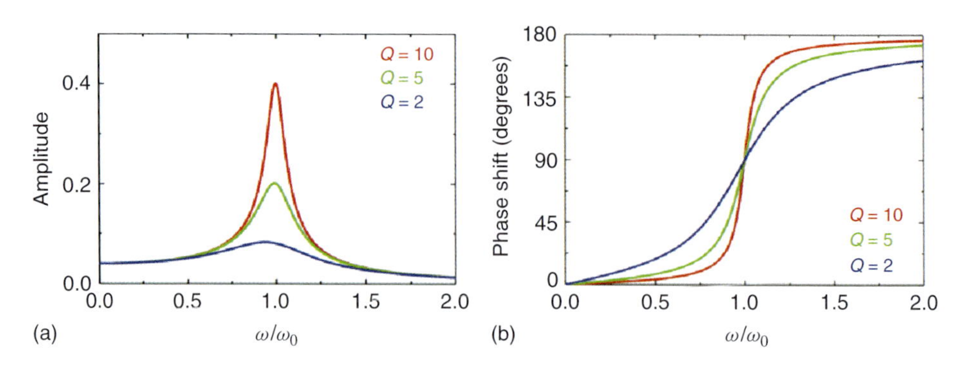

Figure 1.4 Plots of Eqs. (1.9) and (1.10) highlighting the role of the quality factor Q in a harmonic oscillator.

The oscillation of a cantilever in liquid has important differences compared with its oscillation in air or in vacuum. First, as a consequence of the large density of the surrounding liquid compared with the density of air, the cantilever suffers an increase of the effective mass by a factor of 10–40 and a corresponding decrease of the resonant frequency (Eq. (1.3)). Resonance and natural frequencies are related by Eq. (1.8). Therefore as a second consequence, the strong hydrodynamic interaction between the cantilever and the liquid produces a very low quality factor Q. Typically, Q in liquids can be about two orders of magnitude lower than in air (Figure 1.5) [16]. The decrease in resonant frequency and Q has important consequences on the cantilever oscillation and therefore affects the performance of dynamic modes. First, the cantilever oscillation is inharmonic and asymmetric in liquids [17], in contrast with its performance in air where the oscillation is sinusoidal and symmetric [18]. In addition, the low quality factor of the cantilever

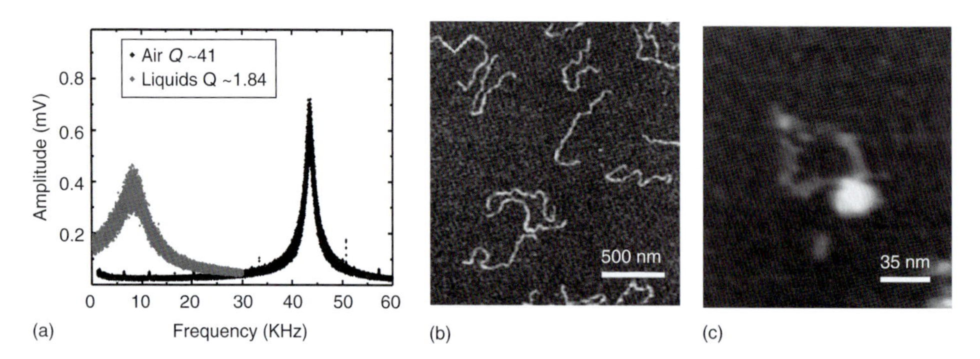

Figure 1.5 (a) Oscillation of the cantilever in air and in liquids. The thermal spectrum of the cantilever (Olympus BL-RC-150VB, $k \sim 0.03\ \mathrm{N\ m^{-1}}$) illustrates the reduction in resonant frequency and Q in liquids. (b,c) Examples of images of biomolecules obtained in liquids using dynamic modes and soft cantilevers (Olympus TR400PSA, $k \sim 0.08\ \mathrm{N\ m^{-1}}$). (b) DNA molecules imaged in 10 mM tris-HCl (pH 8.0), 5 mM MgCl2. (c) Rad50/Mre11 protein complex in 25 mM tris-HCl (pH 8.0), 125 mM KCl and 10% Glicerol. (a) Adapted from Ref. [19].

in liquid implies high forces between the tip and sample [19]. Dynamic modes use as a control signal, the amplitude of the oscillation that reflects the interaction between the tip and the sample. A shift in the resonant frequency of the cantilever due to tip–sample interaction produces an amplitude damping at resonance, which is proportional to the quality factor of the cantilever. Some authors developed an AFM technique for liquids in which the cantilever response is controlled by adding an active feedback system that increases the quality factor up to three orders of magnitude [16]. Alternatively, use of cantilevers of high resonance frequency in liquids should improve the performance of dynamic modes, although high frequencies are achieved at the expense of increasing the force constant, which is not convenient for imaging soft materials. Although cantilevers of low k and high f are preferable, molecular resolution has been obtained in buffer using cantilevers of $k \sim 0.08\ \mathrm{N\ m^{-1}}$ and $f \sim 7$ kHz as can be seen in Figure 1.5 [20].

1.4 Force Detection Methods for Imaging in Liquids

The heart of an AFM is a sharp tip that interacts with a force at the surface of a sample. As seen before, the tip is mounted on a flexible beam whose geometrical and material properties makes it possible to probe the force with high sensitivity. The role of the beam is to translate the force acting on the tip into a deflection that can subsequently be monitored by various means. Among these, tunneling of electrons (the original scheme invented by Binnig and coworkers), capacitance, piezoelectric cantilevers and tuning forks, optical interferometry, and optical beam deflection have been developed to a high degree of sophistication. The interaction force is proportional to the deflection of the cantilever following Hooke's law. Electrical methods such as electron tunneling or capacitance were historically the first used to detect the small movement of the cantilever, but they are not applicable to in-liquid imaging. Interferometry has very high sensitivity and signal-to-noise ratio, but the instrument is difficult to set up and gets quickly misaligned. For this reason, it is not used in liquid AFM. For liquid operation, most AFMs use the laser beam deflection method, but piezoelectric cantilevers and tuning forks can also be employed. In the following section, we introduce both methods.

1.4.1 Piezoelectric Cantilevers and Tuning Forks

The piezoelectric cantilever detection method [21] uses a cantilever with an additional piezoelectric thin film containing electrical connections. As the cantilever bends, the piezoelectric layer is stressed and deformed, altering the charge distribution at both sides of the layer (direct piezoelectric effect). By making proper contacts at both sides of the layer and using a simple preamplifier circuit, it is very easy to obtain a voltage proportional to the cantilever deflection (Figure 1.6a). This method has the advantage that the same connections used to detect deflection can be used

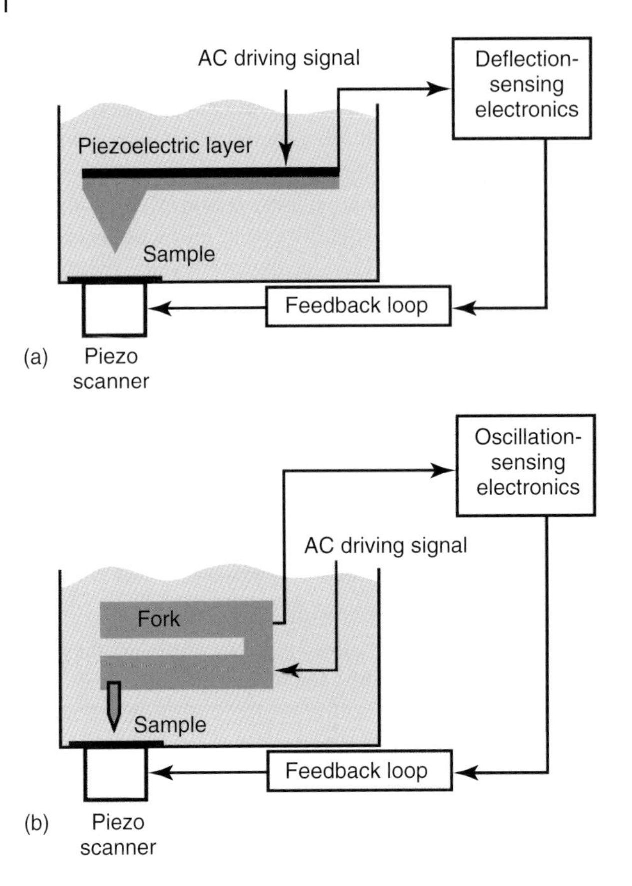

Figure 1.6 (a) Piezoelectric cantilever method and (b) tuning fork method.

to oscillate the cantilever by applying an AC voltage to the piezoelectric film. Piezoelectric cantilevers are also convenient for dynamic operation in liquids. When a cantilever is oscillated in liquid using acoustic excitation, a number of spurious resonances, which are related to the mechanics of the experimental setup, make it very difficult to identify the true cantilever resonant frequency (see discussion in Section 1.5.3.1). In the case of self-oscillated cantilevers, since the only moving object is the cantilever itself, the response of the system does not exhibit the spurious peaks associated with an external drive [22]. On the other hand, electrical contacts in liquids are tricky and may cause unstable oscillation of the cantilever. In addition, this method requires rather expensive cantilevers, often custom manufactured, and their sensitivity is not as good as in optical detection systems.

A similar self-oscillating and detection approach uses a tuning fork for detection of the tip–sample distance (Figure 1.6b). In general, an AFM tip is glued onto one leg of a small quartz tuning fork and it is forced to oscillate. Damping of the amplitude by tip–sample interaction forces is monitored and/or used as a feedback signal. The force resolution of this technique is typically 0.1 pN. This method is employed for NC-AFM [23] (see dynamic modes, Section 1.5.3) and has the advantage that the probe does not touch the sample surface, and therefore damage to the sample by the probe can be avoided. However, many problems must be overcome for the successful application of NC-AFM to biomaterials or biosystems. First, the large oscillation amplitude of the order of 10 or 100 Å used to attain a sufficient signal-to-noise ratio makes interpretation of the force curve difficult. This problem can be solved by using a stiffer force sensor compared to the conventional ones. Second, in liquids, sufficient Q-value required for NC-AFM measurement is hardly implemented because of the viscosity of the liquid.

1.4.2
Laser Beam Deflection Method

Laser beam deflection is the most common detection method used in modern commercial AFMs and was pioneered by Meyer and Amer [24, 25]. The cartoon in Figure 1.7a describes its functioning principle. The cantilever deflection is measured by monitoring the position of a laser beam reflecting from the cantilever and directed to a quadrant photodiode. The photodiode is a semiconductor device that turns the intensity of light falling on it into an electrical voltage signal. The photodiode is usually split into four sections, enabling both vertical and lateral movements of the cantilever to be differentiated. Vertical movement of the cantilever is measured as the difference in voltage between upper and lower quadrants of the photodiode. Similarly, torsional motion of the cantilever is measured as the difference in voltage between the left and right quadrants of

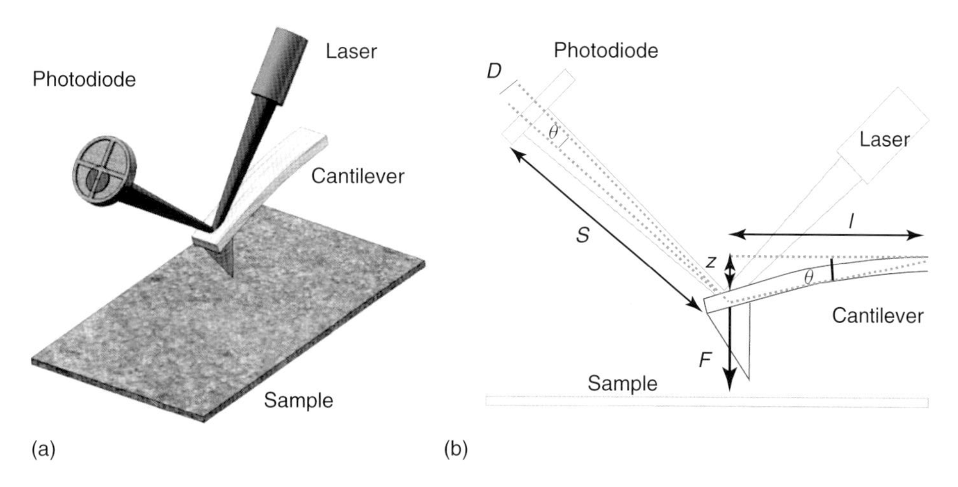

Figure 1.7 Beam deflection method. (a) Cartoon showing the principle of the beam deflection method. A laser beam is focused on the back side of the cantilever and its reflection is directed to a photodiode that records the vertical and lateral movements of the cantilever. (b) A side view of the setup with the relevant dimensions.

the photodiode. Vertical movement of the cantilever originates from the so-called normal forces. The origin of torsional movement of the cantilever arises from frictional forces that originate by the lateral motion of the tip with respect to the sample during the scan [26]. The accuracy of the beam deflection method can be as high as 0.1 Å and is generally limited by random thermal excitation of the cantilever on which the tip is mounted.

It can be shown that the angle at the end of a lever in the presence of a force F acting at this point is [27] (Figure 1.7b)

$$\theta = \frac{Fl^2}{2EI} \tag{1.12}$$

where E is the Young's Modulus and I the area moment of inertia. For a rectangular cantilever, $EI = kl^3/3$, where k is the spring constant fulfilling $F = kz$ and l is the cantilever length. Then we get

$$\theta = \frac{3}{2}\frac{z}{l} \tag{1.13}$$

And since for small deflections $\tan\theta \sim \theta$ and $\tan\theta = D/S$

$$D = \frac{3}{2}\frac{S}{l}z \tag{1.14}$$

This is a large amplification of the movement because for a cantilever of $l \sim 200$ μm at a distance $S \sim 5$ cm, this method amplifies the movement of the cantilever by a factor of 375. In general, the beam deflection method is less sensitive than other electronic methods, but the simplicity of the implementation makes it the preferred method to detect the cantilever deflection in air and in liquids.

1.4.2.1 Liquid Cells and Beam Deflection

The experimental device that allows the laser beam to propagate without suffering scattering in the liquid surface is known as *liquid cell*. The importance of this device is illustrated in Figure 1.8. In almost any air–liquid interface, small mechanical instabilities give rise to surface waves in the liquid that scatter the light coming from a laser beam, producing a noisy spot that is useless for detecting the cantilever deflection (Figure 1.8, left). This problem is solved by creating a well-defined solid–liquid interface with a transparent window. The incoming laser beam (right red line) is transmitted in the liquid without being affected by any surface wave, resulting in a stable spot suitable for detecting the cantilever deflection (Figure 1.8, middle).

The degree of sophistication of a liquid cell depends on the imaging mode employed. We discuss later that there are imaging modes in which the cantilever is oscillated at a particular frequency, and that complicates the design of a liquid cell. Precisely, to oscillate a standard cantilever, most liquid cells incorporate a piezoelectric ceramic, which is isolated to prevent wetting. Some liquid cells use magnetic cantilevers and then there is no need for using piezoelectrics since the cantilever oscillation is achieved with an external magnetic field. Often, the liquid has to be changed in the course of an experiment. Some liquid cells incorporate two

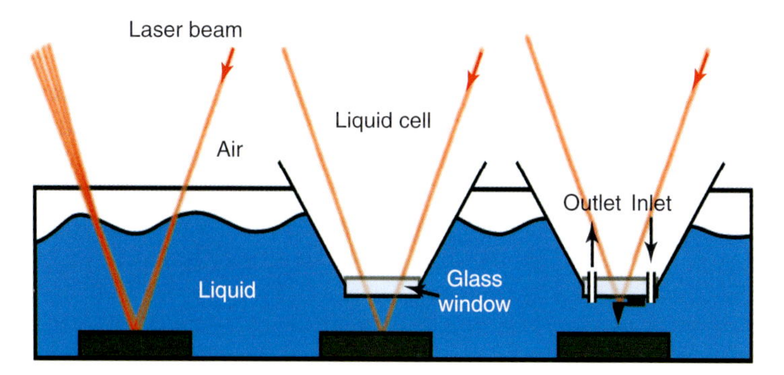

Figure 1.8 Need of a liquid cell device for imaging in liquids. The incoming laser beam (red line) continuously changes direction due to surface waves at the gas–liquid interface. As a consequence, a nonsteady spot is obtained (left). By using a transparent glass window, a well-defined interface is created and the laser light is transmitted without scattering in the liquid (middle). By means of two holes, liquid can be exchanged in the course of an experiment (right).

holes (Figure 1.8, right) at the tip mount. This permits to flow the required solution in and out of the bath, allowing a constant renewal of the liquid environment. The exchange of buffer may cause turbulences in the surrounding of the cantilever. Therefore, to prevent tip damaging, it is recommended to perform this operation with the tip out of contact. Once the old buffer is replaced, the tip can again be approached and set in range for imaging.

1.5 AFM Operation Modes: Contact, Jumping/Pulsed, Dynamic

Imaging modes in AFM are generally classified as static or dynamic modes. This classification is related to the oscillation of the tip during imaging. In static mode, the tip does not oscillate, and in dynamic mode, the tip is forced to oscillate at or near its resonant frequency. Static modes mainly include contact and jumping or pulsed force modes, while dynamic modes include, among others, amplitude-modulation atomic force microscopy (AM-AFM) and frequency-modulation atomic force microscopy (FM-AFM). In the following sections, a general description of the most important ones is given.

1.5.1 Contact Mode

Contact mode atomic force microscopy (CM-AFM) is the oldest and simplest AFM imaging mode in which the tip is brought into direct contact with the surface, deflecting the cantilever (repulsive force) [2]. This deflection is measured in liquid by any of the methods described in Section 1.4 and controlled by a feedback system that keeps it constant. In practice, most AFMs use the beam deflection method, and therefore, the signal used as control is the one given by the photodiode that corresponds to vertical deflection of the cantilever. The value at which images are taken (usually known as the *feedback set point*) is chosen by the operator depending on the conditions of the experiment. As the tip scans the surface, the Z-scanner is automatically adjusted maintaining the normal signal from the photodiode equal to the set point (see later the feedback loop Section 1.6). First applications of AFM in liquids used this operation mode [5]. AFM was used in contact mode to investigate membrane-bound proteins ordered in 2D arrays [28] and to image the chaperonins GroEL and GroES from *Escherichia coli* [29]. However, these first approaches to AFM in biology soon evidenced that contact mode imaging had several drawbacks. First, the set point given by the operator is related to a certain position of the deflected beam in the photodiode and may not reflect a constant force applied to the sample. This is because the photodiode signal that corresponds to free cantilever deflection (out of contact, zero force) may drift with time. As a consequence, while images are taken at constant deflection, they may not be taken at constant force. Second, the normal force in combination with the lateral motion during scanning introduces high lateral forces (friction) that can damage or move the sample. This is particularly relevant for the case of soft biological materials weakly attached to a surface (DNA, proteins, viruses, etc.). In order to minimize shear forces and to accurately control the force applied while imaging, several methods have been developed, and a brief description is given below.

1.5.2
Jumping and Pulsed Force Mode

Jumping mode atomic force microscopy (JM-AFM) [30] combines features of contact and dynamic modes (Section 1.5.3) and is very similar to pulsed force microscopy [31]. It was originally developed as a scanning mode to minimize shear forces and to accurately control the force applied on an image, while the tip is in contact. In JM-AFM, a force-extension measurement is performed on each of the pixels of the image. JM-AFM mode operation can be described as a cycle repeated at each image point with the following four steps: (1) tip–sample separation (moving from point A to B and C in Figure 1.9), (2) lateral tip motion at the largest tip–sample distance, (3) tip–sample approach (moving from point C to D and A in Figure 1.9), and (4) feedback enabled, which is generally performed on the cantilever deflection (point A in Figure 1.9). From this cycle it is clear that shear forces are minimized because lateral motion is always performed out of contact and that the applied force is controlled because the zero-force level is known and adjusted for each cycle.

JM-AFM solved some of the technical problems occurring in contact mode, for instance, the drift in the zero-force level, but its performance is still seriously affected by the environment used for imaging because of the presence of contact between the tip and sample. In air, the strong adhesion force arising from van der Waals and capillary forces makes it difficult to obtain reproducible images of biomolecules because forces of hundreds of picoNewtons are induced by the mere

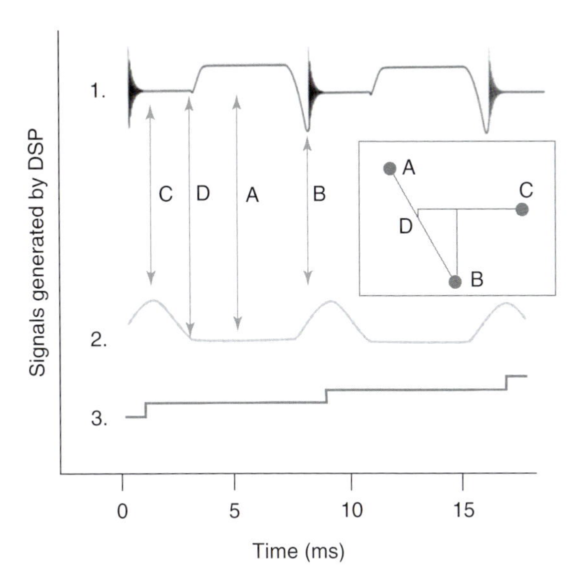

Figure 1.9 Jumping mode oscilloscope signals. (1) Normal force signal or cantilever deflection. (2) Piezoelectric Z (vertical) signal. (3) Piezoelectric (X) lateral displacement as a function of time. In the inset, a force (signal 1) versus extension (signal 2) curve is shown including main points of interest. (A) Region with feedback on. (B) *Jump-off* point or point of maximum negative deflection of the cantilever (adhesion force). (C) Largest tip–sample distance. (D) *Jump-in* point or position where tip–sample contact is established. Adapted from Ref. [30].

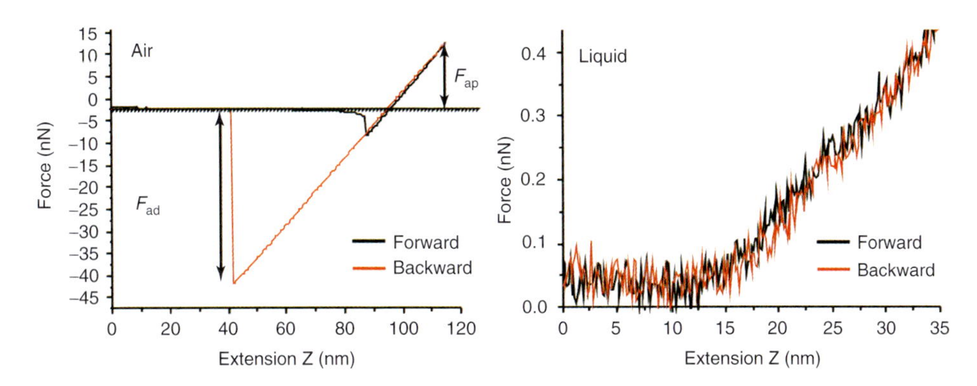

Figure 1.10 Force-extension curves in air and in liquids, enhancing the role of van der Waals and capillary forces in both environments.

contact with the sample (Figure 1.10). In liquids, however, the absence of both capillary forces and low van der Waals forces dramatically decreases the magnitude of the applied forces. This benefits imaging in JM-AFM in liquids, as it has been shown for imaging DNA molecules and viral capsids [17, 32, 33]. Moreover, this method can be used to get information on the mechanical properties of the sample by performing force-extension curves at each image point. For instance, mechanical properties of virus [33, 34] and unfolding of proteins [35] have been investigated. However, JM-AFM has some drawbacks as, for example, the low imaging rate because of the time consumed during its operating cycle.

1.5.3 Dynamic Modes

The common feature of dynamic modes is that the cantilever is driven at, or close to, its free (far from the surface) resonant frequency f_0. There are two major dynamic AFM modes, AM-AFM and FM-AFM, which are classified according to the signal used as feedback: amplitude and frequency of the oscillating cantilever, respectively. In AM-AFM, the cantilever is excited at $\sim f_0$ with a given oscillation amplitude A_0 (Figure 1.11a,b). Then, the oscillating tip and sample are approached as the amplitude signal is monitored. The amplitude of the cantilever can be measured with a lock-in amplifier or with a much simpler root-mean-square (RMS) detector. At some point, the tip starts *feeling* the interaction with the surface and as a consequence, the amplitude decreases linearly with the distance between the tip and surface. A naive way of surface *feeling* is to imagine the tip to be in intermittent contact with the surface, as if one taps the surface with a finger (Figure 1.11c,d). This is why AM-AFM is usually known as *tapping mode* [36–38]. Generally speaking, the degree of reduction of the amplitude in the tapping region with respect to the free amplitude defines the force applied on the sample and in many occasions, the quality of the image.

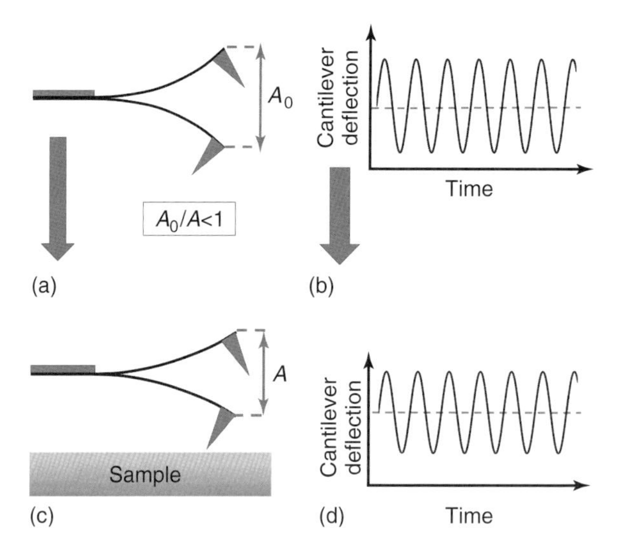

Figure 1.11 Principles of dynamic-mode AFM. (a) Cantilever driven at its free resonance frequency f_0 with amplitude A_0. (b) Rendering of the cantilever deflection as a function of time using, for instance, the beam deflection method. (c) The cantilever is approached to the sample surface and the oscillation amplitude is reduced to A. (d) Same as in (b) but with the cantilever near the surface.

In FM-AFM, the cantilever is kept oscillating with a fixed amplitude at its resonant frequency [39, 40]. An image is formed by scanning at a constant frequency shift (difference between current frequency and the free resonant frequency). FM-AFM is the preferred imaging mode in ultrahigh vacuum (UHV), but recently, true atomic resolution on mica has been reported using this method in liquids [41]. Some authors are now using FM-AFM in liquids as a method to obtain the highest resolution [42].

Despite the superior performance of dynamic modes in air with respect to contact or jumping/pulsed mode, in liquids, both methods yield comparable results. In liquids, van der Waals forces are very weak and there are no capillary forces. As a consequence, in dynamic modes in liquid, contact between tip and sample takes place and forces involved in the working cycle of dynamic and jumping modes are very similar. Nevertheless, there is still an obvious difference between both operating modes, namely, the time consumed in performing each cycle per image point: approximately milliseconds in the case of jumping/pulsed and <0.1 ms in dynamic modes. This affects the imaging rate and gives advantage to dynamic modes compared to others. However, there is still room for improvement because owing to the large effective mass of the cantilever in liquid, its resonant frequency drops and the high damping strongly reduces the Q factor of the system, which in turn means a reduction of the sensitivity of the technique.

1.5.3.1 Liquid Cells and Dynamic Modes

For working in static modes (contact, jumping, or pulse force mode), a liquid cell is almost nothing else but what is shown in Figure 1.8. However, working in liquids in dynamic modes, where the cantilever has to oscillate at or near its resonant frequency, involves a higher degree of sophistication in the liquid cell design. The most common way to oscillate a cantilever in air or vacuum is to use acoustic driving, in which a small piezoelectric element with a very high resonant frequency is located right below the cantilever chip. When working in liquid, this is not so simple because the liquid can easily get in contact with the piezoelectric, creating shortcuts and potential leaks. A simple way to avoid this problem is to locate the small piezo element far away from the liquid. In the liquid cell shown in Figure 1.12, the piezoelectric is located under one of the supporting balls. This solution has an important drawback as the piezo element excites many different mechanical resonances of the liquid cell, thus producing a *forest of resonances*, in which, in many cases, it is almost impossible to locate the real resonance of the cantilever [43]. For amplitude modulation, the right selection of the peak that corresponds to the resonant frequency of the cantilever seems to be not so important, and in many cases, a resonance peak chosen close to the one assumed for the cantilever works fine. For more sophisticated working modes such as frequency modulation a clean resonance and phase lag are required.

There are several remedies for solving the *peak forest* problem. The most common one is to use magnetic driving, in which a coil is placed close to a cantilever covered with a magnetic material (usually cobalt) [44–46]. In this design, the cantilever is driven by a magnetic field induced by an AC current that passes through the coil.

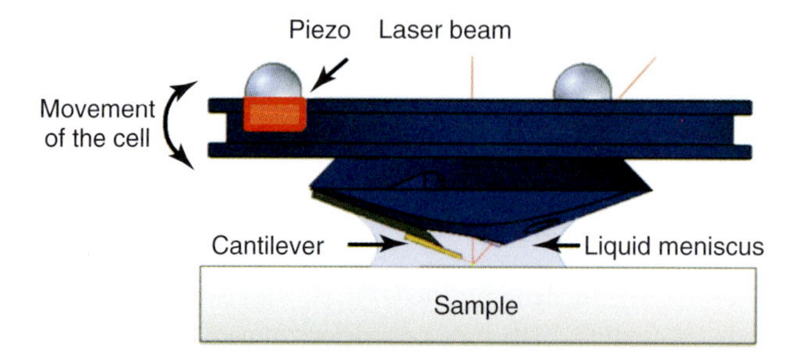

Figure 1.12 Side view of a design of a liquid cell showing a possible mechanism to oscillate the cantilever. A piezoelectric ceramic is located below one of the supporting balls and far away from the liquid meniscus. This solution also moves the liquid, and that excites many other resonances of the liquid cell.

Obviously, this method requires magnetic cantilevers, but the frequency spectrum is free of spurious resonances and the cantilever resonance is very easy to identify. A somewhat similar approximation is the electrostatic excitation, which also requires covering the cantilever with a metal thin film [47]. This method is not as restrictive as the magnetic excitation, but still, not all types of cantilevers can be used. There are also attempts to improve the acoustic driving methods in liquids by moving the piezoelectric closer to the cantilever and selecting a good combination of materials with different impedances. The results are still not as good as in magnetic or electrostatic driving, but cantilevers do not need any special coverage [48, 49]. Recently, a new approach using a secondary laser beam to excite the cantilever has been described [50].

1.6 The Feedback Loop

AFM operation is based on the ability to keep constant a given parameter: deflection, amplitude, phase shift, and so on. This task is done by the AFM control system, which should be as accurate and fast as possible in order to give it a sufficient bandwidth. Control theory provides an extensive arsenal of schemes and procedures to carry out this task. The most common one introduces three mathematical terms to the control signal: proportional (P), integral (I), and differential (D). This control method is commonly known as *PID*. While in early AFM designs, this feedback system was completely analog, the advantages of computer-controlled digital feedback were soon realized, and at present, most SPMs use digitally controlled systems [51]. In most digital controllers, a built-in timer provides the time base for the so-called interruption, which is a piece of code in the control algorithm that is executed at a given frequency. Execution of this code ensures a well-defined and consistent timing (dt) for the control subroutines. In modern systems, the interruption frequency can be up to 100 kHz. A PID feedback loop consists of three correcting terms, whose sum constitutes the Z-scanner position at a given time Z_t following the equation

$$Z_t = P\varepsilon_t + I\,dt\sum_{i=0}^{t}\varepsilon_i + D\frac{\varepsilon_t - \varepsilon_{t-1}}{dt} \tag{1.15}$$

where $t-1$ means $t - dt$, ε_t is the error signal at time t (i.e., the difference between the measured value of the control parameter at time t and the set point value), and P, I, and D are the proportional, integral, and derivative gains, respectively. Currently, most AFMs employ a simplified version containing only two parameters P and I because differential control tends to introduce instabilities. Considering a PI feedback, one can evaluate the Z-scanner position at time $t - dt$

$$Z_{t-1} = P\varepsilon_{t-1} + I\,dt\sum_{i=0}^{t-1}\varepsilon_i \tag{1.16}$$

and after some manipulations

$$Z_t = Z_{t-1} + (P + I\,dt)\varepsilon_t + P\varepsilon_{t-1} \tag{1.17}$$

This can be rewritten as

$$Z_t = Z_{t-1} + a\varepsilon_t + b\varepsilon_{t-1} \quad \text{with} \quad a = P + I\,dt \quad \text{and} \quad b = -P \tag{1.18}$$

Proper feedback control requires fine tuning of the PI parameters. This is usually accomplished by the user by looking at the forward and backward scan signals and aiming to make them overlap. For low a and b values, the response of the system is very slow and the backward and forward profiles are very different. As one increases a and b, the response is faster, but if these parameters are too high, the system becomes unstable and the Z-piezoelectric oscillates at high frequency. In general, the working point is in-between these two situations and it is found following a trial-and error-procedure. In many commercial systems, there is an algorithm to optimize a and b.

1.7 Image Representation

The most outstanding feature of an AFM is its capability to produce three-dimensional, high-resolution images of a surface. A conventional AFM topography image is a surface function $z = f(x, y)$ that can be rendered in different ways. The most common rendering method consists of using a pseudo-color/gray scale map, in which each color/gray tone corresponds to a certain height (Figure 1.13). While this representation is quite intuitive, it does not provide a straightforward visual measurement of heights and distances. To quantify in-plane distances and heights, the usual procedure is to draw a profile line on the color map, resulting in a simple 2D representation (Figure 1.13b).

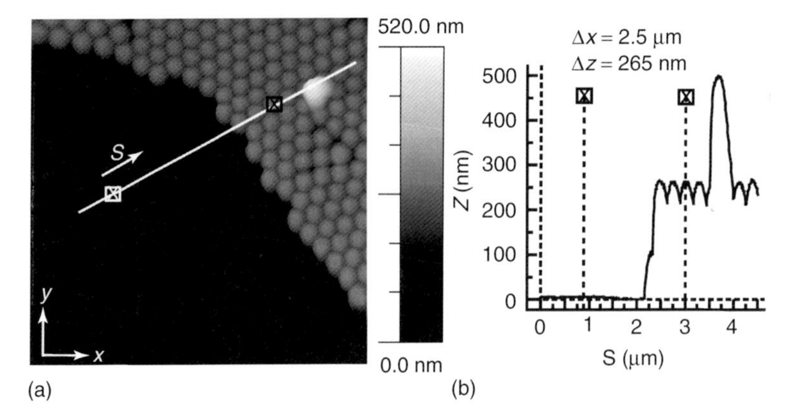

Figure 1.13 (a) AFM topography image of a single crystal made of polystyrene nanospheres. The height information is represented according to a color code included on the right of the image. (b) Height profile marked in the image.

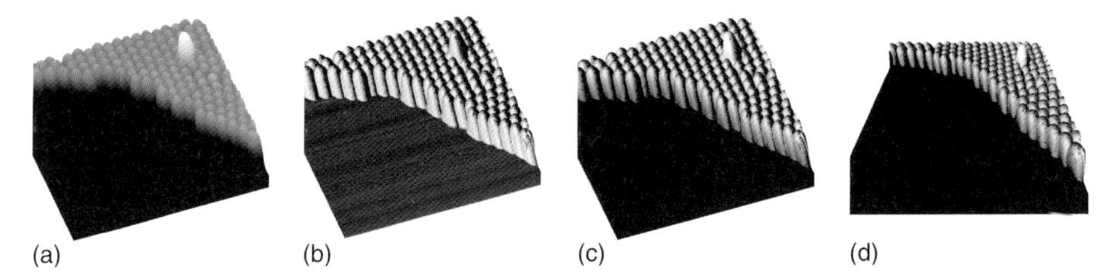

Figure 1.14 Three-dimensional rendering of the AFM topography shown in Figure 1.13. (a–c) Orthogonal views using different textures and shadows. (d) Change of the perspective of the projection.

A more intuitive representation is to render the image as a three-dimensional view, which can be displayed in several ways. For instance, the user can change the pseudo-color texture; apply shadows; change the angle of the illumination, the surface reflectivity, and the amount of ambient light; or change the view point of the image. All these different representation settings are used to enhance a certain detail of an image. Sometimes, 3D rendering must be done with caution because length scales of vertical and horizontal dimensions may be different. For instance, in the examples shown in Figure 1.14, the vertical height of the nanospheres is 265 nm, in contrast with the size of the image, which is 5 μm. Moreover, lateral and vertical dimensions may be distorted as a consequence of the projection used to display the data.

In addition to the surface representation methods described above, there are many filters that can be applied to an AFM image. Almost all AFM images are subjected to a filter known as *plane subtraction* because tip and sample planes are never parallel (Figure 1.15a). As a consequence, the angle between planes will

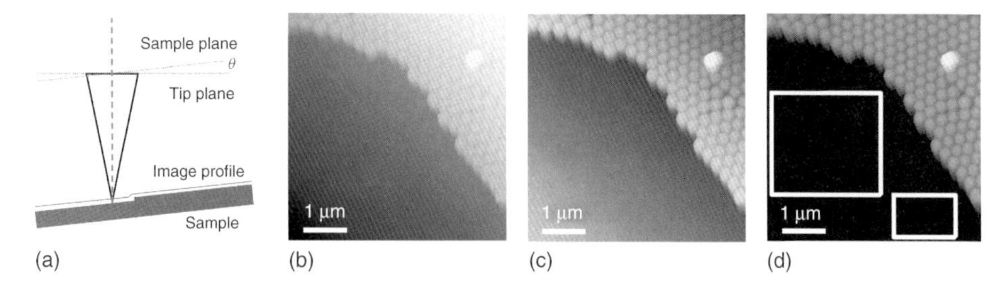

Figure 1.15 Plane subtraction filter. (a) Cartoon showing the angle difference between the tip plane and the sample plane. (b) Raw data showing a contrast caused by the surface tilt. (c) Data processed with a general plane subtraction filter. (d) Local plane subtraction filter. The squared areas are now used to calculate the plane that will be subtracted from the original image.

introduce a spurious height difference along the image (Figure 1.15b). This can be solved by calculating a general plane of the image and then subtracting it from the image raw data (Figure 1.15c). In general, this filter will flatten the image, but sometimes, the peculiarities of the data require some extra procedures. For instance, the data used to illustrate this example is clearly divided in two main planes: the basal surface and the plane given by the polystyrene nanospheres. A *local plane subtraction* considering only the data marked in squares in Figure 1.15d (basal plane) will give much better results, yielding the image shown in Figure 1.13.

Sometimes, raw data must be processed to reduce high-frequency noise or to enhance the edges of objects. This is done with low-pass or high-pass filters, respectively. High frequency noise can be removed by replacing the value of each image point by the average of next neighbors (nine data points). The *smoothing filter* is illustrated in Figure 1.16a,b with an image showing atomic periodicity of a highly oriented pyrolytic graphite (HOPG) surface. On the contrary, to enhance step edges of the graphite terraces in Figure 1.16c, the *derivative filtering* of the data along the x-direction should be applied (Figure 1.16d). Results similar to those shown in Figure 1.16 can be obtained in the reciprocal space using fast Fourier transform methods.

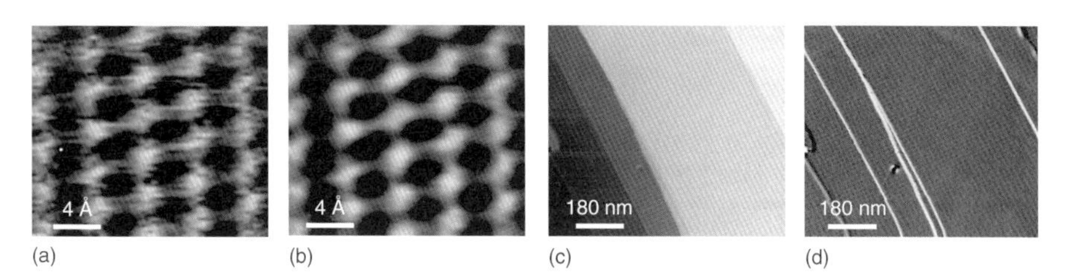

Figure 1.16 Smooth and derivative filters. (a) Raw data obtained showing the atomic periodicity of a HOPG surface. (b) Processed data of (a) with a low-pass filter (smoothed). (c) Atomically flat terraces of HOPG. (d) Derivative filter applied to (c) to enhance the step edges of the terraces.

A detailed description of data analysis for AFM is beyond the scope of this introduction. There are a number of powerful and flexible software applications allowing a variety of image filtering techniques, and some of them can be easily downloaded from the Internet. For instance, images shown in this section have been processed using WSxM [52], a freeware that can be downloaded from *www.nanotec.es*.

1.8
Artifacts and Resolution Limits

1.8.1
Artifacts Related to the Geometry of the Tip

The finite dimensions of the apex of an AFM tip produce images of objects that appear wider than they really are. This phenomenon known as *tip dilation* or convolution is particularly relevant when object dimensions are smaller that tip dimensions, and it affects the X and Y dimensions of an image (Figure 1.17). For instance, a typical AFM tip of 12 nm radius will produce an image of a DNA molecule (2 nm diameter) with a full width at half maximum (FWHM) of 10 nm. Some simple geometric models have been proposed to simulate the broadening

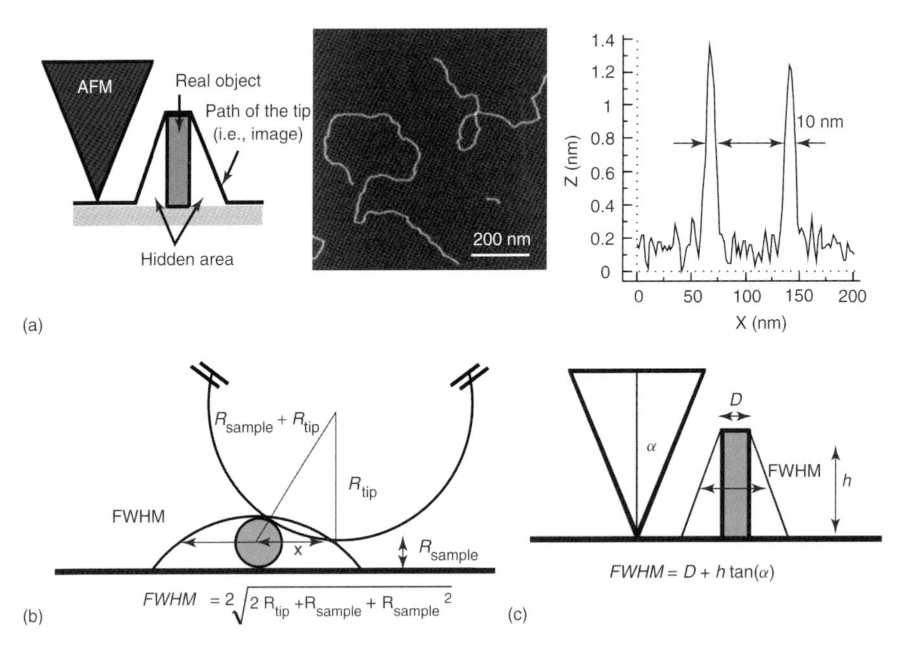

Figure 1.17 AFM broadening of objects. In the example shown in (a), the real object (a DNA molecule of around 2 nm diameter) appears five times wider because of the finite size of the AFM tip. This phenomenon is known as *tip dilation* or *convolution*. (b,c) Two models used to estimate tip dimensions from measurements of full width at half maximum considering a certain geometry of the tip and sample.

of the image by the tip. Figure 1.17b considers the tip as a spherical object of radius R_{tip} and a cylindrical object of radius R_{sample}. A simple calculation based on geometry yields a simple formula that allows an estimation of the radius of the tip from measurements of objects of known dimensions or geometry.

$$\mathrm{FWHM} = 2\sqrt{2R_{tip}R_{sample} + R_{sample}^2} \tag{1.19}$$

An alternative model considers a conical tip with an apex angle of 2α and a square object of dimensions $h \times D$ (Figure 1.17c). From geometry, one can derive the following equation:

$$\mathrm{FWHM} = \sqrt{D + h\tan(\alpha)} \tag{1.20}$$

A common artifact also related to the geometry of the tip is an effect known as *double tip*. The tip may possess two or more apexes, usually as a result of damage or contamination. This produces two or more copies of the image separated by a distance equal to the gap between the apexes (Figure 1.18). The appearance of a double-tip image depends on the size and height of the objects subjected to scan. Sometimes, the secondary apex is tens of nanometers above the main imaging apex and therefore, it only becomes obvious when imaging objects of that height. In general, when imaging small objects such as DNA or proteins (1–2 nm of height), no double-tip effect appears, but rather, broadening and loss of resolution due to blunting of the tip or contamination occurs.

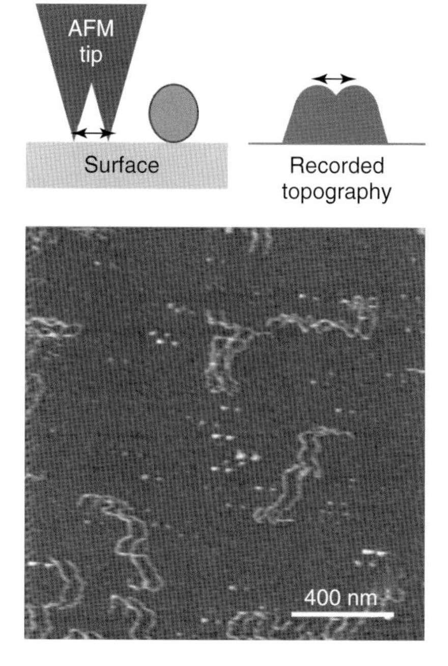

Figure 1.18 Double-tip effect.

1.8.2
Artifacts Related to the Feedback Loop

One of the most common sources for artifacts in AFM images is the feedback. The magnitude of the proportional and integral parameter (Section 1.6) depends on many factors, which include the type of piezoelectric scanner, the particular type of cantilever used, the state of the tip, the vibration level of the building, and so on. While some modern systems include software utilities that help tuning the feedback, in many other cases, the feedback must be adjusted by the user. The number of possible artifacts is too broad to be described here, but there are two kinds that are particularly common. Correct adjustment of a and b (Section 1.6) leads to similar images regardless of the direction of the scan used for acquisition (Figure 1.19a,b). If a and b are too small, the response of the system is very slow, and that results in asymmetric images as shown in Figure 1.19c,d. The effect of a slow feedback is obvious when recording the backward and forward profiles because surface protrusions are elongated along the scan direction. On the other hand, if a and b are too high, the system becomes unstable and a high-frequency oscillation is observed in the image (Figure 1.19e,f). Often, this high-frequency oscillation, present in both forward and backward profile directions, can be wrongly attributed to real surface features.

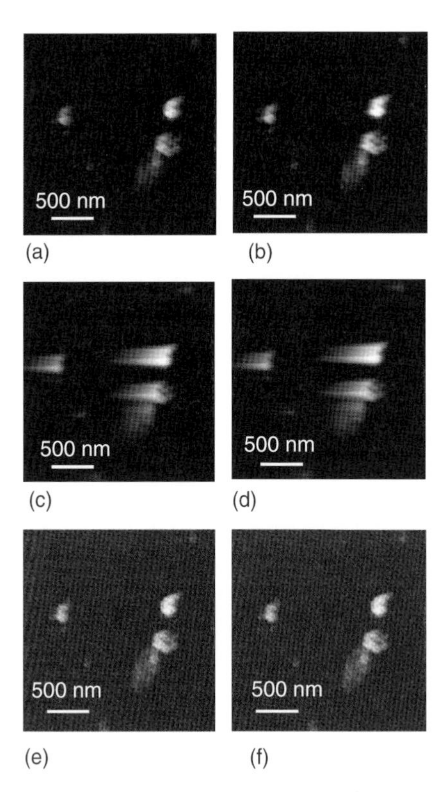

Figure 1.19 Artifacts related to the feedback loop. Left-column images were acquired from left to right (forward direction) and right-column images were acquired from right to left (backward direction). (a,b) represent examples of correct use of feedback parameters, (c,d) are examples of a too low feedback, and (e,f) of a too high feedback.

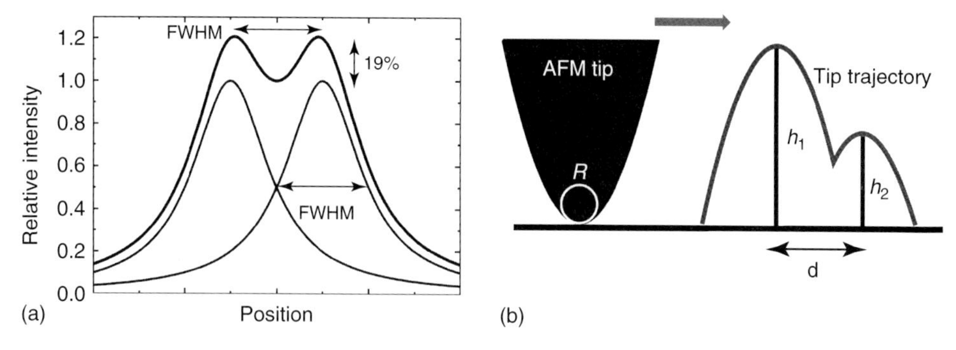

Figure 1.20 (a) Resolution criterion for a Fabry–Perot interferometer: the peaks are separated by one FWHM of each. (b) Spatial resolution in AFM. Two objects need a minimum distance *d* to be resolved. This distance *d* is dependent on the dimensions and shape of the tip and on the relative height of the objects.

1.8.3 Resolution Limits

Spatial resolution in AFM is a complex subject to address because AFM imaging is a three-dimensional imaging technique and tip convolution is not linear. Lateral resolution is usually defined by the ability to distinguish two separate points on an image. In radiation-based microscopies, spatial resolution is limited by diffraction, but in AFM, lateral resolution is defined by two factors: the pixel size of the image and the radius of the tip. Typical images in air or in liquids have a pixel size of 1–2 nm (500 nm at 512 pixels). For this reason, large images are in general limited by the pixel size.

In the microscopy community, one of the criteria used to define resolution was given by Fabry and Perot: two peaks of equal intensity are resolved if they are at least separated by FWHM of each. The dip height of the combined profile is nearly 0.81 times the maximum height [53] (Figure 1.20a). Similarly, in AFM we may ask: what is the minimum distance we can resolve in a two-point-like object configuration? (Figure 1.20b)

Considering that each object produces an image of the parabolic AFM tip, we can formulate an equation for the image of an object of height h_1

$$y_1 = h_1 - \frac{x^2}{2R} \tag{1.21}$$

For $y_1 = h_2$ we obtain

$$\Delta h = h_1 - h_2 = \frac{x^2}{2R} \tag{1.22}$$

Therefore the distance between these two points must be

$$d \geq \sqrt{2R\Delta h} \tag{1.23}$$

This explains why two objects closely spaced and of similar heights can be distinguished provided the tip is sharp enough, in contrast with radiation-based microscopic techniques, which are limited by the wavelength of the light. This also means that true atomic resolution can be achieved with AFM if interaction between the tip and sample is done only through the atom at the very end of the tip [54]. In contact mode and on hard surfaces, true atomic resolution was achieved in 1993 by Ohnesorge and Binnig [55]. Recently, using FM-AFM also true atomic resolution has been achieved in liquids [41]. For biological samples in liquids, the ultimate resolution is achieved on crystal-like arrays of membrane proteins. In 1995, Muller *et al.* [28] imaged the purple membrane at subnanometer resolution.

Acknowledgments

Fernando Moreno-Herrero acknowledges support from a starting grant from the European Research Council (Grant 206117) and by a grant from the Spanish Ministry of Science and Innovation (Grant FIS2008-0025). Julio Gomez-Herrero acknowledges support from the Spanish Ministry of Science and Innovation (Grants MAT2010-20843-C02-01, CSD2010_00024) and from Comunidad de Madrid (Grant S2009/MAT-1467).

References

1. Binnig, G. and Rohrer, H. (1982) Scanning tunneling microscopy. *Helv. Phys. Acta.*, **55**, 726–735.
2. Binnig, G., Quate, C.F., and Gerber, C. (1986) Atomic force microscope. *Phys. Rev. Lett.*, **56**, 930–933.
3. Giessibl, F.J. and Quate, C.F. (2006) Exploring the nanoworld with atomic force microscopy. *Phys. Today*, **59**, 44–50.
4. Bustamante, C. and Keller, D. (1995) Scanning force microscopy in biology. *Phys. Today*, **48**, 32–38.
5. Bustamante, C., Rivetti, C., and Keller, D.J. (1997) Scanning force microscopy under aqueous solutions. *Curr. Opin. Struct. Biol.*, **7**, 709–716.
6. Binnig, G. and Smith, D.P.E. (1986) Single-tube 3-dimensional scanner for scanning tunneling microscopy. *Rev. Sci. Instrum.*, **57**, 1688–1689.
7. Rubio, G., Agrait, N., and Vieira, S. (1996) Atomic-sized metallic contacts: mechanical properties and electronic transport. *Phys. Rev. Lett.*, **76**, 2302–2305.
8. Cleveland, J.P., Manne, S., Bocek, D., and Hansma, P.K. (1993) A non-destructive method for determining the spring constant of cantilevers for scanning force microscopy. *Rev. Sci. Instrum.*, **64**, 403–405.
9. Sader, J.E., Larson, I., Mulvaney, P., and White, L.R. (1995) Method for calibration of atomic force cantilevers. *Rev. Sci. Instrum.*, **66**, 3789–3798.
10. Seder, J.E., Chon, J.W.M., and Mulvaney, P. (1999) Calibration of rectangular atomic force microscope cantilevers. *Rev. Sci. Instrum.*, **70**(10), 3967–3969.
11. Green, C.P., Lioe, H., Cleveland, J.P., Proksch, R., Mulvaney, P., and Sader, J.E. (2004) Normal and torsional spring constants of atomic force microscope cantilevers. *Rev. Sci. Instrum.*, **75**, 1988–1996.
12. Israelachvili, J. (1991) *Intermolecular and Surface Forces*, Academic Press, London.
13. Sader, J.E. (1998) Frequency response of cantilever beams immersed in viscous fluids with applications to the atomic force microscope. *J. Appl. Phys.*, **84**, 64.
14. García, R. and Pérez, R. (2002) Dynamic atomic force microscopy methods. *Surf. Sci. Rep.*, **47**, 197–301.

15. French, A.P. (1971) *Vibrations and Waves*, W.W. Norton and Co., New York.
16. Tamayo, J., Humphris, A.D.L., Owen, R.J., and Miles, M.J. (2001) High-Q dynamic force microscopy in Liquids and its application to living cells. *Biophys. J.*, **81**, 526–537.
17. Moreno-Herrero, F., Colchero, J., Gomez-Herrero, J., and Baro, A.M. (2004) Atomic force microscopy contact, tapping, and jumping modes for imaging biological samples in liquids. *Phys. Rev. E. Stat. Nonlin. Soft. Matter Phys.*, **69**, 031915.
18. Cleveland, J.P., Anczykowski, B., Schmid, A.E., and Elings, V.B. (1998) Energy dissipation in tapping-mode atomic force microscopy. *Appl. Phys. Lett.*, **72**, 2613–1615.
19. Xu, X., Carrasco, C., de Pablo, P.J., Gomez-Herrero, J., and Raman, A. (2008) Unmasking imaging forces on soft biological samples in liquids when using dynamic atomic force microscopy: a case study on viral capsids. *Biophys. J.*, **95**, 2520–2528.
20. Moreno-Herrero, F., de Jager, M., Dekker, N.H., Kanaar, R., Wyman, C., and Dekker, C. (2005) Mesoscale conformational changes in the DNA-repair complex Rad50/Mre11/Nbs1 upon binding DNA. *Nature*, **437**, 440–443.
21. Manalis, S.R., Minne, C.S., and Quate, C.F. (1996) Atomic force microscopy for high speed imaging using cantilevers with an integrated actuator and sensor. *Appl. Phys. Lett.*, **68** (6), 871–873.
22. Rogers, B., York, D., Whisman, N., Jones, M., Murray, K., Adams, J.D., Sulchek, T., and Minne, C. (2002) Tapping mode atomic force microscopy in liquid with an insulated piezoelectric microactuator. *Rev. Sci. Instrum.*, **73**, 3242–3244.
23. Giessibl, F.J. (1995) Atomic resolution of the silicon(111)-7x7 surface by atomic force microscopy. *Science*, **267**, 68–71.
24. Meyer, G. and Amer, N.M. (1988) Novel optical approach to atomic force microscopy. *Appl. Phys. Lett.*, **53**, 1045.
25. Alexander, S., Hellemans, L., Marti, O., Schneir, J., Elings, V., Hansma, P.K., Longmire, M., and Gurley, G. (1989) An atomic-resolution atomic-force microscope implemented using an optical lever. *J. Appl. Phys.*, **65**, 164.
26. Moreno-Herrero, F., de Pablo, P.J., Colchero, J., Gómez-Herrero, J., and Baró, A.M. (2000) The role of shear forces in scanning force microscopy: a comparison between the jumping mode and tapping mode. *Surf. Sci.*, **453**, 152–158.
27. Sarid, D. (1991) *Scanning Force Microscopy with Applications to Electric, Magnetic, and Atomic Forces*, Oxford University Press, Oxford.
28. Muller, D.J., Schabert, F.A., Büldt, G., and Engel, A. (1995) Imaging purple membranes in aqueous solutions at sub-nanometer resolution by atomic force microscopy. *Biophys. J.*, **68**, 1681–1686.
29. Mou, J.X., Yang, J., and Shao, Z.F. (1996) High resolution surface of E.coli GroES oligomer by atomic force microscopy. *FEBS. Lett.*, **381**, 161–164.
30. de Pablo, P.J., Colchero, J., Gómez-Herrero, J., and Baro, A.M. (1998) Jumping mode scanning force microscopy. *Appl. Phys. Lett.*, **73**, 3300–3302.
31. Rosa-Zeise, A., Weilandt, E., Hild, S., and Marti, O. (1997) The simultaneous measurement of elastic, electrostatic and adhesive properties by scanning force microscopy: pulsed-force mode operation. *Meas. Sci. Technol.*, **8**, 1333–1338.
32. Moreno-Herrero, F., de Pablo, P.J., Fernández-Sánchez, R., Colchero, J., Gómez-Herrero, J., and Baró, A.M. (2002) Scanning force microscopy jumping and tapping modes in liquids. *Appl. Phys. Lett.*, **81**, 2620–2623.
33. Carrasco, C., Carreira, A., Schaap, I.A., Serena, P.A., Gomez-Herrero, J., Mateu, M.G., and de Pablo, P.J. (2006) DNA-mediated anisotropic mechanical reinforcement of a virus. *Proc. Natl. Acad. Sci. U.S.A.*, **103**, 13706–13711.
34. Schaap, I.A., Carrasco, C., de Pablo, P.J., MacKintosh, F.C., and Schmidt, C.F. (2006) Elastic response, buckling, and instability of microtubules under radial indentation. *Biophys. J.*, **91**, 1521–1531.
35. Rief, M., Gautel, M., Oesterhelt, F., Fernandez, J.M., and Gaub, H.E. (1997)

Reversible unfolding of individual titin immunoglobulin domains by AFM. *Science*, **276**, 1109–1112.
36. Martin, Y., Williams, C.C., and Wickramasinghe, H.K. (1987) Atomic force microscopy mapping and profiling on a sub 100-A scale. *J. Appl. Phys.*, **61**, 4723–4729.
37. Zhong, Q., Inniss, D., Kjoller, K., and Elings, V.B. (1993) Fractured polymer/silica fiber surface studied by tapping mode atomic force microscopy. *Surf. Sci. Lett.*, **290**, L688–L692.
38. Putman, A.J., van der Werf, K.O., de Grooth, B., van Hulst, N.F., and Greve, J. (1994) Tapping mode atomic force microscopy in liquid. *Appl. Phys. Lett.*, **72**, 2454–2456.
39. Albrecht, T.R., Grutter, P., Horne, D., and Rugar, D. (1991) Frequency modulation detection using high-Q cantilevers for enhanced force microscope sensitivity. *J. Appl. Phys.*, **69**, 668–673.
40. Higgins, M.J., Riener, C.K., Uchihashi, T., Sader, J.E., McKendry, R., and Jarvis, S.P. (2005) Frequency modulation atomic force microscopy: a dynamic measurement technique for biological systems. *Nanotechnology*, **16**, s85.
41. Fukuma, T. and Jarvis, S.P. (2006) Development of liquid-environment frequency modulation atomic force microscope with low noise deflection sensor for cantilevers of various dimensions. *Rev. Sci. Instrum.*, **77**, 043701–043708.
42. Fukuma, T., Ueda, Y., Yoshioka, S., and Asakawa, H. (2010) Atomic-scale distribution of water molecules at the mica-water interface visualized by three-dimensional scanning force microscopy. *Phys. Rev. Lett.*, **104**, 016101-1–016101-4.
43. Hansma, P.K., Cleveland, J.P., Radmacher, M., Walters, D.A., Hillner, P.E., Bezanilla, M., Fritz, M., Vie, D., Hansma, H.G., Prater, C.B. *et al.* (1994) Tapping mode atomic force microscopy in liquids. *Appl. Phys. Lett.*, **64**, 1738–1740.
44. Florin, E.-L., Radmacher, M., Fleck, B., and Gaub, H.E. (1994) Atomic force microscope with magnetic force modulation. *Rev. Sci. Instrum.*, **65**, 639.
45. O'Shea, S.J., Welland, M.E., and Pethica, J.B. (1994) Atomic force microscopy of local compliance at solid-liquid interfaces. *Chem. Phys. Lett.*, **223**, 336–340.
46. Han, W., Lindsay, S.M., and Jing, T. (1996) A magnetically driven oscillating probe microscope for operation in liquids. *Appl. Phys. Lett.*, **69**, 4111–4113.
47. Umeda, K., Oyabu, N., Kobayashi, K., Hirata, Y., Matsushige, K., and Yamada, H. (2010) High-resolution frequency-modulation atomic force microscopy in liquids using electrostatic excitation method. *Appl. Phys. Express*, **3**, 065205.
48. Carrasco, C., Ares, P., de Pablo, P.J., and Gomez-Herrero, J. (2008) Cutting down the forest of peaks in acoustic dynamic atomic force microscopy in liquid. *Rev. Sci. Instrum.*, **79**, 126106.
49. Asakawa, H. and Fukuma, T. (2009) Spurious-free cantilever excitation in liquid by piezoactuator with flexure drive mechanism. *Rev. Sci. Instrum.*, **80**, 103703–103705.
50. Stahl, S.W., Puchner, E.M., and Gaub, H.E. (2009) Photothermal cantilever actuation for fast single-molecule force spectroscopy. *Rev. Sci. Instrum.*, **80**, 073702.
51. Piner, R. and Reifenberger, R. (1989) Computer control of the tunel barrier width for the scanning tunneling microscope. *Rev. Sci. Instrum.*, **60**, 3123.
52. Horcas, I., Fernandez, R., Gomez-Rodriguez, J.M., Colchero, J., Gomez-Herrero, J., and Baro, A.M. (2007) WSXM: a software for scanning probe microscopy and a tool for nanotechnology. *Rev. Sci. Instrum.*, **78**, 013705.
53. Sharma, K.K. (2006) *Optics: Principles and Applications*, Elsevier, Oxford.
54. Morita, S., Wiesendanger, R., and Meyer, E. (2002) *Noncontact Atomic Force Microscopy*, Springer, Berlin.
55. Ohnesorge, F. and Binnig, G. (1993) True atomic resolution by atomic force microscopy through repulsive and attractive forces. *Science*, **260**, 1451.

2
Carbon Nanotube Tips in Atomic Force Microscopy with Applications to Imaging in Liquid

Edward D. de Asis, Jr., Joseph Leung, and Cattien V. Nguyen

2.1
Introduction

Carbon nanotube (CNT) is one of the most researched materials in the world because CNT possesses many highly desirable electrical, mechanical, and thermal properties that can be advantageously utilized in a vast array of diverse applications such as electrical devices, composite materials, and thermal dissipation devices. In the applications of atomic force microscopy (AFM), CNT AFM probes with tips constructed from a single CNT or bundles of CNT have been employed [1]. Compared to conventional microfabricated silicon and silicon nitride AFM probes, CNT AFM tips have been shown to achieve higher resolution while simultaneously providing longer tip lifetime and minimizing interference from spurious nonspecific forces acting on the tip and cantilever [1].

The advances realized with CNT AFM probes result from the CNT atomic structure. CNT consists of a high aspect ratio and a nanometer-diameter cylindrical tube (Figure 2.1a) in which the carbon atoms each make covalent bonds with three nearest neighbors. CNT can be grown using the following processes: (i) arc discharge, (ii) laser ablation, (iii) catalytic decomposition, and (iv) chemical vapor deposition (CVD). For an excellent review of CNT synthesis, we refer the reader to the literature [2]. A single carbon cylinder comprises the single-walled carbon nanotube (SWNT) form, while the multiwalled carbon nanotube (MWNT) form is constructed from multiple concentric cylinders of increasing radius separated by a distance of 0.34 nm (Figure 2.1a).

The intrinsic atomic bonding structure gives CNT impressive structural, mechanical, electrical, thermal, and chemical properties, some of which are exploited in AFM. Structurally, the radius of curvature of the free end of CNT leads to increased resolution and the vertical hydrophobic carbon sidewalls minimize the interference from extraneous nonspecific forces. Electrically, the CNT can be metallic, semimetallic, or semiconducting depending on its helicity [2]. Mechanically, CNTs are strong and ductile. The Young modulus of SWNT is $\sim$1 TPa, while the ductility enables the CNT to elastically deform under large repeated axial tensile and

Atomic Force Microscopy in Liquid: Biological Applications, First Edition.
Edited by Arturo M. Baró and Ronald G. Reifenberger.

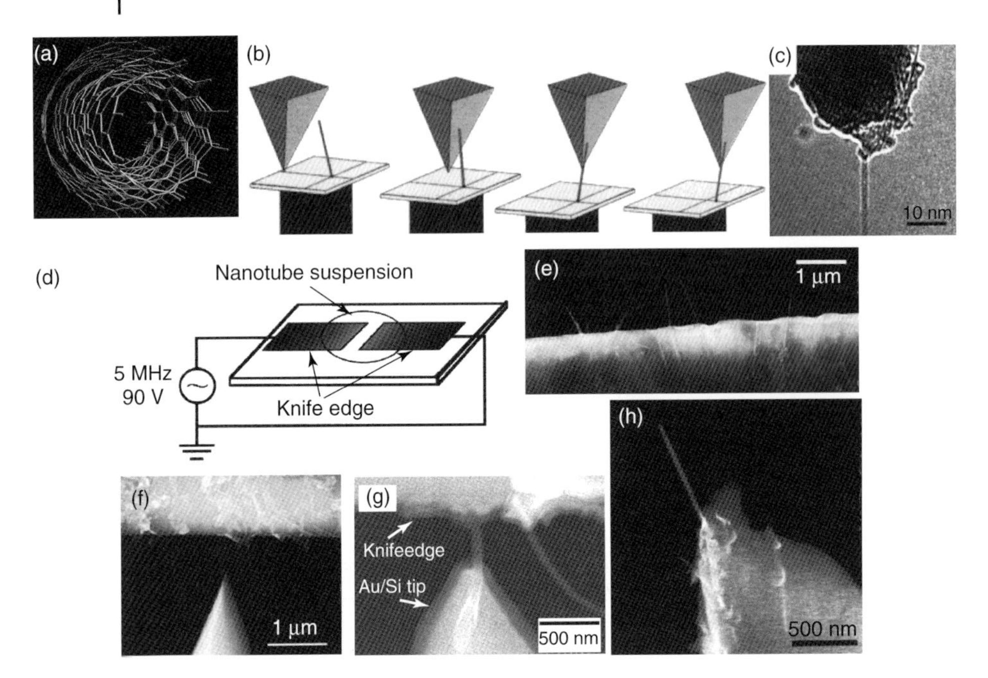

Figure 2.1 (a) Transverse view of MWNT [1]. (b) Schematic illustration of the "pickup method" [3]. (c) TEM image of SWNT tip (scale bar = 10 nm) [3]. Process for attaching single CNT to AFM tip under SEM control: (d) setup for preparation of CNT source cartridge using AC electrophoresis [4]. SEM micrographs of (e) CNT source cartridge [4], (f) AFM tip and CNT cartridge [4], (g) attachment of CNT to AFM tip [5], and (h) completed single MWNT AFM tip [4].

compressive loads [1]. In addition, the carboxylic acid functional group terminating open-ended CNT enables chemical mapping of sample surfaces [6].

While these properties of CNT have improved the performance and extended the capabilities of AFM in air [1], CNT tips can also be used in liquid environments to enable applications that were previously unachievable with conventional silicon and silicon nitride tips. Force microscopy of biological samples submerged in an aqueous liquid medium that mimics the specimen's milieu *in vivo* can be enhanced by CNT AFM tips. CNT promotes less invasive force microscopy of biological specimens because CNT elasticity reduces the risk of damage to delicate, soft samples. CNT's vertical hydrophobic sidewalls minimize the influence of nonspecific adhesion forces on the tip. In addition, the high-aspect-ratio structure of CNT can potentially eliminate interfering fluid drive on the cantilever. These effects allow for the reduction of the forces that are exerted on the sample when generating an image, thereby protecting the sample from damage. CNT tips can also more easily probe complex, contoured, creviced, and three-dimensional biological samples. The advantages of CNT AFM tips can be employed in novel applications such as mechanical characterization of the membrane, drug delivery, and electrophysiology.

In this chapter, we investigate the application of CNT tips to AFM in liquid. First, we survey the various mechanical, electromagnetic, and direct growth processes for fabricating CNT AFM probes. Second, we describe the techniques for functionalizing the CNT open end and coating the CNT sidewalls. Third, we examine the mechanical properties of CNTs in relation to AFM and attribute those mechanical properties to the atomic structure of CNT. In addition, we relate the mechanical strength of CNT to its elastic response to various loads and its susceptibility to vibration noise and nonspecific lateral and adhesion forces that can degrade image quality. Furthermore, we discuss the effect of radius of curvature of the CNT tip on image resolution. Fourth, we investigate the mechanics of CNT tips in liquid. In this treatment, we first study the interaction of the liquid medium with the microfabricated AFM probe, which often serves as the CNT support structure. Then, we examine the CNT tip at each stage of the force microscopy operation in liquid, namely, CNT tip interaction with the air–liquid interface during sample approach, with the liquid–solid interface during imaging, and with the air–liquid boundary during withdrawal. Finally, we present the applications and performance of CNT AFM tips in liquid when imaging DNA and delivering drugs into single cells *in vitro*.

2.2
Fabrication of CNT AFM Probes

In this section, we provide a review of the current methods for fabricating CNT AFM probes for liquid imaging. Afterward, we present two postfabrication steps – shortening and metallic coating of the CNT tip – for adjusting the mechanical behavior of the CNT tip in imaging applications. A review of the current state-of-the-art processes for CNT AFM probe fabrication and CNT tip shortening and coating is necessary because the methods employed for constructing CNT AFM probes directly influence the performance of CNT AFM probes in imaging applications. For example, the methods discussed in this section yield AFM probes with tips consisting of either MWNT or SWNT. The radius of SWNT is typically on the order of single nanometer, while the radius of MWNT is typically on the order of tens of nanometers. Compared to AFM probes consisting of MWNT tips, AFM probes with SWNT tips can potentially achieve higher spatial resolution in imaging applications.

Fabrication of SWNT and MWNT AFM probes can be grouped into three basic methods: (i) mechanical attachment of a single CNT or a bundle of CNTs [3, 4, 7–37], (ii) electromagnetic attachment techniques utilizing an AC magnetic field or AC electric field, and (iii) direct growth of the CNT on the AFM cantilever [33,38–43]. All three techniques utilize a conventional microfabricated AFM probe as the support structure for the CNT tip and require CNT growth via arc discharge, laser ablation, catalyzed decomposition, and CVD. Mechanical and electromagnetic attachment techniques involve CNT synthesis on a suitable substrate before attachment, whereas direct growth entails synthesis of a bundle of CNT on the AFM cantilever.

Following probe fabrication, the CNT tip can be shortened or coated with a thin carbon or metal film to increase the reliability of the CNT tip in penetrating the air–liquid interface and the cell membrane.

2.2.1
Mechanical Attachment

Mechanical attachment utilizes one of the four primary methods: (i) gluing, (ii) the pickup method, (iii) attachment in a scanning electron microscope (SEM), and (iv) attachment under optical microscopy control. Gluing the CNT involves using an adhesive, such as acrylic adhesive or conductive carbon tape, to attach CNT to a conventional AFM tip from a CNT source under optical microscopy control [7, 9, 10, 37]. This technique produces SWNT tips incorporating SWNT ropes [9] and AFM tips with a single MWNT or bundles of MWNT [7, 10, 37, 44]. A closely related method for postgrowth attachment involving adhesion of the CNT to a silicon AFM tip is the "pickup method" [3, 23, 25, 28, 34, 36] (Figure 2.1b,c) [3]. A conventionally microfabricated AFM tip is scanned in tapping mode (TM) along a substrate consisting of vertically freestanding SWNTs. An individual SWNT adheres to the tip due to van der Waals interaction between the CNT and silicon probe [3, 14, 15, 19–21]. The "pickup method" yields an AFM tip consisting of a single SWNT. SWNT tips constructed using the "pickup method" provide superior AFM imaging performance compared to conventional silicon tips [14, 15, 19] and have been functionalized with COOH open-end functionality [3].

A third method for postgrowth mechanical attachment involves attachment of a single CNT to an AFM tip inside the specimen chamber of a modified SEM (Figure 2.1d–h) [4, 5, 11, 12, 16–18, 22, 24, 30, 31, 37, 45–49]. The AFM probe is constructed from silicon or silicon nitride coated with metal such as Au or Os [4]. The fabrication process flow involves five major steps: (1) preparation of the CNT source cartridge via AC electrophoresis (Figure 2.1d), (2) alignment of a single CNT with the AFM tip inside an SEM (Figure 2.1f), (3) contacting a single CNT to the AFM tip (Figure 2.1g), (4) welding the CNT to the AFM tip, and (5) withdrawal of the completed probe. In step 1, two knife edges are placed on a glass plate separated by a 0.5 mm gap, a drop of CNT-alcohol solution is dispersed in between the gap, and an AC voltage is applied between the knife edges to attract CNT to the two knife edges. The applied voltage orients the CNT axis parallel to the electric field lines, and van der Waals forces bind the CNT to the knife edge (Figure 2.1e). This step yields a source cartridge consisting of a single CNT aligned perpendicular to the knife edge. In steps 2 and 3, a CNT cartridge and the AFM probe tip are mounted to two separate and independent translation stages inside a modified SEM. The AFM tip is aligned with and contacted to a CNT. In step 4, the CNT and AFM tip are welded together by application of a DC voltage or a DC current or by deposition of an amorphous carbon film by focusing the electron beam on the CNT/tip junction. In the final step, the CNT/tip is withdrawn from the cartridge.

In the fourth postgrowth mechanical attachment process, a single MWNT is attached to an AFM tip under optical microscopy control. The steps undertaken in this process resemble the process for mechanical attachment under SEM control. Moreover, CNT attachment under optical control requires a microscope equipped with micropositioners (Figure 2.2a) and involves the following steps: (i) synthesis of an MWNT source (Figure 2.2b), (ii) mounting of the MWNT source and a metal-coated Si AFM probe to the micropositioners followed by alignment of a single MWNT with the AFM tip (Figure 2.2c), (iii) contacting a single MWNT to the AFM tip followed by welding of the CNT to the AFM tip via application of a DC voltage (Figure 2.2d), and (iv) withdrawal of the completed probe (Figure 2.2e). We note that a boule of MWNT attached to a brass plate [27] or a vertically aligned MWNT grown via CVD on Fe-coated wire can serve as the MWNT source. Also, the AFM tip can be coated with a nickel adhesion layer via sputter deposition or some other thin-film deposition technique. This process has been shown to result in a mechanically robust Joule's heat-induced weld between the MWNT and metal adhesion layer. Furthermore, in the case of an MWNT/Ni junction, in which the adhesion layer thickness is at least 25 nm, the junction is ohmic with a resistance no greater than 40 kΩ [50]. Like the attachment process performed under SEM visualization, this method allows for control of the CNT attachment angle.

2.2.2 CNT Attachment Techniques Employing Magnetic and Electric Fields

A novel technique for producing CNT AFM tips utilizes an AC magnetic field to align and attach single arc-discharged synthesized MWNT to metal-coated AFM probes [13]. This process requires a modified beaker constructed from a glass coverslip with a shelf attached to the bottom of a 20 mm diameter glass cylinder, an AFM probe sputter deposited with a 60 nm gold coating, a 340 turn solenoid filled with a high-permeability iron core, and a sonicated homogeneous 5 ml suspension of MWNT in dichloromethane (Figure 2.2f). The attachment process proceeds as follows. The modified beaker is positioned above the solenoid and filled with the MWNT suspension. A gold-coated AFM probe is introduced tip side down onto the submerged glass shelf. The AFM tip is aligned with the solenoid under optical control. Thus, the distance between the AFM tip and the solenoid tip is minimized at 400 μm. Afterward, a 60 Hz, 7 A AC current is applied to the solenoid. The MWNTs align along the field lines of the induced 0.1 T magnetic flux, thereby causing the MWNT to align with the AFM tip. The yield for this method is 50% and results in typical MWNT tip lengths ranging from 100 to 500 nm and average attachment angles of 35° (Figure 2.2g).

An AC electric field can also be used to fabricate CNT AFM probes [26, 51–54]. In an example process (Figure 2.2h,i), a drop of solution consisting of MWNT in a dielectric medium such as ethanol is placed onto a gold counter electrode. An AFM tip is mounted onto a z-axis translation stage providing tilt capabilities for attachment angle control. The AFM probe is lowered until the distance between the counter electrode and AFM tip is ~10 μm. An AC electric field is applied between the AFM probe and the Au electrode, creating a nonuniform electric field in the MWNT solution. This nonuniform field induces a dipole moment on

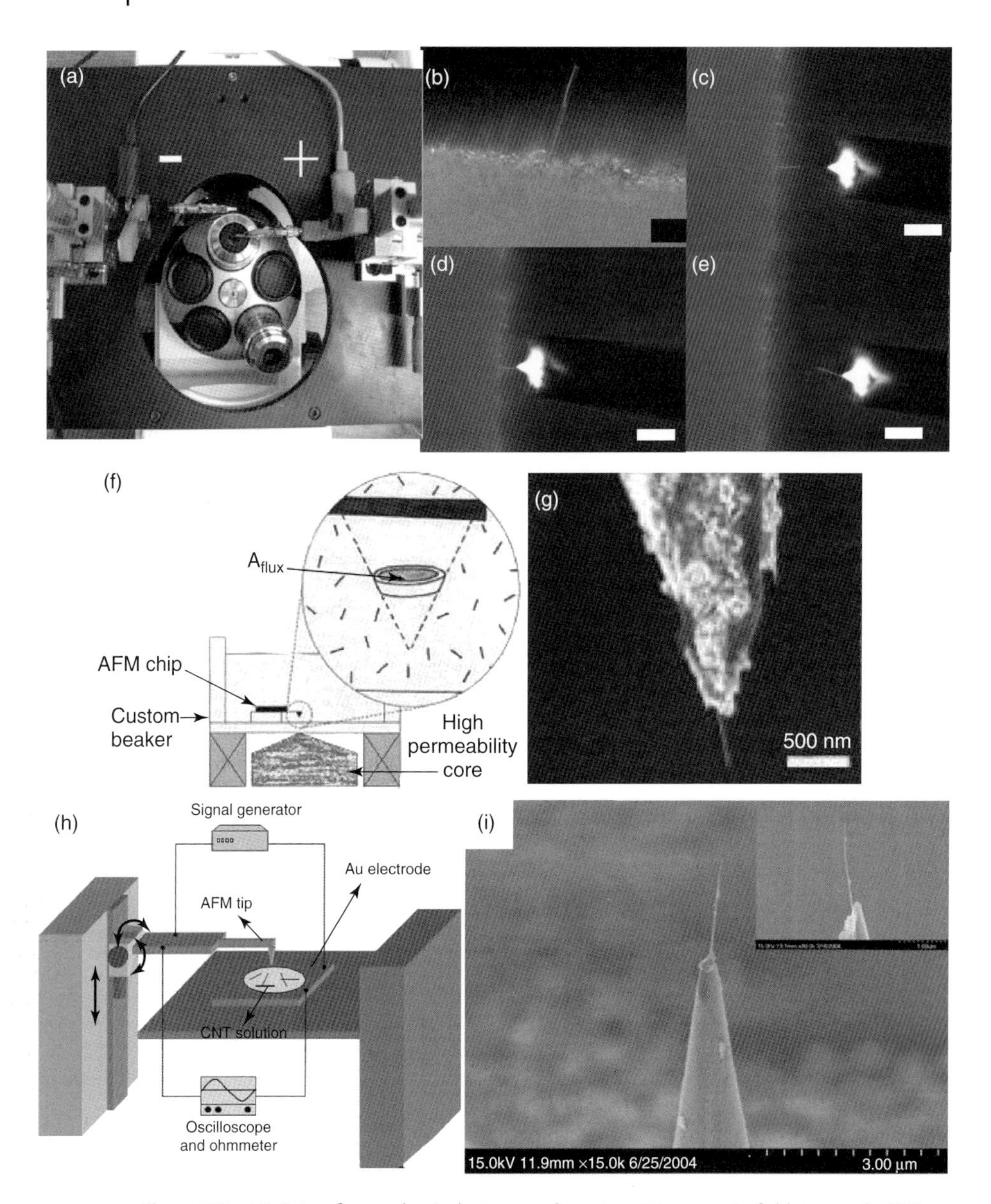

Figure 2.2 (a) Setup for mechanical attachment under optical microscopy control. Optical micrographs at 500× magnification of (b) MWNT grown on a metal wire (scale bar = 40 μm), (c) alignment of MWNT with AFM tip, (d) MWNT in contact with probe tip as DC voltage is applied, and (e) MWNT tip (c–e: scale bar = 20 μm) [1]. (f) Setup for using AC magnetic field to attach CNT tips. (g) Example of CNT AFM probe [13]. (h) Setup for using dielectrophoresis for attaching CNT tips. (i) Examples of MWNT AFM probe fabricated under an ac electric field of 0.61 V_{pp} μm *at* 5MH$_z$. The inset shows a similar tip fabricated using an electric field of 0.70 V_{pp} μm *at* 5MH$_z$[51].

the MWNT, allowing the MWNT to rotate and translate under the influence of a dielectrophoretic force that attracts the MWNT toward the AFM tip. An AC electric field of 0.61 V_{PP} μm^{-1} at 5 MHz produces AFM probes with single MWNT tips and MWNT bundle tips. The yield for single MWNT tips is 35%.

2.2.3 Direct Growth of CNT Tips

Mechanical attachment of CNTs is a time-consuming process necessitating research into the development of batch fabrication processes for the mass production of AFM probes incorporating SWNT and MWNT tips (Figure 2.3) [38, 40–43, 55, 56]. Alternative processes for direct growth of SWNT and MWNT via catalyst-assisted CVD and direct growth of carbon nanofiber tips on carbon-coated AFM tips via Ar irradiation are described in the literature [40, 43, 55, 56]. The SWNT process flow involves: (i) poly(methyl methacrylate) (PMMA) spin coat onto a wafer of AFM probes, (ii) selective removal of PMMA over the AFM tip area, (iii) spin coat of a metal catalyst layer over the entire wafer, (iv) liftoff of the PMMA layer selectively leaving the metal catalyst over the AFM tip only, (v) CVD growth of SWNT onto the AFM tip, and (vi) CNT shortening by mounting the probe in an AFM and passing current through the CNT as it contacts a conducting substrate (Figure 2.3a) [38]. This process yields either tips with single SWNT or bundles of SWNT [38].

The MWNT process flow (Figure 2.3b) consists of a novel mass production scheme combining fabrication of AFM cantilevers and direct growth of MWNT tips into a single process starting from a bare silicon-on-insulator (SOI) wafer. The fabrication process flow involves seven key steps: (i) wafer-scale nanopatterning and registration via electron beam lithography of a PMMA mask layer to define the catalyst areas, (ii) Ti barrier and Ni catalyst deposition via evaporation, (iii) liftoff of PMMA leaving Ti/Ni dots over the CNT tip area, (iv) deposition of a CVD silicon nitride layer to protect the catalyst on the tip area, (v) silicon microfabrication of cantilevers via a dry reactive ion etch of the front side of the wafer to define the cantilever profile and a backside-wet KOH etch to define the cantilever thickness, (vi) release of the silicon nitride protection to expose the catalyst, and (vii) directional growth of CNT on the cantilevers via plasma-enhanced chemical vapor deposition (PECVD). This process achieved individual MWNT growth from 200 nm catalyst dots [41]. Alternative processes for batch fabrication of MWNT AFM tips include fabrication of the AFM probe from a bare silicon wafer but replacing direct CNT growth with dielectrophoresis trapping of SWNTs and MWNTs onto the patterned cantilever [26, 59].

The disadvantages of direct growth of CNT for fabricating AFM probes are the difficulty in achieving high yield and the inability to control CNT tip structural parameters such as diameter, length, and orientation. This is mainly due to the following reasons: (i) the inability to controllably pattern nanoscale catalyst in a highly reproducible manner over an entire wafer and (ii) CVD growth techniques for CNTs with absolutely uniform diameters and lengths are also not highly reproducible over an entire wafer because CVD reactor technology for such a process is currently not available. It should be noted that for CNT AFM probe

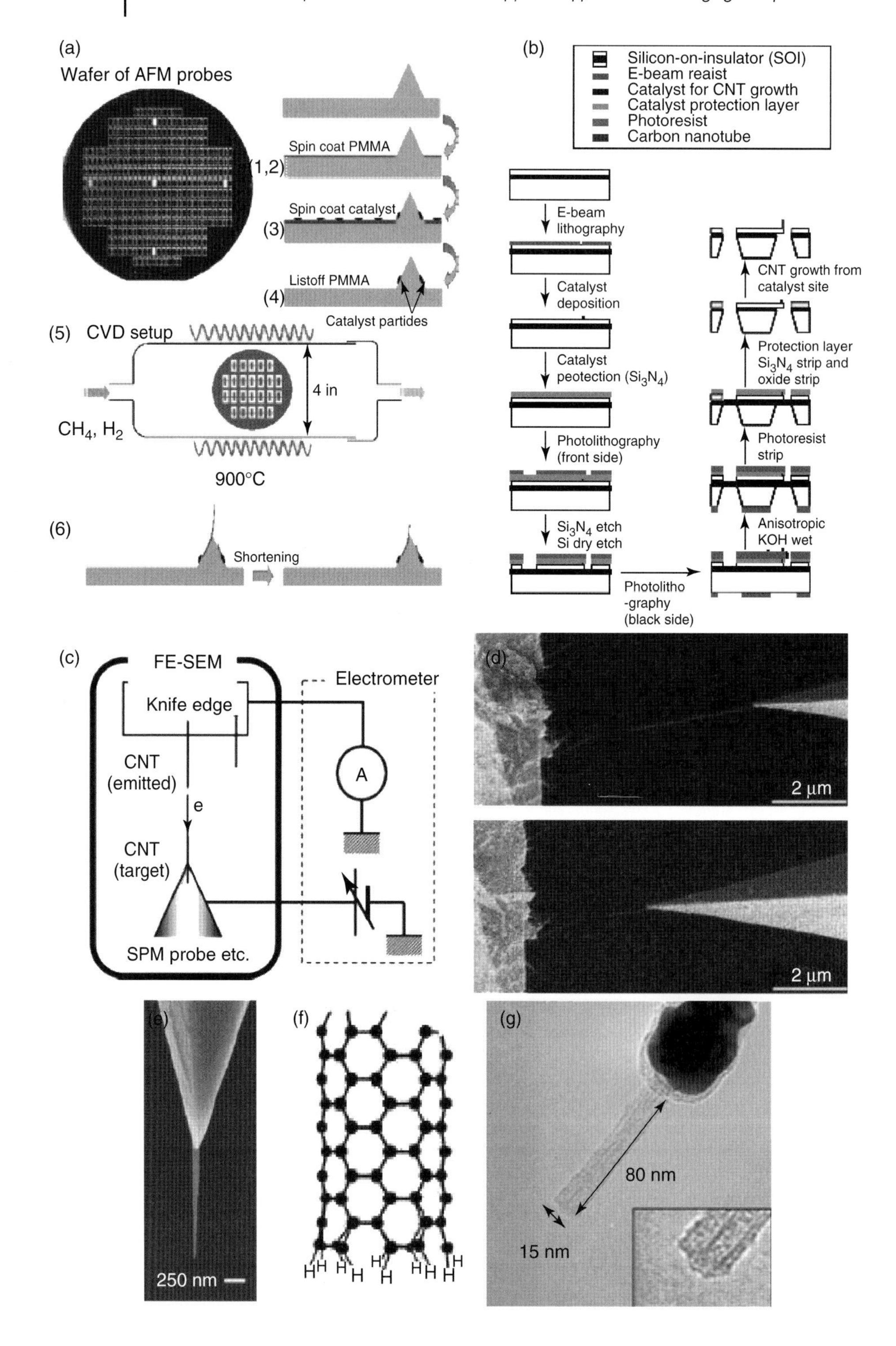
(a)
Wafer of AFM probes
Spin coat PMMA
(1,2)
Spin coat catalyst
(3)
Listoff PMMA
(4)
Catalyst partides
(5) CVD setup
4 in
CH4, H2
900°C
(6)
Shortening
(b)
Silicon-on-insulator (SOI)
E-beam reaist
Catalyst for CNT growth
Catalyst protection layer
Photoresist
Carbon nanotube
E-beam lithography
Catalyst deposition
Catalyst peotection (Si3N4)
Photolithography (front side)
Si3N4 etch
Si dry etch
Photolitho -graphy (black side)
Anisotropic KOH wet
Photoresist strip
Protection layer Si3N4 strip and oxide strip
CNT growth from catalyst site
(c)
FE-SEM
Knife edge
Electrometer
CNT (emitted)
e
A
CNT (target)
SPM probe etc.
(d)
2 μm
2 μm
(e)
250 nm
(f)
H H H H H H H H H H H
(g)
80 nm
15 nm

applications, besides diameter and length control, the control of the orientation of the CNT tip is also extremely important in terms of obtaining CNT probes that would perform well and be easy to use by the AFM end users.

2.2.4 Emerging CNT Attachment Techniques

Lastly, we mention two novel attachment techniques involving an micro-electro-mechanical system (MEMS) microgripper and the Langmuir–Blodgett technique. An MEMS microgripper can be employed to pluck a single CNT from a substrate and place the CNT onto the AFM tip [32]. Subsequent electron beam ligation of the CNT to the support structure yields the completed single CNT AFM probe. Using the Langmuir–Blodgett technique, arc-discharged SWNTs are carboxylated and shortened via chemical oxidation and subsequently thiolated by reaction with 4-aminothiophenol [60]. The thiolated SWNTs are spread onto water and compressed forming a monolayer on the air–water interface. AFM probes are dipped into the monolayer, and SWNT bundles are transferred onto the tip. Drying yields the completed CNT bundle AFM probe. For complete details of these techniques, we refer the reader to the literature [32, 60].

2.2.5 Postfabrication Modification of the CNT Tip

As shown in later sections, the performance of CNT AFM probes in imaging applications is critically dependent on the length and elasticity of the CNT tip. The CNT tip length determines both the maximum attainable resolution in imaging applications and the mechanical reliability of the tip in penetrating the air–water interface. The ability of the CNT tip to elastically deform in response to applied external forces enables imaging of soft, delicate, biological tissue. However, a trade-off exists in that elastic bending of the CNT degrades image resolution. Measuring the minimum force necessary to penetrate the cell membrane may require increasing the rigidity of the CNT tip. Increasing the resolution of the CNT AFM probe is achieved by shortening the CNT tip, while improving the rigidity of the CNT tip is accomplished by coating the CNT tip with a thin metal film.

2.2.5.1 Shortening

The CNT tip length can be shortened *in situ* in an AFM in TM [3] and via a focused electron beam in an SEM [12]. Within an AFM, the CNT tip length is measured using force calibration curves recorded on a clean silicon wafer. The

Figure 2.3 Process flow for batch fabrication of (a) SWNT tips [38] and (b) MWNT tips [41]. (c) Schematic of setup for shortening. SEM micrographs of (d) CNT tip shortening [12] and (e) fortified CNT tip [57]. (f) Schematic of functionalized CNT open end [58]. (g) TEM image of polymer-coated CNT tip. The inset shows micro-structure at the probe apex [25].

CNT is shortened by applying a train of 10–20 V, 50–100 s voltage pulses to the tip in TM. Each voltage pulse trims the CNT length by ~2–5 μm. Measurement of the force calibration curve and application of the voltage pulse train are repeated as needed until the desired CNT tip length is achieved. Inside an SEM (Figure 2.3c,d), the CNT tip is bombarded by electrons emitted by a CNT counter electrode. At high electric fields resulting from an external voltage applied between the CNTs, the electron impact causes Joule's heating-induced thermally assisted evaporation of the tip; consequently, the tip is shortened by 100 nm. The process is repeated iteratively until the desired length is reached. We note that other methods in which the CNT tip is shortened under SEM and optical microscopy control utilizing a PtIr probe or a sharpened Au scanning tunneling microscope (STM) tip are discussed in the literature [50, 61].

2.2.5.2 Coating with Metal

Penetration of the cell membrane has been achieved with bare MWNT [62], MWNT coated with biotin [63], and carbon/gold-coated MWNT [57]. As the length of the CNT tip increases, the force required to bend the CNT is reduced, thus potentially compromising the CNT tip's ability to maintain vertical alignment while penetrating the cell membrane. Fortification of the CNT with a metal coating enhances the rigidity of the CNT tip, possibly enhancing mechanical stability of the CNT tip when penetrating the cell plasma membrane by bolstering the CNT to resist elastic bending and remain upright during insertion into and withdrawal from the cell. A fortifying carbon and gold film is deposited via an ion beam coater providing angstrom-level resolution [57] while rotating the CNT tip to ascertain conformal coating (Figure 2.3e). Failure to rotate the CNT tip during thin-film deposition causes irreversible and permanent bending of the CNT (unpublished results). With this process, CNT tips ranging from 1 to 5 μm in length can be coated with a thin fortifying film of 10–30 nm in thickness.

2.3 Chemical Functionalization

Chemical functionalization of CNT tips improves the mechanical stability of CNT tips, adjusts the bulk electrical conduction properties of the CNT, and potentially enables chemical force microscopy (CFM) of living cells using CNT AFM probes. In CFM applications, CNT tips possess certain advantages over conventional silicon and silicon nitride tips: (i) compared to hard noncompliant tips, CNT ductility potentially reduces the risk of damage to delicate biological samples and (ii) in imaging biological samples with varying hydrophobicity in liquid environments, the hydrophobic nature of the CNT sidewall reduces adsorption of water along the tip, enhancing contrast [48]. Generating image contrast between regions of varying hydrophilicity in a biological sample requires introduction of a hydrophobic or hydrophilic group to the CNT tip. Improving the mechanical stability and adjusting the electrical conduction properties of CNT tips can be achieved through chemical modification of the CNT sidewalls. Thus, in addition to CNT tip shortening and coating, chemical functionalization of the CNT tip is a method that can advantageously be employed to enhance the performance of CNT tips. The methods for CNT chemical functionalization involve chemical modification of the open end of a CNT, conformal coating of the CNT with a fluorocarbon polymer, adsorption of molecules to Au-coated CNT sidewalls, and direct adsorption to bare CNT sidewalls. Application of these methods to CNT AFM probes produces tips for adhesion studies [48, 58, 64], stiff sidewall fluorinated high-resolution SWNT tips [25], and tips with improved mechanical stability when penetrating the air–liquid interface [65]. In addition, pH-sensitive, controlled, intracellular drug delivery is potentially realizable [63].

2.3.1 Functionalization of the CNT Free End

Oxidizing the open ends of CNT results in the CNT sidewalls with terminal carboxyl groups. Carboxy-terminated CNT tips are readily produced with the mechanical attachment processes performed under SEM and optical microscopy control [48, 50]. Shortening closed-end CNT yields open-ended CNT terminated with a carboxy group [58]. The mechanism leading to carboxylic acid termination in the mechanical attachment and shortening processes is unknown. However, one proposed mechanism suggests that the arc discharge produced at the CNT end during application of a current creates reactive carbon sites at the CNT open end while activating surrounding oxygen gas molecules [58]. Subsequently, activated oxygen molecules react with carbon molecules at the CNT terminus, thereby end-functionalizing the CNT with a carboxyl group. This model suggests that the CNT open end can be selectively functionalized by varying the environment in which the shortening process is performed. Indeed, shortening MWNT in an activated H_2/N_2 environment results in H-terminated MWNT ends (Figure 2.3f) [58]. Modifying the functional group terminating the CNT open end can also be achieved by performing subsequent reactions on the carboxyl-terminated CNT. Furthermore, exposing a hydroxyl-terminated MWNT to a reaction solution consisting of 2-aminoethanethiol (AET) and N,N′dicyclohexylcarbodiimide (DCC) converts the termination from carboxylic acid to amide [48]. Analogous results have been demonstrated with shortened SWNT tips [64]. Successful termination of the open end of a CNT AFM tip can be verified through a force titration, in which the adhesion force between the functionalized CNT tip and a chemically treated surface is measured at different pH levels [48, 58].

2.3.2 Coating the CNT Sidewall

Similar to the conformal metal coating described in Section 2.2.5.2, coating CNT with a fluorocarbon Teflon-like polymer also stiffens the CFM probes (Figure 2.3g); however, this technique also alters the electrical conduction properties of the CNT

[25]. For example, thin layer polymer deposition was performed on CVD-grown 5 nm diameter SWNT using an inductively coupled plasma (ICP) reactor [25]. In TM force calibration, SWNTs coated with fluorocarbon Teflon polymer are able to deflect the cantilever when contacting the surface unlike bare SWNTs that elastically bend in response to a load. Thus, the polymer layer stiffens the SWNT. After etching with a 5 V electrical pulse applied to the SWNT tip, electrical testing was performed by contacting the tip with liquid mercury (Hg). For low bias voltages, the etched tip yielded a resistance of 177 kΩ for metallic SWNT, which agrees with the published estimates of SWNT resistivity, suggesting that the deposition process did not affect the electrical conductivity of metallic SWNT. Other polymer-coated nanoelectrodes displayed unique electrical properties as evidenced in current–voltage measurements. Semiconducting SWNT attached to a Au tip showed no rectification. Nanoelectrodes in which SWNT was attached to n-type silicon showed rectification at negative bias, whereas other SWNTs attached to Au tips showed rectification at positive voltages [25]. As with a conformal metal coating, conformal coating of a CNT with a fluorocarbon polymer can stiffen the nanotube as well as change the electrical conduction properties of CNT AFM probes.

Another potential method for functionalizing the tip of a CNT AFM probe involves conjugating the fortified CNT tips presented in Section 2.2.5.2. After coating the CNT tip with a fortifying carbon film, a thin conformal Au layer is deposited; thereafter, the Au can be reacted with a chemical species containing a thiol group, thereby forming a monolayer of linker molecules on the Au layer. While the thiol group covalently bonds to Au, additional molecules can adsorb to the linker molecules. Using this process, 4-ATP has been conjugated to the Au-coated fortified CNT tip and serves as a linker molecule on which gold nanoparticles are adsorbed [57]. This method enables modification of the sidewall chemistry and hydrophobicity of the CNT tip. In addition, incorporation of pH-sensitive enzymes and linker molecules allows for intracellular drug delivery [63].

Lastly, we provide a technique that produces a CNT AFM capable of remaining vertically freestanding while imaging in liquid environments [65]. This method requires deposition of ethylenediamine (ED) on the CNT tip to reduce the hydrophobicity of the CNT sidewalls. The ED coating advantageously obviates conformal coating of a fortifying carbon, Au, or fluorocarbon thin film. Vapor deposition of ED involves exposing the CNT tip to saturated ED vapor in a reaction vessel for 5 min. This treatment time achieves sufficient ED coverage over the CNT. Compared to untreated CNT tips, which irreversibly collapse onto the support structure, the ED-coated CNT tips remain vertically upright on immersion in deionized water.

2.4 Mechanical Properties of CNTs in Relation to AFM Applications

The performance of CNT AFM probes in wet and dry imaging applications is determined by the mechanical properties of CNT. In turn, CNT mechanical properties such as strength and ductility result from the atomic bonds between the carbon atoms in the CNT sidewall. CNT strength and ductility determine the maximum resolution of CNT tips in air and in liquid. When imaging in liquid, the interaction of the CNT with water molecules is influenced by the hydrophobicity of the rolled graphene sheets that comprise CNT. In this section, we relate CNT atomic structure to CNT mechanical and chemical properties, which are then used in Section 2.5 to explain the performance of CNT tips in imaging applications in liquid.

2.4.1
CNT Atomic Structure

CNTs are graphene sheets rolled into a cylindrical tube. SWNTs consist of a single graphene tube while a cylinder of multiple concentric graphene tubes comprises MWNT. In a two-dimensional graphene sheet, each carbon atom forms covalent bonds with three nearest neighbors. Each pair of carbon atoms is sp^2 hybridized with three σ bonds radiating out at 120° bond angles. The mechanical properties of CNT can be understood within the context of the three in-plane σ bonds in each carbon atom within the graphene sheet. With no forces applied on the graphene sheet, the σ bond between each pair of carbon atoms remains at its equilibrium length. Under the influence of an applied external force, the σ bond length increases or decreases from equilibrium. If the resulting strain remains in the elastic deformation regime, the length of the σ bonds between each pair of nearest neighbor carbon atoms returns to equilibrium when the force is removed.

Rolling up the graphene sheet into a cylinder (i.e., an SWNT) changes the mechanical properties of the CNT slightly. The surface curvature and the arrangement of the carbon atoms in CNT cause the graphitic carbon–carbon bonds to be nonplanar, resulting in a less efficient overlapping of the π orbital between carbon–carbon atoms. Hence, the carbon atoms may not be perfectly sp^2 hybridized; furthermore, the three $sp^2\sigma$ bonds which radiate from the carbon atoms are slightly deformed from their preferred trigonal planar orientation. Consequently, these σ bonds are strained and weakened leading to a slight decrease in CNT Young modulus compared to the Young modulus of a graphene sheet [2]. The Young modulus parallel to the plane of a graphite sheet is about 1.3 TPa, and as expected, the Young modulus of SWNT is comparable at about 1 TPa, still making CNT one of the strongest materials. Nevertheless, the bonding of the carbon atoms in CNT maintains the same sp^2-like hybridization characteristics, that is, the carbon–carbon bond as well as the bonding angle can be mechanically distorted in a reversible manner.

Figure 2.4a schematically demonstrates the bending of sp^2-hybridized graphitic carbons when an external force is applied. The bonding angle may be twisted out of plane, with the central C atom becoming more sp^3-like and less sp^2-like and thus losing most of the bonding contribution by the π orbital, that is, transforming from double-bond-like to more of a single-bond-like characteristic. When the strain is removed, the bonding of the C–C atoms would revert to its thermodynamically more stable sp^2 configuration and thus retain the original nanotube structure. The bonding energy of an sp^2 carbon–carbon bond is 612 versus 345 $kJ\,mol^{-1}$

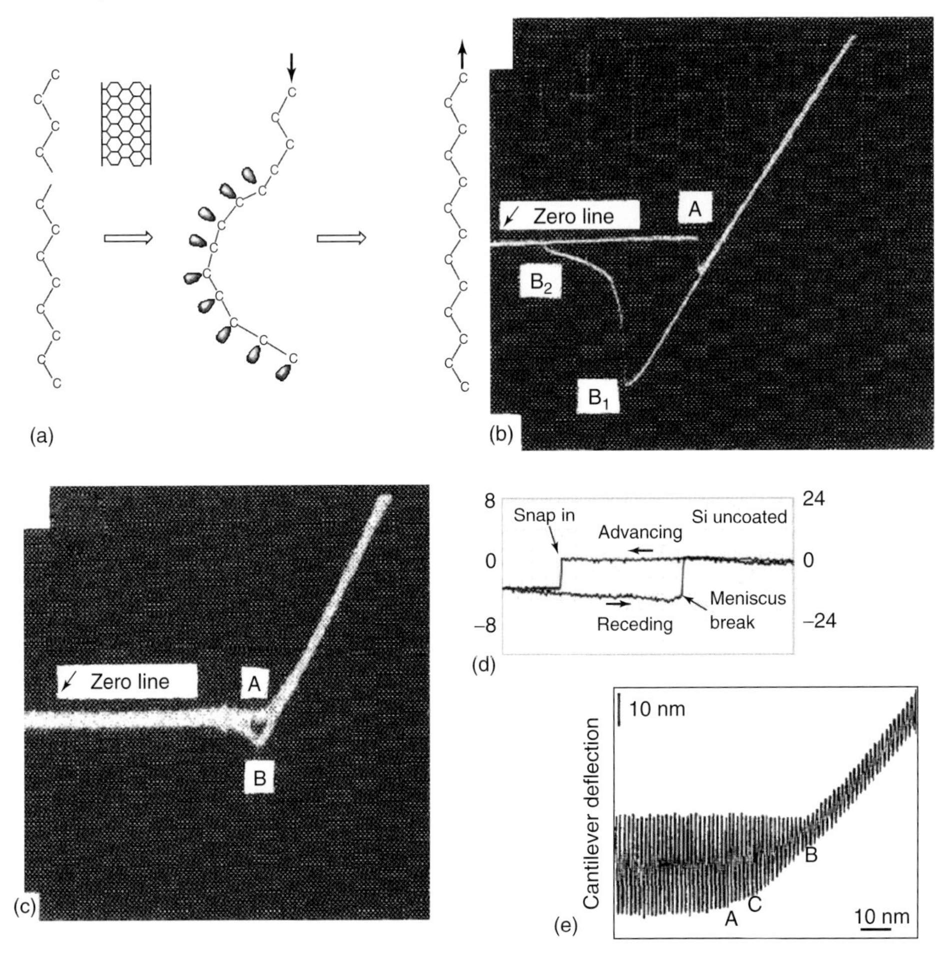

Figure 2.4 (a) The elastic buckling of the CNT is derived from the chemical bonding energy from the sp^2 carbon atoms. Graphitic carbons along CNT at equilibrium (left), under external force (middle), and after removal of external force (right). Force spectroscopy curves in (b) air and (c) water [66]. (d) Force versus distance curve for uncoated silicon AFM tip immersion and retraction from water film [67]. (e) Cantilever deflection versus distance in TM. The decrease in amplitude from A to B enables formation of an image [68].

for an sp^3 carbon–carbon bond. The general chemical nature of the sp^2 carbons and the C–C bonds is the basis of the elastic buckling properties of CNTs. The elastic deformation renders the CNT structurally robust, that is, the CNT will not permanently change its structure when it experiences external forces. Lastly, we note the effect of defects on the structure and mechanical properties of CNT. The existence of defects in the CNT leads to indentations in the CNT sidewalls and kinks in the CNT; as a result, defects prevent CNT from assuming an upright vertically aligned cylindrical shape. Mechanically, defects lead to areas in which the

σ bonds are severely distorted and strained, thus weakening the CNT and leading to a reduction in the Young modulus.

2.4.2
Mechanical Properties of CNT AFM Tips

The ability of CNTs to elastically deform under large tensile and compressive loads, coupled with its nanoscale diameter and high aspect ratio, renders them suitable for application as the tip of scanning probes. Therefore, it is worthwhile to explore some of the elastic responses of the CNT in more detail. The imaging capability of the CNT tips is determined by the thermal vibration of the CNT tip and the external forces acting on the CNT tip. The external forces consist of axial forces applied parallel to the CNT axis, lateral forces applied perpendicular to the CNT axis, and adhesive forces between the substrate and CNT sidewalls. The magnitudes of these forces and amplitude of the thermal vibration are determined by the CNT stiffness, CNT length, and tip radius.

In AFM applications, the CNT tip is likely to encounter a force that causes it to bend while scanning. The bending force constant of a nanotube is given by

$$k_{\mathrm{B}} = \frac{3Y\pi\left(r_{\mathrm{o}}^4 - r_{\mathrm{i}}^4\right)}{4L^3} \tag{2.1}$$

where Y is the Young modulus; r_{o} and r_{i} are the outer and inner diameters of the nanotubes, respectively; and L is the length of the CNT [44, 45]. Elastic deformation also invokes a compression response. The compression force constant is given by

$$k_{\mathrm{c}} = \frac{Y\pi\left(r_{\mathrm{o}}^2 - r_{\mathrm{i}}^2\right)}{L} \tag{2.2}$$

It can be clearly seen from these two equations that the bending response is the larger component of the elastic deformation. Also, the length of the CNT, with the length scale typically an order of magnitude larger than the radius, will dramatically influence the bending response of the CNT tip.

Significantly, Eqs. (2.1) and (2.2) provide a measure of CNT stiffness, in that CNTs characterized by increased k_{B} and k_{c} are more resistant to bending and compression in response to an external load. These equations indicate that long CNT tips are also more prone to elastic deformation in response to applied external axial, lateral, and adhesion forces as has been shown experimentally [69]. Furthermore, in imaging applications, the adhesion force that attracts the CNT to the substrate thereby causing the CNT to bend elastically is counteracted by a repulsive restoring force that causes the CNT to revert back to its original vertically aligned orientation [33]. In addition, k_{B} determines the amplitude of the thermal vibration noise of the CNT tip, thereby influencing the attainable resolution. It has been shown that the spatial resolution of SWNT tips with an aspect ratio greater than 10 is degraded due principally to bucking at the CNT tip in response to lateral forces exerted by the sample on the CNT [14]. We note that to reduce the lateral force components acting on the CNT, the attachment angle must be as close to zero as possible [14, 70]. Single MWNT tips supported with MWNT bundles at the base are stiffer compared to unsupported MWNT tips; as a result, the supported MWNT tip can resist deformation from adhesive forces, thus producing higher-quality images of digital versatile disc (DVD) pits [46]. As mentioned in Section 2.2.3, coating SWNT with a Teflon polymer effectively stiffens the SWNT tip. Compared to uncoated SWNT, the coated SWNTs are more capable of resisting axial compressive forces in force calibration [25]. In addition, long SWNT tips are prone to higher thermal vibration and higher noise [3], as given by Eq. (2.3) where X_{tip} is the amplitude of the thermal vibrations.

$$X_{tip} = \sqrt{\frac{k_{Boltzmann} T}{k_B}} \tag{2.3}$$

where k_B is given by Eq. (2.1), with $r_i = 0$. These equations show that as CNT length increases and CNT radius decreases, the thermal vibration increases. This same trend has been observed for MWNT tips. We note that the attainable image resolution of a CNT tip improves as the radius of curvature of the tip decreases. The radius of curvature is the contact point between the sample and the CNT and determines the image sharpness. The tip radius can be estimated by imaging gold colloid particles on mica and measuring the full width at half maximum (FWHM) height of the gold particles [3].

2.5 Dynamics of CNT Tips in Liquid

The imaging capability of CNT AFM tips in force microscopy of biological samples in liquid environments is influenced by the respective forces exerted by the liquid medium on the CNT tip and cantilever. Therefore, a brief review of the dynamics of CNT tip interaction with liquid is warranted. Because the CNT tip is often attached to a microfabricated support structure, the review begins with a discussion of the mechanics of fluid interaction with the tip and cantilever of conventional AFM probes, which often serve as the CNT support structure. Many of the forces that occur during the interaction of a conventional microfabricated AFM probe also participate in the interaction of a CNT AFM probe with liquid.

2.5.1 Interaction of Microfabricated AFM Tips and Cantilevers in Liquid

Before the development of CNT AFM probes, conventional microfabricated silicon and silicon nitride tips were applied to biological imaging applications in liquid [71]. When imaging in air, the microcontact between the AFM tip and sample is a condensation nucleus for water vapor present in ambient. Formation of a meniscus creates a capillary force of the order of 10–100 nN acting on the tip and sets an absolute minimum to the size of the tip–sample contact. Consequently, resolution is degraded [48]. Induced dipoles in the tip and sample, which arise as a result of dipole fluctuations, create van der Waals forces that also act at the tip. Immersion of the AFM tip in liquid eliminates the capillary force and reduces the van der Waals force [66]. As a result, force spectroscopy curves performed on a mica sample show a dramatic reduction in hysteresis because of the reduction in van der Waals force when the mica is immersed in water (Figure 2.4b,c) [66]. Force spectroscopy of a silicon AFM probe on a water layer elucidates the mechanism underlying AFM submersion (Figure 2.4d) [67]. On approach, attractive van der Waals forces between the tip and water lead to spontaneous wetting of the tip by a meniscus bridge, after which the tip submerges in the liquid with little resistance [67]. Extraction is accompanied by the reformation of a meniscus bridge, the breaking of which requires a measurable finite force [67]. These results indicate that imaging in liquid enables a reduction in the forces applied by the tip on delicate biological samples, thus minimizing damage. In addition, imaging biological material in liquid is less disruptive to the sample compared to imaging performed in air.

While imaging contour-rich, three-dimensional, biological specimens, the sample surface exerts lateral forces on the AFM tip during image scan. As a result, image quality is degraded due to smearing caused by lateral forces [68]. Because the tip is in continuous contact with the sample during a scan, contact mode (CM) AFM on biological samples using hard noncompliant tips is potentially damaging to cells. To overcome these limitations, TM AFM was developed, in which a cantilever oscillating at its resonance frequency is brought into intermittent contact with the sample and the linear change in vibration amplitude is transduced into an image (Figure 2.4e) [68]. The oscillation is characterized by its amplitude, phase, resonance frequency, and quality factor. The quality factor is a measure of the amount of damping of the oscillation due to forces acting on the probe. In addition, TM AFM provides user control of the oscillation amplitude. The free oscillation at resonance is usually large in air; furthermore, when imaging, the oscillation amplitude called as *set-point amplitude* can be selected to a fraction of the free oscillation amplitude. TM AFM is less damaging to biological samples because the intermittent contact between the tip and the sample reduces the lateral force acting on the sample [68]. Hence, in TM, image quality is improved because lateral resolution is determined by tip sharpness and biological samples are less prone to deformation [68].

As the tip is submerged deeper into the liquid, the influence of solvation forces, which are due to the interaction of solvent molecules with the surface of the tip and the sample, becomes significant as the tip–sample spacing approaches molecular dimensions. Solvation forces lead to oscillations in the force detected in dynamic AFM. Using cantilevers (stiffness $\sim$0.65 N m^{-1}) in dynamic AFM to eliminate instability due to spontaneous wetting of the tip on approach enables detection of oscillation in the phase and amplitude as the thickness of the water layers confined between the tip and the mica sample (Figure 2.5a) [72]. As the tip nears the sample surface, the oscillation can be decomposed into a monotonic background due to hydrophilic interactions and due to its increased dampness caused by increased viscosity of the confined water layers [72]. In other studies conducted in TM AFM, this decrease in the quality factor and increased dissipation are attributed to frictional losses due to a slight inclination of the submerged cantilever that lead to lateral displacements of the tip on contact [73].

Having discussed the forces acting on an AFM tip in liquid, we now briefly discuss the interaction of a submerged cantilever with a liquid medium. In TM, a series of resonance peaks are observed in the frequency sweep of the cantilever response amplitude for cantilever chips mounted in glass and plastic modules (Figure 2.5b) [74]. The cantilever response spectrum is at least partially determined by the material and shape of the cantilever module [74] as well as the viscosity and density of fluid [77]. Significantly, the number of resonance peaks in the cantilever response spectrum was increased in fluid compared to air possibly because of additional excitation of the cantilever by acoustic fluid motion driven by the AFM piezoactuator [74]. A theoretical model illustrating the strong dependence of the frequency response of AFM cantilevers on the fluid medium [77] and experimental validation of this model [75] are provided in the literature. In these studies, the thermal noise spectrum, which measures the frequency response of cantilevers thermally excited by Brownian motion in the fluid, possesses fundamental resonance frequencies that shift to lower frequencies when immersed in liquid (Figure 2.5c). This frequency shift is enhanced by the viscous fluid. The resonance peaks in the frequency sweep of the cantilever response amplitude are enhanced at frequency ranges near the maximum of the thermal noise spectrum [74]. These results are applicable to flexural bending of the cantilever; however, the same effects have been observed for cantilevers undergoing torsional bending under the influence of an applied torque [78].

2.5.2 CNT AFM Tips in Liquid

Herein, we examine the mechanics of CNT AFM tips in liquid medium. This treatment begins with a discussion of the CNT interaction with inorganic fluids and with water. Afterward, we focus on the mechanics of CNT AFM tip penetration of the liquid–air interface, the interaction of the CNT AFM tip with the liquid–solid interface during imaging of the submerged sample, and, finally, the mechanics of CNT AFM tip withdrawal from the liquid medium. This discussion is necessary because the performance and reliability of CNT AFM tips are determined by the interaction of CNT tips with the liquid medium during sample approach, image scanning, and tip withdrawal.

2.5.3 Interaction of CNT with Liquids

Fluorescent labeling of SWNT in liquid enables visualization of SWNT dynamics in liquid [79]. Such studies indicate that, mechanically, SWNT of lengths $<1\ \mu m$ behave like rigid rods when immersed in liquid and in the absence of flow [79]. This result suggests that the stiffness of CNT is dependent on length. In addition, only the outer graphitic surface of closed MWNT is wet by the liquid, whereas

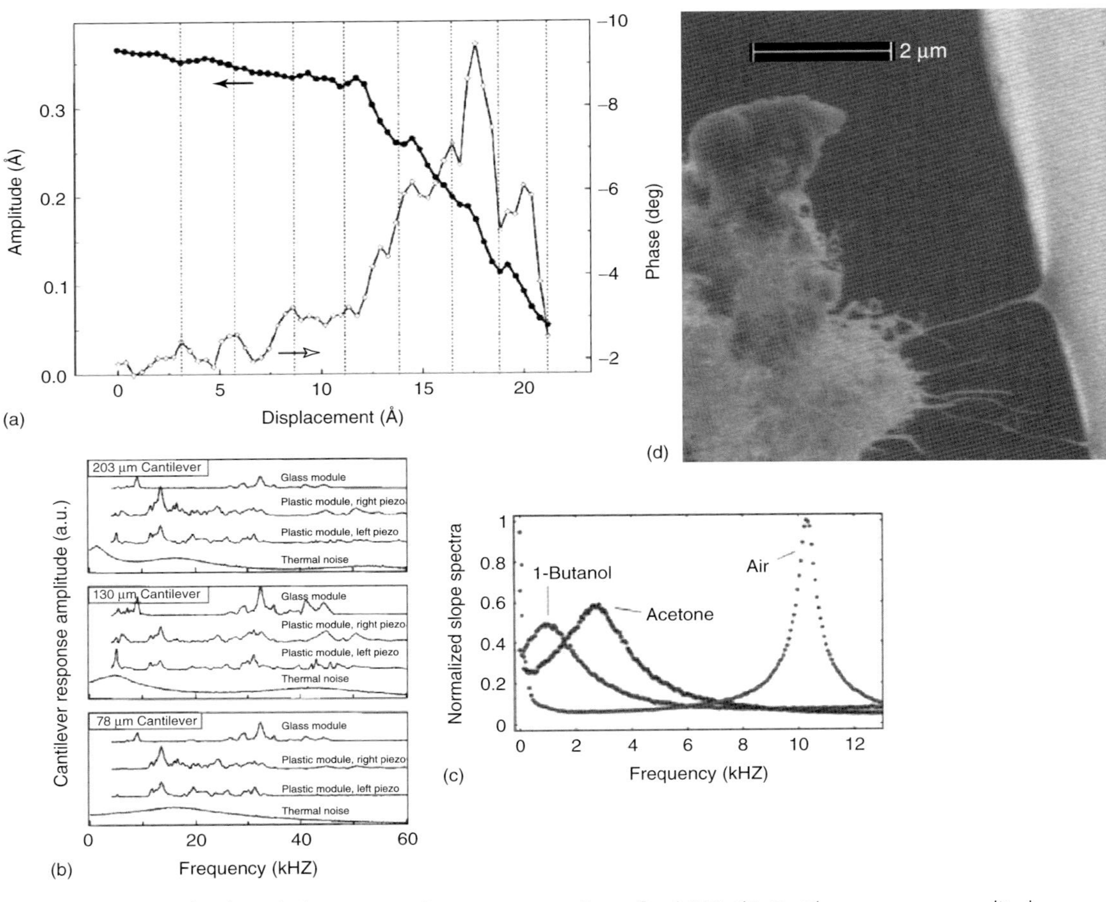

Figure 2.5 (a) Amplitude and phase versus distance as water is confined [72]. (b) Cantilever response amplitude versus frequency [74]. (c) Fundamental resonance peaks in thermal noise spectrum [75]. (d) SEM micrograph of meniscus bridge formation at MWNT tip [76].

for open-ended MWNT, the liquid wets both the interior and exterior surfaces of the MWNT [76]. Wetting of the MWNT surface creates a surface tension force on the order of nanonewton acting on partially submerged MWNT. The presence of defects in the MWNT sidewalls and the use of more polar liquid medium enhance wetting and increase the surface tension force [76]. With increasingly nonpolar, organic liquids and the magnitude of the surface tension, force is reduced [76].

2.5.3.1 CNT Tips at the Air–Liquid Interface During Approach

Successful imaging of biological samples in liquid is dependent on the CNT tip's ability to maintain vertical alignment while penetrating the air–liquid interface during approach to the submerged sample. Similar to conventional AFM tips, spontaneous wetting of CNT AFM tips at the air–water interface by a meniscus bridge has been observed for both water and organic liquids such as polydimethylsiloxane (PDMS) [76]. The extent of spontaneous wetting is dependent on the length of the CNT tip, in that, with CNT tip lengths of $<0.5\ \mu m$, wetting of the entire support structure in addition to the CNT occurs, whereas with CNT tip lengths significantly $>1\ \mu m$, formation of the meniscus bridge is limited to the CNT tip only (Figure 2.5d) [80]. As discussed in the previous section, this wetting leads to a surface tension force that can be equated to the cantilever restoring force [80]. The restoring force corresponds to the cantilever bending force that occurs in response to and tends to resist cantilever deflection. This force can be measured in an AFM, thereby allowing estimation of the contact angle between the liquid and CNT tip [80]. With more polar liquids such as water, the contact angle is increased compared to organic liquids [80].

The length-dependent spontaneous wetting mechanism necessitates an examination of the factors influencing the mechanical reliability of CNT tips at the air–liquid interface. A survey of the literature on force microscopy in liquid using CNT AFM probes suggests that shorter CNT tips are more capable of maintaining vertical alignment on immersion in liquid [81]. A recent study showed that MWNT tips of length $<2\ \mu m$ maintain vertical alignment on penetration into the air–water interface. However, MWNT tips of length $>2\ \mu m$ fail irreversibly, in that the MWNT tip breaks or the MWNT collapses and permanently adheres to its support structure (Figure 2.6) [81]. This observation led to the development of a first-order mathematical model of the mechanics of MWNT tip interaction with the air–water interface [81]. In this model, the MWNT restoring force, which tends to resist MWNT bending due to external forces, acts in opposition to the water surface tension force that tends to bend the MWNT. Both forces are functions of the MWNT length. For MWNT tips longer than $2\ \mu m$, the surface tension force was found to be larger than the restoring force, causing failure. Correspondingly, for MWNT tips shorter than $2\ \mu m$, the surface tension force was calculated to be smaller than the restoring force, causing the MWNT to remain freestanding [81].

The mechanical reliability of MWNT tips is dramatically different with electrolytes. A more recent investigation reported that, on penetration into the interface of air and aqueous solution, MWNT tips remain vertically aligned regardless of the

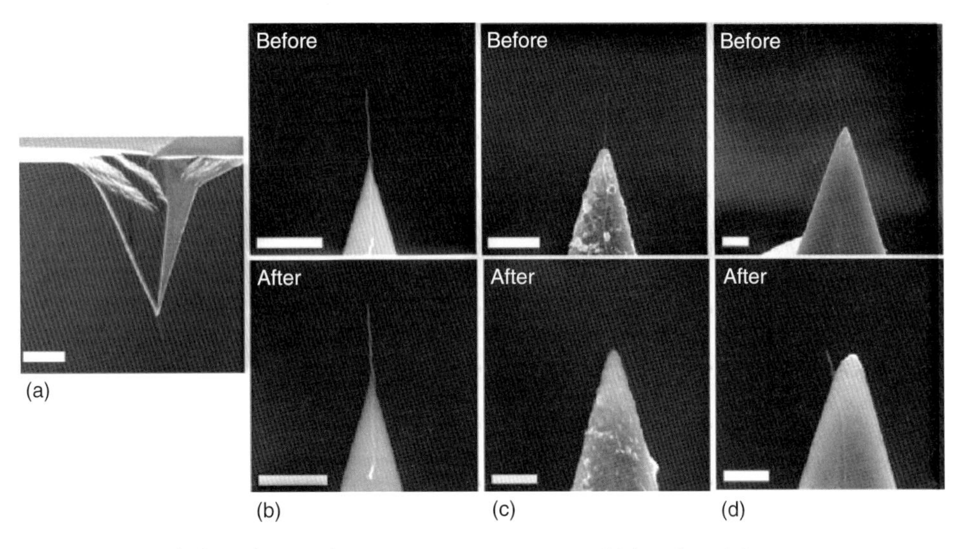

Figure 2.6 Length-dependent mechanical reliability at air–water interface [81]. (a) SEM micrograph of MWNT AFM probe (scale bar = 5 μm). SEM micrographs of three MWNT probes before and after water immersion: (b) length = 1.9 μm, radius = 27 nm; (c) length = 2.1 μm, radius = 22 nm; and (d) length = 4.9 μm, radius = 28 nm (scale bar = 2 μm).

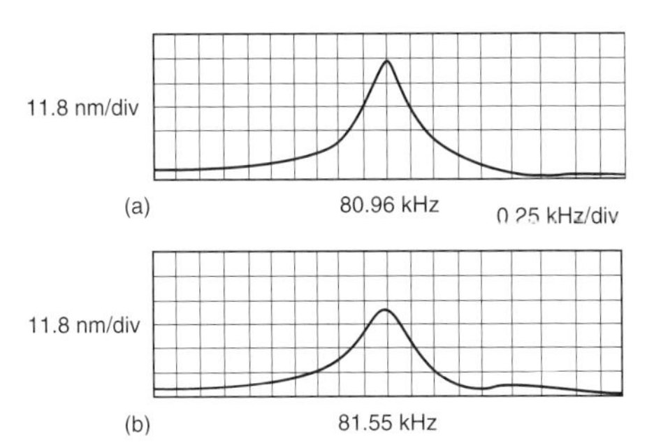

Figure 2.7 Frequency response of CNT AFM probe with soft cantilever in (a) air and (b) water [37].

tip length [82]. Significantly, spontaneous wetting of the MWNT tip is not observed at the air–electrolyte interface [82]. This effect can be attributed to the reduction in van der Waals attractive forces between the tip and electrolyte caused by the presence of dissolved mobile ionic species in the electrolyte.

A TM AFM study of MWNT tips partially submerged in water elucidated the effects of the interaction between the MWNT tip and water on the oscillation of MWNT tips [37]. An increase in the resonance frequency and a reduction in the quality factor that indicates increased damping was observed and attributed to upward directed capillary forces acting on the MWNT, thereby effectively reducing the cantilever's effective mass (Figure 2.7) [37]. This effect is independent of length. As an added advantage, immersing CNT tips minimizes adhesion forces.

2.5.3.2 CNT Tips at the Liquid–Solid Interface

TM AFM with CNT AFM tip is an improved method for imaging samples in liquid. One application of TM AFM with CNT AFM probes takes advantage of the high-aspect-ratio structure of CNT, in that imaging can be performed with the CNT tip partially submerged in liquid. The cantilever remains in air, so that interference imparted to submerged cantilevers by piezoactuator-driven vibration of the liquid, as described in Section 2.5.1, is eliminated [37]. Compared to pyramidal tips in conventional AFM probes, the vertical sidewalls of CNT tips minimize interference from nonspecific van der Waals forces between the probe and the sample and from squeezing of the liquid between the probe and the sample [45]. For open-ended MWNT tips of 2–4 μm in length and 5–20 nm in diameter, the quality factor remains high, preserving the sensitivity of the oscillation amplitude and phase to tip–sample interactions [37]. Nevertheless, similar to conventional AFM tips, the CNT tip also confines liquids such as octamethylcyclotetrasiloxane (OMCTS) at short tip–sample distances, causing the formation of solvation shells [45]. A similar effect is observed with water. Layering of water occurs, and higher dissipation (i.e., more damping) is observed when the confined water is stiffer as indicated by the presence of an integer number of layers between the tip and sample [83].

As with conventional AFM probes, contact between the MWNT tip and the sample causes a linear decrease in oscillation amplitude, which can be transduced into an image (Figure 2.8a). Further decreases in the tip–sample separation forces the oscillation amplitude to zero when the point of the CNT end farthest from the sample surface during free oscillation contacts the sample surface [84]. Further decrease in the separation distance causes the cantilever to snap to the surface and then deflect upward until the CNT reversibly buckles. Thereafter, the cantilever freely oscillates again until the CNT is completely bent away and the support structure contacts the surface. This dynamic interaction of the CNT tip with the surface of a submerged sample in TM AFM is similar to the interaction of a CNT tip with a noncompliant surface in air during force spectroscopy [3, 85].

While the influence of capillary forces is important when imaging with partially submerged CNT tips, hydrodynamic forces are more significant with fully submerged CNT AFM tips because the capillary forces are eliminated [87]. Owing to CNT ductility, imaging in TM AFM with high set-point amplitude causes broadening of the free end of the CNT tip in response to the compressive load that the CNT experiences during intermittent contact [87]. Tip broadening causes an effective increase in the radius of curvature of the CNT, thereby degrading resolution in ambient. Conversely, imaging at lower set-point amplitudes minimizes tip broadening, thereby enhancing resolution [87]. However, imaging soft biological tissue in air with low set-point amplitudes leads to image distortion [87]. TM AFM with CNT tips in liquid lowers the amplitude of the free oscillation. The liquid medium advantageously enables the imaging of biological material with low set-point amplitudes to enhance resolution while reducing distortion. Simulations indicate that these conditions lead to a repulsive sample–CNT tip interaction, with greatly reduced repulsive forces compared to imaging in air [87]. Thus, the risk of damage to delicate biological samples is reduced.

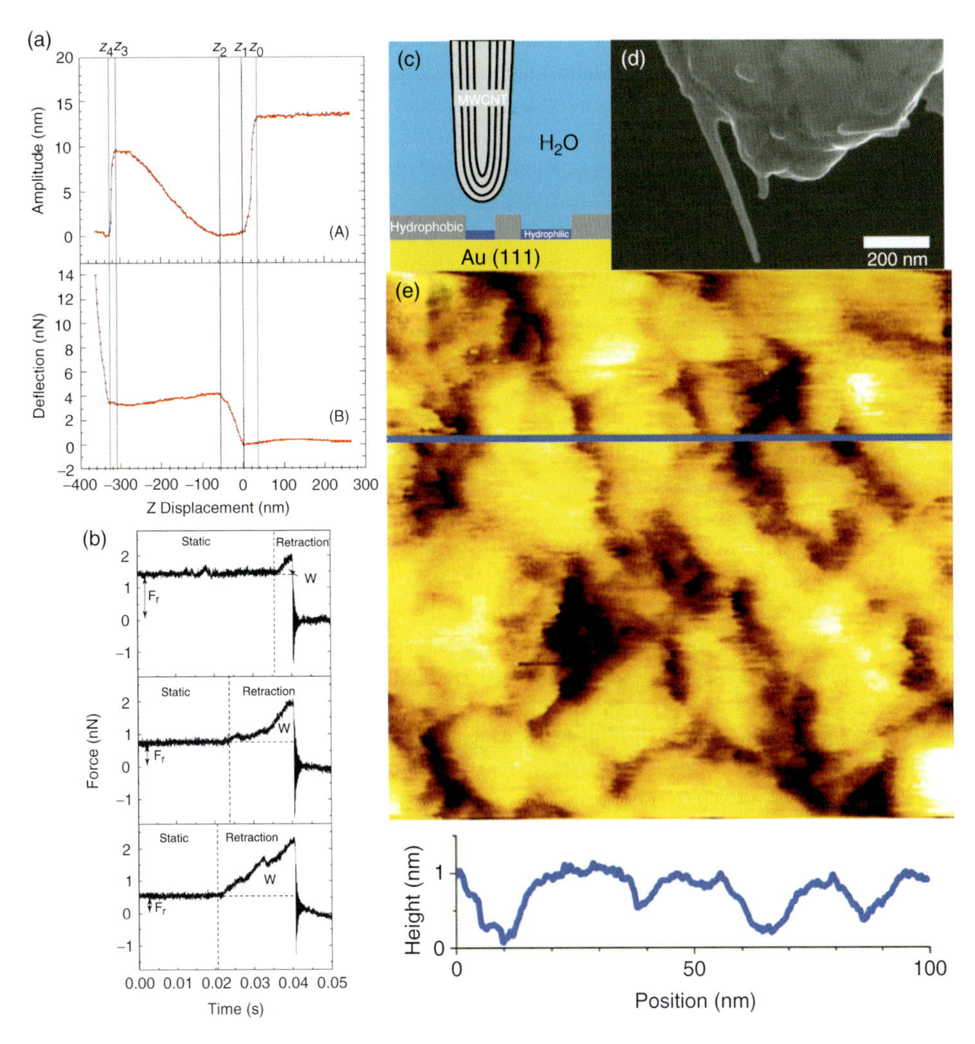

Figure 2.8 (a) Cantilever oscillation amplitude and deflection versus tip–sample separation distance for MWNT AFM probe on mica submerged in water [84]. (b) Force versus time response curves of MWNT on retraction for polyethylene glycol (top), glycerol (middle), water (bottom) [80]. (c) Schematic side view of the arrangement of hydrophilic (blue) and hydrophobic (gray) monolayers on the gold (111) surface, depicted in yellow. The nanotube tip is drawn in gray. (d) Image of the carbon nanotube tip used in the measurement. The image was made at a tilt angle of 45°, with the tilt axis horizontal in this picture. The tip radius was estimated to be 8.5 ± 1.0 nm. (e) Topography and cross section of the same surface in (a), under pure water, image size:100 nm. This image was measured with a miniature cantilever and nanotube tip. A 2 nm high "nanobubble" is visible in the bottom right corner of the image, and two smaller "bubbles" can also be distinguished. The smallest hydrophilic gaps between hydrophobic domains seem less deep because of tip convolution. (From Ref. [86]).

2.5.3.3 CNT Tips at the Air–Liquid Interface during Withdrawal

On withdrawal from liquid, a downward directed surface tension force that increases in magnitude during retraction acts on the CNT sidewalls [80]. This attractive force is accompanied by a meniscus bridge. This effect resembles the interaction of liquids with microfabricated AFM tips as discussed in Section 2.5.1. Clearly, this attractive force should be larger for open-ended CNT tips due to wetting of the interior and exterior CNT surfaces [76]. In addition, the magnitude of this force is larger for more polar liquid (Figure 2.8b) [80]. We note that large surface tension forces may pull off the CNT from the support structure – a failure mechanism that has been observed (unpublished results). Thus, the junction between the CNT and support structure must be mechanically robust so that the CNT probes can be reused.

2.6 Performance and Resolution of CNT Tips in Liquid

CNT AFM tips are used to study biological materials and to perform electrical and chemical characterization experiments involving a variety of force microscopy modes [1]. In air, CNT AFM tips realize a myriad of advantages over conventional microfabricated AFM probes, in that CNT AFM probes achieve nanoscale resolution, allow improved imaging of high-aspect-ratio three-dimensional structures, provide longer tip lifetime, and reduce nonspecific adhesion forces. SWNT AFM tips can resolve 3 nm silicon nitride grains [88]. CNT tips can map near vertical sidewalls when profiling deep trenches in photoresist [26, 88]. When measuring the root mean square (RMS) surface roughness of p-type (100) silicon using dynamic force AFM, the lifetime of the MWNT tip was five times that of supersharp silicon tips [89]. Compared to a silicon nitride tip that is easily wetted, the MWNT tip can more effectively differentiate between hydrophilic and hydrophobic areas because the hydrophobic MWNT sidewalls resist water condensation around the tip, thereby reducing nonspecific adhesion forces [48]. In this section, we examine the performance of CNT tips when imaging in liquid.

2.6.1 Performance of CNT AFM Tips When Imaging in Liquid

In liquid, CNT AFM tips achieve improved resolution [65]; have the capability of probing high-aspect-ratio, three-dimensional, nanoscale structures [37]; and achieve the minimization of spurious nonspecific forces [37, 45]. When imaging a 2 nm ion-beam-sputtered Ir on mica, the resolution of uncoated and ethylenediamine (ED)-treated CNT AFM probes is higher compared to Si AFM probes when imaging in air and in deionized water [65]. CNT tips are able to achieve lateral resolution on the order of single nanometer because of the reduced radius of curvature [87]. Nanometer vertical (height) resolution is also possible [90]. Unlike Si AFM probes, MWNT AFM probes can resolve the 40–50 nm diameter pores of amaranth starch molecules (Figure 2.8c–e). The MWNT tip is capable of imaging contour-rich samples and of resisting adhesion forces on the interior walls of the pore [37]. In addition, the vertical sidewalls of MWNT tips enable the reduction of the magnitude of extraneous background forces to undetectable levels [45]. Using dynamic AFM operating at frequencies >1 MHz, Katan and Oosterkamp [86] were able to obtain force-volume imaging with hydrophobic CNT tips to simultaneously image the surface and measure force–distance with nanometer resolution in three dimensions.

2.6.2
Biological Imaging in Liquid Medium with CNT AFM Tips

CNT's ductility, high aspect ratio, small radius of curvature, and vertical alignment offer myriad advantages toward biological imaging in aqueous environments. CNT AFM tips are robust and durable probes that provide high-resolution images of delicate, three-dimensional, and contour-, crevice-, and cavity-rich biological samples while minimizing the risk of damage. Current biological imaging applications in liquid involve aggressive imaging of DNA with single nanometer resolution [87, 90]. The imaging of DNA on mica and immersed in 20 mM Tris–HCl and 10 mM MgCl buffer solution with a mechanically attached single MWNT tip has been demonstrated [65]. In an analogous study, MWNT bundle tips with a single protruding MWNT imaged lambda DNA with ~2 nm vertical resolution and 5–8 nm lateral resolution (Figure 2.9a) [84, 90]. Imaging DNA with single SWNT tips reveals direct scaling of the DNA width with the tip diameter [87]. For example, when imaging 2 nm wide DNA with a 2.4 nm diameter SWNT tip, apparent width of the DNA was found to be 4.3 nm [87]. Presently, the maximum attainable lateral

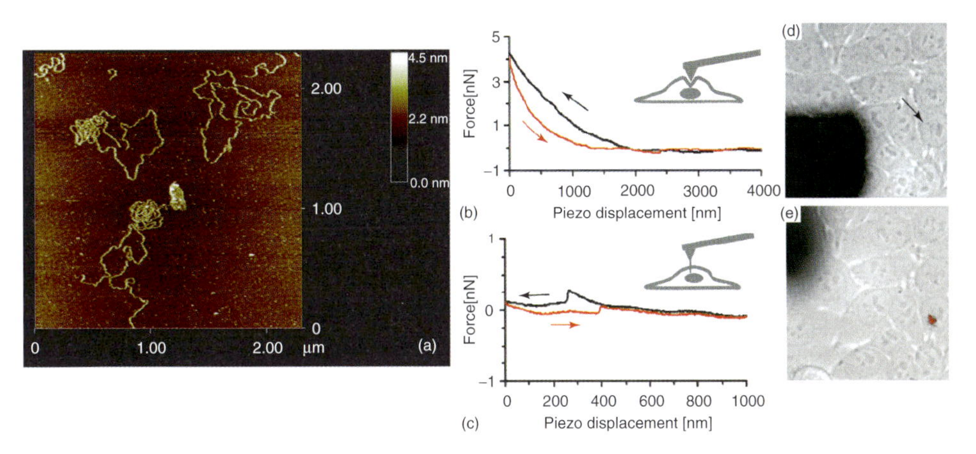

Figure 2.9 (a) Image of plasmid DNA on mica surface in 1 mM $MgCl_2$ taken with MWNT tip [84]. Force curves taken on human mesothelial cells *in vitro* with (b) conventional AFM probe and (c) fortified MWNT tip [57]. Fluorescence microscopy image of HeLa cells (d) before injection and (e) after injection with quantum dots by loaded MWNT tips [63]. The inserted arrow indicates the target cell.

resolution of CNT AFM tips is not sufficient for imaging the DNA double helix structure [87, 90].

2.6.3 Cell Membrane Penetration and Applications of Intracellular CNT AFM Probes

The high-aspect-ratio structure of CNT points to its potential application as a nanoscale needlelike intracellular biological probe for characterizing cells and for drug delivery. The development of CNT AFM tips as intracellular probes requires an understanding of the mechanics of the cell plasma membrane. An excellent review is given in the literature [91]. Here, we summarize the most important points. The cell plasma membrane consists of a phospholipid bilayer connected to an intracellular scaffolding called the *cytoskeleton*. Embedded in the membrane are proteins that perform a wide variety of functions including but not limited to anchoring, cell signaling, and selective permeability to ions. The shape and fluidity of the membrane is partly determined by the extent of carbon–carbon double bonding in the fatty acid chain. The behavior of the phospholipid bilayer transitions along a small temperature range from gel-like at lower temperatures to fluidlike at higher, physiologically relevant temperatures. In addition, the cytoskeleton constituents determine cell membrane elasticity and susceptibility to penetration. Without cytoskeletal support, the membrane bends easily but is difficult to stretch and compress in two dimensions [91]. The rigidity of the cell membrane can be studied through cell indentation experiments, in which the force curve is measured by pushing the AFM probe into the cell membrane [91].

Cell membrane dynamics have been studied with the fortified single MWNT tip protruding from a nanotube bundle as described in Section 2.3 [57]. Probing human mesothelial cells with conventional AFM tips produces a force curve demonstrating the elastic deformation of the cell membrane (Figure 2.9b) [57]. The gradually increasing force in the approach curve (top) indicates simultaneous bending and stretching of the lipid bilayer [57]. For fortified MWNT tips, the force versus distance curve is remarkably different (Figure 2.9c), in that, during approach, the increasing force reaches a maximum after which the force drops to a stable lower magnitude, suggesting successful penetration of the cell membrane [57]. The MWNT tip's ability to penetrate the cell membrane has been used for targeted single cell drug delivery [63]. In a recent study, N-(pyren-1-yl)butanoyl-N′-(biotinyl)cystamine (compound 1) was absorbed onto the MWNT sidewalls and then loaded with streptavidin-coated quantum dots [63]. Compound 1 is connected to biotin through a pH-sensitive disulfide bond that can be cleaved after 15–30 min in the cell's cytosol. Quantum dot (QD) delivery into a single HeLa cell using this method was verified through fluorescence microscopy (Figure 2.9d,e) [63]. Significantly, the labeled HeLa cell remains viable for 10 h after injection [63].

Finally, single MWNT tips have been employed in single cell electrophysiology as a nanometer-diameter solid-state stimulation and recording probe that eliminates the problem of washout in glass micropipettes [62]. Single MWNT tips have been shown to provide lower electrode impedance, thus stimulation efficiency in intracellular excitation experimentation is enhanced, in that lower stimulation voltages are required to elicit action potentials [62, 92]. At low excitation voltages that preclude electrolysis of water, MWNT tips successfully stimulate cells through a capacitive current conduction mechanism thereby preserving cell viability [92]. Furthermore, noise levels in intracellular recording applications can be potentially reduced thereby improving signal-to-noise ratio. Significantly, the high-aspect-ratio MWNT structure has been shown to influence stimulation efficiency. Increasing MWNT length increases the stimulation efficiency through possible electric field enhancement at the tip [92]. This application of single MWNT tips to electrophysiology has demonstrated improved performance and the importance of geometry in electrode design. Although these experiments were not performed in an AFM, they suggest the potential combined use of CNT tips and AFM for simultaneous mechanical, electrophysiological, and membrane surface characterization of living cells *in vitro*.

References

1. de Asis, E.D. Jr., Li, Y., Austin, A., Leung, J., and Nguyen, C.V. (2010) in *Nanoscience and Technology: Scanning Probe Microscopy in Nanoscience and Nanotechnology*,Chapter 5 (ed. B. Bushan), Springer, Heidelberg, Dordrecht, London, New York, pp. 129–168.
2. Di Ventra, M., Evoy, S., and Heflın, J.R. Jr. (2004) *Introduction to Nanoscale Science and Technology*, Kluwer Academic Publishers, Norwell, MA.
3. Hafner, J., Cheung, C.-L., Oosterkamp, T.H., and Lieber, C.M. (2001) *J. Phys. Chem. B*, **105**, 743–746.
4. Nakayama, Y., Nishijima, H., Akita, S., Hohmura, K.I., Yoshimura, S.H., and Takeyasu, K. (2000) *J. Vac. Sci. Technol. B*, **18**, 661–664.
5. Uchihashi, T., Choi, N., Tanigawa, M., Ashino, M., Sugawara, Y., Nishijima, H., Akita, S., Nakayama, Y., Ishikawa, M., Tokumoto, H., Yokoyama, K., Morita, S., and Ishikawa, M. (2000) *Jpn.J. Appl. Phys.*, **39**, L887–L889.
6. Wong, S.S., Joselevich, E., Woolley, A.T., Cheung, C.L., and Lieber, C.M. (1998) *Nature*, **394**, 52–54.
7. Dai, H., Franklin, N., and Han, J. (1998) *Appl. Phys. Lett.*, **73**, 1508–1510.
8. Nagy, G., Levy, M., Scarmozzino, R., Osgood, R.M. Jr., Dai, H., Smalley, R.E., Michaels, C.A., Flynn, G.W., and McLane, G.F. (1998) *Appl. Phys. Lett.*, **73**, 529–531.
9. Wong, S.S., Woolley, A.T., Odom, T.W., Huang, J., Kim, P., Vezenov, D.V., and Lieber, C.M. (1998) *Appl. Phys. Lett.*, **73**, 3465–3467.
10. Arnason, S.B., Rinzler, A.G., Hudspeth, Q., and Hebard, A.F. (1999) *Appl. Phys. Lett.*, **75**, 2842–2844.
11. Nishijima, H., Kamo, S., Akita, S., Nakayama, Y., Hohmura, K.I., Yoshimura, S.H., and Takeyasu, K. (1999) *Appl. Phys. Lett.*, **74**, 4061–4063.
12. Akita, S. and Nakayama, Y. (2002) *Jpn.J. Appl. Phys.*, **41**, 4887–4889.
13. Hall, A., Matthews, W.G., Superfine, R., Falvo, M.R., and Washburn, S. (2003) *Appl. Phys. Lett.*, **82**, 2506–2508.
14. Wade, L.A., Shapiro, I.R., Ma, Z., Quake, S.R., and Collier, C.P. (2004) *Nano Lett.*, **4**, 725–731.
15. Wilson, N.R. and Macpherson, J.V. (2004) *J. Appl. Phys.*, **96**, 3565–3567.
16. Wolny, F., Weissker, U., Muhl, T., Leonhardt, A., Menzel, S., Winkler, A., and Buchner, B. (2006) *J. Appl. Phys.*, **99**, 4304–4309.
17. Choi, N., Uchihashi, T., Nishijima, H., Ishida, T., Mizutani, W., Akita, S., Nakayama, Y., Ishikawa, M., and

Tokumoto, H. (2000) *Jpn. J. Appl. Phys.*, **39**, 3707–3710.
18. Ishikawa, M., Yoshimura, M., and Ueda, K. (2002) *Jpn. J. Appl. Phys.*, **41**, 4908–4910.
19. Wilson, N.R., Cobden, D.H., and Macpherson, J.V. (2002) *J. Phys. Chem. B*, **106**, 13102–13105.
20. Shapiro, I.R., Solares, S.D., Esplandiu, M.J., Wade, L.A., Goddard, W.A., and Collier, C.P. (2004) *J. Phys. Chem. B*, **108**, 13613–13618.
21. Solares, S.D., Esplandiu, M.J., Goddard, W.A., and Collier, C.P. (2005) *J. Phys. Chem. B*, **109**, 11493–11500.
22. Arie, T., Nishijima, H., Akita, S., and Nakayama, Y. (2000) *J. Vac. Sci. Technol. B*, **18**, 104–106.
23. Carnally, S., Barrow, K., Alexander, M.R., Hayes, C.J., Stolnik, S., Tendler, S.J.B., Williams, P.M., and Roberts, C.J. (2007) *Langmuir*, **23**, 3906–3911.
24. Kuwahara, M., Abe, H., Tokumoto, H., Shima, T., Tominaga, J., and Fukada, H. (2004) *Mater. Charact.*, **52**, 43–48.
25. Esplandiu, M.J., Bittner, V.G., Giapis, K.P., and Collier, C.P. (2004) *Nano Lett.*, **4**, 1873–1879.
26. Tang, J., Yang, G., Zhang, Q., Parhat, A., Maynor, B., Liu, J., Qin, L.-C., and Zhou., O. (2005) *Nano Lett.*, **5**, 11–14.
27. Stevens, R.M.D., Frederick, N.A., Smith, B.L., Morse, D.E., Stucky, G.D., and Hansma, P.K. (2000) *Nanotechnology*, **11**, 1–5.
28. Bunch, J.S., Rhodin, T.N., and McEuen, P.L. (2004) *Nanotechnology*, **15**, S76–S78.
29. Lee, S.I., Howell, S.W., Raman, A., Reifenberger, R., Nguyen, C.V., and Meyyappan, M. (2004) *Nanotechnology*, **15**, 416–421.
30. Martinez, J., Yuzvinsky, T.D., Fennimore, A.M., Zettl, A., Garcia, R., and Bustamante, C. (2005) *Nanotechnology*, **16**, 2493–2496.
31. Uchihashi, T., Higgins, M., Nakayama, Y., Sader, J.E., and Jarvis, S.P. (2005) *Nanotechnology*, **16**, S49–S53.
32. Carlson, K., Anderson, K.N., Eichhorn, V., Peterson, D.H., Molhave, K., Bu, I.Y.Y., Teo, K.B.K., Milne, W.I., Fatikow, S., and Boggild, P. (2007) *Nanotechnology*, **18**, 345501–345507.
33. Bernard, C., Marsaudon, S., Boisgard, R., and Aime, J.-P. (2008) *Nanotechnology*, **19**, 035709.
34. Nandhakumar, I.S., Gordon-Smith, T.J., Attard, G.S., and Smith, D.C. (2005) *Small*, **1**, 406–408.
35. Hudspeth, Q.M., Nagle, K.P., Zhao, Y.-P., Karabacak, T., Nguyen, C.V., Meyyappan, M., Wang, G.-C., and Lu, T.-M. (2002) *Surf. Sci.*, **515**, 453–461.
36. Gibson, C.T., Carnally, S., and Clive, C.J. (2007) *Ultramicroscopy*, **107**, 1118–1122.
37. Moloni, K., Buss, M.R., and Andres, R.P. (1999) *Ultramicroscopy*, **80**, 237–246.
38. Yenilmez, E., Wang, Q., Chen, R.J., Wang, D., and Dai, H. (2002) *Appl. Phys. Lett.*, **80**, 2225–2227.
39. Kuwahara, S., Akita, S., Shirakahara, M., Sugai, T., Nakayama, Y., and Shinohara, H. (2006) *Chem. Phys. Lett.*, **429**, 581–585.
40. Tanemura, M., Kitazawa, M., Tanaka, J., Okita, T., Ohta, R., Miao, L., and Tanemura, S. (2006) *Jpn. J. Appl. Phys.*, **45**, 2004–2008.
41. Nguyen, C.V., Yi, Q., and Meyyappan, M. (2005) *Meas. Sci. Tech.*, **16**, 2138–2146.
42. Yi, Q., Cassell, A.M., Liu, H., Chao, K.-K., Han, J., and Meyyappan, M. (2004) *Nano Lett.*, **4**, 1301–1308.
43. Cheung, C.L., Hafner, J.H., Odom, T.W., Kim, K., and Lieber, C.M. (2000) *Appl. Phys. Lett.*, **76**, 3136–3138.
44. Barwich, V., Bammerlin, M., Baratoff, A., Bennewitz, R., Guggisberg, M., Loppacher, C., Pfieffer, O., Meyer, E., Guntherodt, H.-J., Salvetat, J.-P., Bonard, J.-M., and Forro, L. (2000) *Appl. Surf. Sci.*, **157**, 269–273.
45. Kageshima, M., Jensensius, H., Dienwiebel, M., Nakayama, Y., Tokumoto, H., Jarvis, S.P., and Oosterkamp, T.H. (2002) *Appl. Surf. Sci.*, **188**, 440–444.
46. Akita, S., Nishijima, H., and Nakayama, Y. (2000) *J. Phys. D*, **33**, 2673–2677.
47. Ishikawa, M., Yoshimura, M., and Ueda, K. (2002) *Phys. B*, **323**, 184–186.
48. Azehara, H., Kasanuma, Y., Ide, K., Hidaka, K., and Tokumoto, H. (2008) *Jpn. J. Appl. Phys.*, **47**, 3594–3599.

49. Larsen, T., Moloni, K., Flack, F., Eriksson, M.A., Lagally, M.G., and Black, C.T. (2002) *Appl. Phys. Lett.*, **80**, 1996–1998.
50. Ribaya, B.P., Leung, J., Brown, P., Rahman, M., and Nguyen, C.V. (2008) *Nanotechnology*, **19**, 185201.
51. Lee, H.W., Kim, S.H., Kwak, Y.K., and Han, C.S. (2005) *Rev. Sci. Instrum.*, **76**, 046108.
52. Kim, J.-E., Park, J.-K., and Han, C.-S. (2006) *Nanotechnology*, **17**, 2937–2941.
53. Wei, H., Kim, S.N., Zhao, M., Jou, S.-Y., Huey, B.D., Marcus, H.L., and Papadimitrakopoulos, F. (2008) *Chem. Mater.*, **20**, 2793–2801.
54. Wei, Y., Wei, W., Liu, L., and Fan, S. (2008) *Diamond Relat. Mater.*, **17**, 1877–1880.
55. Hafner, J.H., Cheung, C.L., and Lieber, C.M. (1999) *Nature*, **398**, 761–762.
56. Kleckley, S., Chai, G.Y., Zhou, D., Vanfleet, R., and Chow, L. (2003) *Carbon*, **41**, CO1–836.
57. Vakarelski, I.U., Brown, S.C., Higashitani, K., and Moudgil, B.M. (2007) *Langmuir*, **23**, 10893–10896.
58. Wong, S.S., Woolley, A.T., Joselevich, E., Cheung, C.L., and Lieber, C.M. (1999) *Chem. Phys. Lett.*, **306**, 219–225.
59. Tong, J. and Sun, Y. (2007) *IEEE Trans. Nanotechnol.*, **6**, 519–523.
60. Lee, J.-H., Kang, W.-S., Choi, B.-S., Choi, S.-W., and Kim, J.-H. (2008) *Ultramicroscopy*, **108**, 1163–1167.
61. Hirooka, M., Okai, M., Tanaka, H., and Sekino, S. (2008) *Mater. Res. Soc. Symp. Proc.*, **1081**, 1081.
62. de Asis, E.D., Leung, J. Jr., Wood, S., and Nguyen, C.V. (2009) *Appl. Phys. Lett.*, **95**, 153701.
63. Chen, X., Kis, A., Zettl, A., and Bertozzi, C.R. (2007) *Proc. Natl. Acad. Sci. U.S.A.*, **104**, 8218–8222.
64. Wong, S.S., Woolley, A.T., Joselevich, E., Cheung, C.L., and Lieber, C.M. (1998) *J. Am. Chem. Soc.*, **120**, 8557–8558.
65. Stevens, R.M., Nguyen, C.V., and Meyyapan, M. (2004) *IEEE Trans. Nanobio.*, **3**, 56–60.
66. Weisenhorn, A.L., Hansma, P.K., Albrecht, T.R., and Quate, C.F. (1989) *Appl. Phys. Lett.*, **54**, 2651–2653.
67. Tao, Z. and Bhushan, B. (2006) *J. Phys. D: Appl. Phys.*, **39**, 3858–3862.
68. Putman, C.A.J., Van der Werf, K.O., De Grooth, B.G., Van Hulst, N.F., and Greve, J. (1994) *Appl. Phys. Lett.*, **64**, 2454–2456.
69. Dietzel, D., Bernard, C., Maroille, D., Iaia, A., Bonnot, A.M., Aime, J.P., Marsaudon, S., Bertin, F., and Chabli, A. (2006) *J. Scanning Probe Microsc.*, **1**, 39–44.
70. Strus, M.C., Raman, A., Han, C.-S., and Nguyen, C.V. (2005) *Nanotechnology*, **16**, 2482–2492.
71. Pelling, A.E., Li, Y., Shi, W., and Gimzewski, J.K. (2005) *Proc. Natl. Acad. Sci. U.S.A.*, **102**, 6484–6489.
72. Jeffery, S., Hoffmann, P.M., Pethica, J.B., Ramanujan, C., Ozgur Ozer, H., and Oral, A. (2004) *Phys. Rev. B.*, **70**, 054114.
73. Nnebe, I. and Schneider, J.W. (2004) *Langmuir*, **20**, 3195–3201.
74. Schaffer, T.E., Cleveland, J.P., Ohnesorge, F., Walters, D.A., and Hansma, P.K. (1996) *J. Appl. Phys.*, **80**, 3622–3627.
75. Chon, J.W.M., Mulvaney, P., and Sader, J.E. (2000) *J. Appl. Phys.*, **87**, 3978–3988.
76. Barber, A.H., Cohen, S.R., and Wagner, H.D. (2005) *Phys. Rev. B*, **71**, 115443.
77. Sader, J.E. (1998) *J. Appl. Phys.*, **84**, 64–76.
78. Green, C.P. and Sader, J.E. (2002) *J. Appl. Phys.*, **92**, 6262–6274.
79. Duggal, R. and Pasquali, M. (2006) *Phys. Rev. Lett.*, **96**, 246104.
80. Barber, A.H., Cohen, S.R., and Wagner, H.D. (2004) *Phys. Rev. Lett.*, **92**, 186103.
81. de Asis, E.D. Jr., Li, Y., Ohta, R., Leung, J., and Nguyen, C.V. (2008) *Appl. Phys. Lett.*, **93**, 23129.
82. de Asis, E.D. Jr. (2010) Vertically aligned carbon nanofiber microbrush array & single multi-walled carbon nanotube electrode for electrophysiological probing of electrically active cells. Doctoral Thesis. Santa Clara University, Santa Clara, CA.
83. Jarvis, S.P., Uchihashi, T., Ishida, T., Tokumoto, H., and Nakayama, Y. (2000) *J. Phys. Chem. B*, **104**, 6091–6094.

84. Li, J., Cassell, A., and Dai, H. *Application Notes: Carbon Nanotube Tips for MAC Mode AFM Measurements in Liquid*, Molecular Imaging, Phoenix, AZ, *http://cp.literature.agilent.com/litweb/pdf/5989-6376EN.pdf.* accessed 2006
85. Austin, A.J., Ngo, Q., and Nguyen, C.V. (2006) *J. Appl. Phys.*, **99**, 4304–4309.
86. Katan, A.J. and Oosterkamp, T.H. (2008) *J. Phys. Chem. C*, **112**, 9769–9776.
87. Chen, L., Cheung, C.L., Ashby, P.D., and Lieber, C.M. (2004) *Nano. Lett.*, **4**, 1725–1731.
88. Nguyen, C.V., Chao, K.-J., Stevens, R.M.D., Delzeit, L., Cassell, A.M., Han, J., and Meyyappan, M. (2001) *Nanotechnology*, **12**, 363–367.
89. Yasutake, M., Shirakawabe, Y., Okawa, T., Mizooka, S., and Nakayama., Y. (2002) *Ultramicroscopy*, **91**, 57–62.
90. Li, J., Cassell, A.M., and Dai, H. (1999) *Surf. Interface Anal.*, **28**, 8–11.
91. Ikai, A. (2008) *The World of Nano-Biomechanics Mechanical Imaging and Measurement by Atomic Force Microscopy*, Elsevier, Amsterdam.
92. de Asis, E.D. Jr., Leung, J., Wood, S., and Nguyen, C.V. (2010) *Nanotechnology*, **21**, 125101.

3
Force Spectroscopy

Arturo M. Baró

3.1
Introduction

The atomic force microscope (AFM) has demonstrated its capability to provide images with high resolution [1]. To correctly interpret these images, it is important to know what interaction forces are between the tip and sample. This task is done by surface force spectroscopy (SFS). It gives the force curves $F(D)$ as a function of the tip–sample distance D.

Spectroscopy is common to other microscopies. In particular, in the case of scanning tunneling microscope (STM), the first probe microscope, it was found that the topographic images were, in the case of semiconductors, strongly dependent on the tunneling voltage. To understand this observation, another technique was introduced. This was scanning tunneling spectroscopy (STS) where $I(V)$ was measured, I and V being tunneling current and voltage, respectively. Tunneling current $I(s)$ curves are also plotted although their significance is lower [2]. In STM, the tunneling gap is denoted as s.

One of the most challenging objectives of STM was to image biological material [3]. It started at the beginning of STM by the personal decision of H. Rohrer. The need to find appropriate environments was well known. STM images obtained at air ambient pressure were first reported by Baró *et al.* [4], followed by imaging in cryogenic limits [5], under water [6], and in electrolytes [7].

The next important step was to solve the problem of the insulator character of biological material. The introduction of the AFM in 1986 by Binnig *et al.* [8] removed this difficult barrier. AFM breakthrough was the observation that AFM gave not only high-resolution images but was able to study as well the interaction forces between the tip and sample which were responsible of those images. The ability of AFM to work in different liquids was reported by Drake *et al.* [9]. In this chapter, we deal with the operation of AFM in liquid, more precisely in aqueous and buffer solutions [10] avoiding complex sample preparation procedures. From a physicochemical point of view, the liquid is an electrolyte, that is, an aqueous solution containing salts and dissolved ions. In the AFM setup in liquid, the sample and the tip are both immersed in a liquid cell containing the electrolyte

Atomic Force Microscopy in Liquid: Biological Applications, First Edition.
Edited by Arturo M. Baró and Ronald G. Reifenberger.

solution. Concerning intermolecular and surface forces, attractive van der Waals (vdW) forces are always present, although relatively weak. Electrostatic forces are always present at the interface solid liquid due to the electrical charge developed on the solid surface. vdW and electrostatic forces are long-range forces; we also treat short-range forces such as solvation and steric forces. Excellent reviews on FS have been reported by Cappella and Dietler [11] and Butt *et al.* [12].

Most surfaces and adsorbed biomolecules immersed in an electrolyte solution develop a surface charge due to various mechanisms. Charged surfaces will be surrounded by ions of opposite sign so that, from a distance, they appear to be electrically neutral. The arrangement of electric charge around the surface, together with the balancing charge in the solution, is called the electrical double layer (EDL). In AFM, when the surfaces of the sample and tip approach each other, an interaction comes out due to the overlapping of their corresponding EDLs. The surface charge and potential on the sample can be determined from the force between the electrode and the AFM cantilever tip.

We focus our interest in biological systems. In fact, it is well established that EDL interaction plays a crucial role in many biological processes due to its long-range character, to the physiological medium containing a small or large concentration of ions, and to the substantial charge of molecules such as amino and nucleic acids. Examples of these processes are cell adhesion [13], stability of protein structure [14], and DNA condensation [15], as well as protein–membrane [16], protein–protein [17], and protein–DNA [18] interactions. It plays also an important role in colloidal stability. If the force between two dispersed particles in a liquid is repulsive, the particles do not aggregate and the dispersion is stable [19].

From a technical point of view, AFM operation in liquid is more demanding than operation in air. Technical aspects have been treated in Chapter 1. Once the technical questions have been solved, our interest is to illustrate the new and important properties which are associated to the liquid state. We are interested in the forces that take place between biomolecules. Their size covers the 5–100 nm range. In the case of vdW interaction between two single atoms or molecules separated by a distance r, vdW is short range since it depends on distance like $1/r^6$. However, for large bodies (such as biomolecules), the vdW interaction is long range as we show in Section 3.4.

It is known that biological objects are inhomogeneous. For example, the surface charge has patches of uniform value, but these are surrounded by zones having a different charge. A cell membrane is a clear example of this property. An important application of AFM is to use the sharp tip to localize and resolve laterally spectroscopic forces together with the topography. We will illustrate how SFS can provide simultaneously this information in the case of DNA [20].

Although there is an ever-expanding repertoire of single-molecule spectroscopic techniques, atomic force microscopy is one of the most commonly used. Single-molecule force spectroscopy (SMFS) [21] by AFM is based on the measurement of low forces in the range of piconewton and piezoelectric displacements in the subnanometer range. Detailed measurement of the force–extension relationship (elasticity) of individual polymers, in particular of nucleic acids, opened up the possibility to investigate unconventional nucleic acid molecular motors that translocate or otherwise modify DNA or RNA. In parallel, the analysis of rupture force, or force spectra, provides a measure of bond energies, lifetimes, and, more recently, entire energy landscapes [22]. FS is used as a tool to characterize ligand and antibody binding [23] and has been extended to study the complex and multistate unfolding of single protein and nucleic acid structures. Owing to its importance, SMFS is treated separately in Chapter 7.

3.2 Measurement of Force Curves

FS allows probing different kind of attracting as well as repulsing forces. Strictly speaking, a force versus distance curve is a plot of the deflection of the cantilever versus the extension of the piezoelectric scanner. The way to move the sample toward the tip is done by changing the voltage of the piezoelectric attached to the sample. Since the tip and sample are located along the vertical axis, the piezo used for this displacement is the z-piezo. The vertical axis of the force curve records the photodiode signal, which can be transformed in cantilever deflection or in force measured by the cantilever once it is calibrated. On the other hand, the horizontal axis of the force curve records the voltage applied to the z-piezo that moves either the sample or the tip. This depends on the AFM design. In commercial AFMs, both cases are found. Just as an option, we take the case where the voltage applied to the z-piezo moves the sample. The z-piezo voltage can be actually transformed to give the tip–sample distance. To do these transformations, it is necessary to know the spring constant of the cantilever, and three parameters to be deduced from force curves: (i) the point of contact between the tip and sample, (ii) the system sensitivity, and (iii) the photodiode voltage when the tip and sample are far away (zero force).

When there is no interaction between the tip and sample, the photodiode voltage has a constant value that corresponds to zero force. This value is taken as a reference for the remaining points of the curve, so that all measured values of photodiode voltage are modified by subtracting this value. To convert the photodiode voltage signal to cantilever deflection, it is necessary to know the system sensitivity $\Delta z/\Delta V$, which is known if z is known. An easy way to do it is, to push the tip against a hard sample like mica. In this case, the increment of cantilever deflection voltage coincides with the distance traveled by the mica. This draws a straight line and both approach and retraction have to be the same. However, there are a number of complications and disadvantages associated with this measurement. In particular for biological samples, it may be the case that the sample substrate is soft. Force measurements on soft surfaces show a nonlinear response in the tip–sample contact region. An alternative method provides a quick *noncontact* approach to calculate the dynamic and static optical lever sensitivity in both air and liquid [24].

To convert z-displacement to the tip–sample distance, it is necessary to detect the contact point in the force curve; from there tip and sample go together. If we assume that the contact line is a straight line (remember that there are more complex situations, see above), the distance D between tip and sample is zero. This means that z-displacement is zero at the contact point. Therefore, to make the conversion from z-displacement to the tip–sample distance, it is necessary to remove $z_{\text{tip}}-z_{\text{sample}}$ to z-displacement in all the abscissa axis points of the force curve.

3.2.1
Analysis of Force Curves Taken in Air

When tip–sample distance gets smaller, some interaction force starts to develop. In the absence of any other interaction, a weak interaction barely visible is due to an attractive vdW force. Inherent to AFM, tip–sample force is in equilibrium with spring force, $F_{\text{i}} + F_{\text{k}} = 0$. However, in the case of an attractive force, when the sample does approach too much to the tip, the cantilever force cannot counterbalance the tip–sample interaction $F_{\text{i}} > -F_{\text{k}}$ and the tip jumps onto the sample surface (jump in) due to a mechanical instability. This occurs when the gradient of the curve $F(z)$ becomes larger than the spring constant c. This is explained with elegance by Butt *et al.* [12].

A different way to look at the tip–sample instability is to examine the variation of the resonance frequency of the tip–sample system as the tip approaches the surface. According to this strategy, the AFM setup is approximated to a harmonic oscillator. With an interaction potential $V(z)$, the resonance frequency of the free cantilever ($\omega_{00}^2 = c/m_{\text{eff}}$) is modified according to [25]

$$\omega_{00}^2 0\,(z) = \frac{c + V''\,(z)}{m_{\text{eff}}} = \omega_{00}\left(1 + \frac{V''\,(z)}{c}\right) \tag{3.1}$$

where ω_{00} is the free resonance frequency and $V''(z)$ the force gradient of the interaction potential. For attractive potentials, that is, for potentials which induce forces that pull the tip toward the surface, the force gradient is negative and therefore the resonance frequency shifts toward lower values. An interesting feature of Eq. (3.1) is that for $V'' = -c$, the resonance frequency is physically not defined. This coincides with the instability that we have described before, which occurs when the force gradient is more negative than the force constant of the cantilever. The vdW tip–sample interaction represents just one contribution to the cantilever deflection. In the discussion above, we have considered that the jump to contact is due to the force gradient of the surface potential (typically vdW force). However, it has been shown that in the case of AFM operation in air ambient, the attractive force is due to the spontaneous creation of a meniscus formed between the interfaces of tip and sample [26]. There is a way to avoid the meniscus formation, by building a relative humidity control chamber that encloses the AFM. The relative humidity is controlled by passing a ratio of dry and humid nitrogen across the sample [27].

In normal atmospheric air conditions, as a consequence of the instability, the sample jumps into the tip (Figure 3.1a). If the sample is made of a hard material (mica, for example), tip and sample move together. The force curve takes the form of a straight line where the force is positive and the slope is constant and negative, which indicates that it is a repulsive force. As a matter of fact what is really obtained is the photovoltage cantilever deflection. There is a maximum photovoltage value that cannot exceed a reference or trigger value, in order that there is not a crush between sample and tip. Some AFMs do this process in a different way.

Once this process is finished, the maximum "force" has to be as low as possible in order to avoid damage to the sample particularly in the case of biological material. Then, the sample is withdrawn and the force traces its way back. The withdrawn curve does not coincide with the approach curve, but it continues further below the zero force value, because of the presence of several water layers adsorbed on the surface (Figure 3.1a).

This water layer produces a capillary force that is very strong and attractive. As the scanner pulls away from the surface, the water holds the tip in contact with the surface, bending the cantilever strongly toward the surface. At some force value (adhesion force), the scanner retracts enough that the tip springs free. This is known as the *snap-back point*. Since the snap-in and snap-back points do not coincide, the force curve has hysteresis (Figure 3.1a) and this is a characteristic property of this behavior. The adhesion position corresponding to the snap-back point corresponds to the product of the adhesion force and the spring constant of the cantilever [28].

The value of the capillary force in air is large (in the range of nanonewton), generating a problem in the analysis of force curves, by hiding possible peaks

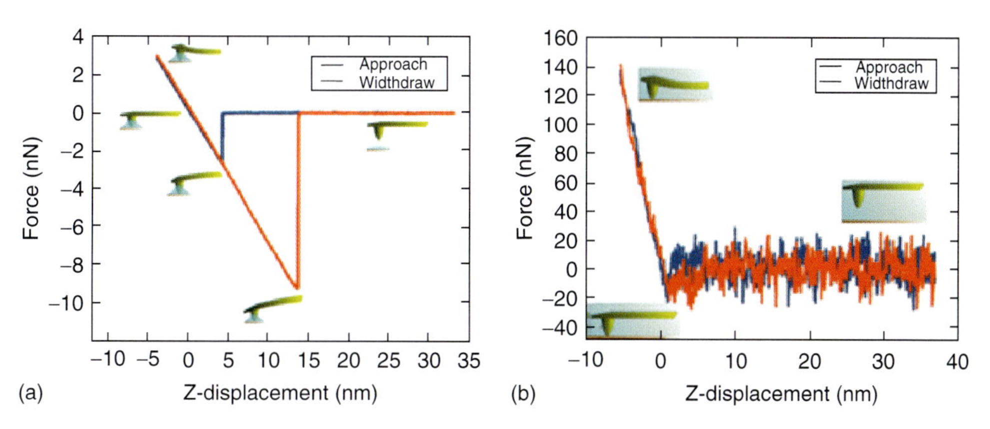

Figure 3.1 Force curves obtained on a mica surface in (a) air ambient atmosphere and (b) in PBS buffer 150 mM pH 7.4. In both curves, the blue color corresponds to approach and the red to withdraw. During the approach at air ambient, a jump in contact to the sample surface is observed due to the formation of a water meniscus between tip and sample. During separation, an intense adhesion peak is generated by the water meniscus. In liquid environment, the jump in as well as the adhesion peak disappear. Cantilever deflection is shown in several points of the force curves. (Reprinted with permission from J. Sotres 2009, Ph. D. Thesis.)

having smaller value. A second consequence is that the large value of the adhesion force contributes to the loading force applied to the sample, since they add together. In total, it can reach a very large value if the AFM is operated in contact, producing a large deformation particularly in the case of soft-matter-like biological material. For DNA, the reduced measured height was attributed to deformation [29]. A quantitative study of this deformation was reported by Moreno-Herrero *et al.* [30] where a height of 1.4 nm was found for an applied force of 300 pN and 1.8 nm for a normal force of 100 pN.

In this chapter, we present the force curves by using the static operation mode of the microscope. In the static mode, the cantilever is not oscillated, but it measures cantilever deflection when there is a force between tip and sample. The force curves are done with the "jumping mode" [31] where during the scanning, the tip is withdrawn from the sample to avoid the tip dragging the sample.

3.2.2 Analysis of Force Curves in a Liquid

The amount of hysteresis in a force versus distance curve is found to be much larger in air (Figure 3.1a) than in liquid (Figure 3.1b). This result was already reported by Weisenhorn *et al.* [32]. They were looking at the differences of $F(D)$ curves when working in air or in water and found that the adhesion peak value was lower (in the piconewton range). This is an important observation that favors AFM operation in liquid versus air. This property is due to the removal of the capillary force between tip and sample when they are both in the liquid cell and consequently completely immersed in the liquid. Under these conditions, capillary forces disappear, which is an extra advantage for liquid operation. Nevertheless, other forces exist such as electrostatic and vdW forces discussed in Sections 3.5 and 3.6.

3.3 Measuring Surface Forces by the Surface Force Apparatus

Another technique to measure the interaction between two macroscopic surfaces in air, vacuum, or a *liquid* environment is the surface force apparatus (SFA), which was invented by Israelachvili 1992 [33]. It allows the measurements of interactions between biological molecules (such as lipids and proteins). The SFA geometry involves two smooth cylindrical curved surfaces at 90° (mostly mica due to its planar surface) with a distance resolution of 0.1 nm and force sensitivity in the 10^{-8} N range. It can be used to measure electrostatic forces and even hydration or solvation forces [34]. SFA is quite demanding and only a handful of laboratories have functional instruments. AFM and SFA have some similarities except that in AFM, the sample surface does not need to be flat. In addition, the AFM sharp tip allows to measure inhomogeneous samples.

3.4
Forces between Macroscopic Bodies

The forces between macroscopic bodies can be calculated from the forces between single atoms or single molecules. They have several common properties: (i) the net interaction energy is proportional to size (e.g., radius) and geometrical shape and (ii) the energy and force decay are much more slowly as a function of the separation [33].

To obtain the interaction energies between two macroscopic surfaces, he starts from the pair potential between two atoms or two small molecules. With the further assumption of additivity, the net interaction energy of a molecule and a planar surface of a solid, made up of like molecules separated by a distance D, is obtained. Following this idea, they get the interaction energy between two solid surfaces like two flat plates.

The next step is to use the Derjaguin approximation [35] that considers the influence of arbitrary geometry on the interaction potential $W(D)$ by reducing it to the simple geometry of two flat surfaces $U(A)$ separated by a distance x, obtaining:

$$U(D) = \int U_A(x)\mathrm{d}x \tag{3.2}$$

For forces, there is an analogous expression

$$F(D) = \int f(x)\mathrm{d}A \tag{3.3}$$

where F is the force between two bodies of arbitrary shape and f the force per unit area between two flat surfaces. This permits to give the force between two spheres in terms of the energy per unit area of two flat surfaces at the same separation.

$$F(D) \approx 2\pi \frac{R_1 R_2}{R_1 + R_2} W(D) \tag{3.4}$$

If one sphere is very large so that $R_2 \gg R_1$, $F(D) = 2\pi R_1\ W(D)$, which corresponds to the limiting case of a sphere near a flat surface.

3.5
Theory of DLVO Forces between Two Surfaces

The DLVO theory which comes from the initials of their authors Derjaguin–Landau–Verwey–Overbeek describes the interaction between similarly charged double layers interacting through a liquid medium at low-surface potentials or when the potential energy of an elementary charge on the surface is much smaller than the thermal energy scale, $k_B T$. It combines the effects of the vdW attraction and the electrostatic repulsion (EDC) due to the so-called double layer of counterions. The central concept of the DLVO theory is that the total interaction energy of two surfaces or particles is given by the sum of the attractive W_{vdW} and repulsive W_{EDC} contributions:

$$F_t(D) = F_{vdW}(D) + F_{EDC}(D) \tag{3.5}$$

The DLVO theory establishes that like-charged colloidal particles in an electrolyte should experience a purely repulsive screened electrostatic (coulombic) interaction. The vdW attraction is greater than the double layer repulsion at very short distances. Experimental data about this point can be seen in Figure 3.3 (Section 3.7).

3.6
Van der Waals Forces – the Hamaker Constant

The total interaction between any two surfaces must include the vdW attraction. vdW is largely insensitive to variations in electrolyte concentration and pH so that it may be considered as fixed. The Derjaguin approximation can be used to obtain the vdW force between two spheres of the same radius R. This is

$$F_{\mathrm{vdW}}(D) = -HR/12D^2 \tag{3.6}$$

where H is the Hamaker constant and D the separation distance. Hamaker constants have been calculated and experimentally measured for two identical phases interacting across the air or across a medium (for example, water). The Hamaker constant depends on the medium. In the case of mica interacting across air, the Hamaker constant is 10^{-19} J, whereas across water it is 2×10^{-20} J (measured with the surface force apparatus). We have measured by AFM the system mica as sample and (Si_3N_4) as tip across electrolytes at various concentrations, and found $H \sim 7 \times 10^{-21}$ J (see Figure 3.3, Section 3.7).

3.7
Electrostatic Force between Surfaces in a Liquid

An EDL is a structure that appears on the surface of a charged object that acquires charge when it is placed in an electrolyte. The object might be a solid surface or a colloidal particle or a porous body. Most surfaces immersed in an electrolyte solution develop a surface charge. Various processes are responsible of the charging. At the interface between a surface and an aqueous solution by the dissociation of surface groups (e.g., the dissociation of protons from carboxylic groups):

$$\mathrm{R-COOH \rightarrow H^+ + R-COO^-} \tag{3.7}$$

where the surface acquires a negative charge. Another mechanism is by the adsorption of ions from the solution onto a previously uncharged surface. Owing to the high dielectric constant of water, surface dissociation or adsorption of a charged surface in water is very common. Whatever the charging mechanism is, the final surface charge is balanced by an equal but oppositely charged region of counterions, some of which are bound, usually transiently, to the surface within the so-called Stern layer, while others form a second layer loosely associated with the charged surface because they are free ions that move in the fluid under the

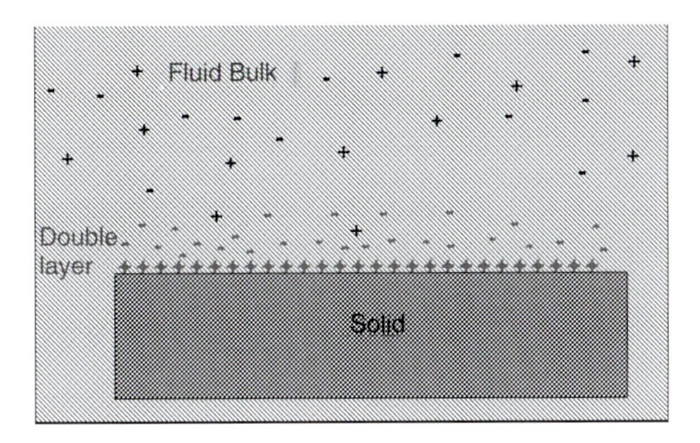

Figure 3.2 Schematic of double layer in a liquid at contact with a positively charged solid. Depending on the nature of the solid, there may be another double layer (unmarked on the drawing) inside the solid (*http://en.wikipedia.org/wiki/Double_layer_(interfacial)*).

influence of electric attraction and thermal motion, known as the *diffuse electric double layer* [33] (Figure 3.2).

In the AFM, there are two charged surfaces immersed in a liquid and each one is surrounded by its own EDL. When sample and tip approach each other, their EDLs overlap, and consequently, an electrostatic force builds up [36].

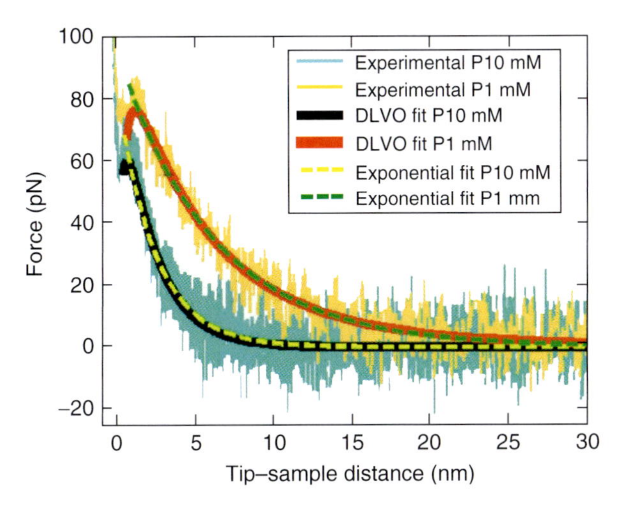

Figure 3.3 Force curves as a function of tip–sample distance in the case of a silicon nitride tip scanning a mica surface in sodium phosphate at 1 and 10 mM concentrations. The fits of both plots to Eq. (3.13) are drawn as continuous lines and those to Eq. (3.12) as dashed lines. From the fits, we attribute the repulsive part of the curves to the EDL force. The attractive part is attributed to the vdW interaction and is analyzed by the DLVO expression (red and black thicker lines correspond to 1 and 10 mM concentrations). This allows obtaining the average value of $H \sim 7 \times 10^{-21}$ J. (Reprinted with permission from J. Sotres 2009, Ph. D. Thesis.)

The most common approach for calculating the EDL force is to take a Boltzmann distribution of counterions. This is given by

$$\rho(x) = \rho_0 \exp\left(-ze\psi(x)/k_B T\right) \tag{3.8}$$

where $\psi(x)$ and $\rho(x)$ are the electrostatic potential and the number density of counterions of valence z at each point between the surfaces. This together with the Poisson equation yields

$$\frac{d^2\psi(x)}{dx^2} = -\frac{ze\rho(x)}{\varepsilon\varepsilon_0} \tag{3.9}$$

ε being the dielectric constant of the liquid. This is the Poisson–Boltzmann (PB) equation. When solved, the PB equation gives the potential ψ, electric field $E = d\psi/dx$, and counterion density ρ. PB equation is a second-order nonlinear differential equation that cannot be solved analytically. In addition, it does not consider ion size and ion interactions.

To prevent this mathematical difficulty, a linearization of the exponential term is often down by neglecting the higher order terms. It is worth to mention that this approximation is in good agreement with experimental results particularly with forces between charged surfaces measured by the SFA technique.

One of these approximations is based on the following expression for the EDL interaction energy per unit area of two parallel plane half-spaces:

$$W_{EDL}(D) = \frac{2\sigma_1\sigma_2\lambda_D}{\varepsilon\varepsilon_0} \exp -\frac{D}{\lambda_D} \tag{3.10}$$

where ε is the dielectric constant of the medium, ε_0, σ_1, and σ_2 are the surface charge densities of both surfaces. λ_D also known as the *Debye length*, is the decay length of the interaction, defined as

$$\lambda_D = \left(\frac{\varepsilon\varepsilon_0 k_B T}{e^2 \Sigma_i c_i z_i^2}\right)^{1/2} \tag{3.11}$$

where $k_B T$ is the Boltzmann factor, c_i and z_i are the concentration and valence values of the ith type of ion present in the solution.

The interaction energy W given by Eq. (3.10) can be related to the force between two curved spheres coming from the Derjaguin approximation. Therefore, the EDL force between an AFM tip and a plane surface with surface charge densities σ_{tip} and σ_{sample}, respectively, separated by a distance D, can be expressed as

$$F_{EDL}(D) = \frac{4\pi R_1 \sigma_{tip}\sigma_{sample}\lambda_D}{\varepsilon\varepsilon_0} \exp -D/\lambda_D \tag{3.12}$$

This derivation is valid under several assumptions, including small surface potentials, tip–sample separations larger than the Debye length, and tip radii larger than the separation, $R \gg D \gg \lambda_D$.

Other studies have been reported. Butt [37] calculated the following expression that includes the vdW force and is valid for tip radius R larger than tip–sample separation D

$$F_{\mathrm{DLVO}}(D) = \frac{-HR}{6D^2} + \frac{2\pi\lambda_{\mathrm{D}}R}{\varepsilon}\left[\left(\sigma_1^2+\sigma_2^2\right)\exp\left(-2D/\lambda_{\mathrm{D}}\right)+2\sigma_1\sigma_2\exp -D/\lambda_{\mathrm{D}}\right] \tag{3.13}$$

It is interesting to apply Eqs. (3.12) and (3.13) to our force curves. Equation (3.13) should be more accurate, but Eq. (3.12) (Derjaguin approximation) is also used because of its simplicity. Although the exponential function of Eq. (3.12) should only be valid for separations $D \gg \lambda_{\mathrm{D}}$, it is often used to adjust force curves even in the case of shorter separations (Figure 3.3).

We are aware that the expressions based on Derjaguin approximation loose accuracy when the Debye length of the medium is comparable to the separation and size of the interacting surfaces, which is likely to be the case in EDL force measurements by AFM. Moreover, our model considers the molecule as a sphere smaller than the AFM tip, when this will not always be the case. Nevertheless, these approximations are commonly used to model EDL forces from AFM measurements [38]. In spite of the reasonable agreement of Derjaguin approximation, more accurate force expressions can be obtained by applying more sophisticated strategies such as the surface element integration (SEI) method [39].

Another study was reported by attaching single-nanoparticle-terminated tips for SPM. When the size of the particles becomes comparable to the range of the interaction force, some deviations can be expected and the Derjaguin approximation might no longer be valid. Force curves scaled by the particle radius of $R = 20$ and 10 nm were formally fitted with standard DLVO theory. However, there was an apparent increase in the Debye length, from about 11.5 nm for $R = 20$ nm to 14.0 nm for the $R = 10$ nm particle. These values are well above the theoretical value of $\lambda_{\mathrm{D}} = 9.6$ nm for 10^{-3} M NaCl and $\lambda_{\mathrm{D}} \ll R$ [40].

The low values of the Hamaker constant in solution (Figure 3.3) contribute to reduce tip–sample interaction in liquid. We can estimate vdW force diminution in a typical AFM experiment. For a tip radius of $R \sim 10$–30 nm, vdW force attains values below AFM resolution (10 pN) for 1–2 nm separation. This means that under these conditions, for $D \geq 1$–2 nm, we may only take into account the EDL force.

It is quite evident that the EDL force depends on the surface charge of tip and sample. Such dependence is shown in Eq. (3.12), which indicates that the EDL force is proportional to the product $\sigma_{\mathrm{tip}}\sigma_{\mathrm{sample}}$, where σ_{tip} and σ_{sample} are the surface charge densities of tip and sample. This means that if tip and sample surfaces are both negatively or positively charged, the EDL force is repulsive. But if tip and sample surfaces have charges of opposite sign, the EDL force is attractive. Notice that Debye length is smaller for higher ion concentrations (blue curve) (Eq. (3.11)) (Figure 3.4).

The measured electrostatic force in an aqueous solution depends on boundary conditions, that is, the surface charge and potential on the electrode. Electrode potential can be controlled by placing it under electrochemical control. The surface charge on the electrode is inferred from the force between the electrode and the tip on the AFM cantilever as a function of the electrode potential.

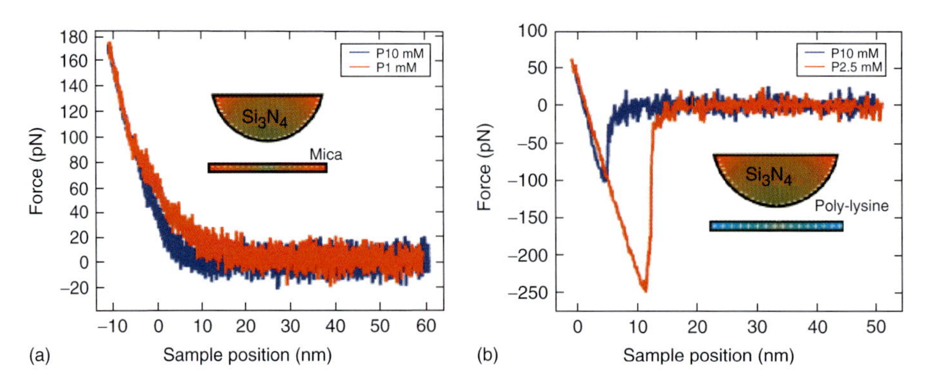

Figure 3.4 (a) Force curves obtained in sodium phosphate pH 7 10 mM (blue curve) and 1 mM (red curve) for silicon nitride tip and mica surface; under these conditions, both surfaces are negatively charged. (b) Force curves taken with a silicon nitride tip in sodium phosphate pH 7 10 mM (blue curve) and 2.5 mM (red curve). On a mica surface covered with polylysine (the sample acquires a positive charge at pH 7). (Reprinted with permission from J. Sotres 2009, Ph. D. Thesis.)

This experiment has been done [41] with a three-electrode electrochemical cell with the gold substrate acting as the working electrode, in the presence of an electrolyte solution with a gold electrode under electrochemical control. The silica (tip)–gold interaction is a strong function of the applied potential and the nature of the electrolyte (silica is negatively charged at the pH used). In aqueous solutions containing KCl, if the Au electrode potential is negative, indicating that the electrode surface is negatively charged, the EDL interaction force between silica and gold at negative potentials is repulsive. When the potential of the electrode is made more positive, the net surface charge becomes positive and the diffusive double layer is formed by repulsion of K^+ and attraction of Cl^-. The measured EDL force is attractive under these conditions.

At constant potential, the charge on both surfaces may change. When the surfaces have a potential of the same sign, the charge on the surface with smaller potential will change sign as the surfaces approach because of an induced charge from the surface of higher potential, which leads to an attractive force. For surfaces with potentials of opposite sign, the surface charges will both increase in magnitude as the surfaces approach, which leads to an even larger attractive force. Experiments suggest that the constant charge model describes the data more accurately than the constant potential model [41].

3.8 Spatially Resolved Force Spectroscopy

We are going to consider now the AFM imaging and EDL force measurement done on a single biomolecule, having a size similar to tip size and Debye length. We have employed a strategy that involves the simultaneous AFM imaging of topography and force curves. This requires the acquisition of a force–distance curve at each pixel of a simultaneously acquired AFM topography image. The method is known by different names: imaging spectroscopy [42], force mapping [43], force volume, [44], and fluid electric force microscopy (FEFM) [45]. Some problems of these techniques, usually associated with long acquisition times, have commonly resulted in a low lateral resolution. We are going to use force spectroscopy imaging (FSI) [46] to emphasize not only the high-resolution AFM imaging of surface topography but also the spectroscopy and imaging of EDL forces. With this method, a resolution of $\sim$12 nm is obtained. To reach such a resolution, an important effort has been devoted to data processing [47] because of the large amount of force values ($>10^6$) produced by a single FSI measurement.

An important characteristic of spatially resolved FS is the possibility to extract interesting physical properties of the sample. The property that can be inferred from EDL force measurements is the surface charge density (Eq. (3.12)). This property can be applied to a single DNA molecule. Likewise, the mechanical properties of a virus [48] or a cell can be solved laterally.

In AFM imaging of DNA deposited on a mica substrate, the tip (negatively charged) is scanning both the DNA molecule and the mica substrate. Since DNA is a small molecule (2.5 nm in diameter), which is lower than the tip radius, it is anticipated that the EDL force will have a substrate contribution. For modeling this situation, we make use of a common strategy where the total EDL force acting on a surface by two or more other surfaces is a linear superposition of the force between each pair of surfaces. This approximation has been used by Müller *et al.* [49]. A more involved calculation by Das and Bhattacharjee [50] in the case of a particle approaching a surface with several particles deposited concludes that the approximation is reasonable when $\lambda_D < R$. Therefore, the force acting on an AFM cantilever close to a molecule is given by the force between the tip and the molecule, F_{tm}, plus the force between the tip and the supporting substrate, F_{ts}. On the other hand, if the molecule is considered as a sphere much smaller than the AFM tip, it is reasonable to approximate, within this component of the force, the surface of the tip as a plane. In this case, Eq. (3.12) can also be used to model F_{tm}. Thus, the total EDL force between the tip and a close molecule deposited on a plane substrate is

$$F_{EDL}(D_{ts}, D_{tm}) = F_{ts}(D_{ts}) + F_{tm}(D_{tm}) = F_{ts0}\exp(-D_{ts}/\lambda_D) + F_{tm0}\exp(-D_{tm}/\lambda_D) \tag{3.14}$$

$$F_{ts0} = \frac{4\pi R_t\sigma_t\sigma_s\lambda_D}{\varepsilon\varepsilon_0} \tag{3.15}$$

$$F_{tm0} = \frac{4\pi R_m\sigma_t\sigma_m\lambda_D}{\varepsilon\varepsilon_0} \tag{3.16}$$

where D_{ts} and D_{tm} are the distances between tip and substrate and between tip and molecule, respectively, σ_m is the surface charge density of the molecule, and R_m its effective radius.

3.9
Force Spectroscopy Imaging of Single DNA Molecules

From Eq. 3.13, we expect that the EDL force exerted on the probe tip near a molecule will show a strong dependence on the separation between tip and molecule D_{tm}. For this reason, the operation in the FSI mode stands out as a promising strategy for studying EDL forces on single biomolecules, as in this mode the position of the tip on the sample where each force curve is performed, is known with high accuracy.

We use a DNA molecule deposited on a mica substrate, the tip (Si_3N_4) being negatively charged. In the case of DNA, mica is covered with polylysine (PL), in order that DNA is strongly fixed on the substrate. As a consequence, the substrate is positively charged. The results are presented in Figure 3.5. Images (a) and (b) are

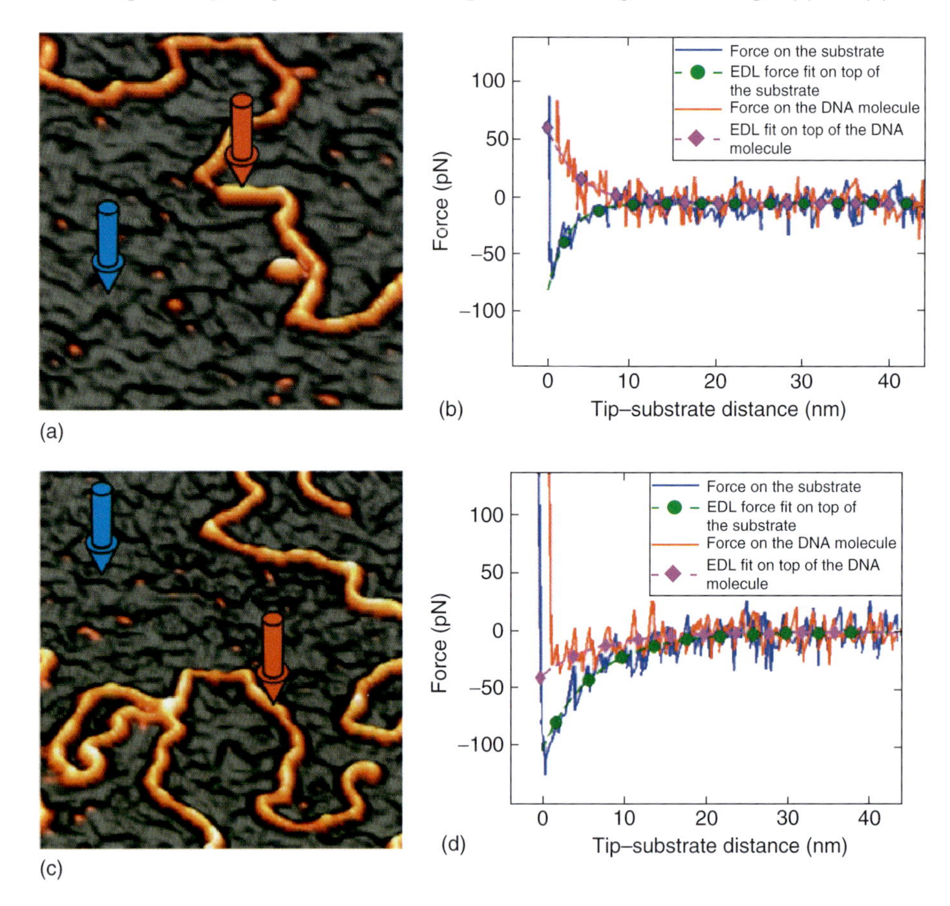

Figure 3.5 Force curves as a function of tip position on DNA or substrate for two different buffer concentrations (see text). (Reproduced from [46] with permission. Copyright 2010, Elsevier.)

taken in a buffer of 10 mM of sodium phosphate that gives a λ_D value = 2.3 nm (Eq. (3.11)), while images (c) and (d) have been taken in a buffer of KCl 1.5 mM, pH 7, which leads to a value of λ_D = 8.3 nm (Eq. (3.11)) (Figure 3.5a). The curves in images (b) and (d) were performed at the positions pointed by arrows of the same color. The curve on top of the PL substrate (blue curve) (Figure 3.5b) exhibits an attractive EDL force as expected for a negative tip approaching a positively charged surface. In the other hand, the EDL force on top of the DNA (red curve) exhibits a repulsive EDL force, as expected for a negative tip approaching a negative molecule. In Figure 3.5d, however, we observe a small attractive contribution which is coming from tip–substrate interaction. This is surprising since images (a) and (c) correspond to regions with the same amount of DNA. The behavior in Figure 3.5d can be explained by the force between the tip and the substrate being superposed to that between the tip and the molecule; this is, in fact, a definite proof of the contribution of the substrate to the EDL value measured on top of the molecule. Another interesting result is due to the different Debye length in images (a) and (c), the force amplitude in Figure 3.5c being smaller than that in Figure 3.5d. This is due to the different Debye lengths in both cases. The most important result of this section is that DNA is resolved in the image of EDL force.

An important characteristic of FSI is the possibility to extract interesting physical properties of the sample. The property that can be inferred from EDL force measurements is the surface charge density (Eq. (3.13)). It has been shown that, by operating in the FSI mode, this property can be measured and imaged down to the single DNA molecule level [46]. The most important difficulty to obtain the surface charge density is the dispersion of the obtained data. This is due to the poor knowledge of the tip characteristics in terms of size and shape and its variability even during a single experiment.

To avoid this problem [51], introduced the so-called colloidal probe technique by gluing a micrometer-sized spherical particle to the AFM cantilever. This has become a powerful tool for quantitative measurements of colloidal forces. However, this is not an option if nanometer lateral resolution is required.

3.10 Solvation Forces

In addition to vdW and electrostatic forces, there are other non-DLVO forces. They take place at small separations between two surfaces (tip and sample surfaces for AFM). The cause of this interaction comes from the fact that the liquid molecules can form one or several ordered layers Therefore, this occurs at small separations, below a few molecular diameters (water diameter about 0.29 nm). These interactions are normally referred to as *solvation forces*. Water is particularly interesting due to its omnipresence in all the cases, but its role is special as the primary medium for biological interactions. Solvation forces are known as *hydration forces* in the case of water molecules.

Short-range forces in liquid were first measured by the SFA [52]. Solvation forces have also been observed between the surfaces of the sample and the AFM tip, which brings the advantages of high spatial resolution to solvation force studies. The solvation structure has been observed for both the OMCTS/graphite and dodecanol/graphite systems [53]. The solvation structure can be seen out to seven shells for both systems. This is comparable with the sensitivity achieved by the SFA. Equally importantly, AFM allows a wider range of solid substrates than used hitherto in SFA.

One of the most interesting systems is that dedicated to the interaction of structured water layers and biological processes. In particular, many processes occur at the lipid/water interface. The unique properties of water may play a critical role in functions such as membrane permeation. It is important to know what are the forces that take place between membrane surfaces and water. The hydration force is that required to remove strongly bound water molecules (Figure 3.6). AFM has increasingly been used to measure oscillatory forces (Figure 3.6) between the probe tip and various substrates in pure water.

Recently, frequency modulation–atomic force microscope (FM–AFM) (treated specifically in Chapter 4) is being used to perform high-resolution imaging before measuring the oscillatory force. FM–AFM measurements reported by Higgins *et al.*

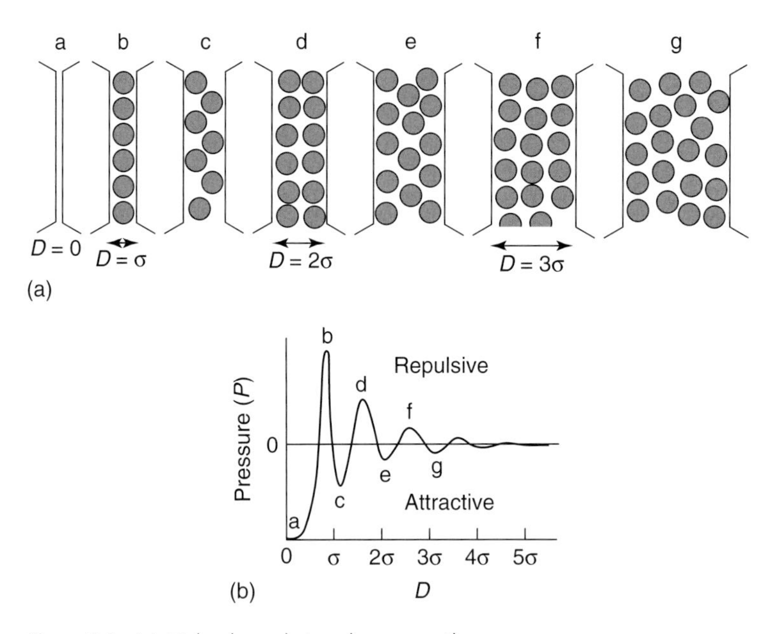

Figure 3.6 (a) Molecular ordering changes as the separation D changes. Note that the density of liquid molecules in contact with the surface varies between maxima and minima. (b) Schematic solvation force. (Reprinted with permission from [33]. Copyright (2011), Elsevier.)

[54] show oscillations that indicate molecular ordering of individual water layers between tip and lipid surface. These findings are of fundamental significance for short-range biological interactions as they highlight the importance of considering water structure as an integral component in cell membrane theory.

3.11 Hydrophobic Forces

A hydrophobic surface usually has no polar groups or hydrogen-bonding sites so that there is no affinity for water and the surface to bond together. The orientation of water molecules in contact with a hydrophobic molecule is entropically unfavorable; therefore, two such molecules tend to come together simply by attracting each other. In spite of these ideas, the intimate mechanisms of hydrophobicity remain elusive. Another source of confusion is the apparent existence of two different force regimes. It has been suggested that the measured force between hydrophobic surfaces is, in fact, a combination of a "truly hydrophobic" shorter range force (<10 nm) and a longer range force (>10 nm) because of an unclear mechanism.

Hydrophobic forces have been investigated in model surfaces by the SFA. Chemical force microscopy (CFM) with hydrophobic, methyl-terminated AFM tips is a valuable tool for measuring the hydrophobicity of bacterial surfaces. The use of CFM has been validated for quantifying short-range hydrophobic forces. CFM has been applied to *Mycobacterium bovis*, the surface of which is believed to be rich hydrophobic mycolic acids. The measurement of spatially resolved force curves recorded with a CH_3-modified tip yielded large adhesion forces uniformly distributed on the surface. Comparison with the data obtained on organic surfaces indicates that the bacterial surface is markedly hydrophobic [55].

3.12 Steric Forces

When attaching at some point to a solid–liquid interface, chain molecules dangle out into the solution where they are thermally mobile. On approach of two polymer-covered surfaces, the entropy of confining these dangling chains results in a repulsive entropic force which, for overlapping polymer molecules, is known as the *steric* or *overlap repulsion*. This repulsion effect has been used to directly measure forces bearing adsorbed flexible polymers. The stabilization of colloidal dispersions by naturally occurring polymers has been exploited continuously by man for almost five millennia. It is normally termed as *steric stabilization*.

AFM offers the opportunity to locally probe molecular forces at hydrated bacterial surfaces by means of FS. For that measurement, the tip can be modified by the immobilization of alkanethiol monolayers. Owing to their dimensions, bacterial cells in solution may be considered as colloidal particles and the adhesion process can be explained by the DLVO theory. However, microbial cells can form specific

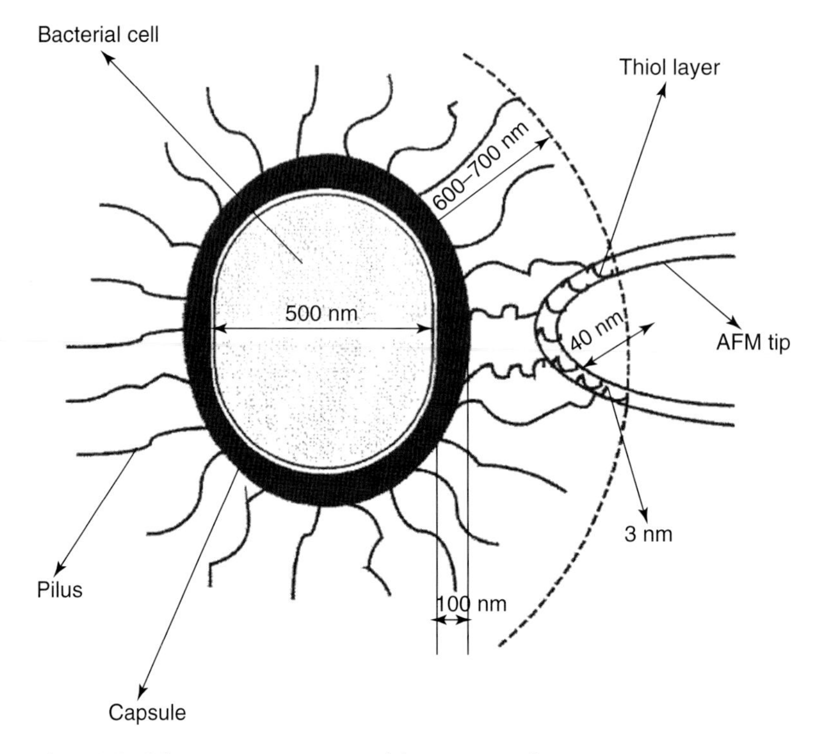

Figure 3.7 Schematic representation of the interaction between an adhering cell possessing exopolymeric capsule and pili- and the thiol-coated AFM probe tip. (Reprinted with permission from [56]. Copyright (2009) American Chemical Society.)

extracellular structures and the cell walls are more complex than the surface of synthetic colloidal particles. Therefore, one would expect that the force curves taken in bacterial surfaces would involve biological polymer interactions. In addition to DLVO theory, other forces including steric interactions have been considered. Surface structures (Figure 3.7) are believed to be involved in bacterial adhesion. The presence of polymeric structures on bacterial cell surfaces has motivated the introduction of polymer interactions in addition to the DLVO force for interpretation of AFM force measurements.

Steric repulsion has been observed to arise from the higher affinity of the polymers for the interacting medium than for the solid surface (e.g., the AFM probe tip) [56]. The importance of AFM is demonstrated because it provides with simultaneous information of local interaction forces. The existence of the various mentioned forces makes difficult to know the origin of different interactions. In the case of several bacteriae, the steric interaction cannot be discarded. The extended models for the bacteria-AFM tip force–distance curves are consistent with the effects of steric interactions.

3.13 Conclusive Remarks

FS is a technique that uses the cantilever tip to probe a sample surface, the deflection system to measure the tip–sample interaction forces and the piezoelectric system to change distances with high precision. The objective is to determine surface and intermolecular interactions. There are interesting changes when AFM is operated in aqueous environments. Specially important is to work in physiological conditions with proteins, nucleic acids, enzymes, cells, and so on. In this chapter, we have dealt with several forces such as DLVO forces including vdW, EDL forces, and short-range forces such as solvation, hydrophobic, and steric forces. FS is specially suited for the study of double layer forces that develop between the diffuse layers formed around the tip and the sample when they are charged. Force values are in the piconewton range and can be easily measured. It is also important that the forces can be laterally solved. We show how the EDL force acting on a single molecule of DNA gives an image of the molecule with the same resolution than that of the AFM topography. Important quantities such as the surface charge density or the elastic constant can be measured.

In future, the technique has to improve the methodology to prepare tips of smaller size and regular shape maintaining the characteristics as a function of time. The modifications of the tip by using appropriate chemicals need a continuous boost. Concerning the force curves, it is important to identify their shape and the proper values of force and length so that the identification of the different interactions can be done with certainty. It is also interesting to fully extend all the features of FS to AFM dynamic operation (see chapter 4).

Acknowledgments

We acknowledge the feedback from the reviewer. This work has been financed by the Ministerio de Ciencia e Innovación through Project No. FIS2009-07657 and Project No. CSD2010-00024.

References

1. Engel, A. and Müller, D.J. (2000) Observing single biomolecules at work with the atomic force microscope. *Nat. Struct. Biol.*, **7**, 715–718.
2. Wiesendanger, R. (1994) *Scanning Probe Microscopy: Methods and Applications*, Cambridge University Press.
3. Binnig, G. and Rohrer, H. (1987) Scanning tunnelling microscopy-from birth to adolescence. *Rev. Mod. Phys.*, **64**, 615–625.
4. Baró, A.M., Miranda, R., Alamán, J., García, N., Binnig, G., Rohrer, H., Gerber, C., and Carrascosa, J.L. (1985) Determination of surface topography of biological specimens at high resolution by scanning tunneling microscopy. *Nature*, **315**, 253–254.
5. de Lozanne, A.L., Elrod, S.A., and Quate, C.F. (1985) Spatial variations in the superconductivity of Nb_3Sn measured by low temperature tunnelling

microscopy. *Phys. Rev. Lett.*, **54**, 2433–2436.
6. Sonnenfeld, R. and Hansma, P.K. (1986) Atomic resolution microscopy in water. *Science*, **232**, 211–213.
7. Sonnenfeld, R. and Schardt, B.C. (1986) Tunnelling microscopy in an electrochemical cell: images of Ag plating. *Appl. Phys. Lett.*, **49**, 1172–1174.
8. Binnig, G., Quate, C.F., and Gerber, C. (1986) Atomic force microscope. *Phys. Rev. Lett.*, **56**, 930–933.
9. Drake, B., Prater, C.B., Weisenhorn, A.L., Gould, S.A., Albrecht, T.R., Quate, C.F., Cannell, D.S., Hansma, H.G., and Hansma, P.K. (1989) Imaging crystals, polymers, and processes in water with the atomic force microscope. *Science*, **243**, 1586–1589.
10. Guthold, M., Bezanilla, M., Eric, D.A., Jenkins, B., Hansma, H.G., and Bustamante, C. (1994) Following the assembly of RNA polymerase-DNA complexes in aqueous solutions with the scanning force microscope. *Proc. Natl. Acad. Sci.*, **91**, 12927–12931.
11. Cappella, B. and Dietler, G. (1999) Force-distance curves by atomic force microscopy. *Surf. Sci. Rep.*, **34**, 1–104.
12. Butt, H.-J., Capella, B., and Kappl, M. (2005) Force measurements with the atomic force microscope: technique, interpretation and applications. *Surf. Sci. Rep.*, **59**, 1–152.
13. Hermansson, M. (1999) The DLVO theory in microbial adhesion. *Colloids Surf. B: Biointerfaces*, **14**, 105–119.
14. Perutz, M.F. (1978) Electrostatic effects in proteins. *Science*, **201**, 1187–1191.
15. Besteman, K., van Eijk, K., and Lemay, S.G. (2007) Charge inversion accompanies DNA condensation by multivalent ions. *Nat. Phys.*, **3**, 641–644.
16. Mulgrew-Nesbitt, A., Diraviyam, K., Wang, J., Singh, S., Murray, P., Li, Z., Rogers, L., Mirkovic, N., and Murray, D. (2006) The role of electrostatics in protein-membrane interaction. *Biochim. Biophys. Acta-Mol. Cell Biol. Lipids*, **1761**, 812–826.
17. Sheinerman, F.B., Norel, R., and Honig, B. (2000) Electrostatic aspects of protein-protein interactions. *Curr. Opin. Struct. Biol.*, **10**, 153–159.
18. Misra, V.K., Hecht, J.L., Yang, A.S., and Honig, B. (1998) Electrostatic contributions to the binding free energy of the lambdacl repressor to DNA. *Biophys. J.*, **75**, 2262–2273.
19. Liang, Y., Hilal, N., Langston, P., and Starov, V. (2007) Interaction forces between colloidal particles in liquids. Theory and experiment. *Adv. Colloid Interface Sci.*, **134–135**, 151.
20. Sotres, J. and Baró, A.M. (2008) DNA molecules resolved by electric double layer force spectroscopy imaging. *Appl. Phys. Lett.*, **93**, 103903.
21. Neuman, K.C. and Nagy, A. (2008) Single-molecule force spectroscopy, optical tweezers, magnetic tweezers and atomic force microscopy. *Nat Methods*, **5**, 491–505.
22. Hummer, G. and Szabo, A. (2005) Free energy surfaces from single-molecule force spectroscopy. *Acc. Chem. Res.*, **38**, 504–513.
23. Lim, C.T., Zhou, E.H., Li, A., Vedula, S.R.K., and Fu, H.X. (2006) Experimental techniques for single cell and single molecule biomechanics. *Mater. Sci. Eng. C-Biomimetic Supramol. Syst.*, **26**, 1278–1288.
24. Higgins, M.J., Protsch, R., Sader, J.E., Polcik, M., Endoo, S.M., Cleveland, J.P., and Jarvis, S.P. (2006) Noninvasive determination of optical lever sensitivity in atomic force microscopy. *Rev. Sci. Instrum.*, **77**, 013701.
25. Dürig, U., Züger, D., and Stalder, A. (1992) Interaction force detection in scanning probe microscopy: methods and applications. *J. Appl. Phys.*, **72**, 1778–1798.
26. Colchero, J., Storch, A., Luna, M., Gómez-Herrero, J., and Baró, A.M. (1998) Observation of liquid neck formation with scanning force microscopy techniques. *Langmuir*, **14**, 2230.
27. Stukalov, O., Murray, C.A., Jacina, A., and Dutcher, J.K. (2006) Relative humidity control for atomic force microscopy. *Rev. Sci. Instrum.*, **77**, 033704.
28. Eastman, T. and Zhu, D.-M. (1996) Adhesion forces between surface-modified

AFM tips and a mica surface. *Langmuir*, **12**, 2859–2862.
29. Lyubchenko, Y., Shlyakhtenko, L., Harrington, R., Oden, P., and Lindsay, S. (1993) Atomic force microscopy of long DNA: imaging in air and under water. *Proc. Natl. Acad. Sci. U.S.A.*, **90**, 2137–2140.
30. Moreno-Herrero, F., Colchero, J., and Baró, A.M. (2003) DNA height in scanning force microscopy. *Ultramicroscopy*, **96**, 167–174.
31. De Pablo, P.J., Colchero, J., Gómez-Herrero, J., and Baró, A.M. (1998) Jumping mode scanning force microscopy. *Appl. Phys. Lett.*, **73**, 3300–3302.
32. Weisenhorn, A.L., Hansma, P.K., Albrecht, T.R., and Quate, C.F. (1989) Forces in atomic force microscopy in air and water. *Appl. Phys. Lett.*, **54**, 2651–2653.
33. Israelachvili, J. (1992) *Intermolecular and Surface Forces*, 2nd edn, Academic Press, London.
34. Gee, M.L., McGuiggan, P.M., and Israelachvili, J.N. (1990) Liquid to solid-like transitions of molecularly thin films under shear. *J. Chem. Phys.* **93**, 1895–1906.
35. Derjaguin, B.V. (1934) Untersuchungen über die reibung und adhesion. *Kolloid Z.*, **69**, 115.
36. Parsegian, V.A. and Gingell, D. (1979) On the electrostatic interaction across a salt solution between two bodies bearing unequal charges. *Biophys. J.*, **12**, 1192–1204.
37. Butt, H.-J. (1991) Electrostatic interaction in atomic force microscopy. *Biophys. J.*, **60**, 777–785.
38. Rotsch, C. and Radmacher, M. (1997) Mapping local electrostatic forces with the atomic force microscope. *Langmuir*, **13**, 2825–2832.
39. Todd, B.A. and Eppell, S.J. (2004) Probing the limits of the Derjaguin approximation with scanning force microscopy. *Langmuir*, **20**, 4892–4897.
40. Vakarelski, I.U. and Higashitani, K. (2006) Single-nanoparticle-terminated tips for scanning probe microscopy. *Langmuir*, **22**, 2931.
41. Hillier, A.C., Sunghyun, Kim., and Bard, A.J. (1996) Measurement of double-layer forces at the electrode/electrolyte interface using the atomic force microscope: potential and anion dependent interactions. *J. Phys. Chem.*, **100**, 18808–18817.
42. Baselt, D.R. and Baldeschwieler, J.D. (1994) Imaging spectroscopy with the atomic force microscope. *J. Appl. Phys.*, **76**, 33–38.
43. Laney, D.E., Garcia, R.A., Parsons, and Hansma, G.H. (1997) Changes in the elastic properties of collinergic synaptic vesicles as measured by atomic force microscope. *Biophys. J.*, **72**, 806–813.
44. Heinz, W.F. and Hoh, J.H. (1999) Relative surface charge density mapping with the atomic force microscope. *Biophys. J.*, **76**, 528–538.
45. Johnson, A.S., Nehl, C.L., Mason, M.G., and Hafner, J.H. (2003) Fluid electric force microscopy for charge density mapping in biological systems. *Langmuir*, **19**, 10007–10010.
46. Sotres, J. and Baró, A.M. (2010) AFM imaging and analysis of electrostatic double layer forces on single DNA molecules. *Biophys. J.*, **98**, 1995–2004.
47. Horcas, I., Fernández, R., Gómez-Rodríguez, J.M., Colchero, J., Gómez-Herrero, J., and Baró, A.M. (2007) WSXM: A software for scanning probe microscopy and a tool for nanotechnology. *Rev. Phys. Instrum.*, **78**, 013705.
48. Carrasco, C., Carreira, A., Schaap, I.A.T., Serena, P.A., Gómez-Herrero, J., Mateu, M.G., and de Pablo, P.J. (2006) DNA-mediated anisotropic reinforcement of a virus. *Proc. Natl. Acad. Sci.*, **103**, 13706–13711.
49. Müller, D.J., Fotiadis, D., Scheuring, S., Müller, S.A., and Engel, A. (1999) Electrostatically balanced subnanometer imaging of biological specimens by atomic force microscope. *Biophys. J.*, **76**, 1101–1111.
50. Das, P.K. and Bhattacharjee, S. (2005) Electrostatic double layer force between a sphere and a planar substrate in the presence of previously deposited spherical particles. *Langmuir*, **21**, 4755–4764.

51. Ducker, W.A., Senden, T.J., and Pashley, R.M. (1992) Measurement of forces in liquids using a force microscope. *Langmuir*, **8**, 1831.
52. Israelachvili, J.N. and Pashley, R.M. (1982) The hydrophobic interaction is long-range, decaying exponentially with distance. *Nature*, **300**, 341.
53. O'Shea, S.J., Welland, M.E., and Pethica, J.B. (1994) Atomic force microscopy of local compliance at solid–liquid interfaces. *Chem. Phys. Lett.*, **223**, 336–340.
54. Higgins, M.J., Polcik, M., Fukuma, T., Sader, J.E., Nakayama, Y., and Jarvis, S.P. (2006) Structured water layers adjacent to biological membranes. *Biophys. J.*, **91**, 2532–2542.
55. Alsteens, D., Dague, E., Rouxhet, P.G., Baulard, A.R., and Dufréne, Y.F. (2007) Direct measurement of hydrophobic forces on cell surfaces using AFM. *Langmuir*, **23**, 11977–11979.
56. Dorovantu, L.S., Bhattacharjee, S., Foght, J.M., and Gray, M.R. (2009) Analysis of force interactions between AFM tips and hydrophobic bacteria using DLVO theory. *Langmuir*, **25**, 6968–6976.

4
Dynamic-Mode AFM in Liquid

Takeshi Fukuma and Michael J. Higgins

4.1
Introduction

In contact-mode atomic force microscopy (AFM) [1], a tip is laterally scanned with its apex in contact with a surface, which often causes damage or deformation of the sample. In addition, contact-mode AFM requires the use of a soft cantilever, leading to an instability known as *jump-to-contact*. Namely, when a force gradient in the z direction exceeds the spring constant (k) of a cantilever during a tip approach, the tip suddenly jumps to the surface. Thus, it is often impossible either to measure a force (F_t) or control the vertical tip position (z_t) near a surface. This often hinders true atomic-resolution imaging. Besides, owing to the use of a DC deflection of a cantilever for the tip–sample distance regulation, it is often difficult to achieve good long-term stability.

To overcome these problems, dynamic-mode AFM was invented in 1987 [2]. In this method, a stiff cantilever is mechanically oscillated at a frequency near the cantilever resonance frequency (f_0). F_t is detected as a shift of amplitude, frequency, or phase of the cantilever oscillation. The use of a stiff cantilever prevents jump-to-contact. In addition, the vertical motion of a cantilever reduces lateral friction force during imaging. This allows the imaging of a soft material without destruction. Due to an enhancement of the force sensitivity provided by the resonance effect of a cantilever, the force sensitivity is sufficiently high in spite of the high stiffness of a cantilever. The long-term stability is also improved due to the use of an AC response of a cantilever as a feedback signal in the tip–sample distance regulation. These capabilities of dynamic-mode AFM allow detection of F_t and control of z_t even at the vicinity of a sample surface. Therefore, dynamic-mode AFM has been used for atomic-scale studies on various surfaces.

However, vibration of a cantilever is considerably damped in liquid because of the high viscosity of liquid compared to that of air. This leads to a high dissipation of the cantilever vibration energy at its resonance, namely, a low Q factor of the cantilever resonance. Since the high force sensitivity in dynamic-mode AFM is provided by the resonance effect of a cantilever vibration, a low Q factor results in low force sensitivity. In addition, the amplitude and phase versus frequency curves

Atomic Force Microscopy in Liquid: Biological Applications, First Edition.
Edited by Arturo M. Baró and Ronald G. Reifenberger.

obtained in liquid often show large distortion due to the influence from spurious resonances in a cantilever holder and sample holder.

To date, considerable efforts have been made to overcome these problems. Various methods have been proposed for spurious-free cantilever excitation in liquid [3–6]. Low-noise cantilever deflection sensors have been developed for obtaining the optimal performance, which is limited only by the thermal fluctuation of a cantilever [7–9]. These efforts finally made it possible to operate dynamic-mode AFM in liquid with true atomic and molecular resolution [10, 11] Due to the low loading force during imaging, the method allows imaging of isolated biomolecules weakly bound to a substrate with molecular-scale resolution [12, 13]. Furthermore, the high force sensitivity of the method makes it possible to visualize not only the surface structures of solid surfaces but also the distribution of water and ions interacting with a surface [14–16].

There are three major operation modes in dynamic-mode AFM, which are referred to as *amplitude-* [2], *frequency-* [17], *and phase* [18]*-modulation AFM* (AM-, FM- and PM-AFM, respectively). In this chapter, basic principles and experimental setups for these operation modes are described. In the setups, some of the components have to meet special requirements for the operation in liquid. Special designs and techniques developed to meet these requirements are described in detail. Typical applications of liquid-environment dynamic-mode AFM include quantitative force measurements and high-resolution imaging. Recent examples of such applications are also described.

4.2 Operation Principles

4.2.1 Amplitude and Phase Modulation AFM (AM- and PM-AFM)

Figure 4.1a shows an experimental setup for AM- and PM-AFM. In this setup, a cantilever is mechanically oscillated at a frequency ($\omega = 2\pi f$) near its resonance frequency ($\omega_0 = 2\pi f_0$). The cantilever vibration is typically excited by applying an AC voltage signal ($A_{ex}\cos(\omega t)$) to a piezoactuator placed near the cantilever. The cantilever vibration is typically detected with the optical beam deflection (OBD) method as shown in Figure 4.1a. As the cantilever approaches a surface, F_t induces a shift (Δf) of the cantilever resonance frequency. In AM- and PM-AFM, Δf is detected as a shift of amplitude (ΔA) and phase ($\Delta\phi$), respectively.

Figure 4.1b shows amplitude (A) and phase (ϕ) versus frequency curves with and without F_t. In most cases, f is set to the value of f_0 measured at a tip position far from the surface. At this frequency, A takes its maximum value, while ϕ takes a value of -90°. As the tip approaches, F_t induces Δf, which in turn gives rise to ΔA and $\Delta\phi$ as shown in Figure 4.1b. Thus, variation of F_t can be detected as ΔA or $\Delta\phi$.

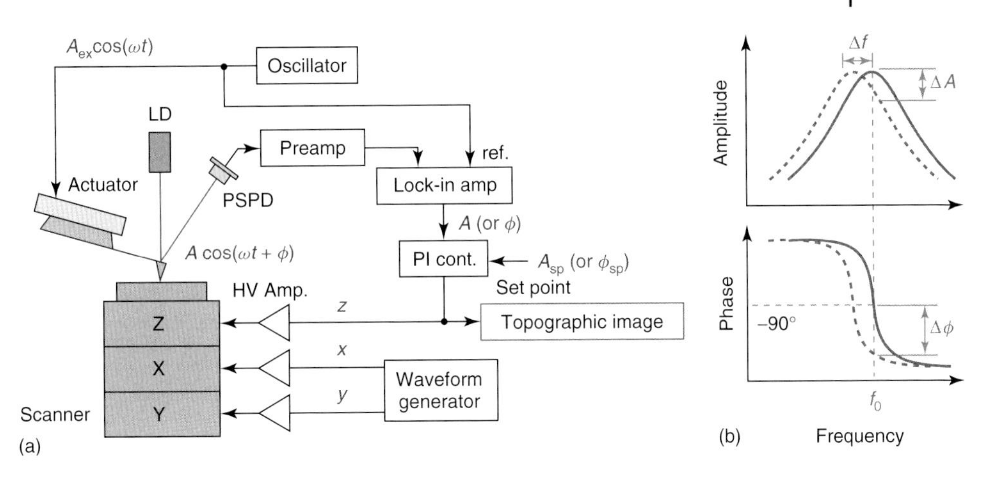

Figure 4.1 (a) Experimental setup for AM- and PM-AFM. (b) Amplitude and phase versus frequency curves with (dotted lines) and without (solid lines) a tip–sample interaction.

A and ϕ signals are typically detected with a lock-in amplifier from the cantilever deflection and excitation signals. Its output is fed into a proportional-integral (PI) controller, which controls the z position of a sample through a high-voltage (HV) amplifier to keep A or ϕ at a set point value (A_{sp} or ϕ_{sp}, respectively). In this way, the tip–sample distance is kept almost constant as long as an A (or ϕ) versus distance curve does not show strong dependence on the lateral tip position.

With the tip–sample distance regulation feedback turned on, a relative position of the tip with respect to a sample is raster-scanned in $x - y$ direction. This lateral scan is typically performed by driving x and y sample scanners. The driving signals are generated by a waveform generator and applied to the scanners through HV amplifiers. As the Z-control signal (z) changes in proportion to the surface corrugations during the tip scan, a topographic image is obtained by recording z as a two-dimensional map.

4.2.2 Frequency-Modulation AFM (FM-AFM)

Figure 4.2 shows the experimental setup for FM-AFM. In FM-AFM, a cantilever is mechanically oscillated at the cantilever resonance frequency using a piezoactuator placed near the cantilever. The detected cantilever deflection signal is routed back to the piezoactuator through a phase shifter and an automatic gain control (AGC) circuit to form a self-oscillation circuit. In this self-oscillation circuit, a cantilever works as a mechanical resonator and determines the oscillation frequency.

Figure 4.2b shows A and ϕ versus frequency curves around f_0. For the self-oscillation circuit to produce a continuous sine wave output, the total phase delay in the self-oscillation circuit should be a multiple of -360°. Thus, ϕ is always kept at -90° if the phase delay at the phase shifter is adjusted to an appropriate

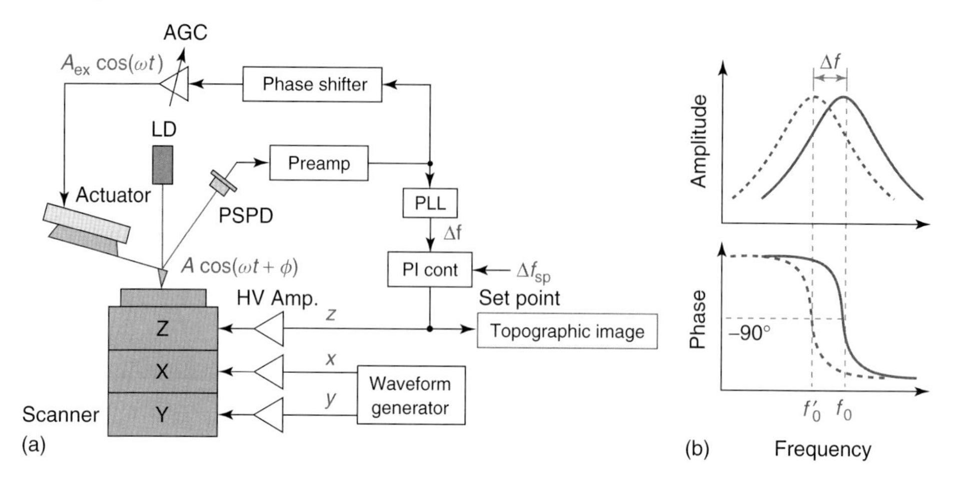

Figure 4.2 (a) Experimental setup for FM-AFM. (b) Amplitude and phase versus frequency curves with (dotted lines) and without (solid lines) a tip–sample interaction.

value. Consequently, the oscillation frequency f is always kept at the cantilever resonance.

As the tip approaches the sample surface, F_t induces Δf. Owing to the self-oscillation circuit, f is also shifted by the same amount. The deflection signal is fed into the frequency detector, which produces a voltage signal proportional to Δf. A phase-locked loop (PLL) circuit is typically used as a frequency detector. The Δf signal is fed into a PI controller, which outputs a z-signal to control the tip–sample distance.

Here we have explained the operation principle of FM-AFM with an assumption that a cantilever is driven by a self-oscillation loop. However, it has recently become common to use a PLL circuit not only for detecting Δf but also for producing a cantilever excitation signal. In this case, the output signal from a voltage-controlled oscillator (VCO) in a PLL circuit is used for the cantilever excitation. In a PLL circuit, the phase difference between the PLL input and VCO output signals is kept constant by adjusting the frequency of the VCO output signal. Therefore, the cantilever vibration frequency is adjusted such that ϕ is always kept at -90°.

4.3 Instrumentation

4.3.1 Cantilever Excitation

Implementation of dynamic-mode AFM typically requires the cantilever to be excited close to one of its resonances. Several methods have been developed to do this, including acoustic excitation, magnetic excitation, photothermal excitation, and electrostatic excitation. Acoustic excitation is the most common method that uses a piezoelectric component, generally integrated into the cantilever holder, to excite the cantilever. However, because the piezoelectric component is not directly coupled to the cantilever, the holder, cantilever chip and surrounding fluid (in the case of operation in liquid) may also be excited and lead to spurious resonances that convolute the true resonance frequency of the cantilever [19]. To overcome this, particularly when it is desired to make quantifiable measurements using dynamic modes, direct excitation of the cantilever is preferred. Magnetic excitation is a method that directly induces cantilever oscillations by subjecting a cantilever, modified by either a magnetic coating or the attachment of a magnetic particle, to an oscillating magnetic field supplied by a solenoid positioned beneath the cantilever [20, 21]. For further integration of components required to control magnetic excitation, an AC current can be generated via an integrated circuit in the cantilever, which is then placed in a DC magnetic field [22]. In contrast to the above methods that mechanically oscillate the cantilever, photothermal excitation uses AC modulation of a laser diode beam focused onto the rear of the cantilever to thermally induce excitation [23]. More recently, excitation of the cantilever has been achieved electrostatically [24]. This was done by applying an oscillating bias voltage between a gold-coated cantilever and an optically transparent electrode integrated into the cantilever holder.

4.3.2 Cantilever Deflection Measurement

A cantilever deflection measurement system is one of the most important elements in AFM. To date, various methods have been proposed for the cantilever deflection measurement [1, 2, 25–30]. Among the most widely used methods is the OBD method [26] owing to its simple setup and high sensitivity. For this method, a focused laser beam is irradiated onto the cantilever backside and the reflected beam is detected with a position-sensitive photodetector (PSPD). The PSPD is typically made of a two- (or four-) divided photodiode array.

The cantilever deflection induces a displacement of the laser spot on the PSPD. Thus, the difference between the photoinduced currents from the upper and lower elements varies in proportion to the cantilever deflection. The current signals are converted to voltage signals using current-to-voltage ($I-V$) converters. These voltage signals are fed into a differential amplifier, which produces a cantilever deflection signal.

There are several noise sources in the OBD method, including Johnson noise from the transimpedance resistors in the $I-V$ converters, the shot noise from the photodiodes, and the noise arising from the laser beam. In most of the designs, the ultimate detection limit is determined by the photodiode shot noise. However, the laser beam noise is often predominant in practice. This is particularly evident in the case of liquid-environment AFM.

In the liquid-environment AFM, the laser beam is reflected or scattered at the interface between the air–glass and glass–liquid as shown in Figure 4.3a. Some of

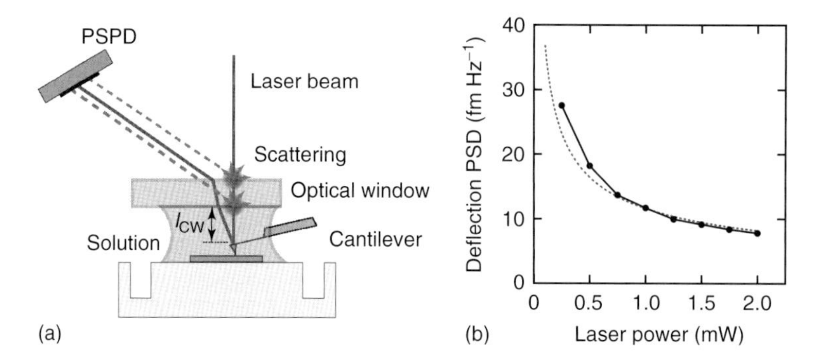

Figure 4.3 (a) Schematic model of an OBD sensor showing a laser beam interference at the PSPD surface caused by the scattering at air–glass and glass–liquid interfaces. (b) Laser power (P_0) dependence of n_{zs} [31].

the reflected beam goes back to the laser diode to induce optical feedback noise, while some of the scattered laser beam goes to the PSPD to interfere with the beam reflected on the backside of the cantilever. These optical feedback and interference noises are often predominant in the case of liquid-environment AFM.

Such laser beam noises can be suppressed by modulating the laser power at a high frequency (typically 300–500 MHz) [7]. The high-frequency laser modulation changes the laser oscillation mode from the single mode to the multimode. As the optical feedback noise arises from the competition between the different laser oscillation modes (i.e., mode hopping), the multimode oscillation is effective in suppressing the optical feedback noise. In addition, the multimode oscillation reduces the coherence length of the laser beam, which is also effective in suppressing the interference noise at the PSPD surface.

The ultimate detection limit in the OBD method is determined by the photodiode shot noise. The PSD of the deflection noise arising from the photodiode shot noise (n_{zp}) is given by, [32]

$$n_{zp} = \frac{a_0 \ell_c}{6\chi n_m \ell_f}\sqrt{\frac{2e}{\eta\alpha P_0}} \tag{4.1}$$

where a_0, χ, ℓ_f, e, η, α, ℓ_c, and n_m are the diameter of the collimated laser beam, the correction factor for a Gaussian laser beam profile, the focal length of the focusing lens, the elementary electric charge, the efficiency of the light-to-current conversion at the PSPD, the efficiency of the optical transmission from the focusing lens to the PSPD, the cantilever length, and the refractive index of the environment, respectively. A similar analysis has been reported by Meyer and Amer [26] in the early stage of the AFM development. While these analyses assume that the laser spot size is much smaller than the cantilever size, Schäffer *et al.* reported an analysis that is applicable to cases where the laser spot size is comparable to the cantilever size [33].

When the laser beam noise is sufficiently reduced to reach the shot noise limit, Eq. (4.1) provides a direct guideline for reducing the deflection noise. For example, the use of a smaller a_0 and a longer ℓ_f gives a better performance in the OBD sensor. Figure 4.3b shows dependence of n_{zs} on laser power (P_0) [31]. The good agreement between the experimentally measured and theoretically calculated values with Eq. (4.1) reveals that the noise performance of this OBD sensor is limited mainly by the photodiode shot noise.

n_{zs} of an OBD sensor typically ranges from 100 to 1000 fm $\sqrt{\text{Hz}}^{-1}$. However, the use of the high-frequency laser modulation technique and a careful design of the optical system made it possible to reduce n_{zs} to a value less than 10 fm $\sqrt{\text{Hz}}^{-1}$ [31, 32]. Such an excellent noise performance in the deflection measurement is particularly important when a relatively stiff cantilever is used, as will be discussed below.

The theoretical limit of the force sensitivity in dynamic-mode AFM is determined by the thermal vibration of a cantilever. To achieve this thermal noise limit, n_{zs} must be sufficiently smaller than the deflection noise arising from the Brownian motion of a cantilever (n_{zB}) around f_0. For typical stiff ($k = 30$ N m^{-1}) and soft ($k = 0.1$ N m^{-1}) cantilevers, $n_{zB}(f_0) = 70$ and 3,000 fm $\sqrt{\text{Hz}}^{-1}$ in liquid, respectively. Thus, for a soft cantilever, the thermal noise limit can be achieved even with a conventional OBD sensor. However, a low-noise cantilever deflection sensor is required for achieving the thermal noise limit with a stiff cantilever.

4.3.3 Operating Conditions

The use of cantilever vibration in dynamic-mode AFM enables more operating parameters than static-mode AFM. For instance dynamic-mode AFM has several operation modes, namely, AM-, FM-, and PM-AFM. The choice of appropriate operation modes and operating conditions is essential for the success of the intended AFM application. Here, we present the principles and typical operating conditions of the different operation modes. We also discuss their implementation and applicability to various applications.

4.3.4 AM-AFM

In AM-AFM, a cantilever is oscillated with a constant excitation signal having a fixed amplitude and frequency. Even if the tip motion is heavily distorted by an accidental tip crash with the surface, the excitation signal is not affected. Thus, the normal motion of the cantilever is recovered after a transient behavior. In addition, A monotonically decreases with decreasing tip–sample distance, except at the point near the tip–sample contact (Figure 4.4). Therefore, the tip–sample distance regulation stably works in both noncontact and contact regimes. These features are useful in applications in which the contact of a tip with a surface is hard to avoid. Typical examples include imaging of a rough surface, dynamic motion of mobile species and fluctuating structures.

On the contrary, it is often difficult to control an atomic-scale contact of a tip and surface in AM-AFM. In an A versus distance curve, there is often a discontinuous

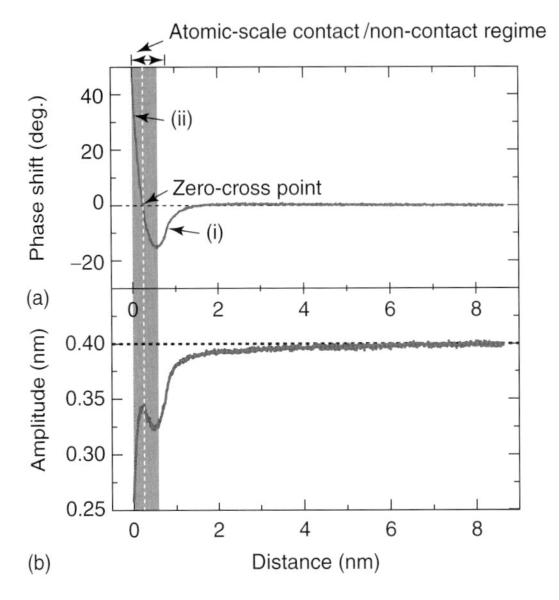

Figure 4.4 (a) $\Delta\phi$ and (b) A versus distance curves measured on mica in water [18]. A Si cantilever (Nanosensors: NCH) with $k = 24.9\,\text{N}\,m^{-1}$, $f_0 = 143.238\,\text{kHz}$, and $Q = 6.2$ was used.

jump at the vertical tip position just before and after an atomic-scale contact, which is referred to as *atomic-scale contact/noncontact regime* as shown in Figure 4.4a. Such an instability hinders precise control of the vertical tip position within this range. Although the position of the discontinuous jump can be shifted to a certain extent by changing the excitation frequency, this makes it difficult to find a set point corresponding to an atomic-scale contact. Depending on the nature of a solution and surface, an A versus distance curve does not show any discontinuity. However, this also makes it difficult to decide on an atomic-scale contact position. The capability of controlling the vertical tip position at an atomic-scale contact/noncontact regime is essential for true atomic-resolution imaging and atomic-scale manipulations. Thus, AM-AFM has not been widely used for such atomic-scale applications.

From these features, AM-AFM has mainly been used for imaging of rough surfaces, dynamic motion of mobile species, and fluctuating structures with nanometer-scale resolution. In such applications, it is very difficult to scan a tip without making tip–sample contact. Thus, a relatively soft cantilever with $k = 0.01–0.1\,\text{N}\,m^{-1}$ is used to avoid damaging a sample. These soft cantilevers show a relatively large thermal vibration. The root-mean-square (RMS) amplitude of the thermal vibration of a cantilever (z_{th}) is given by

$$z_{\text{th}} = \sqrt{\frac{k_{\text{B}}T}{k}} \tag{4.2}$$

where k_{B} and T are the Boltzmann's constant and absolute temperature, respectively. For a cantilever with k of $0.01\,\text{N}\,m^{-1}$, $z_{\text{th}} = 0.64\,\text{nm}$. This corresponds to a peak-to-peak amplitude of about 2 nm. Thus, A is typically set to a value ranging from 5 nm to 20 nm.

The use of large amplitudes is also effective in avoiding an instability known as *jump-to-contact*. If an attractive force shows a rapid increase near the surface, the slope of the force versus distance curve, namely, a force gradient, can exceed the spring constant of a cantilever. In this case, the tip suddenly jumps to a surface, which is referred to as *jump-to-contact*. To avoid a jump-to-contact, one of the following equations has to be met.

$$k > \left| \frac{\partial F_t}{\partial z} \right| \tag{4.3}$$

$$kA > F_{ad} \tag{4.4}$$

where F_{ad} is an adhesion force. In many cases, Eq. (4.3) cannot be satisfied because of the low k. However, Eq. (4.4) can be met even with a soft cantilever by using a large A. This is another reason for using a relatively large A in AM-AFM.

4.3.4.1 **FM-AFM**

In FM-AFM, a cantilever is oscillated with a self-oscillation circuit. In the self-oscillation circuit, a cantilever deflection signal is used for generating a cantilever excitation signal. If an accidental tip crash with the surface causes a large distortion in the cantilever deflection signal, the cantilever excitation signal also becomes distorted, leading to an instability of the cantilever vibration. In addition, Δf versus distance curve shows a steep change at a tip–sample contact position. Thus, an accidental tip–sample contact gives a very large and rapid change in Δf wherein it is often impossible to keep Δf constant because of the insufficient bandwidth of the tip–sample distance regulation feedback. This leads to an instability of the tip–sample distance regulation.

In contrast, FM-AFM is suitable for the precise control of the tip–sample distance at an atomic-scale contact/noncontact regime. Owing to the steep change of the Δf signal near the surface, the tip–sample contact position is very clear in the curve. In addition, the steep change of Δf near the surface makes it easier to choose an appropriate set point value (Δf_{sp}). This feature of FM-AFM makes it suitable for subnanometer-scale imaging and atomic-scale manipulations. Note that a Δf versus distance curve shows a behavior similar to that of a $\Delta \phi$ versus distance curve shown in Figure 4.4a.

In FM-AFM, a relatively stiff cantilever with k of 10–40 N m^{-1} is typically used. For a cantilever with k of 40 N m^{-1}, $z_{th} = 10$ pm. A typical corrugation of an atomic-scale AFM image is 10–100 pm. Thus, a stiff cantilever is desirable to control the vertical tip position with atomic-scale precision. Note that an averaged tip height can be controlled with a precision smaller than z_{th}. However, even in such a situation, the position of the tip apex atom cannot be well controlled due to the thermal vibration. Similarly, the thermal vibration can be suppressed by making gentle contact with a surface. However, this does not allow control of the tip position near the surface in an atomic-scale noncontact regime. The precise control of a tip position becomes possible only when using a stiff cantilever, and such capability is essential for atomic-scale applications.

In FM-AFM, a relatively small oscillation amplitude ($A = 0.1{-}0.5$ nm) is used. With a large oscillation amplitude, the tip apex feels a short-range interaction force only in a fraction of an oscillation period. With a small oscillation amplitude, the tip apex always stays in a distance range where a short-range interaction force is dominant. Consequently, the use of small amplitude enhances the sensitivity to a short-range interaction force and reduces the sensitivity to a long-range interaction force, leading to a high lateral spatial resolution. A short-range interaction force typically has a decay length of 0.2–0.3 nm. To obtain enough sensitivity to such a short-range force, A is typically set to 0.1–0.5 nm.

To use a small oscillation amplitude, a relatively stiff cantilever should be used to prevent a jump-to-contact. To meet Eq. (4.4), either a large A or a high k is required. To perform high-precision and high-resolution imaging, a combination of small A and high k is desirable, which is used in a typical FM-AFM experiment. In contrast, a combination of a large A and low k is effective for stable imaging of rough surfaces or dynamic events. Thus, this condition is used in a typical AM-AFM experiment.

In FM-AFM, a cantilever is always oscillated at its resonance frequency. This enables one to measure conservative and dissipative forces independently and quantitatively. A tip–sample interaction force can be classified into two categories: conservative and dissipative forces. The former is a force component that does not dissipate any vibration energy of a cantilever, while the latter is a component that dissipates it. The conservative force is used for measurements of various elastic forces as well as for the regulation of the tip–sample distance. The dissipative force is used for measurements of inelastic forces, which are related to friction and viscosity of a surface and a surrounding medium. Since viscosity and friction of a surface depend on its material, a two-dimensional map of energy dissipation measured simultaneously with a topographic image is often used for a compositional mapping of a heterogeneous surface.

In FM-AFM, a conservative force varies Δf, while a dissipative force varies A_{ex}. Therefore, from Δf and A_{ex}, these two force components can be independently measured. This is valid only when a cantilever is always oscillated at its resonance frequency. In AM- and PM-AFM, this is not necessarily true thus the quantitative interpretation of measured forces is more complicated.

4.3.4.2 PM-AFM

PM-AFM has features intermediate between those of AM- and FM-AFM. In PM-AFM, the cantilever is always oscillated with a constant excitation signal, which is the same in AM-AFM. This feature is suitable for stable imaging of rough surfaces and dynamic events. However, a $\Delta\phi$ versus distance curve in PM-AFM shows a steep change near the tip–sample contact position (Figure 4.4), which is similar to a Δf versus distance curve in FM-AFM. This may hinder stable tip–sample distance regulation due to a drastic change of $\Delta\phi$ caused by an accidental tip crash. Therefore, PM-AFM is not as stable as AM-AFM but is more stable than FM-AFM, particularly in the case of an application where tip–sample contact cannot be avoided.

Owing to the steep slope of a $\Delta\phi$ versus distance curve (Figure 4.4), PM-AFM is capable of precisely controlling the vertical tip position near the tip–sample contact/noncontact regime. Thus, PM-AFM is suitable for subnanometer-scale imaging. As discussed above, the use of a stiff cantilever and small oscillation amplitude is desirable for such an application.

In principle, PM-AFM and FM-AFM are very similar. In FM-AFM, ϕ is kept constant at -90° by a self-oscillation circuit, while f is kept constant at an arbitrary set point by the tip–sample distance regulation. In PM-AFM, ϕ is kept constant at an arbitrary set point ϕ_{sp} by the tip–sample distance regulation while f is set to an arbitrary value. In both operation modes, ϕ and f are kept constant. In particular, if ϕ_{sp} is set to -90° in PM-AFM and the tip–sample distance is controlled by changing f, the physical behavior of PM-AFM is almost the same as that of FM-AFM [34–36].

The difference between FM-AFM and PM-AFM becomes evident when it is used for force measurements. As mentioned above, the behavior of PM-AFM is very similar to that of FM-AFM as long as ϕ is kept constant by the tip–sample distance regulation. However, during a force versus distance curve measurement, there is no mechanism to keep ϕ constant at -90° and therefore the cantilever may not necessarily oscillate at its resonance frequency. This behavior during force measurements is exactly the same as that in AM-AFM.

At present, PM-AFM applications in liquid [18] have been very limited compared to those of AM- and FM-AFM. There have been no application areas where PM-AFM is predominantly used. However, the method may be the most suitable for high-speed subnanometer-scale imaging [37, 38]. This is because FM-AFM requires a phase feedback loop and frequency feedback loop, while PM-AFM requires only a phase feedback loop. Although the current imaging speed of FM- and PM-AFM is not limited by such factors, PM-AFM may have larger room to improve its imaging speed.

4.4 Quantitative Force Measurements

Owing to the significant progress made in understanding the motion of an oscillating cantilever [39], particularly in response to tip–sample interaction forces that introduce complex nonlinear dynamics and adhesion-viscoelastic effects, it is now relatively straightforward to obtain quantitative force information from dynamic modes of AFM. Progress in this area has seen the use of theoretical models based on perturbed harmonic approximations and numerical simulations [40] of which typically require the involvement of contact elastic models or an *a priori* knowledge of the interaction forces. More recently, the development of analytical approaches and algorithms has enabled the ability to directly quantify measurements from the AFM signals [41, 42]. Perhaps some of the more amenable approaches have effectively implemented what is known as a *force inversion* where there is a direct conversion of the dynamic signals to the interaction force [42–44]. These algorithms allow the interaction force, as a function of the tip–sample separation distance, to be directly calculated from the change in amplitude or frequency signal for any given oscillation amplitude. For example, Eq. (4.5) gives a generic formula for quantifying the interaction force, $F_t(z_t)$, for both AM-AFM and FM-AFM modes.

$$F_t(z_t) = 2k \int_{z_t}^{\infty} \left(1 + \frac{A^{1/2}}{8\sqrt{\pi(t - z_t)}}\right) \Omega(t) - \frac{A^{3/2}}{\sqrt{2(t - z_t)}} \frac{d\Omega(t)}{dt} dt \tag{4.5}$$

$\Omega(z_t)$ in FM-AFM is given by [42]

$$\Omega(z_t) = \frac{\Delta\omega(z_t) - \omega_0}{\omega_0} \tag{4.6}$$

while that in AM-AFM is given by[43]

$$\Omega(z_t) = \frac{\omega}{\omega_0} \left[1 + \frac{A_d}{A(z_t)} \cos(\phi(z_t))\right]^{1/2} - 1 \tag{4.7}$$

A_d is the driving amplitude of a cantilever defined by the following equation:

$$A_d = \frac{A_0}{\omega^2} \sqrt{(\omega_0^2 - \omega^2)^2 + \left(\frac{\omega\omega_0}{Q}\right)^2} \tag{4.8}$$

where A_0 is the free vibration amplitude of a cantilever without an interaction force.

4.4.1
Calibration of Spring Constant

Common to many of the force conversion algorithms is the requirement of knowing the spring constant, k, of the cantilever, as is the case in Eq. (4.5). While much of the emphasis in extracting the force is on determining the relationship between the accessible parameters (i.e., frequency, amplitude, and phase), the determination of the spring constant in the process remains highly critical. Various calibration methods to determine the spring constant have been well established and assessed for their applicability and accuracy [45]. Table 4.1 provides a general description of the most commonly used calibration methods. It is noted that their constraints do not imply any erroneous aspects or arduous requirements but rather the applicability of the method to different types of cantilevers and experimental setups.

In terms of practicality, the most commonly used is the thermal calibration method [46] because of its versatility and implementation into many of the currently available commercial AFM instruments, the latter feature of which probably self-perpetuates its use. The Sader method [47], which is available as an online calculation (*http://www.ampc.ms.unimelb.edu.au/afm/calibration.html*) and iPhone application, is also relatively easy. It requires only the geometric values (cantilever length and width), which can be sufficiently obtained from the manufacturers specifications, the resonance frequency, and quality factor of the cantilever. The last two parameters are conveniently obtained through the cantilever tuning procedure. Note that the Sader method does not assume highly viscous media thus it is more accurately applied in air rather than liquid. Namely, one should calibrate the spring

Table 4.1 Summary of calibration methods for cantilever spring constant.

Cantilever Calibration Method	Principle/Description	Advantages	Constraints	Accuracy
Analytical	Calculated analytically using the material (Young's modulus, E) and geometrical (length L, width b, and thickness h) properties of the cantilever $k = \frac{Ebh^3}{4L^3}$	Simple formulae to calculate k. No physical perturbation or prior contact with the cantilever tip required. Modified to require resonance frequency of cantilever instead of thickness (Cleveland method)	Suitable for single material (e.g. pure silicon) and rectangular cantilevers. Not suitable for cantilevers with metallic coatings. Limited by uncertainties in Young's modulus and thickness of cantilever	≈20%
Reference spring	The deflection of the cantilever is measured as it is "pushed" down onto a precalibrated reference cantilever. The measured inverse optical lever sensitivity (invOLS) and known spring constant of the reference cantilever, k_{ref}, are used to calculate k $k = k_{ref}\left(\frac{\text{invOLS}_{ref}}{\text{invOLS}} - 1\right)$	Suitable for a majority of cantilever types. Independent of cantilever material and geometry. Easy to implement and simple formula to calculate k. Precalibrated reference cantilevers for calibrating a wide range of k are now commercially available	Initial contact with the cantilever tip is required. Limited by uncertainties in the measurement of the invOLS and positioning of the cantilever over the reference cantilever	≈10–20%
Added Mass	The resonance frequency, f, of the cantilever is measured as micron spheres of different masses, M, are successively attached to the end of the cantilever. A plot of M versus $1/(2\pi f)^2$ gives a slope equal to k $M = \frac{k}{(2\pi f_0)^2} - 1$	Indepndent of cantilever material and geometry	The attachment of spheres to the end of the cantilever is potentially destructive and time consuming. Limited by uncertainties in the mass of the spheres and their positioning on the cantilever	≈10–30%

(continued overleaf)

Table 4.1 *(continued)*

Cantilever Calibration Method	Principle/Description	Advantages	Constraints	Accuracy
Sader	Calculated using the measured resonance frequency, f, quality factor, Q, and geometry (length L and width b) of the cantilever. Also uses the Reynolds number (Re) requiring density/viscosity of the surrounding medium (typically air) $$k = 0.1906pb^2 LQf^2(\mathrm{Re})$$	Quick and easy to implement (calculation now available online and on iPhone app). Does not require prior contact with the cantilever tip. Only requires measurement of the cantilever resonance frequency and quality factor using tuning or thermal procedure on most commercial AFM's	Generally restricted to the use of rectangular cantilevers and cantilevers with higher quality factors (Qgg1). Limited by uncertainties in the cantilever geometry	≈5–20%
Thermal	The thermal noise of the cantilever is measured and used to obtain the mean square deflection $\langle z^2 \rangle$ of the cantilever. $K_B T$ is the Boltzmann constant $$k = \frac{K_B T}{\langle z^2 \rangle}$$	Suitable for a majority of cantilever types. Quick and easy to implement. Available on most commercial AFM's. Independent of cantilever material or geometry	Initial contact with the cantilever tip is required. Limited by uncertainties in the measurement of the invOLS	≈5–20%

constant in air before undertaking measurements in liquid. Other useful methods include the reference spring method [48] and analytical approaches [49, 50]. The reference spring method is simple to implement and requires a set of commercially available precalibrated cantilevers. The analytical approaches are based on the geometrical properties and the cantilever's material properties (e.g., modulus/density) and are typically only suitable for single-material cantilevers (e.g., pure silicon with no coatings). For in depth reviews on the fundamental aspects and development of the various methods to improve their accuracy, please see review [45].

To directly compare the thermal, Sader, and analytical (Cleveland) methods, Burnham *et al.* [51] experimentally measured the spring constant of 10 different cantilevers using the different methods. Figure 4.5 shows the spring constant values reported in the study and replotted to visualize their variation across the different methods. It was found that the calibration methods agreed within 17% of the nominal manufacturer's k for rectangular levers, while the agreement for silicon nitride cantilevers was significantly worse. A KRISS nanoforce calibrator that can determine traceable spring constants with uncertainties $<1\%$ has been used as a standard to compare the analytical, reference spring, Sader, and thermal methods [52]. More recently, an interlaboratory study involving eight laboratories from three different countries was performed to assess reliable calibration protocols for soft cantilevers [53]. The study found that the spring constant of soft rectangular and v-shaped cantilevers could be accurately determined using both the Thermal and Sader methods, with the latter showing improved accuracy. The simultaneous application of both methods proved to be a good approach for reliable calibration. Even with the above information at hand, the question of which calibration method should be used, or is "best", is still a frequent cause of confusion, especially among newcomers to the field. As a general rule, the method chosen should be appropriate to each researcher's individual circumstance, including the experiments being undertaken (e.g., type of cantilever), accessibility of the method, and personal preference based on the potential destructiveness or user friendliness of the method. Of importance is that if one is unsure of the accuracy of their calibration method, then initially verifying the measurements using at least two other methods is a good approach, such as in Figure 4.5.

4.4.2 Conservative and dissipative forces

The interaction forces acting on the excited tip are generally divided into two classes – conservative forces and dissipative forces. Conservative forces describe the situation in which the tip on approaching a surface experiences a force gradient that is fully reversible during retraction; the net energy lost from the oscillating cantilever in this case is zero. Fundamental forces such as electrostatic and Van der Waals interaction, in addition to pure elastic contact forces, are representatives of conservative forces. In contrast, the profiles for dissipative forces involving the breakage of bonds (i.e., ligand–receptor unbinding) or viscoelastic effects are not fully reversible force profiles and result in energy losses. These definitions, however,

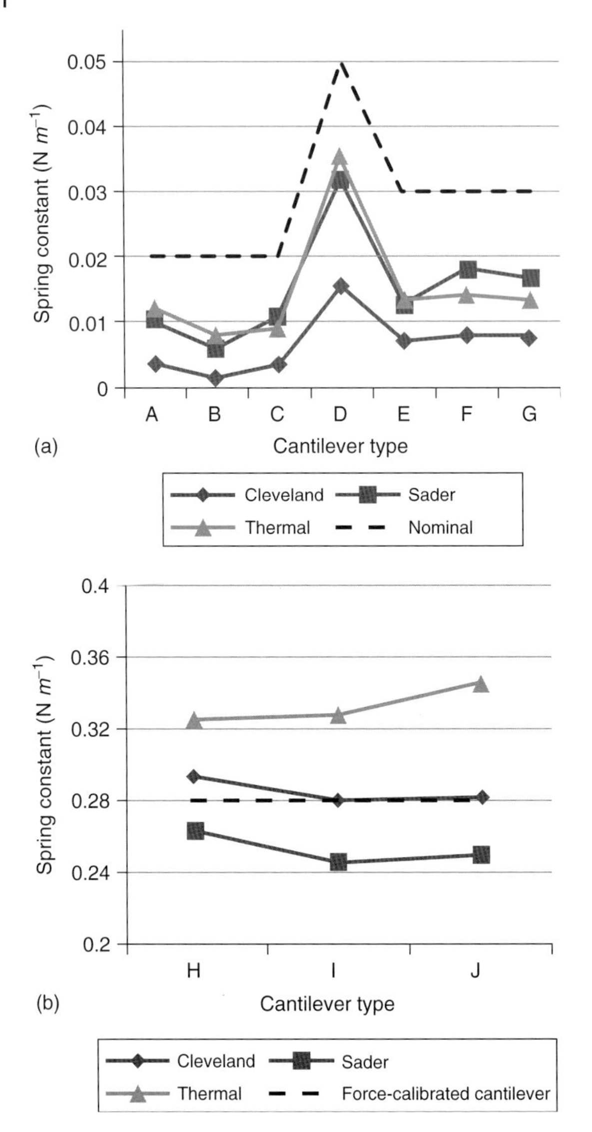

Figure 4.5 Comparison of spring constant (k) values for cantilever types A–J, using three different calibration methods. The measured values are compared with nominal manufacturer's k (A) and Force-calibrated k (B) values of the cantilevers. Cantilevers A–G are silicon nitride and H–J are silicon. Cantilevers A–C and H–J are rectangular and D–G are triangular. All measurements were done in air [51].

are rather simplified and a recent description has pointed out, for example, that conservative forces may be made up of conservative, dissipative, and energy-gaining processes but still result in an overall energy loss of zero [54]. One of the main advantages of dynamic AFM modes is that, in addition to conservative forces, the dissipative forces (e.g., damping/viscoelasticity) can also be quantified and utilized to reveal new information about material properties, liquid structural states, and dynamic molecular processes. Based on the work used to derive Eq. (4.5) for conservative forces [42], a general formula (see Eq. (4.9)) valid for any operating amplitude is also available for quantifying the dissipative forces, $\Gamma(z_t)$, for both AM- and FM-AFM. In this case, $\Gamma(z)$ is given by:

$$\Gamma(z_t) = -b\frac{\partial}{\partial z_t}\int_{z_t}^{\infty}\left(1+\frac{A^{1/2}}{8\sqrt{\pi(t-z_t)}}\Theta(t)-\frac{A^{3/2}}{\sqrt{2(t-z_t)}}\frac{d\Theta(t)}{dt}\right)dt \qquad (4.9)$$

where b is the damping coefficient of the cantilever vibration. $\Theta(z_t)$ in FM-AFM is given by [42]

$$\Theta(z_t) = \frac{\Delta F_0(z_t)}{\overline{F_0}} - \frac{\Delta\omega(z_t)}{\omega_0} \qquad (4.10)$$

where $\overline{F_0}$ and ΔF_0 are the driving force in the absence of an interaction force and its change induced by an interaction force, respectively. $\Theta(z_t)$ in AM-AFM is given by [43]

$$\Theta(z_t) = -\Omega(z_t)\left[\frac{QA_d\omega}{A(z_t)\omega_0}\sin(\phi(z_t))+1\right] \qquad (4.11)$$

There appears to be little experimental work done in liquid using dynamic-mode AFM to investigate fundamental conservative and dissipative forces specifically related to electrostatic, short-range attractive, or contact-repulsive (elastic and nonelastic) interactions (i.e., such as that done with colloidal probes using static AFM). This could be due to the already extensive work done using static-mode AFM and at a time when the operational and quantitative aspects of dynamic-mode AFM in liquid were still under development. Furthermore, dynamic-mode AFM has also found its foray into selective areas where the limitations of static-mode AFM have not been able to accommodate the measurements to the same extent. These areas include measurements on solvation forces, hydration forces, and dissipative interactions of biomolecules, all of which are presented below.

4.4.3 Solvation Force Measurements

The increase in sensitivity of dynamic-mode AFM to weaker, longer-range interaction forces has led to a large body of work focused on the structural forces of ordered liquid layers at solid–liquid interfaces. These forces, commonly known as *solvation forces,* arise as two solid bodies that are brought together to within nanometer distances of each other induce the ordering of the intervening liquid molecules [55]. If adequate pressure between the two surfaces is reached, an ordered liquid layer can be displaced laterally into the bulk, leaving the remaining liquid molecules free to disorder in the newly occupied space. Repeated ordering and disordering of the liquid molecules as the surfaces continue to approach gives rise to a characteristic oscillatory force profile, where the periodicity of the oscillations relates to the size of the liquid molecule (Figure 4.6). These unique forces influence

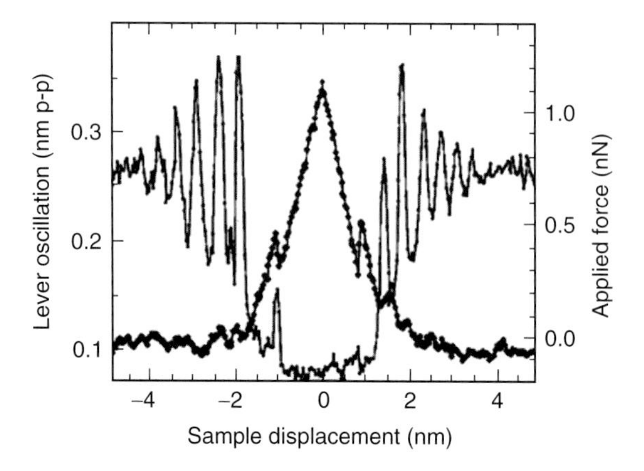

Figure 4.6 Static (heavy line) and dynamic force measurement of a tip approaching and retracting (right to left) a graphite surface immersed in *n*-dodecanol [58]. Oscillatory forces are very clear and reversible in the dynamic force profile.

the friction, lubrication, wear, and adhesion properties of a material and thus are of great interest in tribology applications.

4.4.3.1 **Inorganic Solids in Nonpolar Liquids**

Since the first solvation forces were measured using the surface force apparatus [56], the advantages of using AFM to study these forces became quickly apparent due to its versatility in accommodating different solid substrates and ability to probe the solid–liquid interface over lateral nanometer length scales. Model solid–liquid systems, consisting of an atomically smooth inorganic substrate (mica and graphite) in nonpolar solvent octamethylcyclotetrasiloxane (OMCTS), have typically been studied as their properties are conducive to observing the oscillatory forces [57]. This is because the smooth surface of the inorganic substrate does not provide any topographic feature to disrupt the liquid ordering and the chemically inert nature of the nonpolar liquid molecules, without their complex intramolecular- or substrate interactions, facilitates the ordering. For the earlier static-mode measurements preceding the use of dynamic-mode AFM, O'Shea *et al.* [57] measured solvation forces for approximately five ordered layers on a highly orientated pyrolytic graphite (HOPG) surface immersed in OMCTS and *n*-dodecanol. The periodicity of the oscillations, 0.73 nm and 0.35 nm for OMCTS and *n*-dodecanol, respectively, was commensurate with the size of the liquid molecules and indicated the packing geometry of the ordered layers. For OMCTS, the magnitude of the oscillatory forces were 0.2–0.4 nN corresponding to an energy of 5–25 kT required to displace an ordered layer. However the force profile in these static measurements was not continuous as the tip unstably jumped from the maximum repulsive force of one oscillation to the force minimum of the next oscillation. For attempted measurements in water (HOPG in water), a large attractive force (3 nN) caused a jump-into-contact with the surface at a separation distance of $\approx$6 nm, preventing the observation of solvation forces. These discontinuous jumps occur when the force gradient exceeds the stiffness of the cantilever and are common in static-mode measurements that require the use of low-spring-constant cantilevers to enhance sensitivity. To increase the sensitivity of their measurements, the same researchers implemented a small-amplitude AM-AFM, or "off-resonance" technique by oscillating the tip well below its resonance frequency with constant amplitude [58]. In addition to the force, this approach enabled the force gradient or stiffness of the ordered liquid to be quantified using a simple relationship between the spring constant and amplitude. When studying a dodecanol–graphite system, the small-amplitude AM-AFM measurements showed oscillations with a periodicity of 0.37 nm (as above) and peak stiffnesses ranging from of 0.5–5 N m^{-1}, with the latter values increasing as the tip probed less compliant layers closer to the surface. In contrast to the static deflection signal, the off-resonance mode was capable of detecting more layers with weaker structural forces further away from the surface because of the increased sensitivity of the detection scheme (Figure 4.6).

By extending these small-amplitude AM-AFM measurements to monitor the phase signal, damping effects and viscoelastic properties related to the structuring of OMCTS layers at small separation distances could be measured [59]. In particular, a significant increase in the damping term indicated that a confinement effect between the tip and graphite surface caused the liquid nearest the surface to become more viscoelastic. FM-AFM has also been used to observe an oscillatory force profile, along with a monotonic increase in the effective damping at smaller tip–sample separations, for the OMCTS–graphite system [60]. For these measurements, the forces scaled by the tips radius were of the order of 1–20 mN m^{-1}, depending on their distance from the surface, and were similar to values obtained in previous AM-AFM and SFA measurements. In particular, the quantitative similarities with SFA demonstrated that measurements over the macroscopic (i.e., 1 cm^2 interaction area for SFA) and nanometer-scale interaction areas (i.e., 10–20 nm for AFM) coincided well for this system. A few studies on the damping properties of the confined liquid have reported an oscillatory component in the damping profiles for both OMCTS and *n*-dodecanol–graphite systems [61–63], suggesting phase changes in the viscoelastic properties of ordered and disordered liquid states. However, caution must be taken in the interpretation of both the accuracy of the oscillatory forces and their occurrence in the damping profile, as coupling of the conservative and dissipative interactions has experimentally been shown to introduce such artifacts [64]. Further studies aimed at elucidating the origins of inconsistencies in solvation measurements, including the effect of the solvent contamination, tip properties (e.g., radius, shape, roughness), cantilever excitation modes, coupling of conservative and dissipative interactions, and values set for the different parameters (e.g., frequency and amplitude) remain of particular importance. For example, a closer examination of the effect of tip radius and shape in AM-AFM measurements has revealed that the presence of nanoscale asperities at the tip, and the lack of tip symmetry, may influence the magnitude of the oscillatory forces [65].

While most of the studies involving inorganic-substrate-nonpolar-liquid-systems have revolved around the development of techniques, there is an increasing focus on understanding the effect of different conditions on the fundamental properties of the ordered liquid. The effect of surface epitaxy [66], temperature [66, 67] and different types of liquid [68] has recently been studied. De Beer *et al.* [66] showed that temperature changes close to the melting point of the dodecanol and hexanol can cause changes in the oscillatory force. Atkin *et al.* [68] has investigated different liquids, such as ionic liquids (e.g. ethylammonium nitrate) that are entirely a salt in the liquid state, and revealed oscillatory forces. Moving toward such measurements will be evermore important for studying solvation forces in varying environments under which potential technological applications and devices will operate.

4.4.3.2 **Measurements in Pure Water**

Although earlier attempts of measuring solvation forces in water using static- and dynamic-mode AFM were made, it was not until the introduction of FM-AFM mode in liquid that observing oscillatory forces with a periodicity ($\approx$2.5Å) related to the size of single water molecule was possible [69]. Jarvis *et al.* [69, 70] made several modifications to the FM-AFM setup, including the use of magnetic excitation and attachment of a single multi-wall carbon nanotube (CNT) to the apex of the tip to decrease unwanted hydrodynamic damping effects from the lower aspect ratio of the bulk AFM probe. In doing so, the FM-AFM detection scheme offered seemingly increased local sensitivity in the region of the tip–sample interaction while also avoiding tip–sample instabilities due to the acquiescent use of higher-spring-constant cantilevers. This setup allowed for a direct measurement of $\approx$5 solvation layers with oscillatory forces of 1–50 mN m^{-1} in the vicinity of a COOH-terminated self-assembled monolayer in pure water (Figure 4.7). In contrast to the previous SFA and AFM measurements, a number of significant outcomes were demonstrated in the work, including the ability to (i) measure solvation forces on surfaces other than those of mica or graphite substrates, (ii) localize the force measurements with nanometer lateral resolution by a priori imaging of the surface with the same CNT-functionalized tip, and (iii) detect the solvation forces of water, thus opening up AFM to exploring this phenomenon in biological systems. The implementation of dynamic techniques such as small-amplitude AM-AFM [71], FM-AFM [72], and transverse force microscopy [73] in other studies have since been able to directly measure the oscillatory and damping forces of water on mica.

4.4.3.3 **Solvation Forces in Biological Systems**

With the sensitivity of dynamic-mode AFM reaching levels where the structural forces of water can be detected, it is an exciting time to study the water molecules and layers that are ever-present in cellular systems. Other techniques have experimentally shown that the biological water is intimately dispersed and bound as solvation shells surrounding proteins and discrete layers at membranes, yet little is known about their functional role in biology. With the desire to understand how the interactions of this water may mediate numerous cellular functions, including cytoplasmic phase transformations, membrane permeability, protein–protein

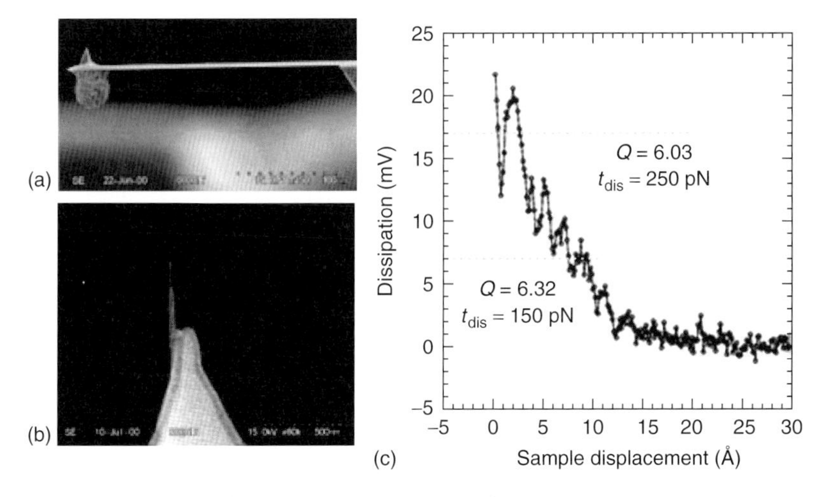

Figure 4.7 Scanning electron microscopy images of a magnetic particle (a) and carbon nanotube (b) attached to the cantilever and end of the tip, respectively. (c) FM-AFM measurement showing an oscillatory profile for the magnetically activated CNT tip approaching a self-assembled monolayer in pure water [69].

interactions, biocatalytic processes, and endocytosis, there is growing interest in probing their biophysicochemical interactions in much the same way protein interactions have been interrogated with AFM.

In applying solvation force measurements to biological substrates, the probe tip is likely to interact with more chemically reactive, softer, nanometer-corrugated, and mobile interfaces, which is in contrast to the inorganic substrates described above. The difference in this environment has questioned the ability to observe oscillatory forces in biological systems, for example, lipid bilayers, where the magnitude of their surface properties (e.g., roughness and mobility of surface groups) is of similar scale to water molecules and will disrupt their ordering. However, the application of FM-AFM measurements by Higgins *et al.* [74] to study lipid bilayers in water demonstrated that oscillatory forces could be detected for a biological system, presumably facilitated by the gel-phase state and hexagonal arrangement of the lipid head groups (Figure 4.8). Up to five solvation layers were present and an analysis of their oscillatory forces (ranging from 1–20 mN m^{-1}) and periodicity indicated that the layers closest to the surface were more tightly packed and difficult to displace. Conversely, lipid bilayers brought into the fluid phase by heating them above their transition temperature did not exhibit oscillatory forces, suggesting that a change in the thermal motion, chemistry, or deformability of the bilayer disrupted or inhibited detection of the water layers.

In the studies to date on water, the question stills remains as to whether the oscillatory forces are due to an induced confinement effect between the tip and surface, intrinsic water layers at the probe, and/or intrinsic layers known to be tightly associated with polar lipid head groups. A recent study using a low-noise

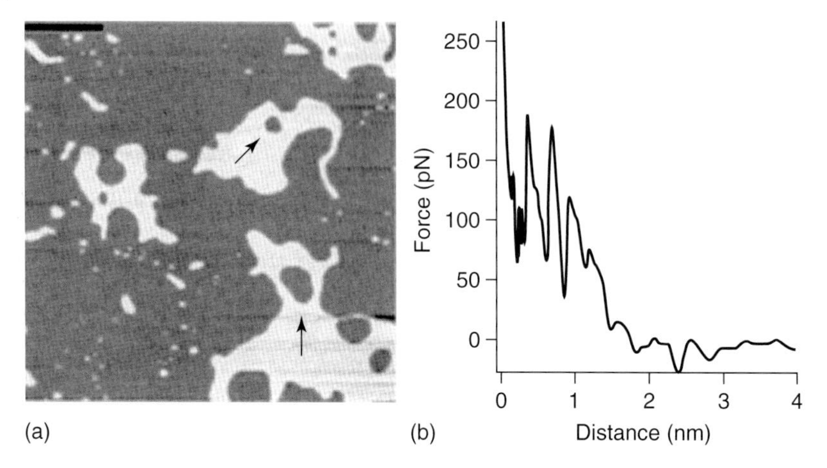

Figure 4.8 (a) AFM image of gel-phase DPPC lipid bilayers in water. The lipid bilayers (light areas) form small 5 nm thick islands on the mica substrate. (b) FM-AFM frequency shift measurement that has been converted to a force showing an oscillatory profile as the tip approaches the lipid bilayer in water [74].

FM-AFM setup designed for the use of subangstrom oscillation amplitudes has come a way to clarifying the origin of oscillatory forces. In this study, Fukuma *et al.* [14] consistently observed two solvation water layers in phosphate buffer solution for a lipid bilayer in water. Significantly, "tip jumps" corresponding to the height of quasi-stable water layers were observed in corresponding molecular-resolution AFM images and provided additional evidence for the detection of intrinsic hydration layers associated with the lipid bilayer (Figure 4.9). These advances in resolution, sensitivity, and stability of dynamic AFM modes represent the new and exciting tools of the future for probing biological systems in liquid.

4.4.4 Single-Molecule Force Spectroscopy

4.4.4.1 Unfolding and "Stretching" of Biomolecules

Traditional "stretching" experiments of proteins, DNA, or biological polyelectrolytes using static-mode AFM have been expanded using dynamic-mode approaches by superimposing a sinusoidal oscillatory component of the cantilever to measure dynamic responses of the molecule under tension. In these experiments, either the sample stage or the tip is oscillated, and as the piezo is retracted to extend the molecule tethered between the tip and substrate, the elastic and dissipative interactions are quantified from the change in the amplitude and phase signals. Lanz *et al.* [75] introduced the use of AM-AFM measurements to measure the force required to stretch a single peptide, cysteine$_3$-lysine$_{30}$-cysteine, from the helical state into a linear chain, in addition to simultaneously measuring the stiffness during elongation of the molecule (Figure 4.10a).

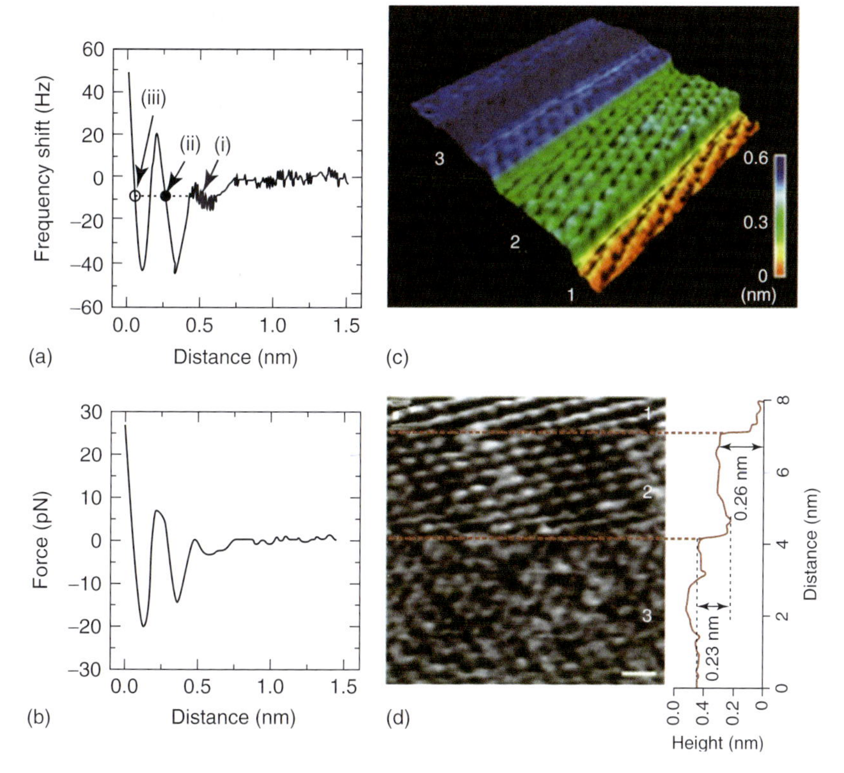

Figure 4.9 FM-AFM frequency shift (a) and converted force measurement (b) showing an oscillatory profile for a tip approaching a gel-phase DPPC bilayer in phosphate buffer solution. Positions (i), (ii), and (iii) denote equivalent set point values for the frequency shift that can be located on each repulsive branch of the oscillations. (c) FM-AFM molecular-resolution height image of the lipid bilayer with steps in the height (terraces 3, 2, and 1) corresponding to the regions (i), (ii), and (iii). The tip can "jump" between the repulsive branches (water layers) for the same set point, causing the height steps in (c). (d) Height image showing the change in the molecular topography as the tip probes the different water layers during scanning. The corresponding height-line profile (right) gives values of 0.23--0.26 nm for the step heights [14].

By using FM-AFM with a feedback to system to increase the quality factor of the cantilever, Humphris *et al.* [76] tracked the frequency shift and change in amplitude to measure the elastic and damping components during the stretching of a single dextran molecule (Figure 4.10b). The effective damping values were used to calculate a viscosity of 6000 Pa for the dextran molecule at high extensions. Various other configurations of dynamic-mode AFM have been applied to measure the elastic and dissipative responses of different biomolecules, including carbonic anhydrase II [77], bacteriorhodopsin [78], polysaccharides [79], I27 titin domains [80], and native titin [81]. In many of these dynamic approaches, the stretching experiments have unveiled new force transitions and unfolding pathways, not previously detected alone in the static-mode AFM measurements.

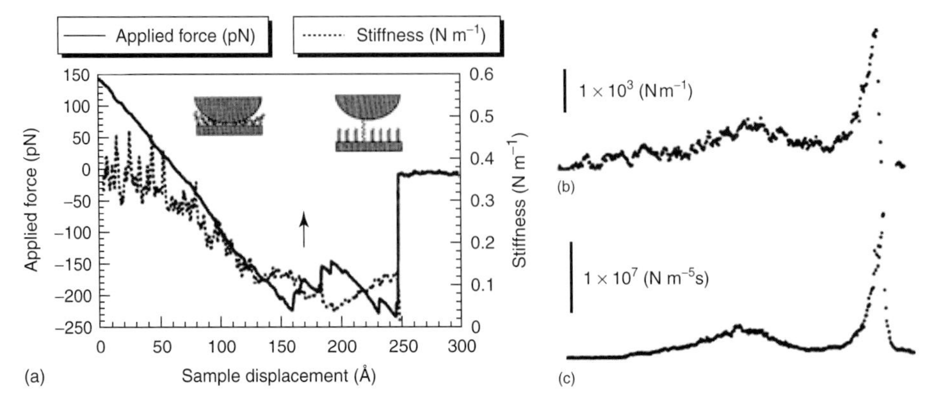

Figure 4.10 (a) Small-amplitude, off-resonance AM-AFM force (solid) and stiffness (dotted) measurements for the stretching of a single α-helical peptide [75]. At the point marked with the arrow, the tip separates from the nonbinding peptides and the stiffness then increases as the peptide that is still attached to the tip is stretched. FM-AFM measurements showing the stiffness (b) and dissipative forces (c), given as the effective damping, for the stretching of a single dextran molecule. The small hump in the profiles corresponds to the chair-to-boat transition of the glucose ring followed by further elongation and then unbinding (peak) [76].

4.4.4.2 **Ligand–Receptor Interactions**

The use of dynamic modes to quantitatively measure the unbinding forces of ligand–receptor interactions has been done to a much lesser extent. For example, Chtcheglova *et al.* [82] measured the interaction between single bovine serum albumin (BSA) and its polyclonal antibody (Ab-BSA) and also fibrinogen–fibrinogen (Fb-Fb) complexes. These researchers applied a dithering voltage to the cantilever, resulting in oscillation amplitudes of 0.5–2 nm, which were monitored to quantify the stiffness of the complexes at the moment of bond rupture (Figures 4.11a,b). In one of the few other studies, Higgins *et al.* [83] have investigated the biotin–avidin complex, via PEG linker functionalization, using FM-AFM measurements with magnetic excitation to oscillate the cantilever. The relationship between the measured frequency shift profile and increase in the stiffness of the biotin–avidin complex leading to bond rupture was described (Figure 4.11c). Like this study, many of the mentioned dynamic approaches for probing single biomolecule interactions are complex but offer new insights into exploring ligand–receptor interactions and protein unfolding – particularly as a function of the frequency dependence and subangstrom oscillation amplitudes that are smaller than their own chemical bonds.

4.5 High-Resolution Imaging

Until recently, subnanometer-scale imaging in liquid has been performed predominantly by contact-mode AFM [84] while AM-AFM is mainly used for

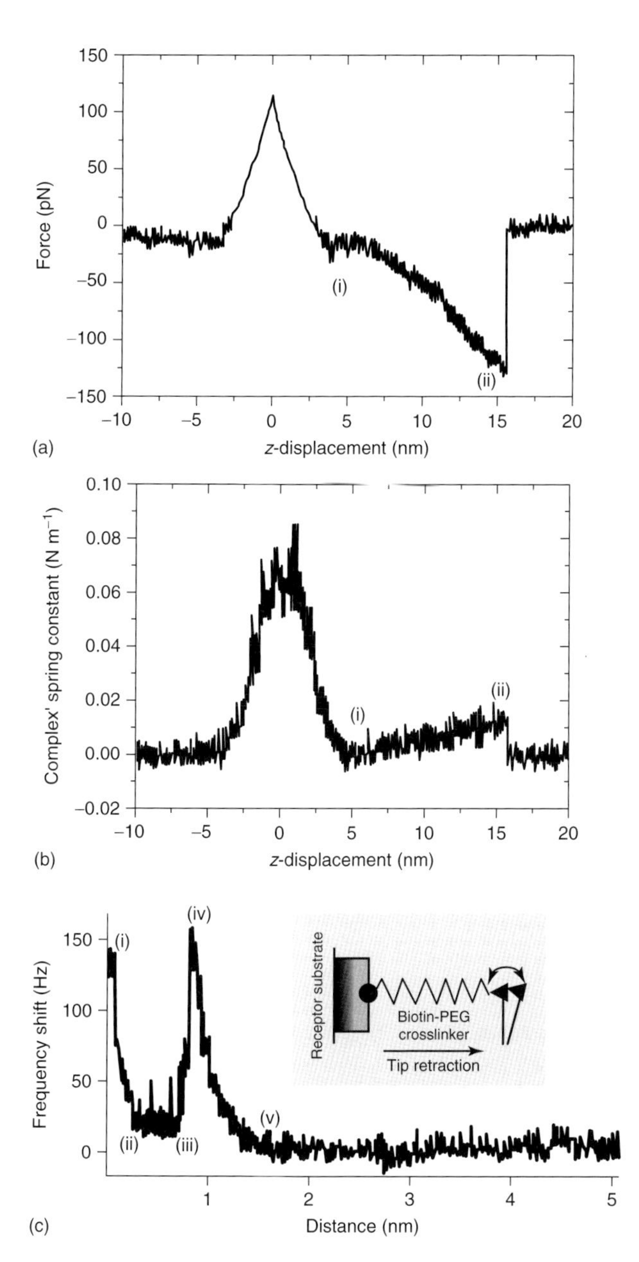

Figure 4.11 (a) Off-resonance AM-AFM static force (a) and stiffness (b) measurements for the unbinding interaction of single BSA– antibody complex [82]. The region between (i) and (ii) corresponds to the extension of the complex followed by bond rupture at point (ii). (c) FM-AFM measurement showing the frequency shift profile for the unbinding of a single biotin– avidin complex [83]. An increase in the effective stiffness due to restrained motion caused by the PEG-linker-biotin-avidin complex causes a positive frequency shift (point iii to iv) followed by bond rupture at point (iv).

nanometer-scale imaging of various samples, including isolated biomolecules that cannot be imaged by contact-mode AFM. Recently, subnanometer-scale imaging by FM-AFM in liquid has become possible, which enabled imaging of surfaces having imaging of surfaces having isolated biomolecules [12] as well as solid crystals [11] with subnanometer-scale resolution. In addition, this technical progress has made it possible to visualize not only surface structures but also distribution of water [14] and ions [15] at a solid–liquid interface. In this section, recent applications of FM-AFM to high-resolution imaging in liquid are summarized.

4.5.1 Solid Crystals

The cleaved surface of a crystal is one of the most suitable samples for high-resolution AFM imaging. A crystal surface has high rigidity so that it is hardly damaged by a loading force. The cleaved surface has a small roughness and periodic structures, which also makes it easier to image by AFM. In addition to the investigations of the crystal surfaces, they are also used as a substrate for imaging isolated molecules, molecular layers, and their dynamic behaviors.

Among the various solid surfaces, the cleaved surface of muscovite mica has been the most widely used for AFM experiments in liquid. Muscovite mica can be easily cleaved with scotch tape to present an atomically flat surface. The cleaved surface has a negative charge, which is useful to adsorb various molecules and their assemblies. The electrostatic interaction between molecules and substrate stabilize the adsorbates, which allows imaging with a high spatial resolution. Indeed, the first true atomic-resolution images obtained by dynamic-mode AFM in liquid were demonstrated by imaging a cleaved mica surface in water (Figure 4.12) [11].

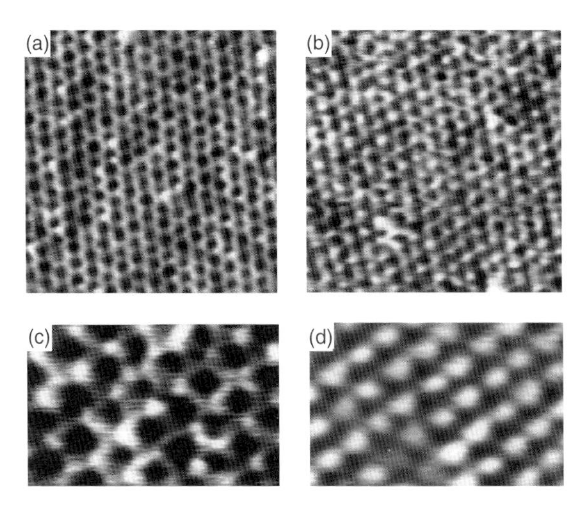

Figure 4.12 FM-AFM images of the cleaved (001) surface of muscovite mica taken in water [11]. (a) 8 nm × 8 nm, (b) 8 nm × 8 nm, (c) 4 nm × 2.5 nm, and (d) 4 nm × 2.5 nm.

To date, FM-AFM has been used for investigating surfaces of other solid crystals. In 2005, Fukuma *et al.* reported submolecular-resolution images of a polymer (polydiacetylene) signal crystal, shortly before the imaging of muscovite mica shown in Figure 4.12. In 2009, Rode *et al.* [85] reported atomic-scale images of a calcite crystal. In 2010, Nagashima *et al.* [86] reported molecular-resolution images of the surface of a protein (lysozyme) single crystal. These results showed that FM-AFM can be used for subnanometer-scale imaging of various crystals such as inorganic, organic, and biomolecular crystals.

4.5.2 Biomolecular Assemblies

One of the major application areas of liquid-environment AFM is biological science. Biological systems are soft and fragile and typically have large corrugations, fluctuations, and inhomogeneity. Therefore, imaging of biological systems imposes special requirements on AFM. The loading force during imaging must be kept as small as possible (typically less than 100 pN) to minimize influence on a biological system. The time response of the tip–sample distance regulation must be kept as fast as possible to follow the large corrugations of biological interfaces. A relatively large scanning area is desirable for investigating the whole structure of a nonperiodic inhomogeneous biological system.

Among various biological systems, one of the most widely used samples in AFM experiments is a purple membrane. A purple membrane consists of a two-dimensional crystal of proteins referred to as bacteriorhodopsins (bRs). Owing to the intermolecular interactions, positions and structures of bRs are stabilized. Thus, they can withstand a lateral friction force during the tip scan in contact-mode AFM [87]. To date, molecular-resolution images of a purple membrane has been obtained by various AFM imaging techniques including contact-mode AFM [87], AM-AFM [88], and FM-AFM [13, 89].

Dynamic-mode AFM is capable of imaging isolated biomolecules weakly bound to a substrate due to the low lateral friction force during imaging. This is one of the major advantages of dynamic-mode AFM over contact-mode AFM. The most widely used model biological system for such experiments is GroEL. GroEL is a molecular chaperone, that helps a protein to be folded into a proper three-dimensional structure. GroEL consists of upper and lower rings with a diameter of 14 nm. Each ring consists of seven molecules so that the structure has a seven-fold symmetry. The individual molecules constituting the ring have been directly imaged by FM-AFM in liquid [13].

Biological systems often have large fluctuations. Thus, it is not necessarily possible to image their subnanometer-scale structures. However, in some cases, a biological system shows a stable position or stable structure that can be imaged with subnanometer-scale resolution. For example, a surface of a lipid bilayer in the gel phase consists of an array of lipid head groups. As the acyl chains of lipid molecules are closely packed to form a crystalline structure, individual head groups can be clearly resolved by AFM imaging. Higgins *et al.* [74] reported a molecular-resolution image of a DPPC bilayer, wherein the individual phospholipid headgroups separated by 0.5 nm are resolved. Asakawa *et al.* [90] reported molecular-resolution images of a DPPC–cholesterol mixed bilayer. They proposed a molecular-scale model of the mixed bilayer based on the FM-AFM images.

Another example is imaging of stable structures constituting a surface of a biomolecular assembly. For example, Fukuma *et al.* reported molecular-resolution images of amyloid fibrils where individual β-strands separated by 0.5 nm were clearly resolved (Figure 4.13). Among the secondary structures of a protein, a loop and a random coil show a large fluctuation, while an α-helix and a β-sheet show a relatively small fluctuation. Therefore, if an α-helix or a β-sheet is stably fixed at a surface of a biomolecular assembly, it is likely that they can be imaged by FM-AFM with subnanometer resolution as shown in Figure 4.13.

4.5.3 Water Distribution

At a solid–liquid interface, a tip interacts with both water molecules and solid surface. Thus, spatial distribution of F_t measured near a surface is strongly influenced by a nonuniform distribution of water. One of the distinctive features

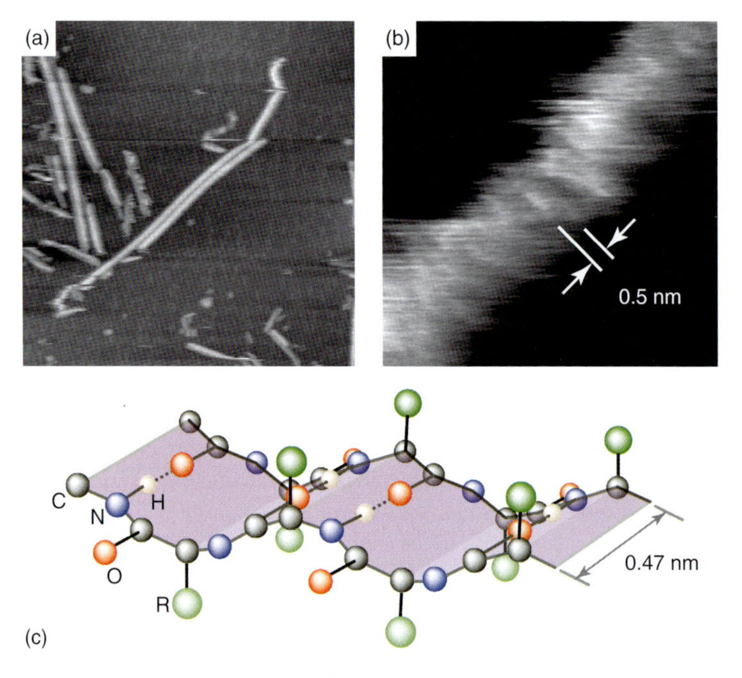

Figure 4.13 (a) FM-AFM image of islet amyloid polypeptide fibrils on mica in PBS solution (800 × 800 nm^2, $\Delta f = -55$ Hz). (b) FM-AFM image taken on one of the fibrils in Figure (a) (10 × 10 nm^2, $\Delta f = +50$ Hz). (c) Schematic model of β-strands [12].

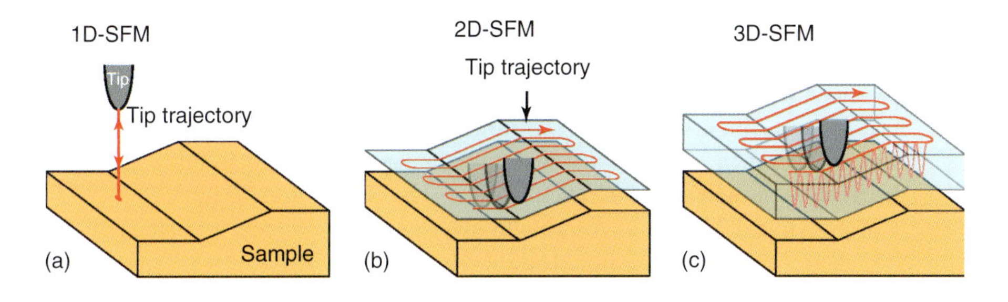

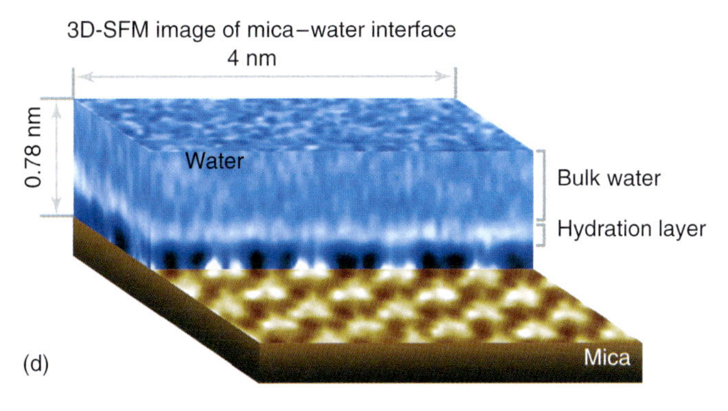

Figure 4.14 Principles of (a) 1D-SFM, (b) 2D-SFM, and (c) 3D-SFM. (d) 3D-SFM image obtained at a mica– water interface in phosphate buffer solution [16].

of dynamic-mode AFM is the capability of force measurements at a tip position in a near-contact region. This capability enables one to investigate nonuniform distribution of water at a solid–liquid interface.

The simplest way of visualizing force distribution at a solid–liquid interface is to take a force curve. In this method, a tip is scanned in the Z-direction, and Δf variation during the scan is recorded (Figure 4.14a). Here, we refer to this mode as one-dimensional scanning force microscopy (1D-SFM) for the purpose of discussion. An example of such a Δf versus distance curve taken at a lipid–water interface is shown in Figure 4.9a. Due to the existence of hydration layers, a force versus distance curve often shows an oscillatory profile with a peak separation corresponding to the diameter of a water molecule (0.2–0.3 nm). As described in Section 4.4, a Δf versus distance curve can be mathematically converted to an F_t versus distance curve (Figure 4.9b) with Eq. (4.5) for a quantitative analysis.

Although 1D-SFM is useful to investigate vertical distribution of water, it does not investigate its lateral distribution. In some limited cases, such lateral distribution of water can be visualized by two-dimensional (2D) imaging, where a tip–sample separation is controlled whereby Δf is kept constant during the lateral tip scan over a surface (Figure 4.14b). Here, we refer to this mode as two-dimensional scanning force microscopy (2D-SFM). If a Δf curve has an oscillatory profile, there are multiple positions for a stable tip–sample distance control as indicated by arrows (i)–(iii) in Figure 4.9a. Therefore, a tip can jump between different feedback positions during imaging. Figure 4.9c shows an example of such imaging. During the imaging, the tip spontaneously jumped twice. The force branches indicated by arrows (i)–(iii) correspond to the surfaces of the secondary hydration layer, the primary hydration layer, and the lipid membrane, respectively. In this way, lateral distribution of individual hydration layers can be directly visualized.

In principle, it should be possible to investigate both vertical and lateral distributions of water using 1D and 2D-SFM. However, this requires very complicated experimental procedures and hence is inefficient. Moreover, the interpretation of the obtained results are complicated by the existence of tip drift. To overcome this problem, a method referred to as three-dimensional scanning force microscopy (3D-SFM) was recently proposed [16] (Figure 4.14c). In the method, a tip is scanned in both vertical and lateral directions to cover the whole interfacial space (Figure 4.14c). While the tip is laterally scanned over a surface, it is also scanned in the vertical direction at a frequency faster than the bandwidth of the tip–sample distance regulation. Δf variation induced by the scan is recorded in real time to construct a 3D Δf image. The Δf image can be mathematically converted to a quantitative 3D force field using Eq. (4.5). As the averaged tip position in the Z-direction is always kept constant by the distance feedback regulation, a surface corrugation or sample tilt does not cause a tip crash with surface.

An example of 3D-SFM images taken at a mica–water interface is shown in Figure 4.14d [16]. Once a 3D force field is obtained, 1D force profiles and 2D force images at an arbitrary position can be extracted from it. In fact, the model shown in Figure 4.14d was constructed with vertical and lateral cross sections derived from the 3D-SFM image. The 3D-SFM image reveals a layer-like force contrast corresponding to a hydration layer as well as localized force contrasts corresponding to adsorbed water molecules. The result showed good agreement with the previously reported results obtained by Monte-Carlo simulation [91] and X-ray reflectometory [92]. As shown by this example, 3D-SFM is much more powerful than 1D- and 2D-SFM for studies on hydration phenomena at a solid–liquid interface which inherently has a 3D extent.

4.6
Summary and Future Prospects

The performance of liquid-environment dynamic-mode AFM has tremendously advanced in the past decade. Quantitative and independent force measurements of conservative and dissipative forces have become possible. Subnanometer-scale imaging has become possible on both rigid inorganic crystals and soft biological systems. Furthermore, direct visualization of 3D force field at a solid–liquid interface has also become possible. These technical advances have enabled investigations of inter- and intramolecular interaction forces of biomolecules, atomic and molecular-scale structures of solid surfaces, and hydration phenomena at a solid–liquid interface.

There are several issues that are currently under investigation or expected to become important subjects in the near future. An example of such subjects is an understanding of atomic-scale interactions between a tip and surface (or water). While experimental techniques reached the level required for an atomic-resolution imaging in liquid, atomic-scale simulation of AFM in liquid has yet to be achieved. Such a study is particularly important for understanding the correlation between the water distribution and 3D force field obtained by 3D-SFM.

Another important subject is an improvement of the force sensor. There are two approaches that have been considered. One of them is to reduce the size of the cantilever, which enhances f_0 without changing k and Q. Another approach is to use a self-sensing quartz sensor, which enhances both Q and k, while f_0 is determined by the size of the sensor. Although both sensors seem to be very promising, their applicability and performance are yet to be experimentally verified.

The major application areas of liquid-environment dynamic-mode AFM include biology and electrochemistry. So far, biological applications have been intensively explored, which has been driven by the rapid growth of the nanobioscience field. On the contrary, applications in electrochemistry are still in their infancy. However, considering the increasing interests in the so-called green science (e.g., solar power and water splitting), we expect that its applications to the studies on catalysis and batteries will rapidly increase in the near future.

References

1. Binnig, G., Quate, C.F., and Gerber, Ch. (1986) *Phys. Rev. Lett.*, **56**, 930.
2. Martin, Y., Williams, C.C., and Wickramasinghe, H.K. (1987) *J. Appl. Phys.*, **61**, 4723.
3. Umeda, N., Ishizaki, S., and Uwai, H. (1991) *J. Vac. Sci. Technol., B*, **9**, 1318.
4. Jarvis, S.P., Oral, A., Weihs, T.P., and Pethica, J.B. (1993) *Rev. Sci. Instrum.*, **64**, 3515.
5. Degertekin, F.L., Hadimioglu, B., Sulchek, T., and Quate, C.F. (2001) *Appl. Phys. Lett.*, **78**, 1628.
6. Asakawa, H. and Fukuma, T. (2009) *Rev. Sci. Instrum.*, **80**, 103703.
7. Fukuma, T., Kimura, M., Kobayashi, K., Matsushige, K., and Yamada, H. (2005) *Rev. Sci. Instrum.*, **76**, 053704.
8. Hoogenboom, B.W., Frederix, P.L.T.M., Yang, J.L., Martin, S., Pellmont, Y., Steinacher, M., Zäch, S., Langenbach, E., and Heimbeck, H.-J. (2005) *Appl. Phys. Lett.*, **86**, 074101.
9. Kawai, S., Kobayashi, D., Kitamura, S., Meguro, S., and Kawakatsu, H. (2005) *Rev. Sci. Instrum.*, **76**, 083703.
10. Fukuma, T., Kobayashi, K., Matsushige, K., and Yamada, H. (2005) *Appl. Phys. Lett.*, **86**, 193108.
11. Fukuma, T., Kobayashi, K., Matsushige, K., and Yamada, H. (2005) *Appl. Phys. Lett.*, **87**, 034101.
12. Fukuma, T., Mostaert, A.S., Serpell, L.C., and Jarvis, S.P. (2008) *Nanotechnology*, **19**, 384010.
13. Yamada, H., Kobayashi, K., Fukuma, T., Hirata, Y., Kajita, T., and Matsushige, K. (2009) *Appl. Phys. Express*, **2**, 095007.
14. Fukuma, T., Higgins, M.J., and Jarvis, S.P. (2007) *Biophys. J.*, **92**, 3603.
15. Fukuma, T., Higgins, M.J., and Jarvis, S.P. (2007) *Phys. Rev. Lett.*, **98**, 106101.
16. Fukuma, T., Ueda, Y., Yoshioka, S., and Asakawa, H. (2010) *Phys. Rev. Lett.*, **104**, 016101.
17. Albrecht, T.R., Grütter, P., Horne, D., and Ruger, D. (1991) *J. Appl. Phys.*, **69**, 668.
18. Fukuma, T., Kilpatrick, J., and Jarvis, S.P. (2006) *Rev. Sci. Instrum.*, **77**, 123703.

19. Revenko, I. and Proksch, R. (2000) *J. Appl. Phys.*, **87**, 526.
20. Jarvis, S.P., Weihs, T.P., Oral, A., and Pethica, J.B. (1993) MRS Symposia Proceedings No 308 (Pittsburgh), p. 127.
21. Han, W., Lindsay, S.M., and Jing, T. (1996) *Appl. Phys. Lett.*, **69**, 4111.
22. Buguin, A., Du Roure, O., and Silberzan, P. (2001) *Appl. Phys. Lett.*, **78**, 2982.
23. Ratcliff, G.C., Erie, D.A., and Suerfine, R. (1998) *Appl. Phys. Lett.*, **72**, 1911.
24. Umeda, K.-I., Oyabu, N., Kobayashi, K., Hirata, Y., Matsushige, K., and Yamada, H. (2010) *Appl. Phys. Express*, **3**, 065205.
25. McClelland, G.M., Erlandsson, R., and Chiang, S. (1988) *Review of Progress in Quantitative Non-Destructive Evaluation*, **vol. 6B**, Plenum, New York, pp. 1307–1314.
26. Meyer, G. and Amer, N.M. (1988) *Appl. Phys. Lett.*, **53**, 1045.
27. Schönenberger, C. and Alvarado, S.F. (1989) *Rev. Sci. Instrum.*, **60**, 3131.
28. Neubauer, G., Cohen, S.R., McClelland, G.M., Horne, D., and Mate, C.M. (1990) *Rev. Sci. Instrum.*, **61**, 2296.
29. Tortonese, M., Yamada, H., Barrett, R.C., and Quate, C.F. (1991) *The Proceedings of Transducers '91*, Publication No. 91 CH2817-5, IEEE, Pennington, NJ, pp. 448–451.
30. Itoh, T. and Suga, T. (1993) *Nanotechnology*, **4**, 218.
31. Fukuma, T. (2009) *Rev. Sci. Instrum.*, **80**, 023707.
32. Fukuma, T. and Jarvis, S.P. (2006) *Rev. Sci. Instrum.*, **77**, 043701.
33. Schäffer, T.E. and Hansma, P.K. (1998) *J. Appl. Phys.*, **84**, 4661.
34. Nishida, S., Kobayashi, D., Sakurada, T., Nakazawa, T., Hoshi, Y., and Kawakatsu, H. (2008) *Rev. Sci. Instrum.*, **79**, 123703.
35. Pham Van, L., Kyrylyuk, V., Thoyer, F., and Cousty, J. (2008) *J. Appl. Phys.*, **104**, 074303.
36. Pham Van, L., Kyrylyuk, V., Plesel-Maris, J., Thoyer, F., Lubin, C., and Cousty, J. (2009) *Langmuir*, **25**, 639.
37. Li, Y.J., Kobayashi, N., Nomura, H., Naitoh, Y., Kageshima, M., and Sugawara, Y. (2008) *Jpn. J. Appl. Phys.*, **47**, 6121.
38. Li, Y.J., Takahashi, K., Kobayashi, N., Naitoh, Y., Kageshima, M., and Sugawara, Y. (2010) *Ultramicroscopy*, **110**, 582.
39. García, R. and Pérez, R. (2002) *Surf. Sci. Rep.*, **47**, 197.
40. Dürig, U. (2000) *Appl. Phys. Lett.*, **76**, 1203.
41. Stark, M., Stark, R.W., Heckl, W.M., and Guckenberger, R. (2002) *Proc. Natl. Acad. Sci. U.S.A.*, **99**, 8473.
42. Sader, J.E. and Jarvis, S.P. (2004) *Appl. Phys. Lett.*, **84**, 1801.
43. Katan, A.J., van Es, M.H., and Oosterkamp, T.H. (2009) *Nanotechnology*, **20**, 165703.
44. Giessibl, F.J. (2001) *Appl. Phys. Lett.*, **78**, 123.
45. Palcio, M.L.B. and Bhushan, B. (2010) *Crit. Rev. Sol. Stat. Mater. Sci.*, **35**, 73.
46. Hutter, J.L. and Bechoefer, J. (1993) *Rev. Sci. Instrum.*, **64**, 1868.
47. Sader, J.E., Chon, J.W.M., and Mulvaney, P. (1999) *Rev. Sci. Instrum.*, **70**, 3967.
48. Ruan, J. and Bhushan, B. (1994) *ASME J. Tribol.*, **116**, 378.
49. Albrecht, T.R., Akimine, S., Carver, T.E., and Quate, C.F. (1990) *J. Vac. Sci. Technol., A*, **8**, 3386.
50. Cleveland, J.P., Manne, S., Bocek, D., and Hansma, P.K. (1993) *Rev. Sci. Instrum.*, **64**, 3789.
51. Burnham, N.A., Chen, X., Hodges, C.S., Matei, G.A., Thoreson, E.J., Roberts, C.J., Davies, M.C., and Tendler, S.J.B. (2003) *Nanotechnology*, **14**, 1.
52. Kim, M.-S., Choi, J.-H., Kim, J.-H., and Park, Y.-K. (2010) *Measurement*, **43**, 520.
53. te Riet, J, Katan, A.J, Rankl, C, Stahl, S.W, van Buul, A.M, Phang, I.Y, Gomez-Casado, A, Schon, P, Gerritsen, J.W, Cambi, A, et al (2011) *Ultramicroscopy*, **111**, 1659.
54. Sader, J.E., Uchihashi, T., Higgins, M.J., Farrell, A., Nakayama, Y., and Jarvis, S.P. (2005) *Nanotechnology*, **16**, S94.
55. Israelachvili, J.N. (1982) *Adv. Colloid Interface Sci.*, **16**, 31.
56. Pashley, R.M. and Israelachvili, J.N. (1984) *J. Colloid Interface Sci.*, **101**, 511.

57. O'Shea, S.J., Welland, M.E., and Rayment, T. (1992) *Appl. Phys. Lett.*, **60**, 2356.
58. O'Shea, S.J., Welland, M.E., and Pethica, J.B. (1994) *Chem. Phys. Lett.*, **223**, 336.
59. O'Shea, S.J. and Welland, M.E. (1998) *Langmuir*, **14**, 4186.
60. Uchihashi, T., Higgins, M.J., Yasuda, S., Jarvis, S.P., Akita, S., Nakayama, Y., and Sader, J.E. (2004) *Appl. Phys. Lett.*, **85**, 16.
61. Maali, A., Cohen-Bouhacina, T., Couturier, G., and Aime, J.-P. (2006) *Phys. Rev. Lett.*, **96**, 086105.
62. De Beer, S., van de Ende, D., and Frieder, M. (2010) *Nanotechnology*, **21**, 325703.
63. O'Shea, S.J., Gosvami, N.N., Lim, L.T.W., and Hofbauer, W. (2010) *Jpn. J. Appl. Phys.*, **49**, 08LA01.
64. Kaggwa, G.B., Kilpatrick, J.I., Sader, J.E., and Jarvis, S.P. (2008) *Appl. Phys. Lett.*, **93**, 011909.
65. Lim, R., Li, S.F.Y., and O'Shea, S.J. (2002) *Langmuir*, **18**, 6116.
66. De Beer, S., Wennink, P., van der Weide-Grevelink, M., and Mugele, F. (2010) *Langmuir*, **26**, 13245.
67. Lim, L.T.W., Wee, A.T.S., and O'Shea, S.J. (2009) *J. Chem. Phys.*, **130**, 134703.
68. Atkin, R. and Warr, G.G. (2007) *J. Phys. Chem. C*, **111**, 5162.
69. Jarvis, S.P., Uchihashi, T., Ishida, T., Tokumoto, H., and Nakayama, Y. (2000) *J. Phys. Chem. B*, **104**, 6091.
70. Jarvis, S.P., Ishida, T., Uchihashi, T., Nakayama, Y., and Tokumoto, H. (2001) *Appl. Phys. A*, **72**, S129.
71. Jeffery, S.P.M., Hoffmann, J., Pethica, J., Ramanujan, C., Ozer, O., and Oral, A. (2004) *Phys. Rev. B*, **70**, 054114.
72. Uchihashi, T., Higgins, M.J., Nakayama, Y., Sader, J.E., and Jarvis, S.P. (2005) *Nanotechnology*, **16**, S49.
73. Antognozzi, M., Humphris, A.D., and Miles, M.J. (2001) *Appl. Phys. Lett.*, **78**, 300.
74. Higgins, M., Polcik, M., Fukuma, T., Sader, J.E., Nakayama, Y., and Jarvis, S.P. (2006) *Biophys. J.*, **91**, 2532.
75. Lanz, M.A., Jarvis, S.P., Tokumoto, H., Martynski, T., Kusumi, T., Nakamura, C., and Miyake, J. (1999) *Chem. Phys. Lett.*, **315**, 61.
76. Humphris, A.D.L., Tamayo, J., and Miles, M.J. (2000) *Langmuir*, **16**, 7891.
77. Okajima, T., Arakawa, H., Alam, M.T., Sekiguchi, H., and Ikai, A. (2004) *Biophys. Chem.*, **107**, 51.
78. Janovjak, H., Muller, D., and Humphris, A.D.L. (2005) *Biophys. J.*, **88**, 1423.
79. Kawakami, M., Byrne, K., Khatri, B.S., Mcleish, T.C.B., Badford, S.E., and Smith, D.A. (2005) *Langmuir*, **21**, 4765.
80. Higgins, M., Sader, J.E., and Jarvis, S.P. (2006) *Biophys. J.*, **90**, 640.
81. Forbes, J.G. and Wang, K. (2004) *J. Vac. Sci. Technol., A*, **22**, 1439.
82. Chtcheglova, L.A., Shubeita, G.T., Sekatskii, S.K., and Dietler, G. (2004) *Biophys. J.*, **86**, 1177.
83. Higgins, M.J., Riener, Ch.K., Uchihashi, T., Sader, J.E., McKendry, R., and Jarvis, S.P. (2005) *Nanotechnology*, **16**, S85.
84. Ohnesorge, F. and Binnig, G. (1993) *Science*, **260**, 1451.
85. Rode, S., Oyabu, N., Kobayashi, K., Yamada, H., and Kühnle, A. (2009) *Langmuir*, **25**, 2850.
86. Nagashima, K., Abe, M., Morita, S., Oyabu, N., Kobayashi, K., Yamada, H., Ohta, M., Kokawa, R., Murai, R., Matsumura, H., Adachi, H., Takano, K., Murakami, S., Inoue, T., and Mori, Y. (2010) *J. Vac. Sci. Technol., B*, **28**, C4C11.
87. Muller, D.J., Schabert, F.A., Buldt, G., and Engel, A. (1995) *Biophys. J.*, **68**, 1681.
88. Möller, C., Allen, M., Elings, V., Engel, A., and Müller, D.J. (1999) *Biophys. J.*, **77**, 1150.
89. Hoogenboom, B.W., Hug, H.J., Pellmont, Y., Martin, S., Frederix, P.L.T.M., Fotiadis, D., and Engel, A. (2006) *Appl. Phys. Lett.*, **88**, 193109.
90. Asakawa, H. and Fukuma, T. (2009) *Nanotechnology*, **20**, 264008.
91. Park, S.-H. and Sposito, G. (2002) *Phys. Rev. Lett.*, **89**, 085501.
92. Cheng, L., Fenter, P., Nagy, K.L., Schlegel, M.L., and Sturchio, N.C. (2001) *Phys. Rev. Lett.*, **87**, 156103.

5
Fundamentals of AFM Cantilever Dynamics in Liquid Environments

Daniel Kiracofe, John Melcher, and Arvind Raman

5.1
Introduction

As discussed in other chapters of this book, the use of atomic force microscopy (AFM) in liquid environments is one of the main growth areas in AFM and is driven by increased interest in biological applications and in the study of solid–liquid interfaces. In particular, the potential of using *dynamic* atomic force microscopy (dAFM) in liquids was recognized in the early 1990s and two important driving modes – the acoustic excitation mode [1, 2] and the magnetic mode [3] – were established.

This transfer of dAFM from vacuum/air to liquid environments brought with it a set of assumptions that do not quite hold in liquid environments. For example, in liquids, the shapes of the resonance peaks strongly depend on the method of excitation used. Further, the cantilever oscillations couple strongly to viscous hydrodynamics; this can significantly change the cantilever eigenmodes, their frequency bandwidths, and force sensitivities in comparison to ambient conditions. The low Q-factors and frequencies in liquid environments also lead to special, often counterintuitive cantilever dynamics that open up new instabilities, while also creating new channels for energy transfer and local material property contrast. Lastly, the very nature of the tip–sample interaction forces can be quite different in liquids, which leads to unique cantilever dynamics. Without a firm understanding of how AFM cantilevers oscillate and interact with samples in liquid environments, it is very difficult to achieve quantitative force spectroscopy or interpret correctly the local material contrast channels while scanning samples in liquid environments.

In this chapter, we aim to review the state of art of this important subject. Following a brief review of the fundamentals of cantilever oscillations, we focus on the following three aspects of cantilever dynamics in liquid environments. First, we consider the *hydrodynamics of cantilevers in liquids*. In this section, we discuss the influence of the surrounding fluid on the cantilever's vibration, without regard to a particular excitation method and in the absence of the tip–sample interactions. The free vibration properties of the cantilever such as eigenmodes, stiffness, optical sensitivity, Q-factors, and natural frequencies can all be altered in liquids compared

Atomic Force Microscopy in Liquid: Biological Applications, First Edition.
Edited by Arturo M. Baró and Ronald G. Reifenberger.

with their values in air. This plays an important role in better understanding the ideal imaging conditions and force sensitivities in liquids. Second, we consider the various *methods of dynamic excitation in liquids*. In this section, we consider the forced vibration of cantilevers but still do not include any tip–sample interactions. The choice of method of excitation can have a significant effect on the dynamic response and can change the nature of the observable quantities in dynamic AFM. We review the details of many common excitation methods including magnetic, acoustic, thermal, and photothermal. Finally, we describe results on the *dynamics of cantilevers interacting with samples in liquids*. We show that the cantilever dynamics are quite different compared with air and that new channels of local material contrast are created in liquids. This has a major influence on how material contrast channels such as phase or higher harmonics are interpreted in terms of local material properties of the sample. We do not discuss in detail the various models of the tip–sample interaction, which are covered in other chapters.

5.2 Review of Fundamentals of Cantilever Oscillation

In air or vacuum environments, it is common to treat the AFM cantilever as if it were a point-mass oscillator. That is, it is assumed that the cantilever can be modeled as a simple spring-mass-dashpot system that is characterized by a single natural frequency, quality factor, and stiffness. However, an AFM cantilever is actually a continuous structure and therefore it possesses an infinite number of eigenmodes, each with an equivalent point-mass model. That is, the cantilever dynamic response can be regarded as a superposition of the dynamic response of an infinite number of point-mass oscillators, each with its own natural frequency, quality factor, and stiffness.

Each eigenmode can be characterized by several frequencies. One is the natural frequency, which is the frequency at which the cantilever's response is exactly 90° out of phase with the excitation force. Another is the resonance frequency, which is the frequency at which a driven cantilever's amplitude response is largest. In other words, this is the frequency at which the peak appears in a tuning curve. In vacuum, the natural frequency and the resonance frequency are virtually identical, but they are typically different in liquids.

Finally, one point that is commonly misunderstood is the difference between higher eigenmodes and higher harmonics. A common source of confusion is that, for some systems (e.g., strings), the natural frequencies of higher eigenmodes may occur at integer ratios of the fundamental and are mistakenly referred to as *harmonics*. However, for cantilever beams, the natural frequencies of higher eigenmodes do not generally occur at integer ratios of the fundamental. For a uniform beam, the second natural frequency is $\sim$6.3 times the fundamental natural frequency.

With this brief introduction to AFM cantilever vibrations, we return to a discussion of how their dynamic response is modified on immersion in liquids.

5.3
Hydrodynamics of Cantilevers in Liquids

As an AFM microcantilever is immersed in liquid, its oscillations couple significantly to the viscous, low-Reynolds number hydrodynamics of the surrounding fluid. The most immediately obvious effect of this coupling is that a cantilever's resonance frequency and quality factor appear to be significantly reduced in liquid environments compared with ambient conditions (sometimes the terms "*wet resonance frequency*" and "*dry resonance frequency*" are used to distinguish these two cases). This effect is known to depend on the frequency of oscillation, eigenmode number, and it also depends sensitively on the distance of the microcantilever from the sample because of the squeeze film that develops between the microcantilever and the rigid substrate.

Predicting the hydrodynamics of cantilevers near a substrate has been a focus of many research groups and has been generally based on

- *Ad hoc*, but intuitive models [4, 5].
- Computational solutions using the boundary element method of the unsteady Stokes equations in two and three dimensions [6–10]. These have been recently expanded to include computations with finite cantilever thickness [11].
- Transient, fully coupled fluid–structure interaction calculations using the Navier–Stokes equations (Figure 5.1a) [12–14].

Many experimental results have also been performed to study this phenomenon [15–17]. These works have clearly shown the reduced cantilever resonance frequency and Q-factors in liquids and the dependence of these quantities on the oscillation frequency and mode number. Moreover, they have shown that when brought close to a surface in a liquid medium, the Q-factors and wet resonance frequencies of the different eigenmodes decrease significantly (Figure 5.1b,c). The rate of decrease with gap depends strongly on the eigenmode of interest and also on the orientation of the cantilever relative to the surface (Figure 5.1b,c). While the distance-dependent damping and added mass effect are relevant for dAFM, sometimes it is possible that as the gap is reduced the cantilever eigenmodes become overdamped ($Q < 0.5$) and no resonance peak is observed [13, 18, 19].

An important implication of this result is that experimental measurements of quality factor and resonance frequency should be made as closer to the sample preferably at a distance comparable to the cantilever–sample gap at which the image is acquired.

The results of many of these studies can be conveniently expressed through the use of the so-called hydrodynamic function. If the cantilever is undergoing a periodic motion at frequency ω, then the force per unit length exerted on the cantilever by the fluid is given by Sader [20]

$$f_{\text{hydro}}(x, \omega) = \frac{\pi}{4}\rho_f \omega^2 b^2 \Gamma(\omega) w(x, \omega)$$

where b and ρ_f are the cantilever width and fluid density, $w(x, \omega)$ is the Fourier transform of the displacement of the beam at location x, and Γ is the (complex-valued)

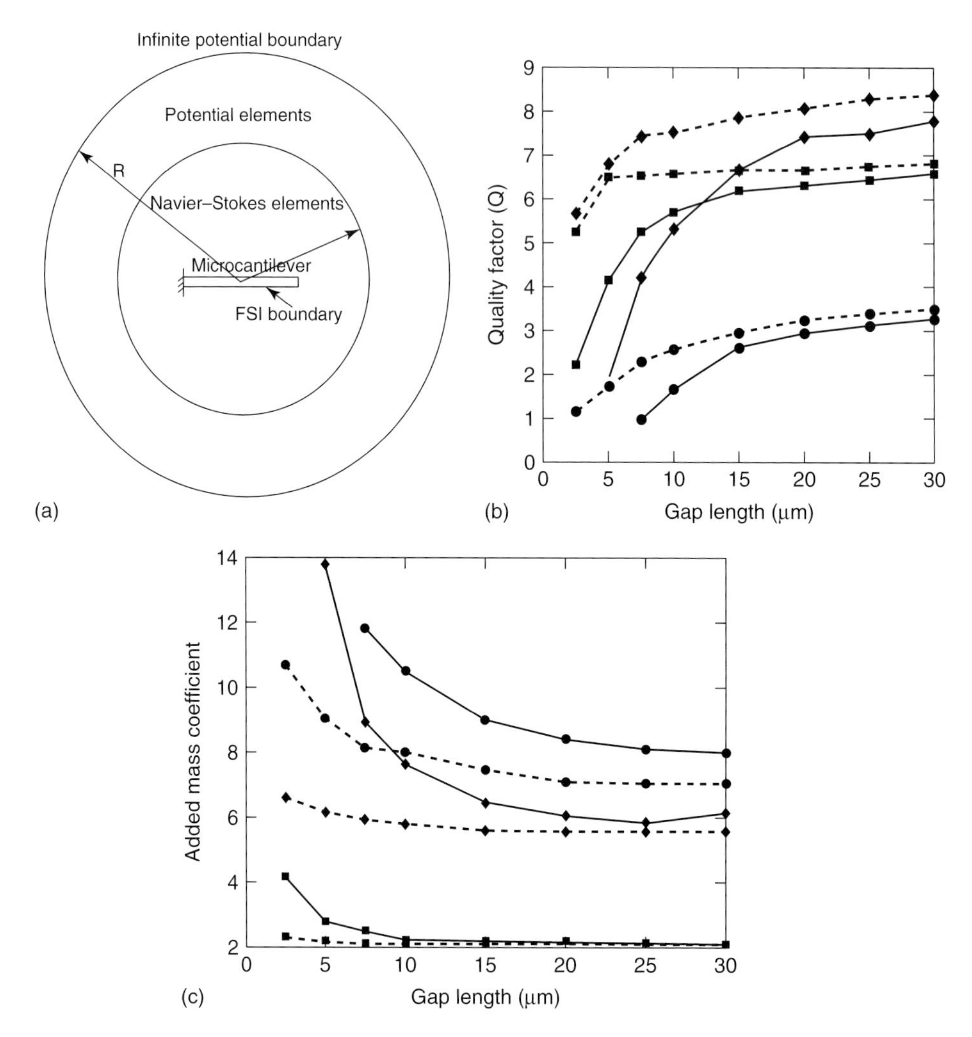

Figure 5.1 (a) Three-dimensional computational fluid–structure interaction modeling of cantilever hydrodynamics in liquids uses three-dimensional finite element models of cantilevers immersed in Navier–Stokes elements that are bounded by potential flow elements, and the cantilever hydrodynamics both far away and near to and inclined to a surface can be studied; (b) computational results for a rectangular Si cantilever ($197 \times 20 \times 2$ μm) close to a surface in water. Computed *Q*-factors of the first bending mode (circles), second mode (diamonds), and first torsion (squares) modes show that the *Q*-factors decrease rapidly while (c) the added mass effect (that leads to reduced wet natural frequency) increases on decrease of gap. The rate of decrease/increase depends on the mode number and on the orientation of the cantilever (dashed lines are for a cantilever oriented at 11° to the sample surface). (Reprinted with permission from Ref. [12]. Copyright 2006, American Institute of Physics.)

hydrodynamic function. The definition of the hydrodynamic function will vary depending on the geometry and situation of interest. For example, see Ref. [21] for the hydrodynamic function for torsional oscillations, Refs. [22, 23] for hydrodynamic functions including squeeze film effects near rigid walls, and Ref. [24] for corrections for higher-order eigenmodes. It may sometimes be more useful to show the results in terms of lumped parameters. In that case, we can rewrite the hydrodynamic force in terms of an effective added mass and added viscosity [25]

$$M = \frac{\pi}{4}\rho_f b^2 \mathrm{Re}\left[\Gamma(\omega)\right] \quad c = -\frac{\pi}{4}\rho_f \omega b^2 \mathrm{Im}\left[\Gamma(\omega)\right] \tag{5.1}$$

where Re and Im refer to the real and imaginary parts.

Recently, the effect of hydrodynamic loading on the eigenmode shapes, modal stiffnesses, and optical sensitivities of AFM microcantilevers were investigated [25] by measuring their vibrations in air and water using a scanning laser Doppler vibrometer (Figure 5.2). It was found that, for rectangular tipless cantilevers, the measured fundamental and higher eigenmodes and their equivalent stiffnesses are nearly identical in air and water. However, for microcantilevers with a tip mass or for picket-shaped levers (as is the case in AFM cantilevers), there is a marked difference in the second (and higher) eigenmode shapes between air and water that leads to a large decrease in the modal stiffness and optical sensitivity in water compared with in air (Figure 5.3).

To understand this interesting result, a mathematical model of a rectangular beam with the tip mass and immersed in a fluid was developed [25]. The fluid dynamics was modeled using unsteady Stokes equations in the two-dimensional cross-sectional plane of the cantilever, through use of semianalytical hydrodynamic loading functions. The combined eigenvalue problem was solved and the following was found:

1) In the absence of fluid loading, the presence of the tip mass significantly alters the second eigenmode shape leading to larger modal curvatures and thus greater modal stiffness and optical sensitivity. Optical sensitivity is the slope per unit transverse deflection at the end of the beam, this being the relevant quantity that is monitored in an AFM.
2) When immersed in a liquid, the added mass effect of the surrounding fluid is very large and is uniformly distributed over the beam (added fluid mass can be 10–20 times the cantilever mass). As a consequence, the effect of the concentrated tip mass is greatly diminished, relatively speaking. Thus, in water, a beam with tip mass possesses eigenmodes that are nearly identical to those of a uniform beam without tip mass.

There are many implications of these results to AFM applications, for instance, the theory in [26] lays out what corrections need to be made when a modal stiffness is calibrated in air and then used for imaging samples in liquids.

As can be seen the effects of hydrodynamic loading on cantilever dynamics can be quite profound and must be well understood before studying the influence of excitation methods on the cantilever dynamics.

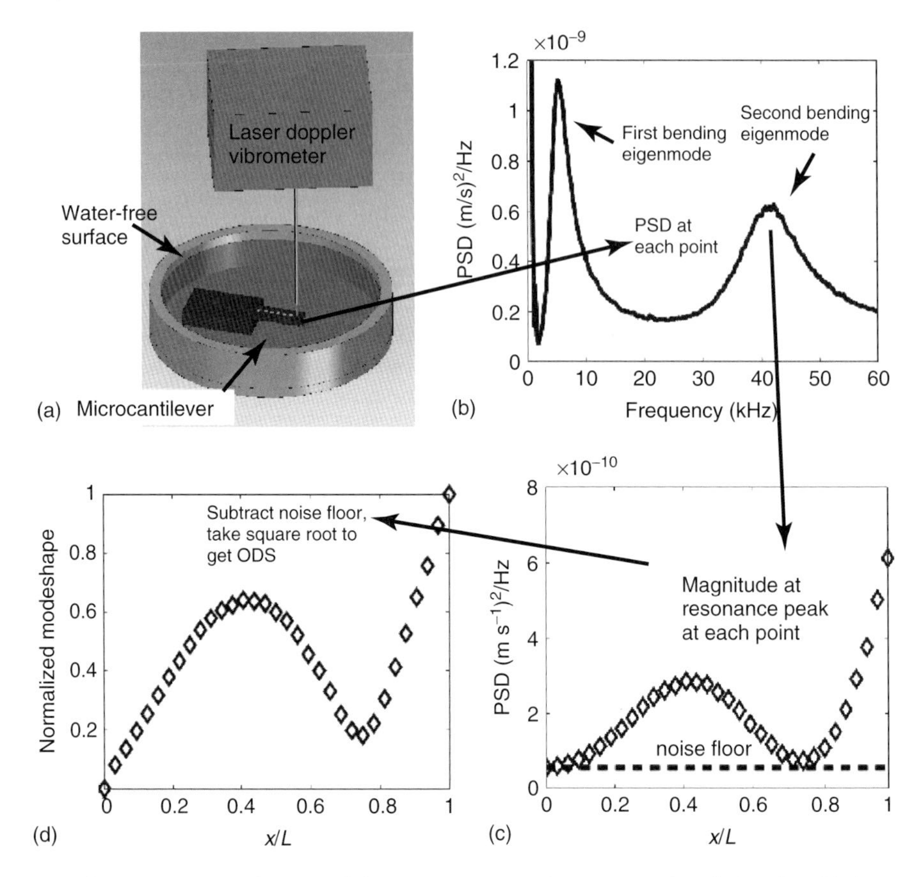

Figure 5.2 (a) A schematic of the experiment to measure the eigenmodes of AFM cantilevers in air and liquids; (b) thermal (Brownian motion excited) vibration time series data were collected at each point along the axis of the lever; (c) from which the operating deflection shape (ODS) at a given frequency is identified; and in (d) from the ODS at the second resonance peak, the first eigenmode is subtracted to yield the second eigenmode shape. (Reprinted with permission from Ref. [25]. Copyright 2010, American Institute of Physics.)

5.4
Methods of Dynamic Excitation

All dAFM techniques require some method of exciting the cantilever in the first place. There are multiple methods currently in use (e.g., piezo/acoustic and magnetic). As far as the dynamics are concerned, all the excitation methods are essentially equivalent in air or vacuum where quality factors are high. However, in liquids, where quality factors are low, there can be significant differences between the various methods.

The quantification of the forces that excite the cantilever is a vital and often implicit component in the quantification and interpretation of experimental

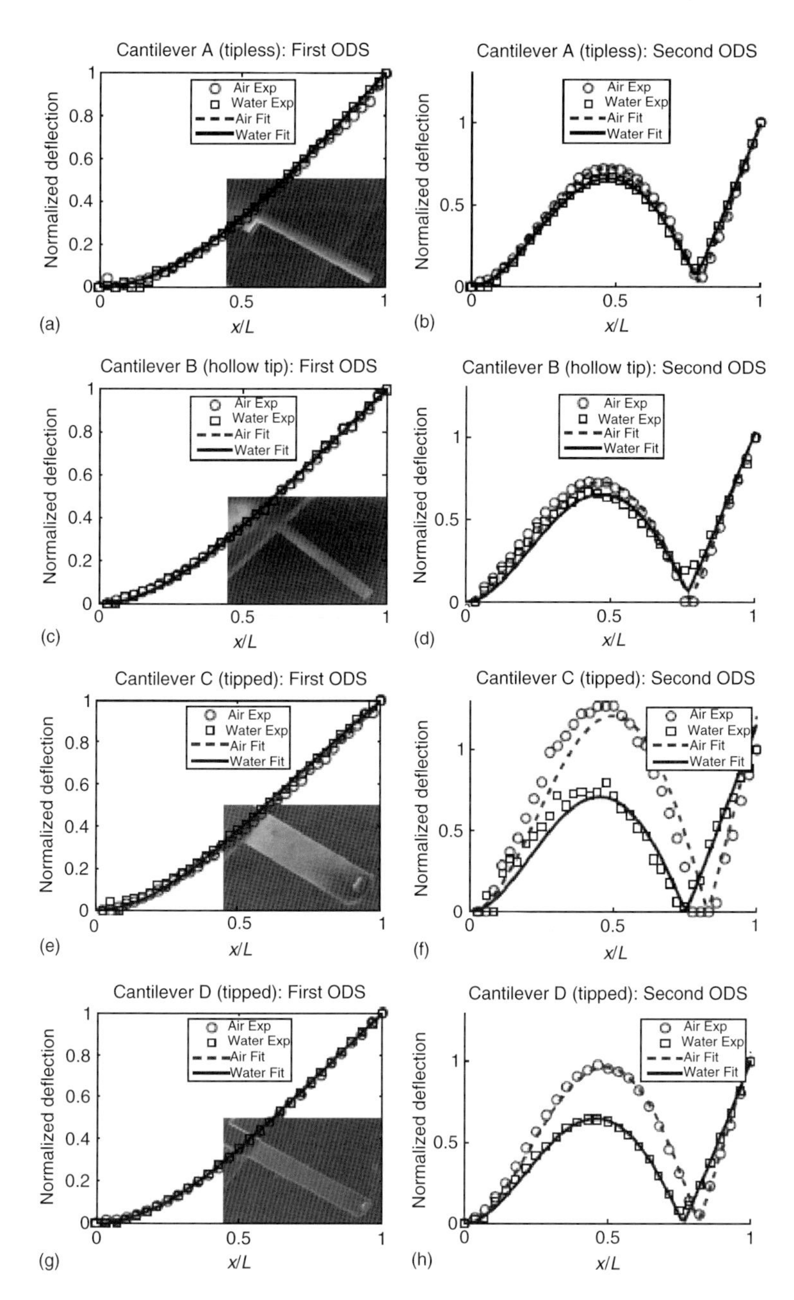

Figure 5.3 Plots of measured first and second eigenmodes in both air and water of two tipped and two tipless cantilevers. The first eigenmode is nearly identical in all cases (a,c,e,g), and for tipless cantilevers, the second eigenmode is the same in air versus water (b,d). However, the cantilevers with a tip mass show a significant change in the second eigenmode shape in water compared with air (f,h). (Reprinted with permission from Ref. [25]. Copyright 2010, American Institute of Physics.)

dAFM data in liquids. In dynamic AFM, the cantilever acts as a force transducer – converting force into displacement. There are two types of forces that are applied to the cantilever, the tip–sample forces and the excitation forces. Neither of these forces can be observed directly, they are only inferred from the cantilever motion. We are interested in measuring the tip–sample forces, but to get the tip–sample forces from the cantilever motion, we must first account for the excitation forces. Therefore, failure to understand the excitation forces can lead to a significant error in calculation of the tip–sample forces.

In this section, we first give a brief review of the various different methods of excitation. Then, we show the fundamental differences between the various classes of methods. Finally, we discuss practical difficulties involved with some of them in liquids, and discuss attempts to overcome those differences. The majority of the discussion is applicable to both amplitude modulation (AM) and frequency modulation (FM) methods. Specific comments for FM are provided at the end of this section.

5.4.1
Review of Cantilever Excitation Methods

Broadly speaking, the different types of excitation can be grouped into three different categories. The first category is *the induced displacement methods*. As the name suggests, a displacement is applied to some component, and this indirectly creates a force that drives the cantilever. The second category is *the direct forcing methods*. In this case, a force, rather than a displacement, is applied to the cantilever. These two methods are shown schematically in Figure 5.4. The final category is the *thermal excitation*, in which random thermal fluctuations of the cantilever and the surrounding fluid cause a random vibration. We now describe each of these categories in more detail:

Induced displacement methods include the following:

1) **Dither piezoelectric/acoustic**: The first AFM [27] used a piezoelectric crystal that vibrated the base of the probe up and down. This method is inexpensive and simple to operate, and it does not require any special cantilevers as some of the other methods do. It comes standard on nearly every commercial AFM and works very well in air or vacuum. However, as we shall discuss later, there are difficulties using it in liquid environments.
2) **Piezoelectric cantilevers**: One problem with the acoustic method is that there are many different mechanical elements between the piezoelectric crystal and the cantilever (e.g., chip, chip holder, etc.). Each of these elements can have resonances and mechanical impedances that affect the transmission of vibration to the cantilever. One solution is to locate the piezoelectric actuator directly on the chip, as close to the cantilever as possible. This method was demonstrated by Rogers *et al.* [28, 29] and greatly reduces the "forest of peaks" problem that we will discuss later, although the difficulty of large base motions may still be an issue. This type of cantilever is commercially available from Bruker (formerly Veeco).

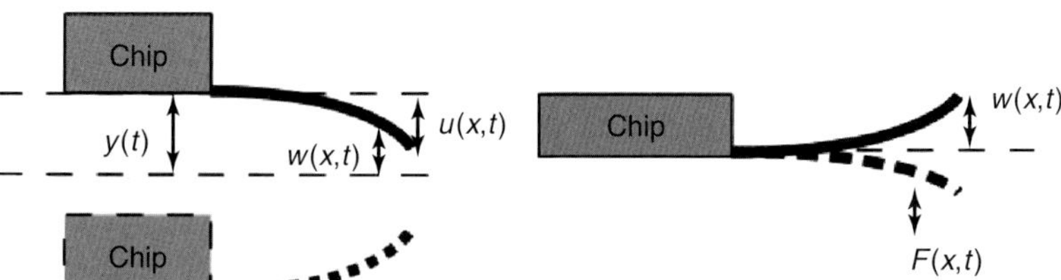

Figure 5.4 Comparison of the motions and forces in the induced displacement (i.e., piezo/acoustic) forcing versus the direct forcing methods (dashed lines show displaced positions). (a) The expansion/contraction of the piezo causes the base of the cantilever to move up and down an amount $y(t)$. In response, the cantilever is displaced relative to the chip by an amount $u(x,t)$. The absolute motion relative to the sample is then $w(x,t)$. The photodiode/optical beam deflection method measures $u(x,t)$, not $w(x,t)$. In the direct forcing case, a force $F(x,t)$ acts directly on the cantilever, causing a displacement $w(x,t)$. In this case, there is no relative motion, so the photodiode is directly measuring $w(x,t)$.

3) **Sample excitation**: Some AFMs hold the cantilever fixed and vibrate the sample up and down. Unfortunately, the dynamics of this configuration are significantly more complicated than shaking the chip [30–32]. Subharmonics and chaotic motions are possible.

There are also many direct forcing methods available including:

1) One of the first direct forcing methods was the magnetic method. In early AFMs, magnetized iron wires were used as probes [33]. As silicon cantilevers became more popular, many groups glued a small magnetic particle onto the free end of a cantilever to create magnetized probes [34–36]. Later Han and coworkers [3], coated an entire silicon cantilever with a thin film of a magnetic material (commercially available as Agilent's MAC™ levers). In all these methods, the probe is a permanent magnet that is driven by an external coil or solenoid. It is also possible to integrate the coil onto the cantilever and has the permanent magnet external [37, 38] (commercially available as Asylum's i-Drive™). A different magnetic mechanism, magnetostriction (the dimensional change of a material when subjected to a magnetic field), was recently proposed [39].
2) Electrostatic excitation was developed for imaging surface dielectric properties and surface potentials [40, 41]. It was only recently that electrostatic actuation was realized in liquid environments [42].
3) The photothermal method uses an intensity-modulated laser focused on a cantilever [43–45]. There are several mechanisms by which this can induce bending, including bimorph bending (i.e., a thin metal coating has a different coefficient of thermal expansion than the silicon cantilever) and thermal gradients through the thickness.
4) A very similar method is electrothermal, in which the heat input is provided by a resistive heater that is integrated onto the cantilever [46, 47].

5) A final method is acoustic radiation pressure, which uses focused acoustic waves at radio frequencies (100–300 MHz) to exert a force on the cantilever [48].
6) Quartz tuning forks. Not all AFM probes are based on cantilevers. A popular method in vacuum AFM is to attach the tip to the end of a quartz tuning fork. Because quartz is a piezoelectric material, a direct force can be applied to the fork with an applied voltage. This method has been used by a few groups in liquids [49, 50].

Thermal excitation refers to random fluctuations of a body in thermal equilibrium with the surrounding bath. The equipartition theorem then states that all of the energy in the system will be distributed equally among all the degrees of freedom (DOF) in the system and quantifies the amount of energy that each should have. A cantilever's oscillation eigenmodes count as DOF in this scheme, such that every cantilever eigenmode must have a energy equal to $\frac{1}{2}k_B T$, where k_B is Boltzmann's constant and T is temperature. The thermally driven vibration is the smallest excitation that is possible at a given temperature. Thermal excitation is regularly used for stiffness calibration [51–53] and has also been used for force spectroscopy by several groups [54, 55].

5.4.2 Theory

Here, we formally derive the equations of motion of an AFM cantilever driven in a fluid. Many times throughout the literature, these equations have been developed in an *ad hoc* manner. This approach generally gets the right form of the equation, but various constants are often neglected or stated incorrectly or inconsistently. In the following sections, we derive the equations for each of the three general cases: direct forcing, base displacement, and thermal. This derivation mainly follows the work of [56] and [57].

5.4.2.1 Direct Forcing

We start with *direct forcing* of cantilevers using the Euler–Bernoulli beam equation [58] subject to a hydrodynamic force and a periodic driving force. The damping is assumed to come from the fluid only. As the book is focused on AFM in liquids, we neglect any tip mass, which has been shown to have little effect in liquid [25]. The hydrodynamic function (Section 5.3) is typically described by its Fourier transform, so we write the equations in frequency domain:

$$EI\frac{\partial^4 w(x,\omega)}{\partial x^4} - \omega^2 \rho_c A w(x,\omega) = f_{\text{hydro}}(w) + f_{\text{drive}}(x,\omega) \tag{5.2}$$

where w, x, ω, E, I, ρ_c, and A are the cantilever deflection, length along the cantilever axis, driving frequency, Young's modulus, area moment of inertia, density, and cross-sectional area. The equation is subject to the boundary conditions

$$w(0,\omega) = 0 \quad \frac{\partial}{\partial x}w(0,\omega) = 0 \quad \frac{\partial^2}{\partial x^2}w(L,\omega) = 0 \quad \frac{\partial^3}{\partial x^3}w(L,\omega) = 0$$

where L is the cantilever length.

Using the definitions of hydrodynamic added mass and added viscosity from (5.1), Eq. (5.2) can be written as

$$EI\frac{\partial^4 w(x,\omega)}{\partial x^4} + ci\omega w(x,\omega) - \omega^2(\rho_c A + M)w(x,\omega) = f_{\text{drive}}(x,\omega) \tag{5.3}$$

Equation (5.3) is a partial differential equation, which describes the motion at every point along the length of the cantilever at every frequency. In AFM, we measure the displacement only at the free end and apply tip–sample forces at the free end. Therefore, it will be convenient to have an ordinary differential equation that describes the motion at the free end.

A Galerkin discretization [58] is used to accomplish this. We write the cantilever deflection as a linear combination of eigenmodes:

$$w(x,\omega) = \sum_{j=1}^{\infty} C_j(\omega)\psi_j(x)$$

where ψ_j is the jth cantilever eigenmode, normalized to $\psi_j(L) = 1$. This transforms Eq. (5.3) into

$$\sum_{j=1}^{\infty} C_j(\omega)\left(EI\frac{\partial^4 \psi_j(x)}{\partial x^4} + ci\omega\psi_j(x) - \omega^2\left(\rho_c A + M\right)\psi_j(x)\right) = f_{\text{drive}}(x,\omega) \tag{5.4}$$

The essence of the Galerkin method is to multiply both sides of the equation by $\psi_k(x)$, integrate over the length of the cantilever, and then use the orthogonality properties of the eigenmodes to simplify. The result is

$$C_j(\omega) = \frac{1}{\alpha_j}\frac{\frac{\omega_j^2}{k_j}F_j}{\omega_j^2 + i\frac{\omega\omega_j}{Q_j} - \omega^2} \tag{5.5}$$

where $k_j = EI\int_0^L \left(\psi_j''(x)\right)^2 \mathrm{d}x$ [57] and $\alpha_j = \frac{1}{L}\int_0^L \psi_j^2(x)\mathrm{d}x$. For $j = 1$ and zero tip mass, k_j is 1.03 times the static bending stiffness and $\alpha_j = 0.25$. Finally, the modal force is

$$F_j(\omega) = \int_0^L \psi_j(x) f_{\text{drive}}(x,\omega)\mathrm{d}x \tag{5.6}$$

All the direct forcing methods are describable by Eq. (5.5) and thus are similar in terms of their dynamics. The only difference between the methods is the size of F_j/k_j for various j. For example, a uniform magnetic coating has $f_{\text{drive}}(x,\omega) = f_{\text{mag}}$ where f_{mag} is a constant. This leads to $F_2/k_2 \ll F_1/k_1$; therefore, the uniform coating method is not efficient at exciting higher eigenmodes. On the other hand, the magnetic bead method has $f_{\text{drive}}(x,\omega) = f_{\text{mag}}\delta(x - L)$ (Dirac delta) and photothermal excitation has a very complicated function for f_{drive} (see [59] and the discussion on photothermal later in this chapter), both of which are efficient at exciting higher eigenmodes.

5.4.2.2 Ideal Piezo/Acoustic

For *dither piezoelectric/acoustic excitation*, we shall first consider an idealized model in this section and then discuss a more realistic situation in the next section. In the ideal piezo method, there is no $f_{\text{drive}}(x, \omega)$. Instead, the base of the cantilever is moved up and down. That is, the boundary conditions are modified to be

$$w(0,\omega) = y(\omega) \quad \frac{\partial}{\partial x}w(0,\omega) = 0 \quad \frac{\partial^2}{\partial x^2}w(L,\omega) = 0 \quad \frac{\partial^3}{\partial x^3}w(L,\omega) = 0$$

However, using a well-known technique [58], we can convert the moving boundary conditions into an *equivalent* force on the cantilever, when we consider it in a moving reference frame (i.e., we consider that we sit on the base of the cantilever and ride up and down along with it and report the force that we would observe in that frame). We do this by making the change of variables $w(x,\omega) = u(x,\omega) + y(\omega)$ where u is the deflection in the moving frame and y is the base motion. This is appropriate because the optical beam method used in most AFMs measures a relative quantity (the slope at the end of the cantilever) instead of an absolute quantity anyway. The coordinates y, u, and w are shown in Figure 5.4.

In this case, the equations of motion (Eq. (5.3)) become

$$EI\frac{\partial^4 u(x,\omega)}{\partial x^4} + ci\omega\left(Y(\omega) + u(x,\omega)\right) - \omega^2\left(\rho_c A + M\right)\left(Y(\omega) + u(x,\omega)\right) = 0$$

Then, following the same procedure as for the direct forcing, we expand $w(x,\omega) = \sum_{j=1}^{\infty} U_j(\omega)\psi_j(x)$, where we use U instead of C to emphasize that this is in the relative frame. Then using the Galerkin discretization and moving the y terms to the right-hand side, we have:

$$U_j(\omega) = \frac{\beta_j}{\alpha_j}\frac{Y(\omega)\left(\omega^2 - i\frac{\omega\omega_j}{Q_j}\right)}{\omega_j^2 + i\frac{\omega\omega_j}{Q_j} - \omega^2} \tag{5.7}$$

where $\beta_j = \frac{1}{L}\int_0^L \psi(x)\mathrm{d}x$.

5.4.2.3 Thermal

Thermally driven vibration differs from the equations above in two important aspects. First, both the motion and the excitation forces are random variables. Therefore, we can describe them only in terms of their power spectral densities [60]. Second, whereas in the direct forcing case, we were given a force per unit length on the beam $f_{\text{drive}}(x,\omega)$ and had to use the Galerkin discretization to obtain a modal force F_j, the thermally driven force is already a modal force and no transformation is needed.

There is one subtlety here, in that any scalar multiple of an eigenmode is still an eigenmode. Therefore, a modal force is only uniquely specified if we also specify the normalization of the eigenmode. Many authors [20, 61] give the following expression for the thermal driving force: $F_{\text{B}}(\omega) = 4k_{\text{B}}T\frac{\pi}{4}\rho_f b^2\omega\,\text{Im}\left[\Gamma(\omega)\right]$ (in units of $\text{N}^2\ \text{Hz}^{-1}$). However, these authors have chosen the normalization $\|\psi_j\| = \int_0^L \psi_j^2(x)\mathrm{d}x = 1$, which leads to $\psi_j(L) = 2$ for an Euler–Bernoulli beam. However, this normalization is not convenient for AFM because we measure the deflection at the free end. Therefore, we must require $\psi_j(L) = 1$. This leads to $\|\psi_j\| = \int_0^L \psi_j^2(x)\mathrm{d}x = \frac{1}{4}$. Therefore, the formulation of the thermal driving force that we require is actually

$$F_{\mathrm{B}}(\omega) = k_{\mathrm{B}} T \frac{\pi}{4} \rho_f b^2 \omega \,\mathrm{Im}\left[\Gamma(\omega)\right]$$

Therefore, the power spectral density of the cantilever motion is

$$\widetilde{C}_j(\omega) = \frac{k_{\mathrm{B}} T \frac{\pi}{4} \rho_f b^2 \omega \,\mathrm{Im}\left[\Gamma(\omega)\right]}{\alpha_j^2 \left(\omega_j^2 + i\frac{\omega\omega_j}{Q_j} - \omega^2\right)^2} \tag{5.8}$$

where the tilde is to emphasize that this is a deflection power spectral density (in units $\mathrm{m}^2\ \mathrm{Hz}^{-1}$) and not a deflection.

5.4.2.4 Comparison of Excitation Methods

The three methods can be compared in a few different ways. First, consider the response of the various methods versus driving frequency, as shown in Figure 5.5. The key observations from this figure are the following: [56]

1) In air, the peak frequencies of all three methods are identical, but in liquid, the peak (resonant) frequencies are different. In air or vacuum, the terms "*natural frequency*" and "*resonant frequency*" are essentially synonymous, but in liquid, they can be quite different.
2) At zero frequency, the response of ideal acoustic mode is zero but the response of magnetic mode is not. Conversely at high frequency, the response of the magnetic mode goes to zero, but the response of the ideal acoustic mode approaches the base motion Y.
3) In magnetic mode, the phase is unique over the frequency range, but for ideal acoustic, there may be two frequencies that have the same phase.

Furthermore, we can compare the two methods at a fixed frequency but decreasing Z (i.e., approaching a sample in tapping mode). Such an experiment is shown in Figure 5.6, which shows an approach curve using both acoustic and magnetic excitation in deionized water on mica. The raw data versus Z is shown on the left, but the more relevant graph is phase versus amplitude ratio (i.e., the amplitude divided by the initial amplitude). It is clear that at high-amplitude ratios, the two methods are fairly similar. But at approximately $A/A_0 = 20\%$, there is a sudden cusp in the magnetic curve and the phase rapidly decreases. This cusp may indicate the onset of a multiple impact regime [62]. However, the acoustic curve goes the opposite direction, with phase increasing up to almost 180°, and the amplitude does not go to zero. The traditional interpretation of AM–AFM phase in air indicates that phase lag $>90°$ is an attractive regime and phase $<90°$ is a repulsive regime [63]. This naturally leads to the question if the acoustic drive somehow created an attractive regime? Fortunately, this quandary is resolved by considering the way the AFMs photodiode measures deflection. The photodiode measures the slope at the free end of the cantilever, not the actual deflection. Normally, the slope is proportional to the actual deflection. But when the cantilever

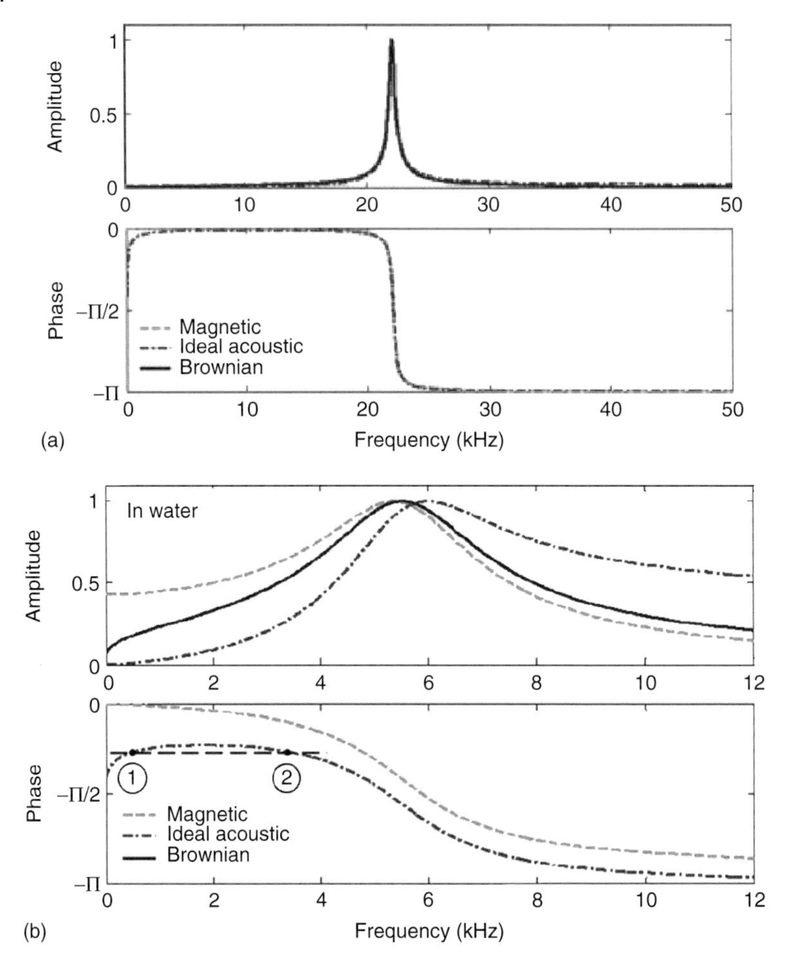

Figure 5.5 This figure shows the theoretical transfer functions of magnetic and ideal acoustic modes and Brownian motion for a cantilever in (a) water and in (b) air. In water, both the Q-factor and the resonance frequencies clearly decrease compared to the case in air; thus, the peak frequencies of the magnetic mode, ideal acoustic mode, and Brownian motion are not the same, and the phase responses are different too. In the ideal acoustic mode, one phase angle can correspond to two frequencies, for example, points 1 and 2. In air, the difference between these transfer functions is nearly indistinguishable. (Reprinted with permission from Ref. [56]. Copyright 2007, American Institute of Physics.)

enters permanent contact with the sample, the tip is no longer moving up and down, but the cantilever is still flexing because the base is moving. The photodiode misinterprets this bending as actual motion. In air, this is not noticed because the base motion is so small. But in liquids, the base motion can be on the same order as the cantilever tip motion.

In light of the fact that the base is moving significantly for dither piezo excited cantilevers in liquids, the force spectroscopy equations (i.e., the equations to convert

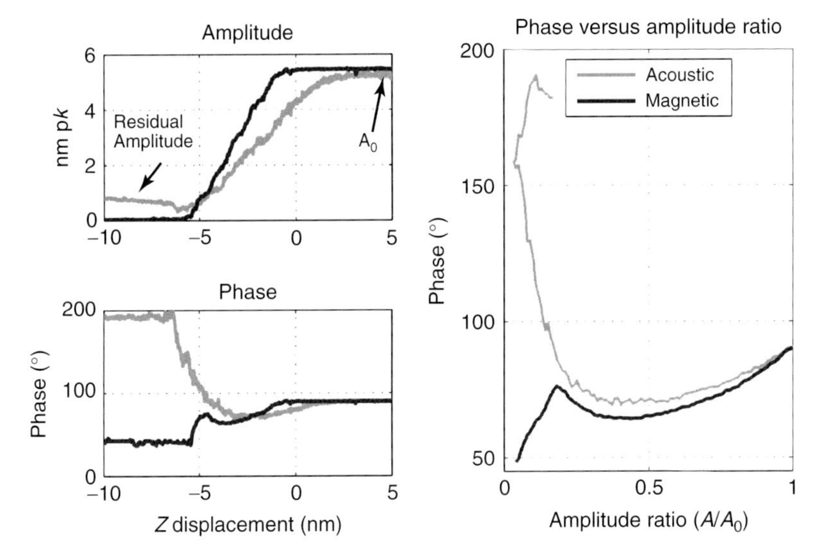

Figure 5.6 AM–AFM approach curves of a magnetically coated CSC 37 cantilever approaching a mica surface in deionized water, comparing the acoustic and magnetic excitation methods. The results are shown as amplitude and phase versus Z and also phase versus amplitude. The magnetic method approaches zero amplitude as the tip reaches permanent contact with the sample; however; the acoustic method does not. Instead, a "residual" amplitude is indicated. This is an artifact of the optical beam deflection method that measures the slope at the free end of the cantilever instead of the actual displacement.

observed amplitude and phase into the tip–sample forces/stiffness, and energy dissipation/damping) for direct forcing cannot be used for acoustic drive in air. For small amplitude (linear) AM–AFM, the equations have been derived [64] and see also [65, 66]. However, to our knowledge, the corresponding equations have not been derived for any other mode (e.g., no frequency modulation formula exists that correctly accounts for the base motion).

5.4.3 Practical Considerations for Acoustic Method

In the above, we have described the ways in which an ideal acoustic (base excitation) forcing differs from a direct forcing in low-Q environments such as in liquids. However, there are two nonideal factors by which a real acoustic drive differs from an ideal one in liquids [64, 65]. The first is the so-called forest of peaks, and the second is fluid borne loading.

What is now called *the forest of peaks* was first identified by Putman *et al.* [1]. That is, there are many peaks in the cantilever frequency response spectrum (i.e., tuning curve) that are not related directly to the cantilever resonance (for example, see Figure 5.7). The ideal acoustic mode would have only one broad peak while a real AFM demonstrates multiple peaks. It was suggested in [1, 2] that the peaks

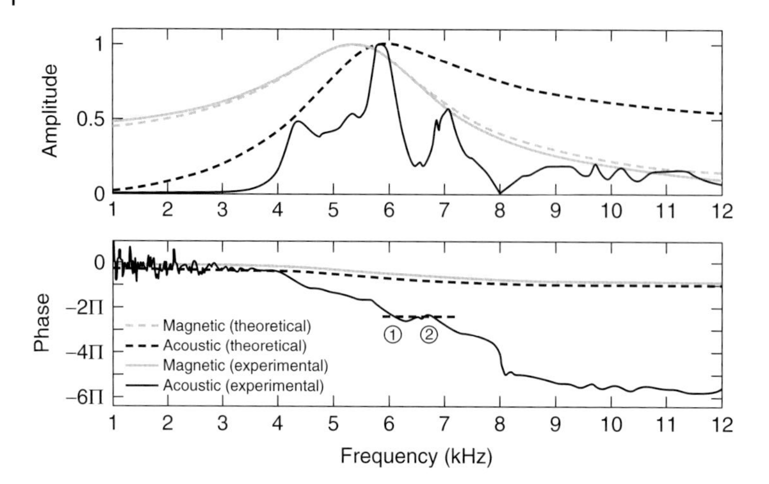

Figure 5.7 A comparison of experimentally measured excitation frequency response with the ideal frequency response for the magnetic (direct forcing) and acoustic (base excitation) cases. (Reprinted with permission from Ref. [56]. Copyright 2007, American Institute of Physics.)

were due to fluid cell resonances (i.e., acoustic resonances of the liquid in the cell). On the other hand, in [67], it was determined experimentally that the peaks in their cell were due to mechanical resonances (piezo, chip, chip holder, etc.) not fluid resonances. Because of the great variability between liquid cell and piezo design between the various AFM manufacturers, it is difficult to make generalizations, but certainly, all AFMs will have some combination of fluid cell resonances and mechanical resonances.

For AM–AFM operation, the primary effect of the forest of peaks is that the cantilever's natural frequency is obscured in the frequency spectrum, so that an operator may accidentally pick to drive at a frequency that is far away from the cantilever's natural frequency. While it is possible to operate AM–AFM off-resonance, the imaging forces applied to the sample will generally be higher [67]. In addition, many standard formulas such as stiffness calibration [68], energy dissipation [69], or force spectroscopy [70] require knowledge of the cantilever's quality factor and natural frequency and not the quality factor of the piezo/chip/chip holder resonance. The solution to these problems is to determine the cantilever's natural frequency and quality factor based on a thermally driven spectrum. This capability is present on most modern commercial AFMs. The affect on FM–AFM operation is discussed later in this section.

Several groups have attempted to build special cantilever holders and liquid cells that reduce the forest of peaks in liquid media [71–74]. Researchers interested in designing such equipment should keep in mind that a smooth amplitude spectrum without any spurious peaks is *not* necessarily the sign of success. It is possible for a piezo resonance and a cantilever resonance to have approximately the same natural frequency and quality factor. In this case, the piezo resonance may not be distinguishable in the amplitude spectrum. To ensure that all piezo resonances are eliminated, one must check that (i) the quality factor of a piezo-driven measurement matches exactly the quality factor from a thermally driven measurement and (ii) the phase changes by 180° (and not more) as the frequency is swept across the resonance.

Fluid borne loading is an effect that has only recently been recognized [56] in the AFM community and relatively few have considered it. In addition to moving the base of the cantilever up and down, the piezo crystal also excites the fluid in the cell (Figure 5.8). This creates a local fluid flow around the cantilever that serves to excite it. That is, in real acoustic mode, there is a direct forcing component coming from the fluid as well as a base excitation component. Therefore, for a given base displacement, the actual tip motion will be larger than predicted using the ideal acoustic theory (Eq. (5.7)). This combined base excitation plus fluid borne loading is not accounted for in any standard AFM formula. Corrections to account for these effects are only beginning to be explored [64]. Further research in this area will be necessary for true quantitative AFM using piezo excitation in liquids. If the excitation forces applied to the cantilever are not well understood, then it is not possible to correctly interpret the data to recover the tip–sample forces.

5.4.4 Photothermal Method

So far, we have reviewed some of the difficulties involved with piezo excitation in liquids. These difficulties have led to the development of alternative excitation methods for liquid applications. We have primarily considered the magnetic method above. However, there are some disadvantages of the magnetic method, primarily the need for special cantilevers as well as limitations at high-frequency and higher eigenmodes. The *photothermal method* does not have these limitations;

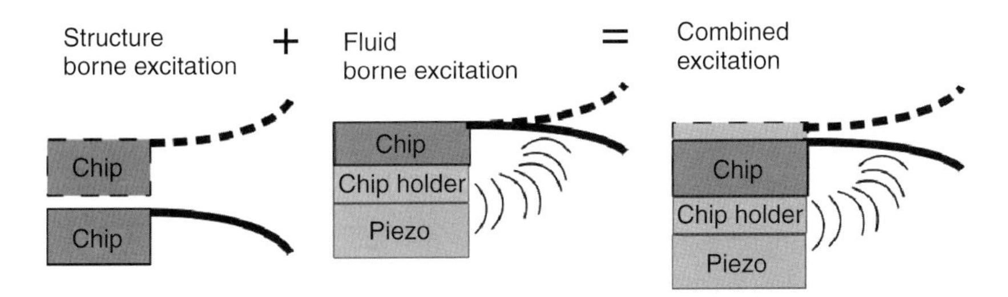

Figure 5.8 Illustration of the two types of loading that are present in real acoustic mode excitation in liquids. The structure borne loading is caused by the expansion/contraction of the piezo that shakes the chip up and down (the dashed lines show displaced position). This is the mechanism described in the section for ideal acoustic. But the piezo also excites the fluid in the cell, causing acoustic waves to travel through the fluid and excite the cantilever without moving the base. This is more like a direct forcing described earlier in this section. The total excitation is a combination of these two effects.

therefore, it gaining popularity, especially in liquids, so we review some of the theory and key results here.

The basis of the photothermal method is to shine an intensity-modulated laser on the cantilever. In other words, the laser power has a mean component and an oscillating component. Care must be taken that this laser does not interfere with the laser used for the photodiode readout. Typically, the two lasers will be at different wavelengths so that the excitation laser can be filtered out of the photodiode signal [51].

There are several mechanisms by which the intensity modulation can cause the cantilever to oscillate. First, the momentum of the photons exerts a pressure on the cantilever [45, 75]. Second, thermal gradients through the thickness (i.e., hotter at the top than at the bottom) cause bending [76]. Third, and typically the largest, effect is the bimorph bending effect. That is, many cantilevers are coated with a thin layer of metal to improve reflectivity. This metal layer has a differing coefficient of thermal expansion and Young's modulus than the silicon body of the cantilever, and this creates bending when the cantilever is heated [77].

An important consideration in photothermal is the placement of the excitation laser spot [59, 70]. In the magnetic and piezo methods, the form of the modal forces (Eq. (5.6)) was fixed, whereas in photothermal, the modal forces can be altered depending on the size and location of the spot. Each of the three types of effects has a different dependence. For light pressure (and assuming a small spot size), the force is localized to the point $x = x_0$ so that the modal force is $F_j = \psi_j(x_0) f_{\mathrm{lp}}$, where f_{lp} is the light pressure force. For the first eigenmode, $\psi_1(x)$ is the largest at the free end, so light pressure forcing is the largest when the laser spot is focused at the free end.

For the bimorph bending effect, the temperature distribution in the cantilever must first be calculated from the one-dimensional unsteady heat equation:

$$\frac{\partial T(x,t)}{\partial t} = K\frac{\partial^2 T(x,t)}{\partial x^2} - \beta T(x,t) + \chi P(x,t) \tag{5.9}$$

where $K = \frac{\kappa_1 d_1 + \kappa_2 d_2}{c_{p1}\rho_1 d_1 + c_{p2}\rho_2 d_2}$ is the thermal diffusivity, $\beta = \frac{h_1 + h_2}{c_{p1}\rho_1 d_1 + c_{p2}\rho_2 d_2}$ is a convective loss coefficient, $\chi = \frac{1}{c_{p1}\rho_1 W d_1} + \frac{1}{c_{p2}\rho_2 W d_2}$ is an absorption coefficient, T, x, t, κ, h, d, W, c_p, ρ, and P are temperature, position along the cantilever axis, time, thermal conductivity, convection heat transfer coefficient, cantilever thickness, cantilever width, specific heat, density, and absorbed optical power per unit length, and the subscripts 1 and 2 refer to the two different material layers (e.g., silicon and a metal coating). For a small spot centered at x_0, the absorbed power term will have the form $P(x,t) = \delta(x - x_0)\left(P_0 + P_1\cos(\omega t)\right)$. The solution to this equation will be an oscillating temperature distribution $T = T_0 + \Delta T(x)\mathrm{e}^{i\omega t}$. In general, the distribution will have a peaked shape with maximum amplitude at the laser spot. Then, the modal forces are calculated from [61] $F_j \propto \int_0^L \Delta T(x)\frac{\mathrm{d}^2\psi(x)}{\mathrm{d}x^2}\mathrm{d}x$. That is, the modal forces will be the highest when the laser spot is focused in the region where the eigenmode has the highest curvature. For the first eigenmode, this will be near the base of the cantilever (Figure 5.9). For the final method, thermal gradients through the cantilever thickness, the situation is more complicated. The

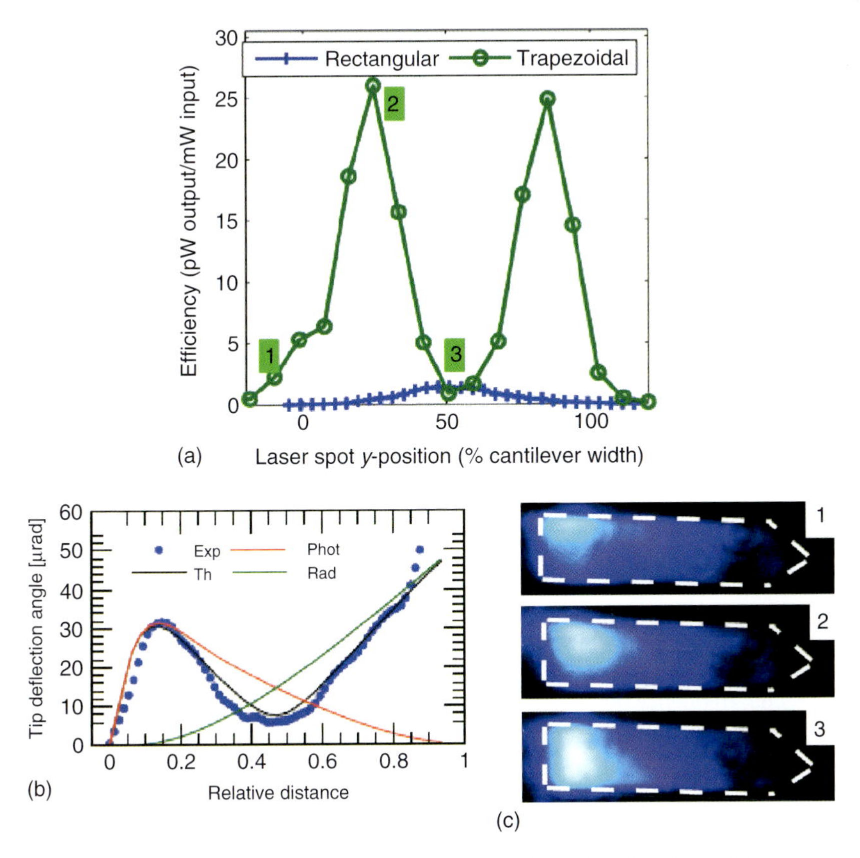

Figure 5.9 The effect of laser spot position on photothermal excitation. (a) For the first eigenmode of an uncoated cantilever, the effect of light pressure (green) is the largest at the free end, whereas the photothermal effect is the greatest near the clamped end. (b) Efficiency (picowatts mechanical response per milliwatts absorbed optical power) as a function of position along the cantilever width for typical uncoated trapezoidal and rectangular cantilevers. The efficiency is greatly enhanced at the edges of the trapezoidal cantilever due to the favorable thermal gradients through the thickness. (c) Photographs of the laser spot on the trapezoidal cantilever for three selected positions (corresponding to the marked positions in (b)) ((a) and (c) Reprinted with permission from Ref. [43] Copyright 2007, American Institute of Physics and (b) Reprinted with permission from Ref. [75]. Copyright 2010, American Institute of Physics.)

temperature distribution along both the length x and the thickness z must be calculated and then $F_j \propto \int_0^L \frac{\partial \Delta T(x,z)}{\partial z} \frac{d^2\psi(x)}{dx^2} dx$. In fact, it has recently been shown [43] that, in some cases, even the temperature distribution along the cantilever width must be taken into account! Some cantilevers do not have a rectangular cross section, but rather a trapezoidal cross section as an artifact of the manufacturing process. In this case, focusing the laser spot at the edge, where the cantilever is the thinnest, enhances the thermal gradient through the thickness, greatly increasing the efficiency of the photothermal method. In this way, the excitation efficiency (mechanical power output per unit optical power input) can be increased by an order of magnitude.

5.4.5
Frequency Modulation Considerations in Liquids

The preceding discussion is all that is necessary to understand the difference between excitation methods for amplitude modulation or phase modulation [78] techniques in which the cantilever driving frequency remains constant throughout the experiment. In frequency modulation and related techniques, the driving frequency changes during the experiment. In this case, there are two additional concerns that must be addressed.

First, frequency modulation tracks the tip–sample dissipation by tracking that amount of driving voltage that is required to maintain a constant cantilever oscillation amplitude. However, the various excitation methods may produce more or less driving forces at different frequencies. This, for example, occurs with the "forest of peaks" phenomenon where the piezo/chip/chip holder resonances make the cantilever base motion depend strongly on the frequency of excitation. In other words, the excitation source may be more efficient at some frequencies than at others. In this case, changes in the excitation efficiency could be mistaken for a change in the tip–sample dissipation.

Second, frequency modulation tracks conservative interaction forces by tracking the cantilever natural frequency, which is the frequency at which the cantilever vibration is 90° out of phase from the driving signal. However, the various excitation methods may have differing phase delays at different frequencies. In other words, there is some time delay between when the driving signal is applied by the electronics and when the actual force is produced. If this delay varies at different frequencies, then this phase difference may be mistaken for a conservative interaction force. These two different effects are examined in detail in [79].

Having described the hydrodynamics and dynamic excitation methods for cantilevers in liquids, we now proceed to discuss the unique dynamics of cantilevers while oscillating and tapping on samples in liquid environments. As has already been elaborated, induced base displacement excitation methods for cantilevers can lead to many challenges in interpreting AFM cantilever dynamics in liquids. For this reason in what follows, we will focus more on directly excited AFM cantilevers in liquids while interacting with samples.

5.5
Dynamics of Cantilevers Interacting with Samples in Liquids

In this section, we review the dynamics of soft cantilever probes interacting with samples in liquid medium. dAFM methods are ultimately based on the ability to sense a nonlinear interaction between a sample and a sharp probe tip in the waveform of an oscillating probe. While the dynamics of cantilever probes interacting with samples in ambient and vacuum environments are relatively well understood, our understanding of dAFM in liquid environments is still developing. As far as the dynamics of the oscillated probe are concerned, two important differences between ambient/vacuum and liquid environments are the following:

- Reduced attractive forces in liquid have popularized the use of relatively soft cantilever probes, on the order of $1\ \mathrm{N\ m^{-1}}$ or less [80], when compared with ambient or vacuum.
- The increased hydrodynamic damping from the liquid medium combined with the low stiffness probes results in quality factors typically on the order of unity.

The use of soft probes enhances the asymmetry of the potential well of the probe in the presence of the sample. The effect of low-quality factors is (i) that the energy of the oscillation is able to occupy large frequency bands and (ii) relatively large base motions for acoustic excitation are needed to sustain the amplitude.

5.5.1 Experimental Observations of Oscillating Probes Interacting with Samples in Liquids

An oscillating dAFM probe was first introduced to a liquid medium by Putman *et al.* [80] in 1994. In this original work, some key observations were made distinguishing cantilever dynamics in liquid environments from that of air or vacuum. However, it was not, for some time, later that the full set of observations made by Putman *et al.* would be reproduced in numerical simulations [81]. In this section, we review key experimental observations of oscillating dAFM probes interacting with samples in liquid medium made by Putman *et al.* [80] and Basak and Raman [81].

Figure 5.10a shows the deflection waveform of a soft triangular cantilever that is acoustically excited and approaches the sample through the extension of the Z displacement of the base. An asymmetry in the envelope of the deflection waveform as the base is displaced between points A and B in Figure 5.10a. If the oscillation waveform were described by a pure tone, as is essentially the case in air and vacuum, the envelope would reduce symmetrically. The asymmetry in the envelope is a result of significant subharmonic (zeroth harmonic) and superharmonic (second and higher harmonic) distortions. Second, a residual amplitude persists from point C onward, which is also unique to liquids. Both these observations are a direct result of the low-quality factors inherent to liquids discussed earlier, which (i) enhance significant higher harmonics and (ii) require significant base motions needed for acoustic excitation in liquid that result in the residual amplitude.

In Figure 5.10b, the experiment in Figure 5.10a is repeated for a rapid approach speed making visible individual, short timescale features in the deflection waveform. These high-frequency components of the waveform are identified as the times of intermittent contact with the sample, which, due to the large base motions, occur at different times in the oscillation period. Similar features in the waveform of a soft triangular cantilever (Figure 5.10c) and soft rectangular cantilever (Figure 5.10d) were demonstrated experimentally in [81]. We note that the high-frequency content is much more pronounced in the case of the rectangular cantilever, which exhibits

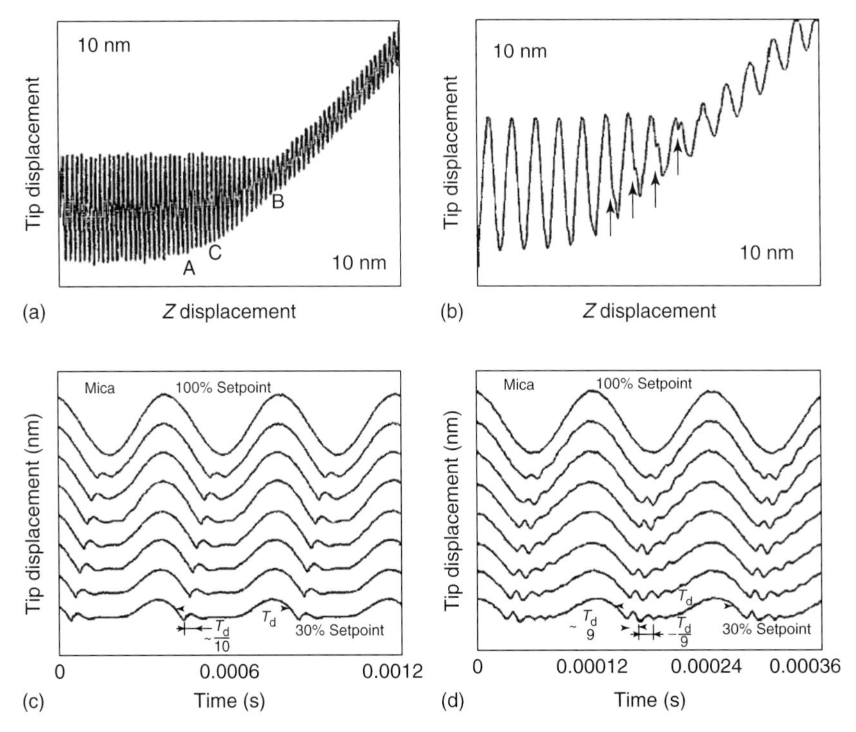

Figure 5.10 Experimental deflection waveform of soft cantilevers interacting with samples in liquid medium. An asymmetry in the deflection envelope, residual amplitude (a), and short timescale features in the waveform (b) were observed by Putman *et al.* [80]. Deflection waveform of a soft triangular cantilever (c) and rectangular cantilever (d). ((a) and (b) Reprinted from Ref. [80], Copyright 1994, with permission from Elsevier, (c) and (d) Reprinted with permission from Ref. [81]. Copyright 2007, American Institute of Physics.)

the signature of an impulse or "ring down" response. The key observation made by Basak and Raman [81] was that the frequency of the ring down corresponded to the resonance of the second eigenmode. This observation leads to the speculation that the features in the waveform are contributed by the second eigenmode that is excited during interaction with the sample and quickly decays. An important difference between the rectangular cantilever and the triangular cantilever used in [81] is that the second eigenmode of the triangular cantilever is more highly damped causing its response decay more rapidly.

5.5.2 Modeling and Numerical Simulations of Oscillating Probes Interacting with Samples in Liquids

The observation that the experimental oscillation waveform appeared includes contributions from the second eigenmode prompted Basak and Raman [81] to propose a two-eigenmode model of the cantilever probe to capture the complex dynamics of oscillating probes interacting with samples in liquid medium. More generally, an N DOF model follows from a Galerkin discretization of the Bernoulli–Euler beam equation [82], which retains the first N normal eigenmodes and yields the lumped-parameter, reduced-order model

$$\begin{aligned}
\frac{\ddot{q}_1}{\omega_1^2} + \frac{\dot{q}_1}{Q_1\omega_1} + q_1 &= \frac{F_1(t)}{k_1} + \frac{F_{\text{ts}}(d,\dot{d})}{k_1} \\
\frac{\ddot{q}_2}{\omega_2^2} + \frac{\dot{q}_2}{Q_2\omega_2} + q_2 &= \frac{F_2(t)}{k_2} + \frac{F_{\text{ts}}(d,\dot{d})}{k_2} \\
&\vdots \\
\frac{\ddot{q}_N}{\omega_N^2} + \frac{\dot{q}_N}{Q_N\omega_N} + q_N &= \frac{F_N(t)}{k_N} + \frac{F_{\text{ts}}(d,\dot{d})}{k_N}
\end{aligned} \tag{5.10}$$

where dots represent temporal derivatives, F_{ts} is the tip–sample interaction force, $d = Z + q_1 + q_2 + \ldots + q_N$ is the instantaneous gap between the tip and sample, where Z is the cantilever-sample separation [81, 83]. The contribution of the *jth* eigenmode to the tip deflection is q_j. k_j, ω_j, Q_j, and F_j are the corresponding equivalent stiffness, natural frequency, quality factor, and modal forcing [84].

Large uncertainties in the cross-sectional geometry of cantilever probes typically make the theoretical prediction of the lumped parameters in Eq. (5.10) unreliable. Instead, these parameters must be calibrated experimentally. The quality factors and natural frequencies of the probe can be determined from the power spectral density of the thermal response of the probe [85]. We note that the frequency dependence of hydrodynamic loading is approximated in Eq. (5.10) locally within the bandwidth of the resonance by the lumped parameters ω_j and Q_j. Additionally, a squeeze film effect of the hydrodynamic loading that occurs when the tip is comparable to the width of the cantilever [12] requires that ω_j and Q_j are measured close to the sample, but not so close that the interactions are significant. For calibration of the equivalent stiffnesses, the two most common techniques are the thermal method of Hutter *et al.* [51], which relates the equivalent stiffness to thermal motion predicted by the equipartition theorem, or the Sader method [86], which uses the predicted hydrodynamic loading to infer the equivalent stiffness. While both methods offer accurate and robust calibration of the equivalent stiffness of the fundamental eigenmode, the thermal method is more robust for the calibration of higher eigenmodes [87].

Numerical simulations of dAFM in liquid media can be performed using Eq. (5.10) [83, 88]. Unlike ambient and vacuum, dAFM in liquids often operates in a highly nonlinear regime and the development of a robust approximate theory becomes complicated. Here, numerical simulations have been instrumental in developing an understanding of the complex nonlinear dynamics of soft probes in liquid medium. In the interests of simplicity and computational expense, the reduced-order model should retain the fewest number of eigenmodes while still capturing the dynamics of the probe. When the excitation frequency is close to the fundamental resonance, a 2-DOF model is typically sufficient [81]. However, driving above the fundamental resonance, such as at the second resonance [89], may require a 3-DOF model that captures the dynamics of the first three eigenmodes.

Figure 5.11 shows numerical simulations of soft probes oscillating in liquids and interacting with a stiff substrate. Figure 5.11a,b compares the simulated deflection waveform for a single DOF (point-mass) model for the fundamental eigenmode to a 2-DOF model that captures the first two eigenmodes of the cantilever probe. The short timescale distortions in the waveform observed are only reproduced in the 2-DOF model. The evidence of multimodal cantilever dynamics in liquids is even more compelling in frequency domain. Figure 5.11c,d compares the amplitude spectrum of experimental oscillation waveform to the simulated amplitude spectrum, which is decomposed into the contributions of the first and

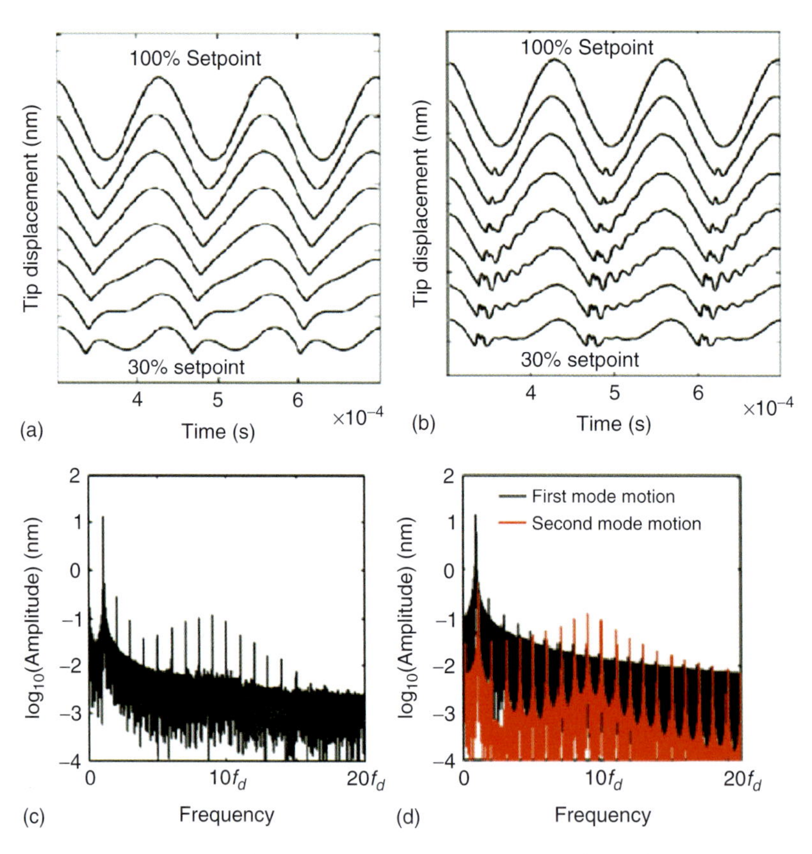

Figure 5.11 Comparison between numerical simulation and experiment of oscillating probes interacting with samples in liquid medium. Simulation of a soft probe interacting with mica in deionized water for (a) a point-mass model and (b) a 2-DOF model [81]. Amplitude spectrum of a soft cantilever interacting with mica in buffer solution (c) as observed in experiment and (d) reproduced in simulation [90].((a) and (b) Reprinted with permission from Ref. [81]. Copyright 2007, American Institute of Physics. (c) and (d) Reproduced from Ref. [90] Copyright 2009 by the American Physical Society.)

second eigenmode. The experimental amplitude spectrum reveals a cluster of enhanced harmonics surrounding the second resonance, which is identified as the second eigenmode in the simulated amplitude spectrum. When the dynamics of the cantilever are captured by a single eigenmode, the higher harmonics decay monotonically with frequency. Higher harmonics that are enhanced around a certain natural frequency indicate the presence of the corresponding eigenmode in the response.

5.5.3
Compositional Mapping in Liquids

Conventional dAFM methods regulate some component of the oscillation waveform to control Z displacement of the base of the probe. The record of the Z displacement during a raster over the sample forms a topographic image. Components of the oscillation that are unregulated by feedback control are free to vary during the raster according to the precise nature of the interactions that are linked to the material properties of the sample. By recording these unregulated components of the oscillation, a compositional map of the sample corresponding to the topography is constructed. In this section, we discuss compositional mapping in liquids and highlight some key differences between compositional maps in liquids versus ambient air or vacuum.

One form compositional mapping in dAFM is higher harmonic imaging where the anharmonic distortions of the waveform are recorded during the imaging raster. The presence of higher harmonics in the oscillation waveform is due to the nonlinear interaction with the sample, and their magnitudes contain information about the composition of the sample. Higher harmonic imaging in ambient is difficult because the higher harmonics are several orders of magnitude smaller than the fundamental harmonic [91]. However, in liquid media, higher harmonics are enhanced by the small quality factors of soft cantilever probes typical in liquid. The first higher harmonic images, taken by Van Noort *et al.* in 1999 [92], included second and third harmonic images of DNA imaged in liquid (Figure 5.12a). Preiner *et al.* [26] have demonstrated molecular resolution of a bacterial S-layer in a second harmonic image (Figure 5.12b), and Turner *et al.* [93] have collected second to fourth harmonics on *Staphylococcus aureus* bacteria cells. On the other hand, instead of using the harmonics close to the fundamental resonance, Xu *et al.* [90] proposed to image with harmonics close to the second eigenmode, which capture some measure of the momentary excitation of the second eigenmode during interaction with the sample. Figure 5.12c shows a nineth harmonic image of a double layer of purple membrane on a mica substrate taken from [90]. The nineth harmonic image is able to distinguish the single layer of purple membrane from the double layer, which is expected to be softer. Finally, it should be noted that instead of measuring higher harmonics, one can excite the cantilever at a frequency considerably higher than the fundamental resonance, say the second resonance, and measure *subharmonics*, that is, amplitudes at fractions of the excitation frequency. This technique is similar to higher harmonic imaging and has been demonstrated on purple membrane [94].

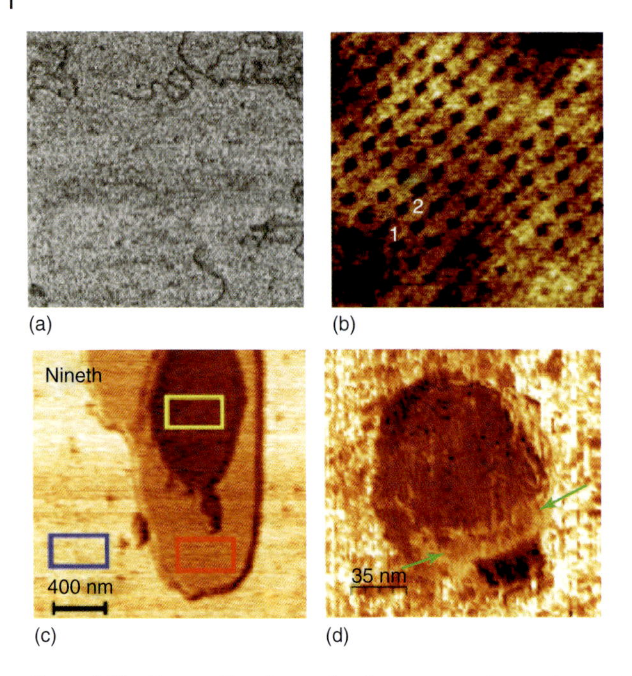

Figure 5.12 Compositional mapping in dAFM in liquids. (a) Second harmonic image of DNA on a mica substrate [92], (b) Second harmonic image of a bacterial S-layer [26], and (c) nineth harmonic image distinguishing a single layer from a double layer of purple membrane on a mica substrate [90]. (c) Nineth harmonic image of purple membrane. (d) Phase contrast image of ϕ29 virion.((a) Reprinted (adapted) with permission from Ref. [92] Copyright 1999 American Chemical Society. (b) Reprinted gure with permission from Ref. [26]. Copyright 2007 by the American Physical Society. (c) Copyright 2009 by the American Physical Society. (d) Reproduced from Ref. [97].)

The relationship between the magnitude of the higher harmonics and the material properties of the sample has been established through numerical simulation of the reduced-order model (Eq. (5.10)) retaining a single eigenmode [26], as well as the first two eigenmodes [90]. The contrast in the second harmonic was attributed to variation in the elastic properties of the sample [26, 90], although the theory is still developing for cells [92]. In addition to their physical interpretation, the sensitivity of the higher harmonics to variations in the sample properties is a primary concern and, more specifically, whether certain harmonics are more sensitive than others. A clear candidate is the second harmonic, proposed by Preiner *et al.* [26], which is typically the largest in magnitude. However, Xu *et al.* [90] have shown conditions where the nineth harmonic amplitude is around one order of magnitude more sensitive than the second to variations in the sample elasticity. The explanation for the improved sensitivity of the nineth harmonic offered by Xu *et al.* [90] is that the natural period of oscillation of the second eigenmode is similar to the duration that the oscillating probe makes contact with the sample during an oscillation period under the conditions studied. As a result, the response of the second eigenmode, which is captured by higher harmonics surrounding the second resonance, is highly sensitive to the properties of the sample, namely local elasticity.

As an alternative to higher harmonic imaging, energy dissipation spectroscopy aims measure small amounts of energy dissipated by nonconservative interaction forces during interaction with the sample. Energy dissipation spectroscopy in ambient air and vacuum can be implemented either in FM–AFM by monitoring the drive signal required to sustain the oscillation or indirectly in AM–AFM by monitoring the phase of the oscillation [69, 93]. Unlike higher harmonic image, energy dissipation spectroscopy can be implemented through the standard dAFM instrumentation. However, the simple formulae that form the basis for energy dissipation spectroscopy in air and vacuum follow from what is essentially a single harmonic description of the probe's oscillation. In liquid media, the significant higher harmonic distortions of the oscillation waveform invalidate formulae for energy dissipation developed assuming a single harmonic. Tamayo [94] proposed a correction to the classic theory for liquids that accounted for the energy lost in the second harmonic of the oscillation. However, the full implications of the energy losses in higher harmonic distortions to energy dissipation spectroscopy were not apparent until recently when the two-eigenmode model was incorporated [95]. When the second eigenmode is excited during interaction with the sample, a relatively large amount of energy is transferred from the driven eigenmode to the second eigenmode. This energy is subsequently dissipated into the surrounding media during the ring down of the second eigenmode. The magnitude of this energy transfer is typically much larger than the actual tip–sample dissipation and changes the meaning of energy dissipation spectroscopy in liquids.

For a magnetically driven cantilever, the energy balance for steady state oscillations in the two-eigenmode, reduced-order model is given by Melcher *et al.* [95]

$$W_1 = E_{\text{ts}} + E_{1-2} + E_{\text{med},1} \tag{5.11}$$

where W_1 is the work of excitation source on the first eigenmode, E_{ts} is the energy dissipated by nonconservative interactions, E_{1-2} is the energy transferred from the first eigenmode to the second eigenmode as a result of the nonlinear interaction with the sample. Finally, $E_{med,1}$ is the energy dissipated by the surrounding liquid medium. Classic formulae relating the energy dissipation to the phase of oscillation start with an energy balance for a point-mass model that omits the term E_{1-2}. From numerical simulations and experiments, it was shown that E_{1-2} is typically much larger than E_{ts} in for soft cantilever operating in liquids [95]. Similar to the higher harmonics, the energy transfer term E_{1-2} is associated with local elasticity of the sample. The stiffer regions of the sample tend to increase the response of the second eigenmode and, in turn, the magnitude of E_{1-2}. On the other hand, the tip–sample dissipation is expected to be larger on soft regions of the sample. As a result, an inversion occurs in the phase contrast in liquids from what is predicted by the classical theory developed in air [69, 93]. Experimental phase contrast images in [95] are consistent with the numerical simulations with the two-eigenmode model, where the bright regions in the phase contrast images correspond to stiffer regions of the sample (Figure 5.12d).

5.5.4
Implications for Force Spectroscopy in Liquids

The most complete form of compositional mapping possible in dAFM is force spectroscopy, where the full tip–sample interaction force is reconstructed from the observed oscillation waveform. Existing methods for force spectroscopy in dAFM can be placed in one of two categories based on their implementation: (i) primary harmonic inversion (PHI) methods that use integral relations involving the primary harmonic of the oscillation as a function of the Z displacement to reconstruct tip–sample interaction forces [96, 97] and (ii) full spectrum inversion (FSI) methods that use the anharmonic distortions in the deflection waveform at a single Z separation for reconstruction [98, 99]. FSI methods are particularly beneficial in liquids where the higher harmonic distortions of the oscillation waveform and significant. Furthermore, FSI methods are attractive since they can be implemented simultaneously while scanning.

In a method described by Stark *et al.* [98] (Figure 5.13a), the cantilever probe is treated as a linear time-invariant (LTI) system subjected to nonlinear feedback, which accounts for the interaction with the sample. Inverting the transfer function of the LTI system allows the measured deflection waveform to be translated into time-resolved interaction forces. The method proposed [98] implemented on the full discrete Fourier transform (DFT) of the deflection signal resulting in a one-to-one correspondence between the measured oscillation and the reconstructed force. Under the assumption that the oscillations of the probe are periodic, a windowed sample of the oscillation waveform can be used to approximate the Fourier coefficients of the periodic signal. In the scanning probe acceleration microscopy (SPAM) method described by Legleiter *et al.* [100], only the Fourier coefficients in a truncated Fourier series are used for force spectroscopy. This method leads to considerable noise reduction compared to the original method of Stark *et al.* [98] because the higher frequency noise is amplified by the inverted transfer function of the LTI system. Simultaneous topography and peak force images are demonstrated on purple membrane in [100] (Figure 5.13b).

While the significant higher harmonic distortions in the oscillations of the first eigenmode in liquids are amiable for force reconstruction via FSI methods, these methods are complicated when the dynamics of the cantilever probe are multimodal [101]. In general, the short timescale features in the waveform contributed by the second eigenmode will be mistaken for the tip–sample interactions by the methods in [98, 100]. One elegant experimental solution for this issue is to focus the laser at the antinode of the second eigenmode (Figure 5.13c). In the traditional optical lever setup, the bending angle of the deflected cantilever is used to infer deflection, focusing the laser at the antinode of the second eigenmode effectively screens its contribution from the measured waveform. This simple technique allows standard FSI methods developed for a single eigenmode to be implemented in liquids incurring force artifacts from the response of the second eigenmode (Figure 5.13d).

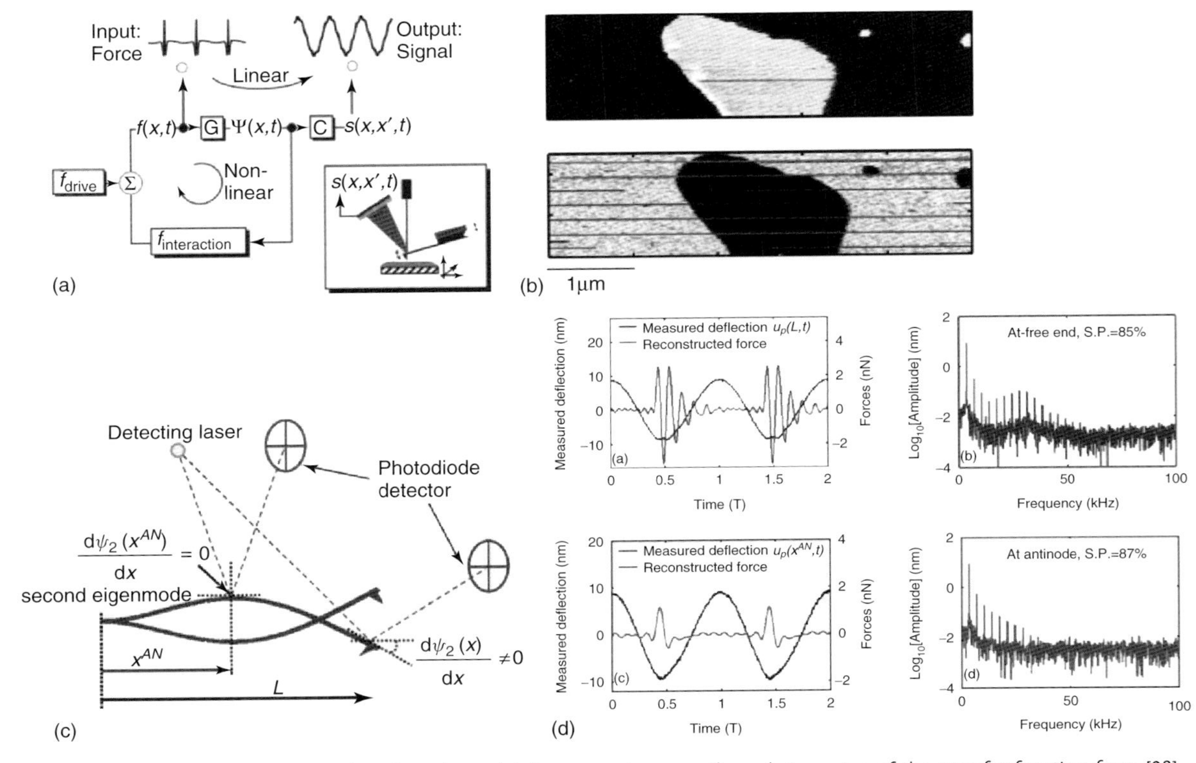

Figure 5.13 Force spectroscopy in liquid medium. (a) Force spectroscopy through inversion of the transfer function from [98]. (b) SPAM method implemented while scanning a bilayer membrane [100]. The top image is the topography, and the bottom image is the peak interaction force. (c) A method where the laser beam focused at the antinode of the cantilever beam to screen the contribution to the second eigenmode from the bending signal in the photodiode [70]. (d) Force spectroscopy on PTFE (polytetrafluoroethylene) showing artifacts in the reconstructed force when the laser spot is focused at the free end but not at the antinode [70].((a) Reproduced from Ref. [100] Copyright 2002 National Academy of Sciences, USA. (b) Reproduced from Ref. [102] Copyright 2006 National Academy of Sciences, USA. (c) and (d) Reproduced from Ref. [70] Copyright (2010) by the American Physical Society.)

5.6 Outlook

In this chapter, we have attempted to describe a range of important phenomena related to cantilever dynamics that occur when dAFM is used in liquid environments. In particular, we have tried to make a strong case for the fact that achieving quantitative dAFM in liquid environments requires a deep understanding of cantilever dynamics. We encourage experimentalists working in the area to consider using simulation tools such as the freeware Virtual Environment for Dynamic AFM (VEDA) which is available on the nanoHUB (*www.nanohub.org/resources/adac*) to develop a deeper intuition for cantilever dynamics in liquid environments.

In spite of the tremendous advances made in the past five years in the field, several open and key challenges remain in understanding AFM cantilever dynamics in liquid environments:

1) The interactions of oscillating AFM cantilevers in liquids with very soft viscoelastic objects such as live cells remain a topic of great interest but very little is understood about the physics of image and contrast channel formation for such samples.
2) The dynamics of small, very high-frequency cantilevers in liquids also has important implications for high-resolution AFM and remains poorly understood.
3) Cantilever dynamics in high-speed AFM is a major topic of interest. At high scanning speeds, cantilever hydrodynamics and cantilever response to perturbation can become more complex compared to conventional speed operation and remains an open topic of interest.
4) The work described in this chapter clearly points out that even if a single eigenmode is excited in liquids that the response often contains higher harmonics and higher eigenmodes. In this scenario, the use of bimodal AFM in liquids where both eigenmodes are simultaneously excited remains an important topic of research.
5) Finally, we have not discussed in this chapter the major role that the nature of the tip–sample interaction forces plays in effecting cantilever dynamics. The presence of electric double layer forces, van der Waals forces in liquids, specific adhesion, and solvation shells have a major influence on cantilever dynamics. This remains a very important and relatively unexplored research area in the field.

We hope that this chapter encourages the reader to use simulations to gain an understanding of the nonintuitive cantilever dynamics that occur in liquid environments and provides an overview of the current state of understanding and future opportunities in the field.

References

1. Putman, C.A.J., Vanderwerf, K.O., Degrooth, B.G., Vanhulst, N.F., and Greve, J. (1994) Tapping mode atomic-force microscopy in liquid. *Appl. Phys. Lett.*, **64** (18), 2454–2456.

2. Schäffer, T., Cleveland, J., Ohnesorge, F., Walters, D., and Hansma, P. (1996) Studies of vibrating atomic force microscope cantilevers in liquid. *J. Appl. Phys.*, **80**, 3622.
3. Han, W.H., Lindsay, S.M., and Jing, T.W. (1996) A magnetically driven oscillating probe microscope for operation in liquids. *Appl. Phys. Lett.*, **69** (26), 4111–4113.
4. Rankl, C., Pastushenko, V., Kienberger, F., Stroh, C.M., and Hinterdorfer, P. (2004) Hydrodynamic damping of a magnetically oscillated cantilever close to a surface. *Ultramicroscopy*, **100** (3–4), 301–308.
5. Nnebe, I. and Schneider, J.W. (2004) Characterization of distance-dependent damping in tapping-mode atomic force microscopy force measurements in liquid. *Langmuir*, **20** (8), 3195–3201.
6. Green, C.P. and Sader, J.E. (2005) Small amplitude oscillations of a thin beam immersed in a viscous fluid near a solid surface. *Phys. Fluids*, **17**, 073102.
7. Green, C.P. and Sader, J.E. (2005) Frequency response of cantilever beams immersed in viscous fluids near a solid surface with applications to the atomic force microscope. *J. Appl. Phys.*, **98** (11), 4913.
8. Clarke, R.J., Cox, S., Williams, P., and Jensen, O. (2005) The drag on a microcantilever oscillating near a wall. *J. Fluid Mech.*, **545**, 397–426.
9. Clarke, R.J., Jensen, O., Billingham, J., and Williams, P. (2006) Three-dimensional flow due to a microcantilever oscillating near a wall: an unsteady slender-body analysis. *Proc. R. Soc. A: Math., Phys. Eng. Sci.*, **462** (2067), 913.
10. Clarke, R.J., Jensen, O., Billingham, J., Pearson, A., and Williams, P. (2006) Stochastic elastohydrodynamics of a microcantilever oscillating near a wall. *Phys. Rev. Lett.*, **96** (5), 50801.
11. Brumley, D.R., Willcox, M., and Sader, J.E. (2010) Oscillation of cylinders of rectangular cross section immersed in fluid. *Phys. Fluids*, **22** (5), 052001.
12. Basak, S., Raman, A., and Garimella, S.V. (2006) Hydrodynamic loading of microcantilevers vibrating in viscous fluids. *J. Appl. Phys.*, **99** (11), 114906.
13. Basak, S., Beyder, A., Spagnoli, C., Raman, A., and Sachs, F. (2007) Hydrodynamics of torsional probes for atomic force microscopy in liquids. *J. Appl. Phys.*, **102** (2), 024914.
14. Clark, M.T. and Paul, M.R. (2008) The stochastic dynamics of rectangular and V-shaped atomic force microscope cantilevers in a viscous fluid and near a solid boundary. *J. Appl. Phys.*, **103** (9), 094910.
15. Roters, A. and Johannsmann, D. (1996) Distance-dependent noise measurements in scanning force microscopy. *J. Phys.: Condens. Matter*, **8** (41), 7561–7577.
16. Harrison, C., Tavernier, E., Vancauwenberghe, O., Donzier, E., Hsu, K., Goodwin, A.R.H., Marty, F., and Mercier, B. (2007) On the response of a resonating plate in a liquid near a solid wall. *Sens. Actuator. A: Phys.*, **134** (2), 414–426.
17. Naik, T., Longmire, E.K., and Mantell, S.C. (2003) Dynamic response of a cantilever in liquid near a solid wall. *Sens. Actuator. A: Phys.*, **102** (3), 240–254.
18. Xu, X., Carrasco, C., De Pablo, P.J., Gomez-Herrero, J., and Raman, A. (2008) Unmasking imaging forces on soft biological samples in liquids when using dynamic atomic force microscopy: a case study on viral capsids. *Biophys. J.*, **95** (5), 2520–2528.
19. Decuzzi, P., Granaldi, A., and Pascazio, G. (2007) Dynamic response of microcantilever-based sensors in a fluidic chamber. *J. Appl. Phys.*, **101** (2), 024303.
20. Sader, J.E. (1998) Frequency response of cantilever beams immersed in viscous fluids with applications to the atomic force microscope. *J. Appl. Phys.*, **84** (1), 64–76.
21. Green, C.P. and Sader, J.E. (2002) Torsional frequency response of cantilever beams immersed in viscous fluids with applications to the atomic force microscope. *J. Appl. Phys.*, **92** (10), 6262–6274.
22. Green, C. and Sader, J. (2009) Frequency response of cantilever beams

immersed in viscous fluids near a solid surface with applications to the atomic force microscope. *J. Appl. Phys.*, **98** (11), 114913.

23. Tung, R.C., Jana, A., and Raman, A. (2008) Hydrodynamic loading of microcantilevers oscillating near rigid walls. *J. Appl. Phys.*, **104** (11), 114905.
24. Van Eysden, C. and Sader, J. (2007) Frequency response of cantilever beams immersed in viscous fluids with applications to the atomic force microscope: arbitrary mode order. *J. Appl. Phys.*, **101** (4), 044908.
25. Kiracofe, D. and Raman, A. (2010) On eigenmodes, stiffness, and sensitivity of atomic force microscope cantilevers in air versus liquids. *J. Appl. Phys.*, **107** (3), 033506.
26. Preiner, J., Tang, J.L., Pastushenko, V., and Hinterdorfer, P. (2007) Higher harmonic atomic force microscopy: imaging of biological membranes in liquid. *Phys. Rev. Lett.*, **99** (4), 046102.
27. Binnig, G., Quate, C.F., and Gerber, C. (1986) Atomic force microscope. *Phys. Rev. Lett.*, **56** (9), 930–933.
28. Rogers, B., York, D., Whisman, N., Jones, M., Murray, K., Adams, J.D., Sulchek, T., and Minne, S.C. (2002) Tapping mode atomic force microscopy in liquid with an insulated piezoelectric microactuator. *Rev. Sci. Instrum.*, **73** (9), 3242–3244.
29. Rogers, B., Sulchek, T., Murray, K., York, D., Jones, M., Manning, L., Malekos, S., Beneschott, B., Adams, J.D., Cavazos, H., and Minne, S.C. (2003) High speed tapping mode atomic force microscopy in liquid using an insulated piezoelectric cantilever. *Rev. Sci. Instrum.*, **74** (11), 4683–4686.
30. Salapaka, S., Dahleh, M., and Mezic, I. (2001) On the dynamics of a harmonic oscillator undergoing impacts with a vibrating platform. *Nonlinear Dyn.*, **24** (4), 333–358.
31. Berg, J. and Briggs, G.A.D. (1997) Nonlinear dynamics of intermittent-contact mode atomic force microscopy. *Phys. Rev. B*, **55** (22), 14899–14908.
32. Burnham, N.A., Kulik, A.J., Gremaud, G., and Briggs, G.A.D. (1995) Nanosubharmonics–the dynamics of small nonlinear contacts. *Phys. Rev. Lett.*, **74** (25), 5092–5095.
33. Martin, Y. and Wickramasinghe, H.K. (1987) Magnetic imaging by force microscopy with 1000-a resolution. *Appl. Phys. Lett.*, **50** (20), 1455–1457.
34. Jarvis, S.P., Oral, A., Weihs, T.P., and Pethica, J.B. (1993) A novel force microscope and point-contact probe. *Rev. Sci. Instrum.*, **64** (12), 3515–3520.
35. Florin, E.L., Radmacher, M., Fleck, B., and Gaub, H.E. (1994) Atomic-force microscope with magnetic force modulation. *Rev. Sci. Instrum.*, **65** (3), 639–643.
36. Lindsay, S.M., Lyubchenko, Y.L., Tao, N.J., Li, Y.Q., Oden, P.I., Derose, J.A., and Pan, J. (1993) Scanning tunneling microscopy and atomic force microscopy studies of biomaterials at a liquid–solid interface. *J. Vac. Sci. Technol. A*, **11** (4), 808–815.
37. Enders, O., Korte, F., and Kolb, H.A. (2004) Lorentz-force-induced excitation of cantilevers for oscillation-mode scanning probe microscopy. *Surf. Interface Anal.*, **36** (2), 119–123.
38. Buguin, A., Du Roure, O., and Silberzan, P. (2001) Active atomic force microscopy cantilevers for imaging in liquids. *Appl. Phys. Lett.*, **78** (19), 2982–2984.
39. Penedo, M., Fernandez-Martinez, I., Costa-Kramer, J.L., Luna, M., and Briones, F. (2009) Magnetostriction-driven cantilevers for dynamic atomic force microscopy. *Appl. Phys. Lett.*, **95** (14), 143505.
40. Martin, Y., Abraham, D.W., and Wickramasinghe, H.K. (1988) High-resolution capacitance measurement and potentiometry by force microscopy. *Appl. Phys. Lett.*, **52** (13), 1103–1105.
41. Weaver, J.M.R. and Abraham, D.W. (1991) High-resolution atomic force microscopy potentiometry. *J. Vac. Sci. Technol. B*, **9** (3), 1559–1561.
42. Umeda, K., Oyabu, N., Kobayashi, K., Hirata, Y., Matsushige, K., and Yamada, H. (2010) High-resolution frequency-modulation atomic force microscopy in liquids using electrostatic

excitation method. *Appl. Phys. Expr.*, **3** (6), 065205.
43. Kiracofe, D., Kobayashi, K., Labuda, A., Raman, A., and Yamada, H. (2011) High efficiency laser photothermal excitation of microcantilever vibrations in air and liquids. *Rev. Sci. Instrum.*, **82** (1), 3702.
44. Umeda, N., Ishizaki, S., and Uwai, H. (1991) Scanning attractive force microscope using photothermal vibration. *J. Vac. Sci. Technol. B*, **9** (2), 1318–1322.
45. Marti, O., Ruf, A., Hipp, M., Bielefeldt, H., Colchero, J., and Mlynek, J. (1992) Mechanical and thermal effects of laser irradiation on force microscope cantilevers. *Ultramicroscopy*, **42**, 345–350.
46. Lee, J. and King, W. (2007) Microcantilever actuation via periodic internal heating. *Rev. Sci. Instrum.*, **78**, 126102.
47. Fantner, G.E., Schumann, W., Barbero, R.J., Deutschinger, A., Todorov, V., Gray, D.S., Belcher, A.M., Rangelow, I.W., and Youcef-Toumi, K. (2009) Use of self-actuating and self-sensing cantilevers for imaging biological samples in fluid. *Nanotechnology*, **20** (43), 434003.
48. Degertekin, F.L., Hadimioglu, B., Sulchek, T., and Quate, C.F. (2001) Actuation and characterization of atomic force microscope cantilevers in fluids by acoustic radiation pressure. *Appl. Phys. Lett.*, **78** (11), 1628–1630.
49. Rensen, W.H.J., Van Hulst, N.F., and Kammer, S.B. (2000) Imaging soft samples in liquid with tuning fork based shear force microscopy. *Appl. Phys. Lett.*, **77** (10), 1557–1559.
50. Kageshima, M., Jensenius, H., Dienwiebel, M., Nakayama, Y., Tokumoto, H., Jarvis, S.P., and Oosterkamp, T.H. (2002) Noncontact atomic force microscopy in liquid environment with quartz tuning fork and carbon nanotube probe. *Appl. Surf. Sci.*, **188** (3–4), 440–444.
51. Hutter, J.L. and Bechhoefer, J. (1993) Calibration of atomic-force microscope tips. *Rev. Sci. Instrum.*, **64** (7), 1868–1873.
52. Butt, H.J. and Jaschke, M. (1995) Calculation of thermal noise in atomic-force microscopy. *Nanotechnology*, **6** (1), 1–7.
53. Lozano, J.R., Kiracofe, D., Melcher, J., Garcia, R., and Raman, A. (2010) Calibration of higher eigenmode spring constants of atomic force microscope cantilevers. *Nanotechnology*, **21** (46), 465502.
54. Kawakami, M., Byrne, K., Khatri, B., Mcleish, T.C.B., Radford, S.E., and Smith, D.A. (2004) Viscoelastic properties of single polysaccharide molecules determined by analysis of thermally driven oscillations of an atomic force microscope cantilever. *Langmuir*, **20** (21), 9299–9303.
55. Gannepalli, A., Sebastian, A., Cleveland, J., and Salapaka, M. (2005) Thermally driven non-contact atomic force microscopy. *Appl. Phys. Lett.*, **87** (11), 111901.
56. Xu, X. and Raman, A. (2007) Comparative dynamics of magnetically, acoustically, and Brownian motion driven microcantilevers in liquids. *J. Appl. Phys.*, **102** (3), 34303.
57. Melcher, J., Hu, S.Q., and Raman, A. (2007) Equivalent point-mass models of continuous atomic force microscope probes. *Appl. Phys. Lett.*, **91** (5), 053101.
58. Meirovitch, L. (1997) *Principles and Techniques of Vibrations*, Prentice Hall, New Jersey.
59. Ramos, D., Tamayo, J., Mertens, J., and Calleja, M. (2006) Photothermal excitation of microcantilevers in liquids. *J. Appl. Phys.*, **99** (12), 124904.
60. Wirsching, P.H., Paez, T.L., and Ortiz, K. (2006) *Random Vibrations: Theory and Practice*, Dover Publications.
61. Clark, M.T., Sader, J.E., Cleveland, J.P., and Paul, M.R. (2010) Spectral properties of microcantilevers in viscous fluid. *Phys. Rev. E*, **81** (4), 046306.
62. Melcher, J., Xu, X., and Raman, A. (2008) Multiple impact regimes in liquid environment dynamic atomic force microscopy. *Appl. Phys. Lett.*, **93** (9), 093111.
63. Garcia, R. and San Paulo, A. (1999) Attractive and repulsive tip-sample interaction regimes in tapping-mode

atomic force microscopy. *Phys. Rev. B*, **60** (7), 4961–4967.

64. Kiracofe, D. and Raman, A. (2011) Quantitative force and dissipation measurements in liquids using piezo excited atomic force microscopy: a unifying theory. *Nanotechnology*, **22**, 485502.
65. Jai, C., Cohen-Bouhacina, T., and Maali, A. (2007) Analytical description of the motion of an acoustic-driven atomic force microscope cantilever in liquid. *Appl. Phys. Lett.*, **90** (11), 113512.
66. Beer, S., Ende, D., and Mugele, F. (2010) Dissipation and oscillatory solvation forces in confined liquids studied by small-amplitude atomic force spectroscopy. *Nanotechnology*, **21**, 325703.
67. Kokavecz, J. and Mechler, A. (2007) Investigation of fluid cell resonances in intermittent contact mode atomic force microscopy. *Appl. Phys. Lett.*, **91** (2), 023113.
68. Sader, J.E., Chon, J.W.M., and Mulvaney, P. (1999) Calibration of rectangular atomic force microscope cantilevers. *Rev. Sci. Instrum.*, **70** (10), 3967–3969.
69. Anczykowski, B., Gotsmann, B., Fuchs, H., Cleveland, J.P., and Elings, V.B. (1999) How to measure energy dissipation in dynamic mode atomic force microscopy. *Appl. Surf. Sci.*, **140** (3–4), 376–382.
70. Xu, X., Melcher, J., and Raman, A. (2010) Accurate force spectroscopy in tapping mode atomic force microscopy in liquids. *Phys. Rev. B*, **81** (3), 035407.
71. Asakawa, H. and Fukuma, T. (2009) Spurious-free cantilever excitation in liquid by piezoactuator with flexure drive mechanism. *Rev. Sci. Instrum.*, **80** (10), 103703.
72. Maali, A., Hurth, C., Cohen-Bouhacina, T., Couturier, G., and Aimé, J.P. (2006) Improved acoustic excitation of atomic force microscope cantilevers in liquids. *Appl. Phys. Lett.*, **88**, 163504.
73. Carrasco, C., Ares, P., De Pablo, P., and Gómez-Herrero, J. (2008) Cutting down the forest of peaks in acoustic dynamic atomic force microscopy in liquid. *Rev. Sci. Instrum.*, **79**, 126106.
74. Motamedi, R. and Wood-Adams, P.M. (2008) Influence of fluid cell design on the frequency response of AFM microcantilevers in liquid media. *Sensors*, **8** (9), 5927–5941.
75. Vassalli, M., Pini, V., and Tiribilli, B. (2010) Role of the driving laser position on atomic force microscopy cantilevers excited by photothermal and radiation pressure effects. *Appl. Phys. Lett.*, **97** (14), 143105.
76. Nishida, S., Kawakatsu, H., and Nishimori, Y. (2009) Photothermal excitation of a single-crystalline silicon cantilever for higher vibration modes in liquid. *J. Vac. Sci. Technol. B: Microelectron. Nanomet. Struct.*, **27**, 964.
77. Ratcliff, G.C., Erie, D.A., and Superfine, R. (1998) Photothermal modulation for oscillating mode atomic force microscopy in solution. *Appl. Phys. Lett.*, **72** (15), 1911–1913.
78. Fukuma, T. and Jarvis, S.P. (2006) Development of liquid-environment frequency modulation atomic force microscope with low noise deflection sensor for cantilevers of various dimensions. *Rev. Sci. Instrum.*, **77** (4), 043701.
79. Labuda, A., Kobayashi, K., Kiracofe, D., Suzuki, K., Grutter, P.H., and Yamada, H. (2011) Comparison of photothermal and piezoacoustic excitation methods for frequency and phase modulation atomic force microscopy in liquid environments. *AIP Adv.*, **1**, 022136.
80. Basak, S. and Raman, A. (2007) Dynamics of tapping mode atomic force microscopy in liquids: theory and experiments. *Appl. Phys. Lett.*, **91** (6), 064107.
81. Merirovitch, L. (1997) *Principals and Techniques of Vibrations*, Prentice Hall, New Jersey.
82. Kiracofe, D., Melcher, J.,and Raman, A. (2011) Gaining insight into the physics of dynamic Atomic Force Microscopy in complex environments using the VEDA simulator. *Rev. Sci. Instrum.*,in press.

83. Melcher, J., Hu, S., and Raman, A. (2007) Equivalent point-mass models of continuous atomic force microscope probes. *Appl. Phys. Lett.*, **91** (5), 053101.
84. Xu, X. and Raman, A. (2007) Comparative dynamics of magnetically, acoustically, and brownian motion driven microcantilevers in liquids. *J. Appl. Phys.*, **102** (3), 034303.
85. Sader, J.E., Larson, I., Mulvaney, P., and White, L.R. (1995) Method for the calibration of atomic-force microscope cantilevers. *Rev. Sci. Instrum.*, **66** (7), 3789–3798.
86. Lozano, J.R., Kiracofe, D., Melcher, J., Garcia, R., and Raman, A. (2010) Calibration of higher eigenmode spring constants of atomic force microscope cantilevers. *Nanotechnology*, **21** (46), 465502.
87. Melcher, J., Hu, S., and Raman, A. (2008) Invited article: veda: a web-based virtual environment for dynamic atomic force microscopy. *Rev. Sci. Instrum.*, **79** (6), 061301.
88. Kiracofe, D. and Raman, A. (2010) Microcantilever dynamics in liquid environment dynamic atomic force microscopy when using higher-order cantilever eigenmodes. *J. Appl. Phys.*, **108** (3), 034320.
89. Xu, X., Melcher, J., Basak, S., Reifenberger, R., and Raman, A. (2009) Compositional contrast of biological materials in liquids using the momentary excitation of higher eigenmodes in dynamic atomic force microscopy. *Phys. Rev. Lett.*, **102** (6), 060801.
90. Rodriguez, T.R. and Garcia, R. (2002) Tip motion in amplitude modulation (tapping-mode) atomic-force microscopy: comparison between continuous and point-mass models. *Appl. Phys. Lett.*, **80** (9), 1646–1648.
91. Van Noort, S.J.T., Willemsen, O.H., Van Der Werf, K.O., De Grooth, B.G., and Greve, J. (1999) Mapping electrostatic forces using higher harmonics tapping mode atomic force microscopy in liquid. *Langmuir*, **15** (21), 7101–7107.
92. Thomson, N.H., Turner, R.D., Kirkham, J., and Devine, D. (2009) Second harmonic atomic force microscopy of living staphylococcus aureus bacteria. *Appl. Phys. Lett.*, **94** (4), 043901.
93. Cleveland, J.P., Anczykowski, B., Schmid, A.E., and Elings, V.B. (1998) Energy dissipation in tapping-mode atomic force microscopy. *Appl. Phys. Lett.*, **72** (20), 2613–2615.
94. Tamayo, J. (1999) Energy dissipation in tapping-mode scanning force microscopy with low quality factors. *Appl. Phys. Lett.*, **75** (22), 3569–3571.
95. Melcher, J., Carrasco, C., Xu, X., Carrascosa, J.L., Gomez-Herrero, J., De Pablo, P.J., and Raman, A. (2009) Origins of phase contrast in the atomic force microscope in liquids. *Proc. Natl. Acad. Sci. U.S.A.*, **106** (33), 13655–13660.
96. Sader, J.E. and Jarvis, S.P. (2004) Accurate formulas for interaction force and energy in frequency modulation force spectroscopy. *Appl. Phys. Lett.*, **84** (10), 1801–1803.
97. Katan, A.J., Van Es, M.H., and Oosterkamp, T.H. (2009) Quantitative force versus distance measurements in amplitude modulation AFM: a novel force inversion technique. *Nanotechnology*, **20** (16), 165703.
98. Stark, M., Stark, R.W., Heckl, W.M., and Guckenberger, R. (2002) Inverting dynamic force microscopy: from signals to time-resolved interaction forces. *PNAS*, **99** (13), 8473–8478.
99. Legleiter, J. and Kowalewski, T. (2005) Insights into fluid tapping-mode atomic force microscopy provided by numerical simulations. *Appl. Phys. Lett.*, **87** (16), 163120.
100. Legleiter, J., Park, M., Cusick, B., and Kowalewski, T. (2006) Scanning probe acceleration microscopy (spam) in fluids: mapping mechanical properties of surfaces at the nanoscale. *Proc. Natl. Acad. Sci. U.S.A.*, **103** (13), 4813–4818.
101. Xu, X., Melcher, J., and Raman, A. (2010) Accurate force spectroscopy in tapping mode atomic force microscopy in liquids. *Phys. Rev. B*, **81** (3), 035407.

6
Single-Molecule Force Spectroscopy

Albert Galera-Prat, Rodolfo Hermans, Rubén Hervás, Àngel Gómez-Sicilia, and Mariano Carrión-Vázquez

6.1
Introduction

This chapter aims to introduce the main concepts of single-molecule force spectroscopy (SMFS) by atomic force microscopy (AFM), essentially focusing on the basic methodology and its most relevant biological applications. Special emphasis will be paid to the criteria used to identify true single-molecule events (the basis of obtaining meaningful data and interpreting it correctly), as well as the main achievements of the technique. More specialized information with details of the experiments can be found in other reviews [1–4].

Since AFM-based SMFS has been so far mainly applied to proteins, this chapter focuses mostly on protein mechanics, although a short account on its application to other biopolymers can be found in Section 6.4.

The field of protein nanomechanics has made progress in three different fronts: intramolecular interactions (protein folding and unfolding), intermolecular interactions (protein–biomolecule), and membrane protein extraction (where intramolecular and intermolecular interactions occur simultaneously). This chapter is mainly focused in intramolecular studies, for which unambiguous single-molecule markers have been developed. In the case of intermolecular interactions, internal single-event markers are not yet available, while in mechanical extraction of membrane proteins, their unfolding and membrane unbinding cannot be easily decoupled.

6.1.1
Why Single-Molecule Force Spectroscopy?

Single-molecule experiments are direct measurements of the behavior of one molecule at a time. These methods present several advantages over ensemble assays [2] in which only average values are typically observed. Single-molecule measurements allow accessing the whole distribution of observable events so

Atomic Force Microscopy in Liquid: Biological Applications, First Edition.
Edited by Arturo M. Baró and Ronald G. Reifenberger.

that heterogeneities, otherwise hidden in the ensemble average, can be resolved [5]. The loss of information in bulk experiments is a consequence of the central limit theorem [6] that states that the average of a sufficiently large number of equally distributed quantities distributes asymptotically as a Gaussian and the information carried by variance and higher moments of the original distribution is lost within the uncertainty of the observation (Figure 6.1). Another advantage of studying single molecules is the ability to follow processes in real time and reveal dynamic information about them. Thus, the possibility of capturing rare events, heterogeneous populations or hidden trajectories makes single-molecule experiments a unique choice to reveal fundamental new information about the dynamics and structure of a molecule.

6.1.2
SMFS in Biology

A fundamental question is whether it is relevant or not to use force to study biological systems. To begin with, force can be used as a perturbation to obtain information from the system of interest. Most biochemical experiments have been performed using other physical or chemical agents in bulk; therefore, the use of a mechanical force may provide new insights. Furthermore, many biomolecules are subjected to or develop mechanical forces during their functioning [1, 7]. Hence, the mechanical study of these biomolecules is not just a different method to study these systems but a physiologically relevant approach to previously unattainable conditions.

6.1.3
SMFS Techniques and Ranges

SMFS can be performed using different techniques, the most relevant being optical tweezers (OTs), magnetic tweezers (MT), and AFM. All these techniques use a force probe to stretch the sample while measuring the response of an individual molecule, in this way obtaining stress–strain or length *versus* time curves [3, 4].

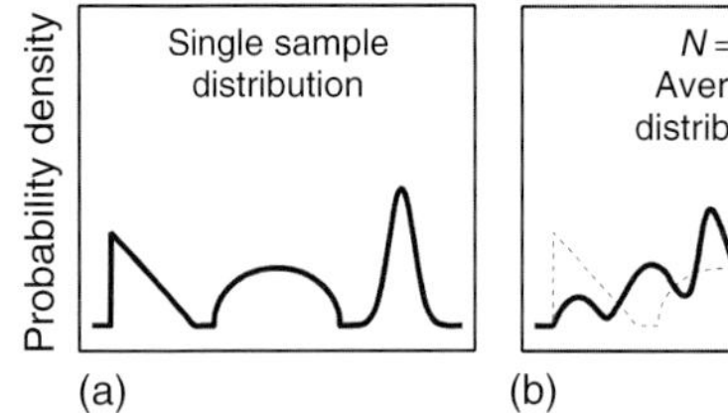

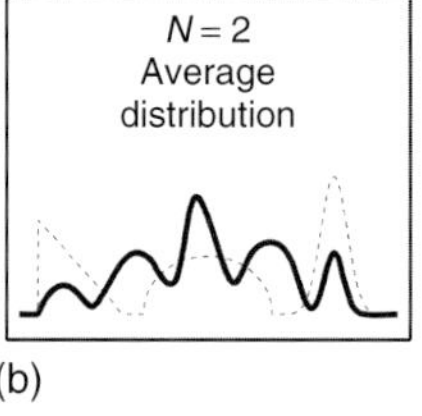

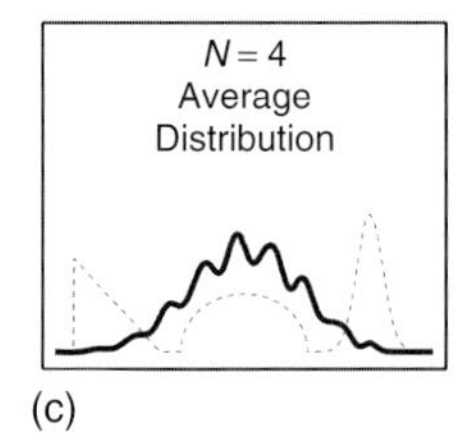

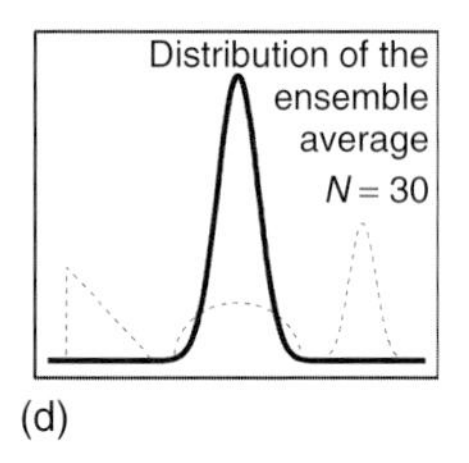

Figure 6.1 Probability density distribution of a random variable x for a single sample (a) and for N-averaged samples (b–d). A hypothetical property x with a probability density with three subdistributions is presented here. This distribution can be clearly observed when dealing with single samples (a). However, when multiple samples are observed at the same time (b–d), the shape of the original distribution (dashed line) is lost and the distribution observed tends toward a Gaussian distribution as predicted by the central limit theorem [6].

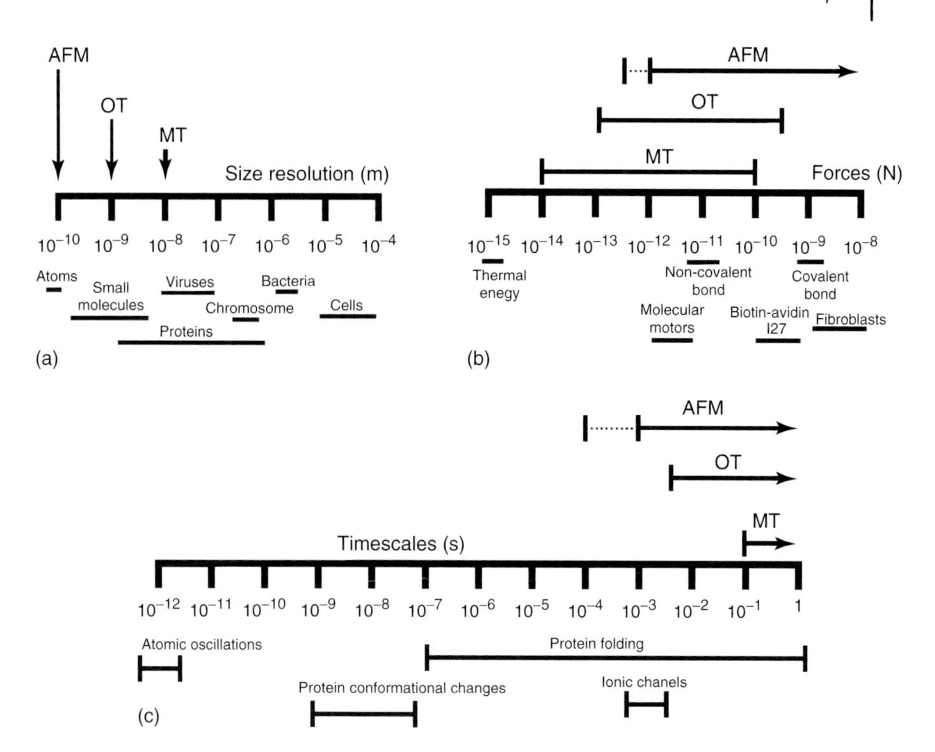

Figure 6.2 Relevant parameters and the ranges covered by different SMFS techniques. (a) AFM has better spatial resolution than OT or MT. (b) OT and MT are more sensitive to lower forces than AFM, although some technical improvements [9] have recently extended the force resolution of AFM close to that of OT. Finally, (c) AFM-SMFS instrumentation has been recently optimized (dashed line) [20] reducing the temporal resolution of this technique to values below the current millisecond values, which was yet faster than that achieved with other SMFS techniques. When compared to the relevant scales for proteins, AFM-SMFS seems to be the most suitable technique.

The small mass of the probe used in OT and MT experiments makes these techniques very sensitive to weak forces (Figure 6.2) [2, 3], while the heavier and stiffer probes used in AFM (cantilevers) are more suitable to apply higher forces to the sample and to resolve shorter distances. Furthermore, the smaller curvature radius of a cantilever tip limits the number of molecules that can be attached, thereby facilitating single-molecule experimentation. For these reasons, AFM has, in practice, mostly been used to study proteins and to measure interaction forces between two biomolecules [8], while OT and MT have been more commonly employed to study DNA, RNA, and molecular motors.

6.2 AFM-SMFS Principles

Apart from being a high-resolution imaging technique, AFM can be used as a very precise tool to study the mechanical properties of different biomolecules in the high-force regime [11]. These experiments are based on measuring the behavior of an individual molecule while it is subjected to mechanical stress. Accordingly, in a typical AFM-SMFS experiment, a sample that usually consists of a purified biopolymer is somehow attached (Section 6.2.1) to a surface (substrate) and then brought in contact with the cantilever tip to establish a mechanical link between these two elements. The substrate is then withdrawn by an actuator and the sample is stretched, while force and distance are measured in the piconewton and angstrom range, respectively. In these conditions, the force response measured depends on the structural and thermodynamic properties of the molecule attached and on the solvent that surrounds it. Depending on how the actuator is controlled, two basic operation modes can be defined: *length clamp* and *force clamp*.

6.2.1
Length-Clamp Mode

The experiment starts with the cantilever tip far from the substrate. Using a calibrated actuator the tip is approached at a constant velocity (typically in the order of 1 nm ms^{-1}), over a range of several hundred nanometers. Once the cantilever tip reaches the surface, it continues to push; subsequently, the actuator trajectory is reversed withdrawing the surface from the cantilever. This increases the tension and stretches the sample until it detaches (from either the substrate or the cantilever tip) such that the force measured returns to zero, marking the end of the experiment.

The measurement of the stretching force is based on the cantilever bending, while the actuator position is estimated from the voltage applied to a piezoelectric actuator or by means of a capacitive sensor or strain gauge. The sample extension (i.e., the distance between the tip and substrate) is calculated as the relative position of the actuator minus the bending displacement of the cantilever. After setting the contact point as the position "zero" of the experiment, it is possible to represent the curve of interest by plotting force versus sample extension, in which phase transitions characteristic of each biopolymer can be identified (Section 6.4).

Length-clamp experiments yield a characteristic elasticity curve for the biopolymer under study. As the tip–surface distance increases, the entropic elasticity of the polymer opposes to its extension [12] by generating a nonlinear restoring force. This is a general property of polymers, which results from the maximization of their conformational entropy in a fluid media where they tend to recoil instead of adopting the linear conformation imposed on stretching (very low in entropy). For protein unfolding, the entropic response produces an increase in force that ends when the critical intramolecular bonds are broken. Once this force is reached, the protein module unfolds and is unraveled. This sudden increase in the length of the molecule typically drops the force to basal levels, which results in the typical force peak (Figure 6.3). This maximum in the trace is enthalpic in nature and related to the mechanical stability of the protein (Section 6.3). As the mechanical circuit is preserved through peptide bonds, further extension results in additional peaks originated from either other unfolding events or the detachment of the molecule. In contrast, when studying a protein–biomolecule interaction only a single peak

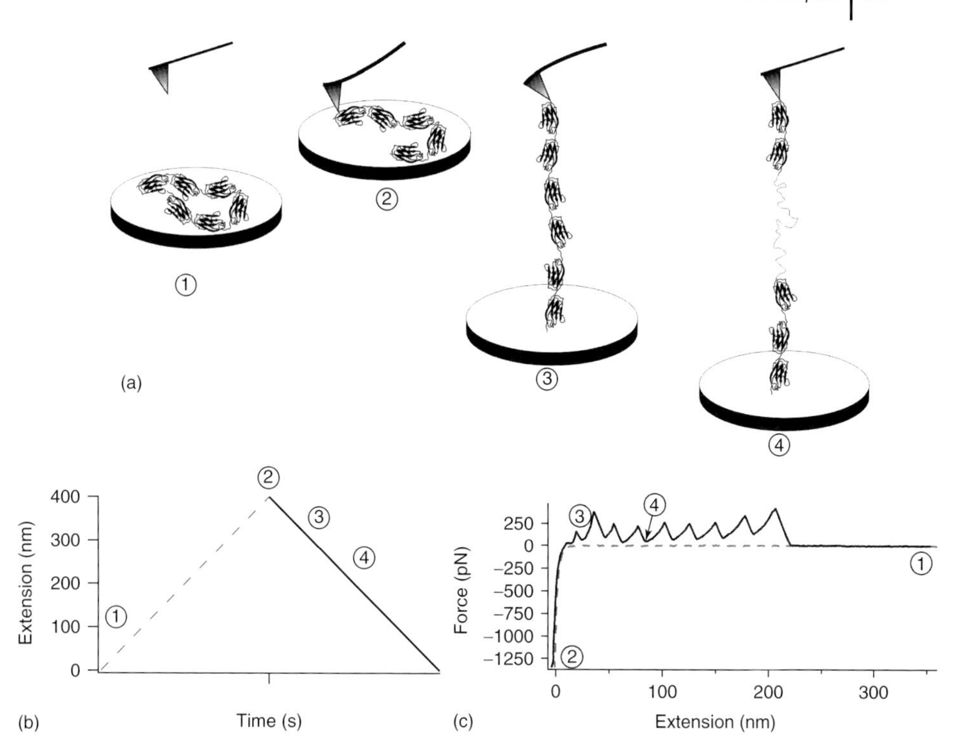

Figure 6.3 *Length-clamp* unfolding experiment. Cartoon representation of a *length-clamp* experiment (a), showing the movement of the actuator (b), and a force-extension curve (c) obtained. In these experiments, the sample is deposited on a substrate. The tip–surface distance is then reduced at a constant speed (dashed line; 1) so that both elements are brought in contact (2), thereby allowing the protein to adsorb to the tip. Then the moving direction of the actuator is reversed (black line) and the protein is stretched at the same pulling speed while a force–extension trace is recorded. As can be seen, the proximal region of the spectrum is noisy (c3) due to uncontrolled interactions. When a protein module is unfolded in a *length-clamp* experiment, force values drop to 0 due to the unraveling of the module, resulting in a force peak (4).

is typically observed (Figure 6.4) resulting from the rupture of the intermolecular interactions, which breaks the tip–surface link.

Length-clamp experiments have also been used to study protein refolding. In this case, the sample is first partially stretched before the actuator direction is reversed to avoid the detachment of the molecule (from the tip or the surface). This reduces the tip–surface distance allowing the protein to autonomously refold with or without force. Repeating the stretch–relax cycle, several times can provide information about the protein folding kinetics and fidelity. This approach has revealed the existence of unusual misfolding events that, because of their low prevalence in the population (2–4%), might be beyond the resolution of other techniques [13].

The initial distribution of the sample over the substrate is unknown and "fishing" for samples is usually done randomly by systematic scanning of the substrate. Setups have been developed that are capable of selectively positioning the tip over

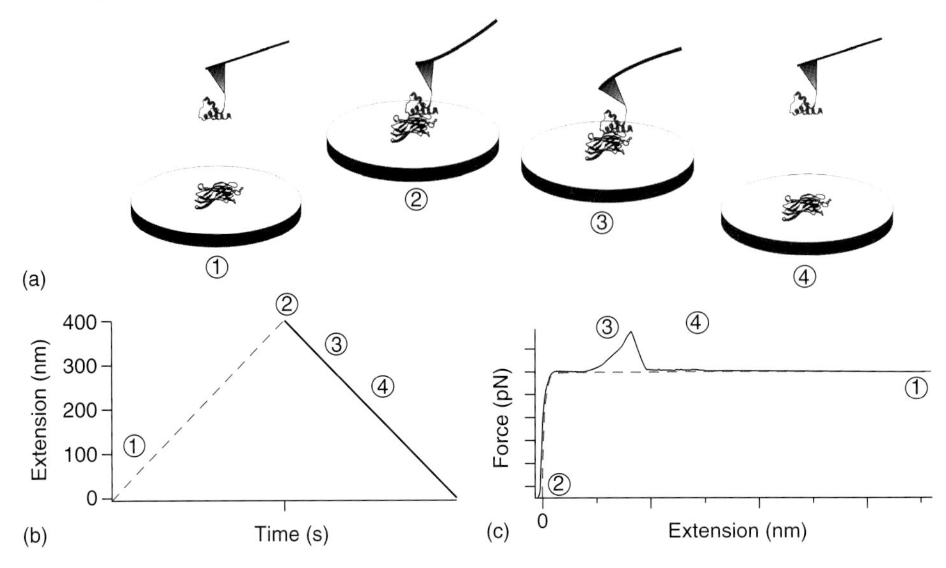

Figure 6.4 *Length-clamp* unbinding experiment. Cartoon representation of a *length-clamp* unbinding experiment (a) showing the movement of the actuator (b) and the force–extension curve (c) obtained. In contrast with unfolding experiments, when studying intermolecular forces, one of the interacting molecules is placed on the substrate and the other one is functionalized on the tip. On approaching the tip and substrate (dashed line in b and c), both elements are brought into contact and the interaction is established (2). Then, the distance is increased and force signal rises (3) until the interaction is broken, restoring the force to its basal level (4). As there is no mechanical continuity between the tip and the substrate, once the interaction is broken no additional force peaks are observed.

chosen molecular species [14], which should enhance the picking-up efficiency. Unfortunately, the compatibility of imaging and SMFS is not always possible due to their opposing technical requirements. SMFS usually uses softer probes and rougher substrate surfaces where a thick layer of sample (rather than isolated molecules) is not attached as firmly as for imaging. Ideally, in SMFS experiments, the sample attachments are strong but limited to specific locations so that sample–surface interactions do not interfere during the experiment. Hence, specific anchoring methods have been developed such as the substrate functionalization with stands for NitriloTriAcetic acid (NTA)-Ni^{2+} or avidin, which bind noncovalently to either the popular histidine tag typically engineered in proteins for affinity purification or biotin molecules, respectively. Furthermore, covalent methods have been used to this end taking advantage of the capability of cysteine residues engineered in proteins to establish S-Au bonds with gold substrates [15]. Other immobilization methods used include silanization, where alkoxysilane molecules self-assemble to cover commonly used substrates (mica, glass, or metal oxide) exposing different functional groups capable of binding proteins [16].

Other challenges in probing single molecules are specific to the AFM technique. Thermal drift limits the duration of an experiment, and consequently, the pulling speed accessible with an AFM is limited to $\sim$0.01 nm ms^{-1}, while viscous drag of the solvent and cantilever response introduce errors when stretching at speeds above 10 nm ms^{-1}. Unfortunately, these pulling speeds are often higher than those relevant for some physiological processes [17].

Furthermore, the tip and cantilever used strongly influence the outcome of the experiment, since they constitute the force sensor contacting the sample. When choosing the cantilever, the aim should be to keep the contribution of thermal noise as low as possible, while minimizing the elastic response of the probe. Thus, given that the fluctuations are proportional to the stiffness of the cantilever, and the elastic response is inversely proportional to it, the choice of a cantilever is a compromise in the stiffness: neither too rigid to become insensitive nor too slack that would have a large elastic response. Typical stiffness values of a cantilever for SMFS range between 1 and 100 pN nm^{-1} [18], while for imaging, stiffer probes are preferred due to their faster response.

Moreover, force sensitivity is limited by thermal fluctuations and could be increased by reducing the cantilever size [19]. A different approach to increase force resolution has recently been reported in which, by oscillating the sample surface at a low frequency and small amplitude while pulling at a low speed, better force resolution (estimated to be about 400 fN) can be achieved at the expense of reducing the bandwidth from the typical 5 kHz to 5 Hz [9].

6.2.2 Force-Clamp Mode

Although the *length-clamp* mode provides a wealth of information, the interpretation of the recordings is not trivial as both the end-to-end length and the force vary during the experiment (Section 6.3). The *force-clamp* mode provides an alternative approach where the position of the actuator is controlled by a feedback system such that the force exerted on the sample is kept constant to a predefined value (force-step) or adjusted to a constantly increased value (*force-ramp*), while recording the tip–surface distance as a function of time [10].

The implementation of this mode is analogous to the standard contact mode protocols for AFM imaging whereby the cantilever deflection (force) signal is fed into a control loop feedback mechanism (e.g., a PID controller), which corrects the actuator position to adjust the force exerted. During a *force-clamp* experiment, the sample length is the only free variable, resulting in highly controlled experimental conditions that permit the kinetic parameters of the process to be extracted directly (Section 6.3), as the lifetime of the molecular conformation is recorded in real time. Since the transition rate is force dependent, the *force-clamp* mode allows this rate to be controlled, in contrast with the situation in *length clamp*.

The sudden length increase in an extension time trace during a *force-clamp* experiment (step) is indicative of a structural phase transition of the sample. For proteins (Figure 6.5), this extension corresponds to the unfolding of the tertiary structure. The time lapse between different events varies reflecting the stochastic

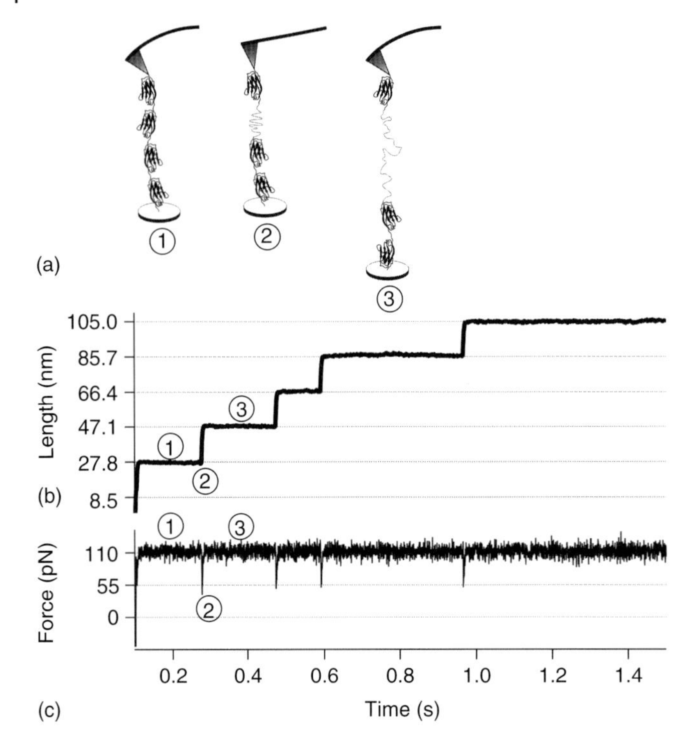

Figure 6.5 *Force-clamp* experiment. Once the protein is bound to the tip (a1), a constant force (c1) is applied to the sample, which initially extends to its folded length (b1). The traces of both molecule length (b) and force (c) versus time are recorded during the experiment. Once the molecule is subjected to a force, it remains folded and maintains its length for a period of time (point 1 in b and c). The applied force promotes the unfolding of any module trapped between the tip and the surface. At a given time point, one of the modules will unfold (a2) resulting in an increase in contour length (b2) that corresponds to the length of the sequence trapped within the mechanical clamp at a given force. As the length suddenly increases, the force drops to lower values (c2) until the feedback mechanism restores it to the set point (c3), which marks the resolution time of the experiment. The same process is repeated for the remaining protein modules or until the detachment of the molecule.

nature of the process (Figure 6.5) and its probability distribution relates to the shape of the potential that gives stability to the module at the set stretching force [10].

Most SMFS instrumentation currently being used is adapted from imaging devices; hence, it is not optimized for the particular requirements of this technique. When performing *force-clamp* experiments, the main limitation comes from the time resolution imposed by the instrument and the feedback mechanism, which is usually in the millisecond range (Figure 6.2). By optimizing an AFM specifically designed for SMFS, it has been possible to reduce the feedback response to about 150 μs [20], expanding the available time window of the technique. This achievement opens the way to obtaining further insight into different processes, such as protein folding.

6.3 Dynamics of Adhesion Bonds

On the basis of the experimental results from bulk assays, protein unfolding and unbinding have typically been considered to be well described as two-state processes [2]. In SMFS, where proteins usually unfold or unbind in an all-or-none manner, it also seems suitable to model these processes as two-state systems as a first approximation. Such processes are characterized by an energy diagram with two minima that define the stable states (folded/unfolded or bound/unbound), separated by an energy barrier characterized by two parameters: height and distance from the folded or bound state (Figure 6.6), the maximum of which is known as the transition state (TS).

6.3.1 Bond Dissociation Dynamics in Length Clamp

Thermal fluctuations allow a protein to explore different conformations and eventually cross the unfolding energy barrier that defines the bound state. In an AFM–SMFS experiment, the application of force to a protein reduces its energy barrier (increasing the crossing probability) and stabilizes the unfolded state. Hence, the unfolding force (the maximum of a force peak in a *length-clamp* trace) of

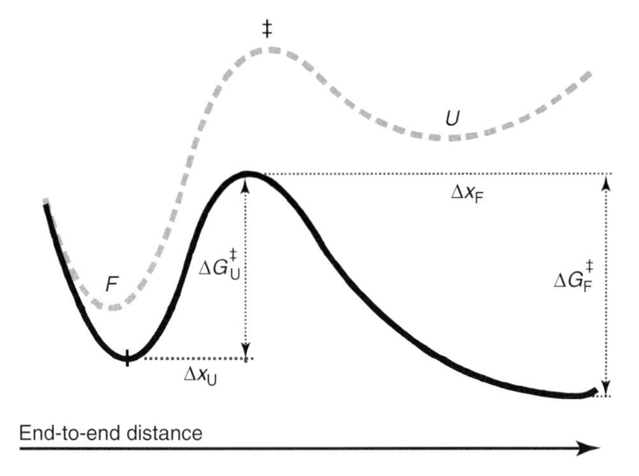

Figure 6.6 Energy diagram for a two-state process. When no force is applied (dashed line), the folded (*F*) and unfolded (*U*) states of a protein are located in two energy minima separated by an energy barrier. The passing structure at the maximum of the energy barrier corresponds to the transition state (‡). The application of force to stretch the protein (black line) "tilts" its energy landscape and reduces the height of the energy barrier ($\Delta G_U^{\ddagger}$) located at a distance Δx_U in the reaction coordinate, which corresponds to the end-to-end distance of the molecule. As a consequence, the unfolded state presents lower energy, making the unfolding of the module more probable. For the folding process under force, the opposite effect leads to an increase in the folding energy barrier ($\Delta G_F^{\ddagger}$) located at a distance Δx_F.

a protein varies stochastically as the energy of each molecule differs and varies over time. As a result, the width of the unfolding force distribution is directly correlated with temperature and inversly to the distance to the TS [21]. Consequently, the *mechanical resistance of a protein* is defined as the most probable unfolding force at a specific loading rate (see below). Important information such as the kinetic parameters of the process could be obtained from the shape of the distribution [12, 22], which enables the mechanical properties of different proteins to be compared [23]. Unfortunately, experimental errors result in histogram broadening that prevents this analysis. Hence, a different approach based on the pulling speed dependence of the unfolding forces is more commonly used, although other methods based on the analysis of raw data are also available [24].

The aforementioned method is based on a model that was developed in 1978 to analyze the increase in the rate of bond dissociation when an external force is applied [25]. On the basis of the transition state theory introduced by Eyring and others, in this model, the energy barrier separating the folded and unfolded states is lowered by the magnitude of the mechanical energy applied.

Later on, as more precise instrumentation was developed, a more detailed model was proposed that allows the kinetic constants and the energy barrier to be calculated from the pulling experiments [26]. In this model [25, 26], during its mechanical unfolding, the protein molecule crosses a single high energy barrier with a relative height $\Delta G^{\ddagger}$ after a deformation Δx_{U} (distance between the native state and the TS in the reaction coordinate, which is the end-to-end distance of the molecule). The unfolding process happens at a rate k_U that depends on the applied force, F, and the thermal energy, $k_{\mathrm{B}}T$, as described by the following equation:

$$k_{\mathrm{U}}(F) = A\exp\left[-\left(\Delta G^{\ddagger} - F\Delta x_{\mathrm{U}}\right)/k_{\mathrm{B}}T\right] \tag{6.1}$$

where A is the frequency factor, k_{B} is the Boltzmann constant, and T is the absolute temperature. Similarly, the rate constant for the opposite process k_{F} (rate of folding under force. Eq. (6.2)) considers Δx_{F} as the distance to the transition state.

$$k_{\mathrm{F}}(F) = A\exp\left[-\left(\left(\Delta G^{\ddagger} + F\,\Delta x_{\mathrm{F}}\right)\right)/k_{\mathrm{B}}T\right] \tag{6.2}$$

according to the above equations, the energy derived from the application of force ($F\Delta x_{\mathrm{U}}$ or $F\Delta x_{\mathrm{F}}$) increases the unfolding rate while it lowers the refolding rate by lowering the energy barrier so that the unfolded state is stabilized in the new tilted energy landscape (Figure 6.6).

The most likely unfolding force F_{U}, representing the mechanical stability of the molecule, can be predicted by calculating the probability density for unfolding in terms of the unfolding rate constant [1, 25, 26]:

$$F_{\mathrm{U}} = \left(k_{\mathrm{B}}T/\Delta x_{\mathrm{U}}\right)\,\ln\left(r\Delta x_{\mathrm{U}}/k_{\mathrm{U}}{}^{0}k_{\mathrm{B}}T\right) \tag{6.3}$$

where $k_{\mathrm{U}}{}^{0}$ indicates the spontaneous unfolding rate (in the absence of force) and r refers to the loading rate, defined as:

$$r = \mathrm{d}F/\mathrm{d}t \approx v_{\mathrm{P}}K \tag{6.4}$$

where K is the spring constant of the cantilever (the stiffness contribution of the molecule is not considered here) and v_P the pulling speed. A dependence of the mechanical stability on the pulling speed indicates that the former is a kinetic property rather than a thermodynamic one.

The expression in Eq. (6.3) cannot be directly applied because the protein stiffness also depends on the applied force. Therefore, to obtain the kinetic parameters, the mechanical stability of a protein measured experimentally at different pulling speeds is matched with the stability observed on Monte Carlo simulations [22, 27, 28]. These simulations are performed at different pulling speeds using different values of Δx_U and k_U so that they fully reproduce the force dependence of the mechanical stability observed experimentally. The procedure models the entropic elasticity of the protein by using the wormlike chain equation (WLC, Section 6.4.1.1) and assumes a two-state unfolding process with a probability P of occurrence that depends on the unfolding rate k_U, the number of folded modules N, and the polling interval Δt [28] as follows:

$$P = N k_U \Delta t \tag{6.5}$$

Even though this model has been presented in the context of protein unfolding, it is also applicable to unbinding studies where the cantilever is used to mechanically separate two interacting molecules. In fact, the model was originally developed to study intermolecular interactions [25, 26] in order to obtain the k_{on} and k_{off} rate constants of bond formation and dissociation, respectively.

It is worth noting that, although the effect of force on bonds (both intra- and intermolecular) is basically to increase the rupture rates, the so-called catch bonds constitute an exception to this general rule [29]. In this case, stabilization of the bond is observed at low forces, while at higher forces, the behavior of the bond is shifted toward the usual one.

Even though the kinetic parameters obtained with this method provide a description of the energy landscape of the protein subjected to a force, some authors suggest the use of other kinetic models [1, 7]. Thus, while Eyring's model (which assumes a transition state similar to the native one) is applicable only to covalent bonds, other models such as Kramers rate theory should be more accurate in the case of proteins where noncovalent bonds are involved in the process [30].

6.3.2
General Considerations

As the mechanical stability of a protein is a kinetic parameter, it is not expected to correlate with its thermodynamic stability [23, 31]. When the force is applied to a protein, the unfolding process follows a different pathway to that in chemical unfolding [32, 33] crossing different unfolding barriers [32, 34]. Both the unfolding pathway and the energy barrier depend on the pulling geometry [35, 36]. This is due to the vectorial nature of force, which makes the mechanical stability of a protein dependent on the direction and point of application of the force [36], which allows the protein to explore different regions of its energy landscape through the imposed reaction coordinate, the end-to-end distance [2]. For simplicity, an orthogonal pulling geometry has been assumed in the above model. Although geometrical errors might be introduced in a real experiment when the sample is stretched at a slight angle, this assumption is valid since the expected effect would only be negligible in the typical systems studied (Section 6.4.1.2) [37].

Most SMFS experiments are carried out under nonequilibrium conditions as the typical pulling speed is faster than the spontaneous rate of equilibration of the molecule. Consequently, hysteresis can be observed when comparing stretching and relaxation traces, indicating that not all the mechanical work applied is directly converted into a free energy change, but rather, part has been dissipated during the process. Even in these nonequilibrium conditions, the behavior observed for some molecules is still relevant to *in vivo* processes in which relatively high pulling speeds are involved [7], although it may not be directly extrapolated when slower pulling speeds are present [38]. Nevertheless, using the Crooks fluctuation theorem, the unfolding free energy in nonequilibrium conditions [39, 40] can also be obtained, providing further insight into protein dynamics, including processes such as folding cooperativity [41].

In a mechanical unfolding experiment of a membrane protein, the molecule is usually picked up from its accessible regions (termini and/or loops) and extracted from the membrane by the applied force [42]. This method can detect numerous unfolding barriers, which surprisingly have been found not only in structured transmembrane elements but also in loops. These proteins can be picked up from either side of the membrane; thus, both sequences of events can be compared. The fact that the location and the presence of these barriers often depend on the pulling direction reflects different unfolding pathways, providing additional and complementary information of the process. Still, molecular dynamics (MD) simulations have shown that these unfolding barriers are contributed by both intramolecular (the unfolding of the protein structure) and protein–membrane intermolecular interactions and their specific effect cannot be decoupled [43].

6.3.3
Bond Dissociation Dynamics in Force Clamp

Because the rate of transition from the folded to the unfolded state is force dependent, maintaining the force parameter constant significantly simplifies the analysis and the understanding of the unfolding data. The addition of a constant force F generates an extra term $-F \cdot x$ in the free energy that lowers the magnitude of the energy barrier increasing the unfolding rate.

For constant force and in a two-state approximation, the probability of an unfolding event before a time t after the force is applied is $P(t) = 1 - \exp(-\alpha t)$ with $\alpha = \alpha_0 \exp(F \cdot \Delta x / k_B T)$.

For a simple estimation of the parameters α_0 and Δx, the unfolding experiment is repeated several hundred times over a range of forces and the time of each individual unfolding event τ_i is recorded. The unfolding rate $\alpha = 1/\langle\tau\rangle$ is then plotted against the force F and fitted to the model. This model is in rough agreement with early experiments [44].

6.3.3.1 The Need for Robust Statistics

The simplest means to examine the data produced by single-molecule experiments is to bin experimental results into a histogram and fit this to a given data model. However, this approach has been shown to introduce unnecessary bias. To take advantage of the extra information contained in single-molecule observations, it is highly recommendable to apply strict considerations to the analysis of the data in order to avoid the loss of information associated with binning and averaging as well as the possible censoring of the data that may introduce bias [45]. Data censoring refers to the possibility that the experimenter has access only to a subset of the existing events, due to the limited resolution or duration of the experiment. The bias generated by the censoring can be easily removed by incorporating the information regarding the experimental limitations into the data model. Nonparametric estimators have been developed to correct these biases on the single-molecule data, such as maximum likelihood [46].

To estimate the subtle characteristics of the unfolding process accurately, a more sophisticated method has been developed that addresses the experimental limitations and that takes advantage of the single-molecule nature of the data [46]. The main experimental limitation is the stability of the experiment (which does not typically allow the calibration of the force to be maintained for a time span longer than several seconds) and the interruption of the experiment by the spontaneous detachment of the sample from the probe or substrate. These two conditions limit the duration of the experiment to a time t_d and induce the censoring of any event of longer duration. Because it is only possible to observe events before t_d, the inverse of the average time $1/\langle\tau\rangle$ is not a reliable estimator for the unfolding rate. In a similar way, events that occur too fast at the beginning of the experiment may not be detected and would be excluded from the data set, inducing a bias.

The experimenter should assess the particular experimental limitations and implement an estimator that takes into account that the collected data corresponds to the probability of an event given that this event is observable. For example, if a protein unfolding event occurs before time t with probability $P_U(t) = 1 - \exp(-\alpha t)$, but the sample spontaneously detaches with probability $P_D(t) = 1 - \exp(-\beta \cdot t)$, then only a fraction $(\alpha/(\alpha + \beta))$ of the unfolding events will be observed with probability $P_O(t) = 1 - exp(-(\alpha + \beta)\ t)$, before the experiment is interrupted by the detachment. It is clear in this simple example that a naïve interpretation of the data could wrongly attribute $(\alpha + \beta)$ as the unfolding rate instead of α [20].

6.4 Specific versus Other Interactions

In SMFS experiments, it is obviously crucial to identify the intramolecular interactions present in the system under study (i.e., individual molecules) and rule out other unwanted interactions monitored by the sensor. Some recordings involve specific interactions but from multiple molecules while others involve nonspecific interactions originated from a variety of sources, such as desorption or stretching of denatured, aggregated, entangled proteins, or contaminants. Even if it is assumed that the sample is pure, which, in practice, is unfeasible, surface attachment could lead to partial or complete unfolding, misfolding, or aggregation [47, 48], which can mask the true single-molecule process under study. Thus, data selection is a must. In order to do this, specific criteria should be taken to avoid biasing the data. This imposes a challenge as a small number of adhesion events, with features that may resemble single-molecule interactions, are observed even in clean substrates [49]. To overcome this difficulty, samples with a measurable property that distinguishes them from other molecules or altered states are used. This distinctive property is known as a "*fingerprint*" and allows single-molecule events to be unequivocally selected [8].

6.4.1
Intramolecular Single-Molecule Markers

Intramolecular interactions of proteins, polysaccharides, and DNA show characteristic fingerprints because of their own structural phase transition. Most of the biopolymers in solution have been shown to behave as predicted by polymer elasticity models such as WLC. This feature by itself can help distinguish between a polymer being stretched from its ends or peeled off from a surface.

6.4.1.1 The Wormlike Chain: an Elasticity Model

The nonlinear force–extension relationship of a polymer being stretched, such as that obtained from an SMFS *length-clamp* experiment is well described by elasticity models of semiflexible polymers. The most widely used is the interpolated approximation to the WLC [50, 51], which has been successfully applied to model the elasticity of proteins, DNA, and RNA on stretching.

The WLC model considers the polymer as a flexible chain of length L_C to predict the restoring force F. This model also includes a parameter p known as the *persistence length* that is related to the rigidity of the molecule, that is, the distance through which the orientation of the polymer remains correlated [52] (for unfolded proteins $p \sim 0.4$ nm, the average value of the length of an amino acid [53]). Equation (6.6) shows the analytical expression of the interpolated approximation to the WLC model [50]:

$$F = \frac{k_B T}{4p}\left(1 - \frac{x}{L_0}\right)^{-2} - \frac{1}{4} + \frac{x}{L_0} \tag{6.6}$$

It is important to note that, when analyzing protein unfolding, multiple peaks can be observed (Figure 6.7). Thus, if the WLC equation is adjusted to consecutive peaks, the increase in contour length (ΔL_C) of the molecule corresponds to the increase in length between its folded and unfolded state (assuming a two-state unfolding process). This corresponds to the number of amino acids found in the unfolded region multiplied by the length contribution of each amino acid [37].

6.4.1.2 **Proteins**

MD simulations have shown that the application of force to a protein from its ends, first acts on a specific region of the protein. This is due to the existence of local secondary structure elements capable of withstanding the force applied and that oppose the unfolding of the whole module. These regions are usually connected by hydrogen bonds, and their number [23] (Figure 6.7) and side-chain packing in the hydrophobic core [54] are the major determinants of the mechanical stability of the module. Another key factor influencing the mechanical stability of a protein is the relative orientation of the force bearing region and the applied force direction [35, 36, 55]. Thus, a series of hydrogen bonds orthogonal to the force (shear topology) tends to show higher mechanical resistance than those arranged in the same direction as force (zipper topology).

A high degree of cooperativity seems to be present during the mechanical unfolding of a typical protein where the stability of secondary structures depends on the tertiary interactions. These tertiary interactions tend to be brittle as they need higher forces but smaller deformations to break, given their short distances to the transition state [1]. Thus, when the force-bearing region, known as a *mechanical clamp*, is broken, the protein usually unfolds completely (Figure 6.7).

The increase in contour length on the unfolding of a module is characteristic of each structured protein and it depends on the number of force-hidden amino

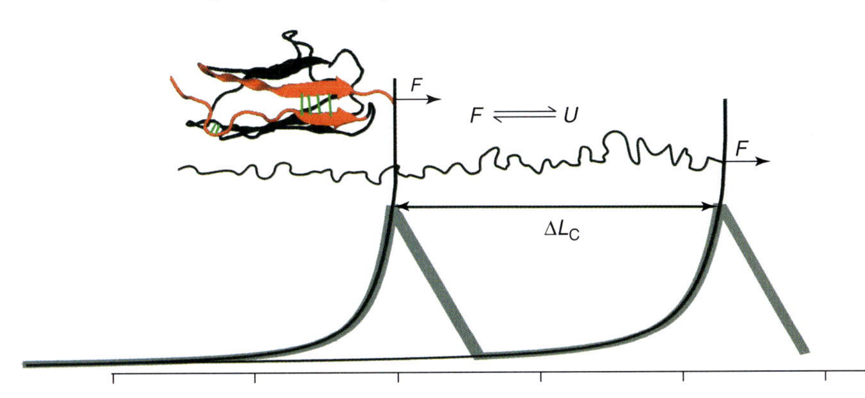

Figure 6.7 WLC model of polymer elasticity fitted on a force–extension trace. A protein fold can typically resist force due to the existence of local secondary structural elements (red strands), sustained by backbone hydrogen bonds (green lines), that form a mechanical clamp (the module depictured shows a shear mechanical clamp, which usually results in high mechanical resistance). Once the resistance of the mechanical clamp is overcome, the fold is broken and the protein unravels, resulting in a peak in a force–extension trace (gray curve). The contour length increment ΔL_C, which is related to the length of the sequence hidden inside the mechanical clamp (black strands of the module shown), can be measured by fitting the WLC (Eq. (6.6) black lines) to two consecutive peaks.

acids in its fold. Hence, it is a useful feature to identify a protein that has a well defined/unique 3D structure. Furthermore, a defined unique persistence length has also been used for single-molecule identification [56]. Although these two parameters are useful to identify single-molecule events originated from a specific protein, they should not be used alone. In such experiments, when the AFM tip is near the surface, many uncontrolled interactions tend to appear that result in a variety of force peaks in this noisy region of the recordings (first 30–70 nm). As the expected contour length of a typical protein of 100 amino acids is about 40 nm, its unfolding events can be masked by uncontrolled interactions that contaminate the data. In addition to this, different proteins can perfectly have similar contour lengths. Moreover, the persistence length can be different for different proteins and even a single protein can display several persistence lengths [57, 58].

A successful strategy to overcome these problems is based on the use of proteins containing multiple repeats of the desired module or whole protein [22] known as

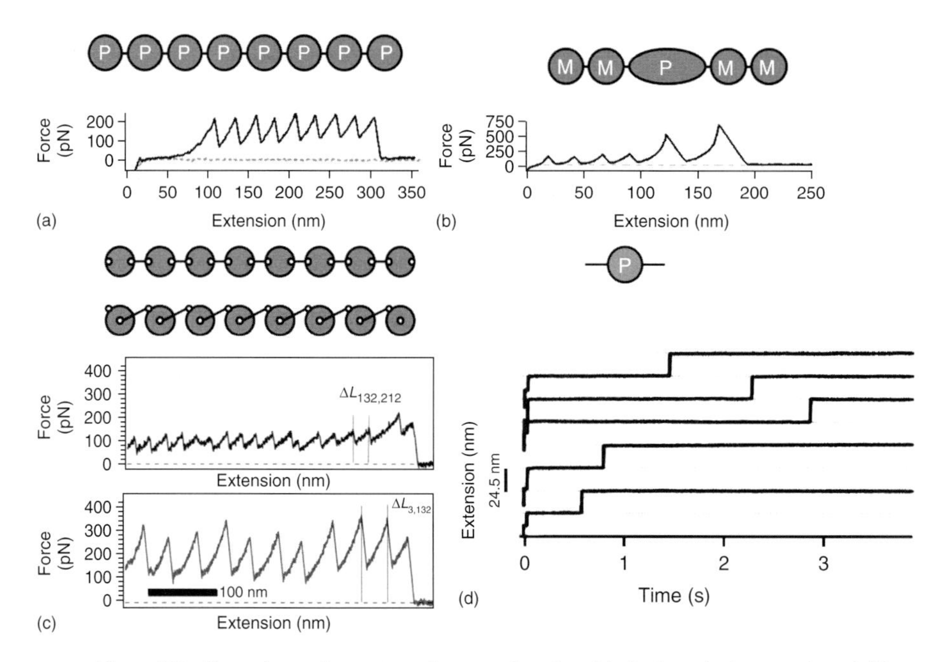

Figure 6.8 The polyprotein strategy. A polyprotein with eight repeats of the protein of interest (a) produces a saw-tooth pattern in which the force peaks are equally spaced and they present a similar maximum force. When a chimeric polyprotein is used (b), the protein under study (P) is flanked by repetitions of a marker (M) protein with known mechanical properties. Different stretching directions can be achieved by synthesizing a polyprotein at the protein level by disulfide bonding (c). By introducing cysteines (white circles in the schematics) into a protein monomer, it is possible to build a polyprotein where the modules are linked through different residues, which results in different mechanical behaviors. A single protein monomer can be studied (d) when the sample is properly engineered and the analysis is carefully designed. ((c) Figure taken from [60] with permission and (d) Modified from [61] with permission.)

polyproteins (Figure 6.8a). Each unfolding event provides a similar force peak and increases the total protein length in a periodic manner; thus, the saw-tooth pattern (fingerprint) produced allows the protein to be unambiguously identified. In the case of *force clamp*, the fingerprint is a staircase pattern of identical length steps over time. Furthermore, long modular proteins are ideal systems as most unfolding events occur at a considerable distance from the surface, avoiding the initial region of the recordings which is typically dominated by uncontrolled interactions. Moreover, although surface interactions may destabilize some repeats, the ones that remain folded can readily be analyzed. An additional advantage of polyproteins is that several data points are obtained from a single molecule, which augments the yield in terms of data acquisition. Finally, their repetitive nature allows some spatial features to be amplified that might otherwise be overlooked [59].

There are a few drawbacks associated with the polyprotein strategy which have been addressed using different approaches. Firstly, the synthesis of a polyprotein at the DNA level is laborious and requires a long time. Furthermore, if the protein of interest is insoluble, the corresponding polyprotein may complicate its production even more because of the increase in the local concentration of the monomer. Both problems can be resolved by introducing the protein sequence of interest into a cloning vector that contains the DNA sequence for a polyprotein (Figure 6.8b). This results in a chimeric protein containing the desired protein flanked by repetitions of a known module that acts as an internal standard and allows single molecules to be identified [62]. This strategy may also promote the solubility of the protein under study [62], although it reduces the yield to a single data point from each molecule.

On the other hand, the synthesis of polyproteins by recombinant DNA methods imposes the N- and C-termini of the protein as the points of force application. As the mechanical response depends on the pulling direction, this restricts SMFS analysis to a limited region of the energy landscape. By producing monomeric protein modules with cysteine residues added to the desired pulling points, a polyprotein can be generated at the protein level by inducing the formation of disulfide bonds between different monomers [60, 63]. This allows the protein to be stretched from virtually any given pair of residues (Figure 6.8c). Finally, although one could think that spurious interactions between the protein modules of a polyprotein may give rise to artifacts [64, 65], in experiments using the monomer (Figure 6.8d), instead of the previously characterized polyprotein, these effects were found negligible [61].

It should be noted that in a polyprotein or modular protein, since all protein modules are exposed to the force, the unfolding events are typically ordered in increasing stability when using *length-clamp* mode. Thus, when a chimeric polyprotein is used, if the protein is less stable than the marker its unraveling falls into the initial region; hence, it may be masked by uncontrolled interactions. Although this issue has not yet been optimally resolved (under some precautions, it allows the right selection of single-molecule recordings [57, 58]), a new strategy has recently been developed to study the mechanical behavior of low-stability and mechanically polymorphic proteins [66]. By protecting the protein of interest inside the force-hidden region of another protein, the order of the stabilities in the recording can readily be reverted [66, 67].

6.4.1.3 DNA and Polysaccharides

SMFS on DNA and polysaccharides also originates phase transitions, which constitute their fingerprint. In both cases, as the tip–surface distance increases, the force rises until a plateau can be observed. This feature is associated to different structural transitions that are characterized by large end-to-end variations at a given force value. In the case of double-stranded DNA in the B conformation, the plateau is reached at about 65–70 pN, which is associated with a transition to an overstretched DNA conformation [68] (Figure 6.9a). This was also seen using optical tweezers [69, 70]. Similarly, a plateau in the force–extension trace of certain polysaccharides has been related to a chair-boat or boat-inverted chair conformational transition [71–73] (Figure 6.9b).

6.4.2 Intermolecular Single-Molecule Markers

When dealing with intermolecular interactions, the general approach differs slightly from the one already described (Section 6.2). The usual procedure requires the specific functionalization of both the substrate and the tip, each with a different element of the interacting pair, so that the interaction can be established when both the tip and the surface are brought into contact and is broken on retraction of the actuator [74–76]. In these experiments, it is crucial to immobilize (see below) the interacting molecules in the correct orientation and with enough mobility to ensure that the interaction occurs.

As mentioned, on stretching two interacting molecules, a single force peak is obtained from each recording. The lack of a repetitive pattern as a fingerprint in these recordings as an internal control makes necessary the use of external controls

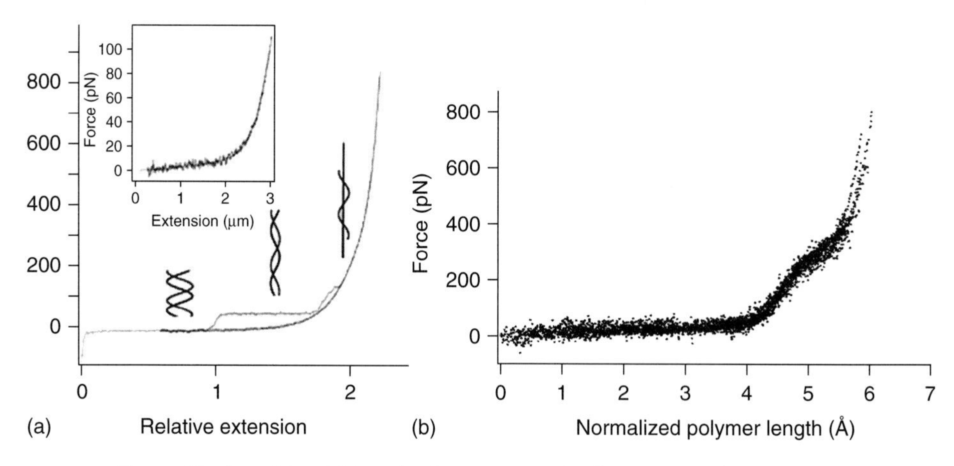

Figure 6.9 Stretching DNA and polysaccharides. Both these biopolymers yield a force–extension trace with a plateau. This characteristic fingerprint originates from the structural transition suffered by these molecules corresponding to a B–S transition of overstretched DNA (a) or chair-boat transition of polysaccharide rings (b). (Figures obtained from [68] and [71] with permission.)

to identify the real interaction events. These include the use of structural analogs, inhibitors, or mutants, which by comparison with the unmodified system allow to assess the specificity of the events. Furthermore, the only feature that is used to identify the rupture of specific events is the contour length of the rupture peak, which corresponds to the stretched length of the interacting pair.

However, even with the use of external controls, it is hard to distinguish single from multiple (acting simultaneously) binding events. The analysis of the distribution of events provides information to this end. Thus, when the force values obtained from these experiments are represented in a histogram of rupture forces, a multimodal distribution is typically obtained (Figure 6.10). These peaks tend to appear at integer multiples of the lowest force value (assumed to be the elementary event); thus, they are interpreted as simultaneous disruption events involving different numbers of interacting molecules [77–79]. Nevertheless, it should be noted that the lowest force peak might not actually represent the elementary event but rather the smallest resolvable one, which could still result from multiple interacting pairs.

Because of these limitations, immobilization techniques are critical for these studies. Therefore, different protocols usually involving bifunctional groups are used in combination with spacer molecules in order to properly functionalize the tip as well as to minimize the possible unspecific interactions between the tip and surface.

The hydrophilic polymer polyethylene glycol (PEG) is one of the most used spacers for the tip functionalization. PEG ends are usually chemically modified in order to provide two different reactive groups that allow to establish directed covalent bonds with specific molecules. Thus, PEG can be used to bind other bifunctional molecules that directly bind both the tip and the molecule to be functionalized on it. A further advantage of PEG is that a range of lengths are available, based on the molecular weights of the different number of monomeric units present in each linear polymer.

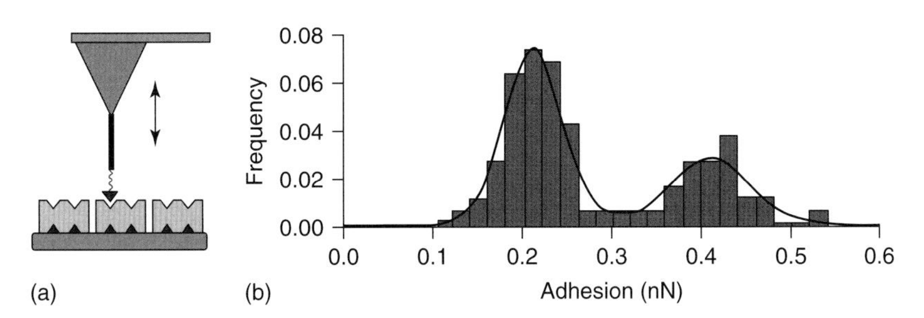

Figure 6.10 Analysis of intermolecular interactions by AFM–SMFS. The usual experimental approach (a) is based on the functionalization of both the tip and the surface with the interacting molecules. When the tip carrying one of the interacting molecules is approached to its pair on the substrate, the interaction is established, while the binding force is measured as a peak when the tip is pulled away. After plotting in a histogram the force peaks obtained, a multimodal distribution is often seen (b). These modes are attributed to different numbers of interacting molecules, and the lowest force value is assumed to be the single interaction. (Figures obtained from [78] with permission.)

The nature of the bonds established between the tip, the spacer, and the molecule under the study depend on the reactive groups present in each element. Thiol groups from cysteine residues present in some proteins can react with a double bond present in maleimide, while its amino groups can bind ester groups such as those present in *N*-hydroxysuccinimide (NHS) [80]. Other methods also include the use of monofunctional PEG molecules in order to saturate the reactive groups present in the tip, which passivates the tip interactions reducing unspecific interactions [81].

6.5 Steered Molecular Dynamics Simulations

MD is a technique that allows the movement of individual particles to be simulated on the basis of their motion equations (Newton's equation for classical particles, Schrödinger's for quantum mechanics). As MD can also be applied to molecules (with the right potential functions), it has largely been applied to the field of single-molecule studies since these simulations reproduce (and sometimes even predict [82]) quite well, the results obtained experimentally, while providing atomic (or even subatomic) detail of the processes. However, this technique is known to be efficient only for short timescales and for a small number of particles, typically one molecule. Thus, single-molecule experiments are ideal for such comparisons.

MD simulations originally monitor every atom in a system, compute the forces acting on each of them, and allow them to move in space accordingly. Nevertheless, this technique is very demanding on resources and it is therefore only applicable to small molecules. Thus, in order to simulate the complete unfolding of a protein module in a reasonable amount of computational time, very large pulling speeds are used that result in overestimated forces [82]. In order to extend the simulation times as well as the molecule sizes, approximations are needed. One of the first approximations developed was the removal of the solvent and consequent modification of the force field to take into account the solvent effects. The generalized born surface area [83] is one of the most commonly used implicit solvent approximations, which was recently used to simulate the largest molecule studied to date (the C-cadherin ectodomain, PDB code 1L3W [84]).

Further simplifications can be found in the so-called coarse-grained simulations, where amino acids are simplified to a single particle at the position of the α-carbons [85]. Considering the connections in the crystallized (assumed to be native) structures, pulling is simulated over shorter times than in the former all-atom approaches (both with explicit and implicit solvents). Hence, one can simulate lower velocities and reach conditions closer to those of the experimental approaches. Using this method, a survey of the PDB ungapped structures with no more than 250 amino acids has recently been reported ([85]: periodically updated at *http://info.ifpan.edu.pl/BSDB/*).

Nevertheless, apart from using good models and approximations, computer speed must increase in the future to be able to simulate real experimental pulling speeds and times and to obtain results that are not only qualitatively but also quantitatively comparable [86].

6.6 Biological Findings Using AFM–SMFS

Far from being just a technological feat, AFM–SMFS is currently used to tackle problems from different scientific fields. In this section, a few examples of important contributions of AFM–SMFS studies are discussed.

6.6.1 Titin as an Adjustable Molecular Spring in the Muscle Sarcomere

The molecular determinants of a macroscopic biological property, the passive elasticity of myofibrils, were described at the single-molecule level through a reductionist approach (Figure 6.11a). When muscles are stretched under physiological conditions, the sarcomere (the basic unit of muscle) can restore its initial dimensions after muscle relaxation [87] thanks to a property called *passive elasticity*. Titin (the third filament of the sarcomere) is responsible for this behavior, acting as a molecular spring [88, 89]. This spring has a multistage nonlinear behavior [58], which has been studied in detail and described using an additive model, as follows.

Titin is a giant protein ($\sim 1\,\mu m$ long) that spans half a sarcomere, from the *Z*-disk to the M-line, although only its I-band displays elastic behavior *in vivo* (Figure 6.11a). This elastic region has two types of components. The first one is formed by unique unstructured segments known as *N2B* [58] and *PEVK* [57], random coil structures that show featureless force-extension curves on stretching. These are true elastic elements that act as reversible springs that store/restore elastic energy with very high efficiency, with no heat dissipation. The second component involves the proximal and distal immunoglobulin (Ig) domain regions of the I-band. This region has also been studied by SMFS; in particular, the mechanics of the I27–I34 region showed a series of reversible unfolding events with mechanical stabilities ranging from 150 to 300 pN at $1\,nm\,ms^{-1}$ [92]. Furthermore, the mechanics of the I27 module have been studied in great detail to the extent that it is now considered a model in the field of protein nanomechanics.

The I27 module is 89 amino acids long and it has a typical Ig topology with a β-sandwich fold composed of seven strands (from A to G, all adjacent strands are antiparallel, except for A′G) that fold into two face-to-face β-sheets through hydrophobic core interactions [93]. SMFS of $(I27)_{12}$, a polyprotein-containing 12 I27 repeats, using the *length-clamp* mode revealed force–extension curves with force peaks of 204 ± 26 pN (at $0.6\,nm\,ms^{-1}$) [22]. The saw-tooth pattern obtained displayed a periodic increase in the polymer length (ΔL_C) of 28.4 ± 0.3 nm, which corresponds to the length of the force hidden region located after the A′G patch. Remarkably, all experimental data obtained is very consistent with the theoretical predictions done by MD simulations [82], which identified two unfolding barriers

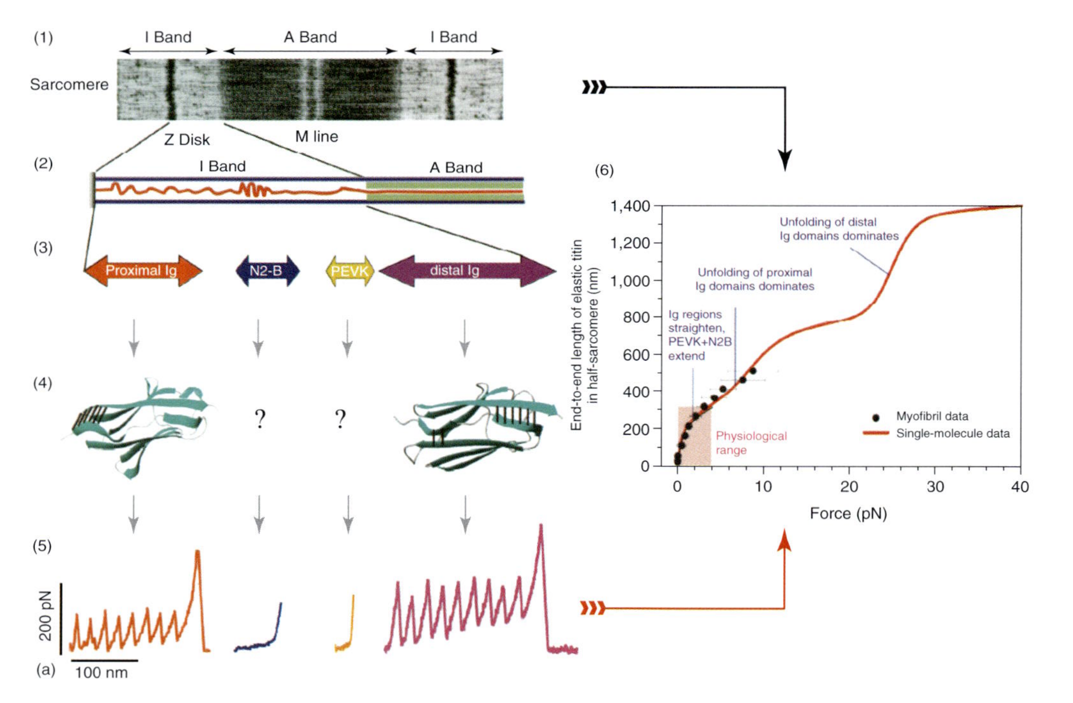

Figure 6.11 SMFS applications in biology. (a) Reverse engineering of titin: scaling-up the mechanical properties of representative elements from the titin elastic region I-band readily accounts for the passive elasticity: proximal Ig domains (left); N2B; PEVK; and distal Ig domains (right, I27 crystal structure as a representative module). In the sarcomere, the physiological force range for titin elasticity results from the unstructured N2B/PEVK entropic springs and straightening of the Ig region. (b) Monitoring the folding pathway of single polyubiquitin molecules. In force clamp mode an initial high-force pulse unfolds the protein (characterized by step increases of 20 nm). (c) Exploring single disulfide bond reduction catalyzed by single Trx molecules. (c1) S–S bond is exposed to the solvent after unfolding the modified $I27_{SS}$ in a first force pulse (~10.5 nm). Later on, in the presence of Trx, length increase steps of ~13.5 nm are observed, corresponding to the residues trapped behind S–S bonds. (c2) Typical recording showing the unfolding and subsequent S–S reduction by single human Trxs. ((a) Figure obtained from [4] with permission; (b) Figure taken from [90] with permission; and (c) Figure obtained from [91] with permission.)

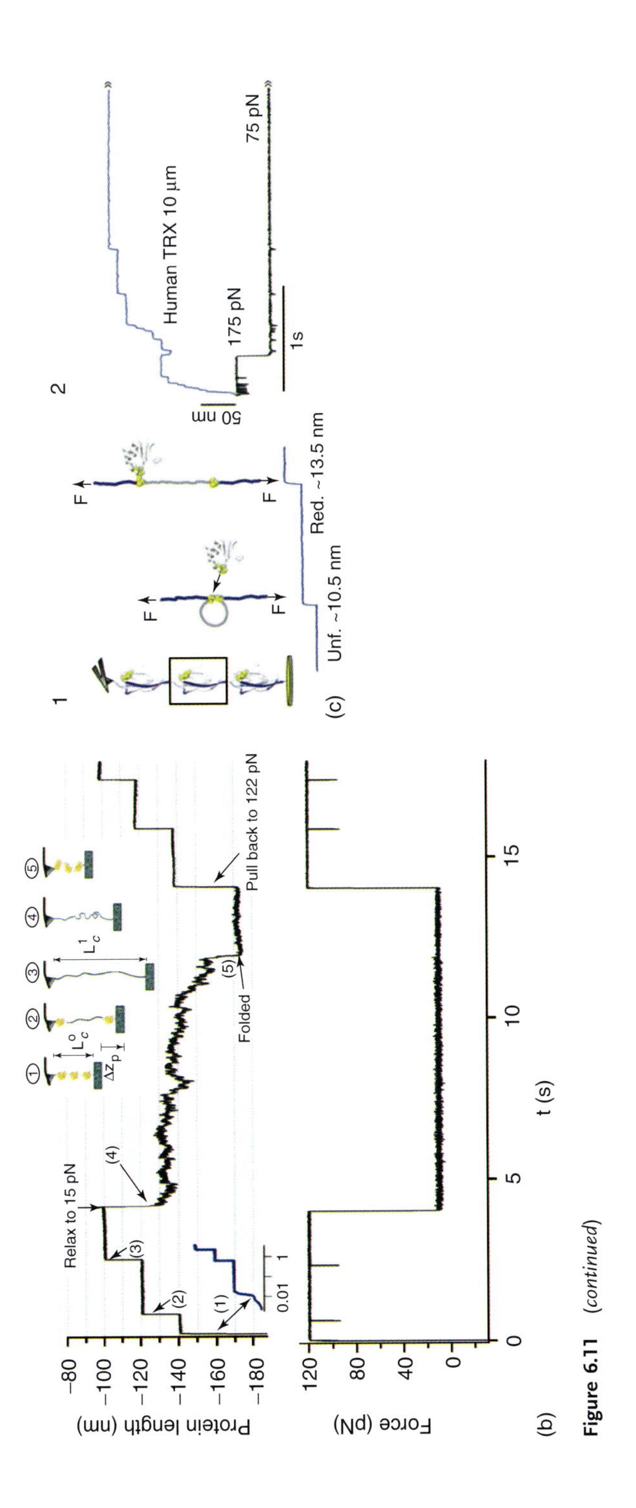

Figure 6.11 *(continued)*

on stretching. First, a minor mechanical barrier (A-B patch composed by two hydrogen bonds) that is experimentally detected as a deviation of the expected WLC ("hump") and that corresponds to an intermediate that refolds after each major unfolding event, when the force drops [59]. Second, the major barrier (that corresponds with the force peak in each force–extension trace) is composed of six hydrogen bonds linking strands A′ and G. These bonds are arranged in a "shear" topology so that to unfold the module, the six hydrogen bonds must be broken simultaneously. Once this barrier is broken, the remaining structure, including the hydrophobic core, unravels without much resistance [82].

Scaling-up from the mechanical behavior of the representative components of the elastic region of titin [58] the passive elasticity, a macroscopic property, has been recomposed. At the physiological force regime of sarcomere extension, the straightening of the Ig region together with the entropic spring effect of PEVK and N2B predominates, while at higher, nonphysiological forces (overextension), the reversible unfolding of Ig domains would act as a "shock absorber," a safety mechanism to maintain the structural integrity of the sarcomere [58] (Figure 6.11a).

6.6.2
Monitoring the Folding Process by Force-Clamp Spectroscopy

Resolving the protein folding problem is one of the major challenges in biology. Recently, *force-clamp* SMFS was used to explore the folding pathways of different proteins, such as ubiquitin. In a pioneering work, constant force was applied to a single polyprotein composed of nine ubiquitin repeats [90]. In this study, polyprotein molecules were initially unfolded by applying a high force. In a following step, the force was relaxed so the refolding of ubiquitin domains could be directly monitored by measuring the changes in the end-to-end length of the protein (Figure 6.11b). The folding time was found to be dependent on both the force applied and the length of the unfolded protein. It was observed that ubiquitin folding in *force clamp* occurs through a series of continuous stages, challenging the classic view of folding through well-defined discrete stages. It should be noted that the restraints imposed by the attachments of the molecule in SMFS experiments could influence the folding process so that it may actually proceed along different pathways from that in solution.

6.6.3
Intermolecular Binding Forces and Energies in Pairs of Biomolecules

Many cellular functions (including cell adhesion, motility, immunological antigen–antibody reactions, or pathogen entrance into target cells) are mediated by specific interactions between two biomolecules. In some of them, mechanical forces are involved. A classic example of intermolecular interactions is the study of the unbinding process of the biotin–(strept)avidin pair [77, 94]. Using AFM–SMFS (Figure 6.10), the force of interaction between these systems was measured providing a different point of view to this classical model. On the basis of the analysis of forced unbinding recordings from these molecules, average unbinding forces of 160 ± 20 and 257 ± 25 pN [77, 94] were found for biotin–avidin and biotin–streptavidin, respectively.

In a more recent study, the interaction force between the catalytic core of a proteasome and an unstructured substrate (tagged β-casein) has been measured [95]. The proteasomes were deposited over a mica substrate while different tips were functionalized with two specific protocols: an engineered cysteine residue in β-casein used to establish a thiol-Au bond with gold-coated tips and PEG-based functionalization with SiN tips. In order to avoid the direct interaction between the tip and the proteasomes while recording the force curve, a three-step procedure was developed. First, the tip is brought into contact with the sample in order to measure its relative position. This is taken as position zero and used in the following approach step as a reference to bring the tip to a known distance from the substrate (the proteasome entrance) while the force curve is recorded. Finally, the first step is repeated in order to ensure that no significant drift affected the measurement. This protocol not only minimizes the unspecific interactions while recording the force curve but also allows to measure protein–proteasome interaction.

Using this approach, the reported interaction forces between the proteasome and an unfolded protein were higher than 100 pN for both active and inactive mutant proteasomes. Furthermore, only active proteasomes showed an additional interacting force of about 250 pN, which may reflect a dynamic structural change occurring in the interior of the proteasome.

Finally, further biological systems studied include the SNARE [96], antigen–antibody [97], protein–peptide [98], and protein–DNA [99] complexes.

6.6.4
New Insights in Catalysis Revealed at the Single-Molecule Level

Single-molecule techniques are also emerging as an important complementary approach to study the dynamics of enzyme–substrate interactions at the sub-angstrom scale. In these studies, a constant force was applied to an engineered substrate (a polyprotein of I27 domains with a nonnative disulfide bond between two exogenous cysteines). The stretching of this substrate produces the partial unfolding of the module and exposes the disulfide bond to the solvent. In the presence of the enzyme thioredoxin (Trx), this individual bond is reduced by the action of one molecule of the enzyme, which is identified by the extension of the residues trapped behind the disulfide bond (Figure 6.11c). Interestingly, the different catalytic mechanisms that exist can be readily distinguished by their sensitivity to the force applied [91] allowing the evolution of Trx catalysis to be studied by comparing different Trx enzymes. In the low-force regime, both prokaryotic and eukaryotic Trxs present characteristic Michaelis–Menten kinetics. However, when a disulfide bond is stretched at higher forces, eukaryotic Trxs follow a single-electron transfer mechanism (SET) whereby prokaryotic Trxs show SET and nucleophilic substitution (SN2), which might be related to the capacity of these organisms to live in extreme environments [100].

6.7 Concluding Remarks

In little more than a decade, AFM–SMFS has been revealed as a powerful biophysical technique that can be used to tackle many different problems in a wide range of scientific areas. A clear example can be found in the studies of the giant protein titin, where this tool has been used to elucidate the molecular determinants of passive elasticity of muscle (a macroscopic physiological process) at the single-molecule level. Furthermore, AFM–SMFS is becoming essential to obtain new information on classic problems such as protein folding or enzyme catalysis at single-bond resolution. Nevertheless, some limitations should be overcome to fully exploit the potential of this new tool. On one hand, force and temporal resolution as well as general stability should be improved in order to fully cover the ranges relevant to most biological processes. On the other hand, better functionalization protocols and single-event markers to study intermolecular interactions should be developed in order to obtain more robust data in these systems.

Acknowledgments

We thank MC-V laboratory for their insights and helpful discussion. This work was funded by grants from the Ministerio de Ciencia e Innovación **(BIO2007-67116)**, Consejería de Educación de la Comunidad de Madrid **(S-0505/MAT/0283)**, and Consejo Superior de Investigaciones Científicas **(200620F00)** to MC-V. AG-P and R. Hervás are recipients of fellowships from the Fundación Ferrer (Severo Ochoa's fellowship) and AG-S is supported by a JAE-pre fellowship from Consejo Superior de Investigaciones Científicas.

Disclaimer

For the sake of simplicity and in order to keep the list of references short, often reviews have been preferred to primary sources. We apologize to those authors whose primary work was not cited here.

References

1. Bustamante, C., Chemla, Y.R., Forde, N.R., and Izhaky, D. (2004) Mechanical processes in biochemistry. *Annu. Rev. Biochem.*, **73**, 705–748.
2. Borgia, A., Williams, P.M., and Clarke, J. (2008) Single-molecule studies of protein folding. *Annu. Rev. Biochem.*, **77**, 101–125.
3. Ritort, F. (2006) Single-molecules experiments in biological physics: methods and applications. *J. Phys. Condens. Matter*, **18**, 531–583.
4. Oberhauser, A.F. and Carrión-Vázquez, M. (2008) Mechanical biochemistry of proteins one molecule at a time. *J. Biol. Chem.*, **283**, 6617–6621.
5. Basché, T., Nie, S., and Fernandez, J.M. (2001) Single molecules. *Proc. Natl. Acad. Sci. U.S.A.*, **98**, 10527–10528.

6. Hoel, P.G. (1984) *Introduction to Mathematical Statistics*, Wiley Series in Probability and Mathematical Statistics Includes Index, 5th edn, John Wiley & Sons, Inc., New York.
7. Howard, J. (2001) *Mechanics of Motor Proteins and the Cytoskeleton*, Sinauer Associates, Sunderland, MA.
8. Carrión-Vázquez, M., Oberhauser, A.F., Díez, H., Hervás, R., Oroz, J., Fernández, J., and Martínez-Martín, D. (2006) in *Advanced Techniques in Biophysics* (eds J.L.R. Arrondo and A. Alonso), Springer-Verlag, Heidelberg, pp. 163–245.
9. Schlierf, M., Berkemeier, F., and Rief, M. (2007) Direct observation of active protein folding using lock-in force spectroscopy. *Biophys. J.*, **93** (11), 3989–3998.
10. Oberhauser, A.F., Hansma, P.K., Carrión-Vázquez, M., and Fernández, J.M. (2001) Stepwise unfolding of titin under force-clamp atomic force microscopy. *Proc. Natl. Acad. Sci. U.S.A.*, **98**, 468–472.
11. Burnham, N.A. and Colton, R.J. (1989) Measuring the nanomechanical properties and surface forces of materials using an atomic force microscope. *J. Sci. Technol. A*, **7**, 2906–2913.
12. Evans, E. and Ritchie, K. (1999) Strength of a weak bond connecting flexible polymer chains. *Biophys. J.*, **76**, 2439–2447.
13. Oberhauser, A.F., Marszalek, P.E., Carrion-Vazquez, M., and Fernandez, J.M. (1999) Single protein misfolding events captured by atomic force microscopy. *Nat. Struct. Biol.*, **6**, 1025–1028.
14. Valbuena, A., Oroz, J., Vera, A.M., Gimeno, A., Gómez-Herrero, J., and Carrión-Vázquez, M. (2007) Quasi-simultaneous imaging/pulling analysis of single polyprotein molecules by atomic force microscopy. *Rev. Sci. Instrum.*, **78**, 113707.
15. Rief, M., Gautel, M., Oesterhelt, F., Fernandez, J.M., and Gaub, H.E. (1997) Reversible unfolding of individual titin immunoglobulin domains by AFM. *Science*, **276**, 1109–11112.
16. Dufrêne, Y. and Hinterdorfer, P. (2007) Recent progress in AFM molecular recognition studies. *Eur. J. Physiol.*, **256**, 237–245.
17. Merkel, R., Nassoy, P., Leung, A., Ritchie, K., and Evans, E. (1999) Energy landscapes of receptor-ligand bonds explored with dynamic force spectroscopy. *Nature*, **397**, 50–53.
18. Bustamante, C., Macosko, J.C., and Wuite, G.J.L. (2000) Grabbing the cat by the tail: manipulating molecules one by one. *Nat. Rev. Mol. Cell Biol.*, **1**, 130–136.
19. Gittes, F. and Schmidt, C.F. (1998) Thermal noise limitations on micromechanical experiments. *Eur. Biophys. J.*, **27**, 75–81.
20. Hermans, R. (2010) Experimental study of single protein mechanics and protein rates of unfolding. PhD thesis. Department of Graduate School of Arts and Sciences, Columbia University.
21. Ng, S.P., Randles, L.G., and Clarke, J. (2007) Single molecule studies of protein folding using atomic force microscopy. *Methods Mol. Biol.*, **359**, 139–167.
22. Carrion-Vazquez, M., Oberhauser, A.F., Fowler, S.B., Marszalek, P.E., Broedel, S.E., Clarke, J., and Fernandez, J.M. (1999) Mechanical and chemical unfolding of a single protein: a comparison. *Proc. Natl. Acad. Sci. U.S.A.*, **96**, 3694–3699.
23. Valbuena, A., Oroz, J., Hervás, R., Vera, A.M., Rodríguez, D., Menéndez, M., Sulkowska, J.I., Cieplak, M., and Carrión-Vázquez, M. (2009) On the remarkable mechanostability of scaffolding and the mechanical clamp motif. *Proc. Natl. Acad. Sci. U.S.A.*, **106**, 13791–13796.
24. Oberbarnsheit, L., Janissen, R., and Oesterhelt, F. (2009) Direct and model free calculation of force-dependent dissociation rates from force spectroscopy data. *Biophys. J.*, **97**, L19–L21.
25. Bell, G.I. (1978) Models for the specific adhesion of cells to cells. *Science*, **200**, 618–627.
26. Evans, E. and Ritchie, K. (1997) Dynamic strength of molecular adhesion bonds. *Biophys. J.*, **72**, 1541–1555.

27. Oberhauser, A.F., Marszalek, P.E., Erickson, H.P., and Fernandez, J.M. (1998) The molecular elasticity of tenascin, an extracellular matrix protein. *Nature*, **393**, 181–185.
28. Rief, M., Fernandez, J.M., and Gaub, H.E. (1998) Elastically coupled two-level systems as a model for biopolymer extensibility. *Phys. Rev. Lett.*, **81**, 4764–4767.
29. Marshall, B.T., Long, M., Pieper, J.W., Yago, T., McEver, R.P., and Zhu, C. (2003) Direct observation of catch bonds involving cell-adhesion molecules. *Nature*, **423**, 190–193.
30. Bieri, O. and Kiefhaber, T. (2000) in *Frontiers in Molecular Biology: Mechanisms of Protein Folding*, 2nd edn (ed. R. Pain), Oxford University Press, Oxford, pp. 34–64.
31. Carrion-Vazquez, M., Oberhauser, A.F., Fisher, T.E., Marszalek, P.E., Li, H., and Fernandez, J.M. (2000) Mechanical design of proteins studied by single-molecule force spectroscopy and protein engineering. *Prog. Biophys. Mol. Biol.*, **74**, 63–91.
32. Best, R.B., Li, B., Steward, A., Daggett, V., and Clarke, J. (2001) Can non-mechanical proteins withstand force? Stretching barnase by atomic force microscopy and molecular dynamics simulation. *Biophys. J.*, **81**, 2344–2356.
33. Fowler, S.B., Best, R.B., Toca-Herrera, J.L., Rutherford, T.J., Steward, A., Paci, E., Karplus, M., and Clarke, J. (2002) Mechanical unfolding of a titin Ig domain: structure of unfolding intermediate revealed by combining AFM, molecular dynamics simulations, NMR and protein engineering. *J. Mol. Biol.*, **322**, 841–849.
34. Brockwell, D.J., Beddard, G.S., Clarkson, J., Zinober, R.C., Blake, A.W., Trinick, J., Olmsted, P.D., Smith, D.A., and Radford, S.E. (2002) The effect of core destabilization on the mechanical resistance of I27. *Biophys. J.*, **83**, 458–472.
35. Brockwell, D.J., Paci, E., Zinober, R.C., Beddard, G.S., Olmsted, P.D., Smith, D.A., Perham, R.N., and Radford, S.E. (2003) Pulling geometry defines the mechanical resistance of a beta-sheet protein. *Nat. Struct. Biol.*, **10**, 731–737.
36. Carrion-Vazquez, M., Li, H., Lu, H., Marszalek, P.E., Oberhauser, A.F., and Fernandez, J.M. (2003) The mechanical stability of ubiquitin is linkage dependent. *Nat. Struct. Biol.*, **10**, 738–743.
37. Carrion-Vazquez, M., Marszalek, P.E., Oberhauser, A.F., and Fernandez, J.M. (1999) Atomic force microscopy captures length phenotypes in single proteins. *Proc. Natl. Acad. Sci. U.S.A.*, **96**, 11288–11292.
38. Williams, P.M., Fowler, S.B., Best, R.B., Toca-Herrera, J.L., Scott, K.A., Steward, A., and Clarke, J. (2003) Hidden complexity in the mechanical properties of titin. *Nature*, **422**, 446–449.
39. Liphardt, J., Dumont, S., Smith, S.B., Tinoco, I., and Bustamante, C. Jr. (2002) Equilibrium information from nonequilibrium measurements in an experimental test of Jarzynski's equality. *Science*, **296**, 1832–1853.
40. Collin, D., Ritort, F., Jarzynski, C., Smith, S.B., Tinoco, I. Jr., and Bustamante, C. (2005) Verification of the Crooks fluctuation theorem and recovery of RNA folding free energies. *Nature*, **437**, 231–234.
41. Shank, E.A., Cecconi, C., Dill, J.W., Marqusee, S., and Bustamante, C. (2010) The folding cooperativity of a protein is controlled by its chain topology. *Nature*, **465**, 637–640.
42. Engel, A. and Gaub, H. (2008) Structure and mechanics of membrane proteins. *Annu. Rev. Biochem.*, **77**, 127–148.
43. Cieplak, M., Filipek, S., Janovjak, H., and Krzyœko, A. (2006) Pulling single bacteriorhodopsin out of a membrane: comparison of simulation and experiment. *Biochem. Biophys. Acta*, **1758**, 537–544.
44. Schlierf, M., Li, H., and Fernandez, J.M. (2004) The unfolding kinetics of ubiquitin captured with single-molecule force-clamp techniques. *Proc. Natl. Acad. Sci. U.S.A.*, **101**, 7299–7304.
45. Koster, D.A., Wiggins, C.H., and Dekker, N.H. (2006) Multiple events

on single molecules: unbiased estimation on single-molecule biophysics. *Proc. Natl. Acad. Sci. U.S.A.*, **103**, 1750–1755.

46. Brujic, J., Hermans, R.I., Walther, K.A., and Fernandez, J.M. (2006) Single-molecule force spectroscopy reveals signatures of glassy dynamics in the energy landscape of ubiquitin. *Nat. Phys.*, **2**, 282–286.
47. Buijs, J., Norde, W., and Lichtenbelt, J.W.T. (1996) Changes in the secondary structure of adsorbed IgG and F(ab')(2) studied by FTIR spectroscopy. *Langmuir*, **12**, 1605–1613.
48. Hlady, V. and Buijs, J. (1998) in *Biopolymers at Interfaces* (ed. M. Malmsten), Dekker, New York, pp. 181–220.
49. Weisenhorn, A.L., Maivald, P., Butt, H.J., and Hansma, P.K. (1992) Measuring adhesion, attraction, and repulsion between surfaces in liquids with an atomic-force microscope. *Phys. Rev. B. Condens. Matter*, **45**, 11226–11232.
50. Bustamante, C., Marko, J.F., Siggia, E.D., and Smith, S. (1994) Entropic elasticity of lambda-phage DNA. *Science*, **265**, 1599–1600.
51. Marko, J.F. and Siggia, E.D. (1995) Stretching DNA. *Macromolecules*, **28**, 8759–8770.
52. Grosberg, A.I.U. and Khokhlov, A.R. (1994) *Statistical Physics of Macromolecules*, AIP Series in Polymers and Complex Materials, AIP Press, New York, translated by Y.A. Atanov.
53. Ainavarapu, S.R.K., Brujic, J., Huang, H.H., Wiita, A.P., Lu, H., Li, L., Walther, K.A., Carrión-Vázquez, M., and Fernández, J.M. (2007) Contour length and refolding rate of a small protein controlled by engineered disulfide bonds. *Biophys. J.*, **92**, 225–233.
54. Ng, S.P., Rounsevell, R.W.S., Steward, A., Geierhaas, C.D., Williams, P.M., Paci, E., and Clarke, J. (2005) Mechanical unfolding of TNfn3: the unfolding pathway of a fnIII domain probed by protein engineering, AFM and MD simulation. *J. Mol. Biol.*, **350**, 776–789.
55. Rohs, R., Etchebest, C., and Lavery, R. (1999) Unraveling proteins: a molecular mechanics study. *Biophys. J.*, **76**, 2760–2768.
56. Watanabe, K., Nair, P., Labeit, D., Kellermayer, M.S.Z., Greaser, M., Labeit, S., and Granzier, H. (2002) Molecular mechanics of cardiac titin's PEVK and N2B spring elements. *J. Biol. Chem.*, **277**, 11549–11558.
57. Li, H., Oberhauser, A.F., Redick, S.D., Carrion-Vazquez, M., Erickson, H.P., and Fernandez, J.M. (2001) Multiple conformations of PEVK proteins detected by single-molecule techniques. *Proc. Natl. Acad. Sci. U.S.A.*, **98**, 10682–10686.
58. Li, H., Linke, W.A., Oberhauser, A.F., Carrion-Vazquez, M., Kerkvliet, J.G., Lu, H., Marszalek, P.E., and Fernandez, J.M. (2002) Reverse engineering of the giant muscle protein titin. *Nature*, **418**, 998–1002.
59. Marszalek, P.E., Lu, H., Li, H., Carrion-Vazquez, M., Oberhauser, A.F., Schulten, K., and Fernandez, J.M. (1999) Mechanical unfolding intermediates in titin modules. *Nature*, **402**, 100–103.
60. Dietz, H. and Rief, M. (2006) Protein structure by mechanical triangulation. *Proc. Natl. Acad. Sci. U.S.A.*, **103**, 1244–1247.
61. Garcia-Manyes, S., Brujic, J., Badilla, C.L., and Fernandez, J.M. (2007) Force-clamp spectroscopy of single-protein monomers reveals the individual unfolding and folding pathways of I27 and ubiquitin. *Biophys. J.*, **93**, 2436–2446.
62. Steward, A., Toca-Herrera, J.L., and Clarke, J. (2002) Versatile cloning system for construction of multimeric proteins for use in atomic force microscopy. *Protein Sci.*, **11**, 2179–2183.
63. Dietz, H., Bertz, M., Schlierf, M., Berkemeier, F., Bornschlögl, T., Junker, J.P., and Rief, M. (2006) Cysteine engineering of polyproteins for single-molecule force spectroscopy. *Nat. Protoc.*, **1**, 80–84.

64. Sosnick, T.R. (2004) Comment on "Force-clamp spectroscopy monitors the folding trajectory of a single protein". *Science*, **306**, 411.
65. Best, R.B. and Hummer, G. (2005) Comment on "Force-clamp spectroscopy monitors the folding trajectory of a single protein". *Science*, **308**, 498.
66. Oroz, J., Hervás, R., and Carrión-Vázquez, M. Unequivocal single-molecule force spectroscopy by AFM using pFS vectors. *Biophys. J.*, (in press).
67. Peng, Q., and Li, H.. (2009) Domain insertion effectively regulates the mechanical unfolding hierarchy of elastomeric proteins: toward engineering multifunctional elastomeric proteins. *J. Am. Chem. Soc.*, **131**, 14050–14056.
68. Rief, M., Clausen-Schaumann, H., and Gaub, H.E. (1999) Sequence-dependent mechanics of single DNA molecules. *Nat. Struct. Mol. Biol.*, **6**, 346–349.
69. Cluzel, P., Lebrun, A., Heller, C., Lavery, R., Viovy, J.L., Chatenay, D., and Caron, F. (1996) DNA: an extensible molecule. *Science*, **271**, 792–794.
70. Smith, S.B., Cui, Y., and Bustamante, C. (1996) Overstretching B-DNA: the elastic response of individual double-stranded and single-stranded DNA molecules. *Science*, **271**, 795–799.
71. Rief, M., Oesterhelt, F., Heymann, B., and Gaub, H.E. (1997) Single molecule force spectroscopy on polysaccharides by atomic force microscopy. *Science*, **275**, 1295–1297.
72. Marszalek, P.E., Li, H., Oberhauser, A.F., and Fernandez, J.M. (2002) Chair-boat transitions in single polysaccharide molecules observed with force-ramp AFM. *Proc. Natl. Acad. Sci. U.S.A.*, **99**, 4278–4283.
73. Walther, K.A., Brujic, J., Li, H., and Fernandez, J.M. (2006) Sub-angstrom conformational changes of a single molecule captured by AFM variance analysis. *Biophys. J.*, **90**, 3806–3812.
74. Zlatanova, J., Lindsay, S.M., and Leuba, S.H. (2000) Single molecule force spectroscopy in biology using the atomic force microscope. *Prog. Biophys. Mol. Biol.*, **74**, 37–61.
75. Weisel, J.W., Shuman, H., and Litvinov, R.I. (2003) Protein-protein unbinding induced by force: single-molecule studies. *Curr. Opin. Struct. Biol.*, **13**, 227–235.
76. Hinterdorfer, P. and Dufrêne, Y.F. (2006) Detection and localization of single molecular recognition events using atomic force microscopy. *Nat. Methods*, **3**, 347–355.
77. Florin, E.L., Moy, V.T., and Gaub, H.E. (1994) Adhesion forces between individual ligand-receptor pairs. *Science*, **264**, 415–417.
78. Willemsen, O.H., Snel, M.M., Cambi, A., Greve, J., De Grooth, B.G., and Figdor, C.G. (2000) Biomolecular interactions measured by atomic force microscopy. *Biophys. J.*, **79**, 3267–3281.
79. Lee, C.K., Wang, Y.M., Huang, L.S., and Shiming, L. (2007) Atomic force microscopy: determination of unbinding force, off rate and energy barrier for protein-ligand interaction. *Micron*, **38**, 446–461.
80. Ebner, A., Wildling, L., Kamruzzahan, A.S.M., Rankl, C., Wruss, J., Hahn, C.D., Hölzl, M., Zhu, R., Kienberger, F., Blaas, D., Hinterdorfer, P., and Gruber, H.J. (2007) A new, simple method for linking of antibodies to atomic force microscopy tips. *Bioconjug. Chem.*, **18**, 1176–1184.
81. Geisler, M., Pirzer, T., Ackerschott, C., Lud, S., Garrido, J., Schiebel, T., and Hugel, T. (2008) Hydrophobic and Hofmeister effects on the adhesion of spider silk proteins onto solid substrates: an AFM-based single-molecule study. *Langmuir*, **24**, 1350–1355.
82. Lu, H., Isralewitz, B., Krammer, A., Vogel, V., and Schulten, K. (1998) Unfolding of titin immunoglobulin domains by steered molecular dynamics simulations. *Biophys. J.*, **75**, 662–671.
83. Still, W.C., Tempczyk, A., Hawley, R.C., and Hendrickson, T. (1990) Semianalytical treatment of solvation for molecular mechanics and dynamics. *J. Am. Chem. Soc.*, **112**, 6127–6129.
84. Oroz, J., Valbuena, A., Vera, A.M., Mendieta, J., Gómez-Puertas, P., and

Carrión-Vázquez, M. (2010) Nanomechanics of cadherin ectodomain. *J. Biol. Chem.*, **11**, 9405–9418.
85. Sikora, M., Sulkowska, J.I., and Cieplak, M. (2009) Mechanical strength of 17134 model proteins and cysteine slipknots. *PLoS Comput. Biol.*, **5**, e1000547.
86. Galera-Prat, A., Gómez-Sicilia, A., Oberhauser, A.F., Cieplak, M., and Carrión-Vázquez, M. (2010) Understanding biology by stretching proteins: recent progress. *Curr. Opin. Struct. Biol.*, **20**, 63–69.
87. Tskhovrebova, L. and Trinick, J. (2003) Titin: properties and family relationships. *Nat. Rev. Mol. Cell Biol.*, **4**, 679–789.
88. Wang, K., McClure, J., and Tu, A. (1979) Titin: major myofibrillar components of striated muscle. *Proc. Natl. Acad. Sci. U.S.A.*, **76**, 3698–3702.
89. Maruyama, K., Kimura, S., Ohashi, K., and Kuwano, Y. (1981) Connectin, an elastic protein of muscle. Identification of "titin" with connectin. *J. Biochem. (Tokyo)*, **89**, 701–709.
90. Fernandez, J.M. and Li, H. (2004) Force-clamp spectroscopy monitors the folding trajectory of a single protein. *Science*, **303**, 1674–1678.
91. Witta, A.P., Perez-Jimenez, R., Walther, K.A., Gräter, F., Berne, B.J., Holmgren, A., Sanchez-Ruiz, J.M., and Fernandez, J.M. (2007) Probing the chemistry of thioredoxin catalysis with force. *Nature*, **150**, 124–127.
92. Rief, M., Gautel, M., Oesterhelt, F., Fernandez, J.M., and Gaub, H.E. (1997) Reversible unfolding of individual titin immunoglobulin domains by AFM. *Science*, **276**, 1109–1112.
93. Improta, S., Politou, A.S., and Pastore, A. (1996) Immunoglobulin-like modules from titin I-band: extensible components of muscle elasticity. *Structure*, **4**, 323–337.
94. Moy, V.T., Florin, E.L., and Gaub, H.E. (1994) Intermolecular forces and energies between ligands and receptors. *Science*, **266**, 257–259.
95. Classen, M., Breuer, S., Baumeister, W., Guckenberger, R., and Witt, S. (2011) Force Spectroscopy of substrate molecules en route to the proteasome's active site. *Biophys. J.*, **100**, 489–497.
96. Liu, W., Montana, V., Bai, J., Chapman, E.R., Mohideen, U., and Parpura, V. (2006) Single molecule mechanical probing of the SNARE protein interactions. *Biophys. J.*, **2**, 744–758.
97. Hinterdorfer, P., Baumgartner, W., Gruber, H.J., Schilcher, K., and Schindler, H. (1996) Detection and localization of individual antibody-antigen recognition events by atomic force microscopy. *Proc. Natl. Acad. Sci. U.S.A.*, **93**, 3477–3481.
98. Lehenkari, P.P. and Horton, M.A. (1999) Single integrin molecule adhesion forces in intact cells measured by atomic force microscopy. *Biochem. Biophys. Res. Commun.*, **259**, 645–650.
99. Bartels, F.W., Baumgarth, B., Anselmetti, D., Ros, R., and Becker, A. (2003) Specific binding of the regulatory protein ExpG to promoter regions of the galactoglucan biosynthesis gene cluster of *Sinorhizobium meliloti* – a combined molecular biology and force spectroscopy investigation. *J. Struct. Biol.*, **143**, 145–152.
100. Perez-Jimenez, R., Li, J., Kosuri, P., Sanchez-Romero, I., Wiita, A.P., Rodriguez-Larrea, D., Chueca, A., Holmgren, A., Miranda-Vizuete, A., Becker, K., Cho, S.H., Beckwith, J., Gelhaye, E., Jacquot, J.P., Gaucher, E.A., Sanchez-Ruiz, J.M., Berne, B.J., and Fernandez, J.M. (2009) Diversity of chemical mechanisms in thioredoxin catalysis revealed by single-molecule force spectroscopy. *Nat. Struct. Mol. Biol.*, **16**, 890–896.

7
High-Speed AFM for Observing Dynamic Processes in Liquid

Toshio Ando, Takayuki Uchihashi, Noriyuki Kodera, Mikihiro Shibata, Daisuke Yamamoto, and Hayato Yamashita

7.1
Introduction

Biological macromolecules are dynamic in nature and only active in aqueous solutions. Reflecting this unequivocal fact, optical microscopy (fluorescence microscopy in particular) is frequently used in biological research. To improve its low spatial resolution, several new types of fluorescence microscopes that overcome the diffraction limit have been recently devised [1]. However, even super-resolution fluorescence microscopy does not allow one to directly observe the entity itself that is labeled with fluorophores, that is, it provides indirect sample images consisting of fluorescent spots, whose positions are, however, determined within 1–20 nm accuracy. On the other hand, atomic force microscopy (AFM) [2] can directly visualize a sample in liquids with submolecular resolution. Nevertheless, AFM is unable to match fluorescence microscopy in the dynamic and noninvasive imaging of live biological samples. This is because of its low imaging rate and the tip–sample contact required in AFM imaging in liquids.

If these weak points of AFM can be overcome, AFM will surpass fluorescence microscopy in some respects, and hence, these microscopy techniques will be complements of each other. High-resolution AFM movies, which directly capture successive images of a biological sample, should contain a large amount of information on the dynamic process and the functional mechanism that cannot be accessed by conventional approaches or can only be reached through a large amount of study from various angles. Thus, high-speed AFM will greatly accelerate our understanding of functional mechanisms. This was what a few researchers including us thought more than 15 years ago when they started to develop high-speed AFM [3, 4]. After long-term efforts, high-speed tapping-mode AFM was finally realized [5–7]. The highest possible imaging rate has now reached 10–25 frames per second for a 250 × 250 nm^2 scan area and 100 scan lines. Remarkably, even delicate protein–protein interactions are not disturbed by the high-frequency intermittent contact with an oscillating cantilever tip [8–10].

Atomic Force Microscopy in Liquid: Biological Applications, First Edition.
Edited by Arturo M. Baró and Ronald G. Reifenberger.

In the following sections, we first theoretically derive the imaging rate and feedback bandwidth, which is followed by descriptions of technical components in the high-speed AFM instrument. We then describe some substrate surfaces that can be used for dynamic imaging and some imaging studies on dynamic biomolecular events and conclude by discussing the future prospects of high-speed AFM.

7.2
Theoretical Derivation of Imaging Rate and Feedback Bandwidth

Feedback control to maintain the tip–sample distance (hence, tip–sample interaction force) at a constant value during scanning is mandatory in bio-AFM imaging. The highest possible imaging rate largely depends on how quickly the feedback control can be performed (i.e., feedback bandwidth) as well as on the imaging conditions and sample fragility.

7.2.1
Imaging Time and Feedback Bandwidth

When a sample is scanned over an area of $W \times W$, with scan velocity V_s in the x-direction and N scan lines in the y-direction, the image acquisition time T is given by $T = 2WN/V_s$. Supposing that the sample surface is characterized with a single spatial frequency $1/\lambda$, then the feedback scan is executed in the z-direction with frequency $f = V_s/\lambda$ to trace the sample surface. The feedback bandwidth f_B of the microscope should be equal to or higher than f, and thus, we obtain the relationship $T \geq 2WN/(\lambda f_B)$. The feedback bandwidth of conventional AFM apparatuses is $\sim$0.1 kHz, and therefore, $T \sim 50$ s for $W = 250$ nm, $N = 100$, and $\lambda = 10$ nm.

The bandwidth of closed-loop feedback control is determined by the sum of the time delays (τ_{total}) caused by devices contained in the feedback loop and the time delay due to "parachuting." Here, parachuting means that the oscillating tip becomes completely detached from the sample surface in the steep downhill regions of the sample and lands on the surface again after a time delay. The phase delay in the closed loop (φ_{closed}) is approximately twice than that in the open loop (φ_{open}), provided the feedback gain is maintained at $\sim$1 [7]. The feedback bandwidth is usually defined by the feedback frequency that results in a phase delay of $\pi/4$ in the closed loop. From the relationship $\varphi_{open} = 2\pi f \tau_{total}$, f_B is thus given by $f_B = 1/(16\tau_{total})$. Considering the phase compensation factor α gained by the (P + D) operation of the proportional-integral-derivative (PID) feedback controller, this relationship is modified to $f_B = \alpha/(16\tau_{total})$. Note that this feedback bandwidth does not guarantee an image acquisition time of $T = 32WN\tau_{total}/(\alpha\lambda)$ for very delicate samples that would be disturbed by the tip force resulting from the phase delay of $\pi/4$ when tracing the sample surface.

7.2.2
Time Delays

The main delays in the feedback loop are as follows: (i) the time required to measure the oscillation amplitude of a cantilever (τ_a), (ii) the response time of a cantilever (τ_c), (iii) the response time of the z-scanner (τ_s), (iv) the time required to integrate the error signals using the PID controller (τ_I), and (v) the parachuting time (τ_p). The minimum τ_a is half of the cantilever oscillation period $1/(2f_c)$, where f_c is the first resonant frequency of the cantilever. The response time of a cantilever is given by $\tau_c = Q_c/(\pi f_c)$, where Q_c is the quality factor. The response time of the z-scanner is given by $\tau_s = Q_s/(\pi f_s)$, where f_s and Q_s are the first resonant frequency and the quality factor of the z-scanner, respectively. The integral time of the PID controller and the parachuting time cannot be expressed explicitly. However, by comparing theoretical analyses and experimental results, we obtained the following relationships [7]:

$$\tau_I = \frac{4\left(\frac{h_0}{A_0}\right)\sin\left(\frac{\phi_{closed}}{2}\right)}{f_c} \tag{7.1}$$

$$\tau_p = \frac{(\tan\beta/\beta - 1)}{f_c} \tag{7.2}$$

where

$$\beta = \cos^{-1}\left[\frac{A_0(1-r)}{5h_0\sin\left(\frac{\phi_{closed}}{2}\right)}\right] \tag{7.3}$$

h_0 is the maximum sample height, A_0 is the free-oscillation amplitude of a cantilever, and r is the dimensionless set point. The dimensionless set point is given by $r = A_s/(2A_0)$, where A_s is the peak-to-peak set-point amplitude. Thus, the feedback bandwidth can be expressed as

$$f_B = \frac{\alpha\frac{f_c}{8}}{\left(1 + \frac{2Q_c}{\pi} + \frac{2Q_s f_c}{\pi f_s} + 2f_c\left(\tau_p + \tau_I + \delta\right)\right)} \tag{7.4}$$

where δ represents the sum of time delays other than those mentioned above.

Figure 7.1 shows the feedback bandwidth as a function of r and $h_0/(2A_0)$. The feedback bandwidth decreases with increasing r, particularly when r is >0.85. This is because tip parachuting is significantly promoted with increasing r [11]. The feedback bandwidth also decreases steadily with increasing $h_0/(2A_0)$. Thus, to increase the feedback bandwidth, a small r and cantilever free oscillation with a large amplitude value are required, which are conditions contradictory to low-invasive imaging. Overcoming this issue is essential for realizing high-speed AFM applicable to delicate biological samples. As described later, this problem can be solved to a large extent by a modification of the conventional PID controller.

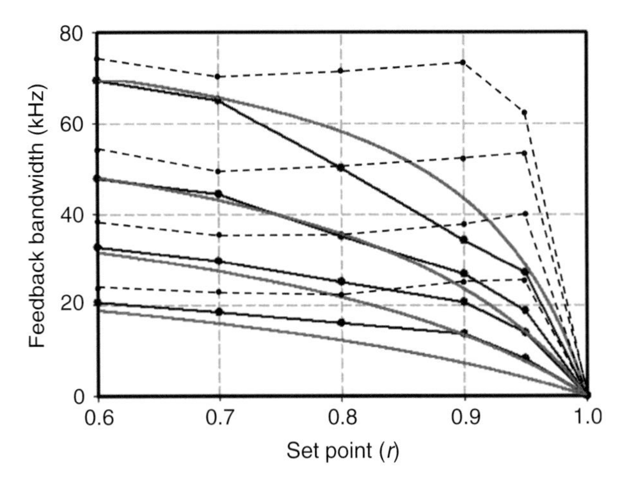

Figure 7.1 Feedback bandwidth as a function of the dimensionless set point r and the ratio of the sample height to the cantilever free-oscillation peak-to-peak amplitude $h_0/(2A_0)$. The feedback bandwidths were obtained under the following conditions: cantilever resonant frequency, 1.2 MHz; quality factor of the cantilever oscillation, 3; resonant frequency of the z-scanner, 150 kHz; quality factor of the z-scanner, 0.5. Solid black lines, experimentally obtained feedback bandwidths using a conventional PID controller; solid gray lines, theoretically derived feedback bandwidths; broken lines, experimentally obtained feedback bandwidths using a dynamic PID controller. Each set of curves corresponds to the ratios $h_0/(2A_0) = 0.2$, 0.5, 1, and 2 from top to bottom.

7.3 Techniques Realizing High-Speed Bio-AFM

There are three key factors for realizing high-speed AFM applicable to biological samples: (i) high feedback bandwidth based on fast devices, (ii) active damping techniques to suppress the mechanical vibrations of the scanner, and (iii) techniques that make high-speed imaging compatible with low-invasive imaging. These factors are related to each other. In the following sections, details of key techniques that we have developed for high-speed AFM are described.

7.3.1 Small Cantilevers

The resonant frequency f_c and spring constant k_c of a rectangular cantilever with length L, width w, and thickness d are, respectively, given by

$$f_c = 0.56\frac{d}{L^2}\sqrt{\frac{E}{12\rho}} \qquad (7.5)$$

and

$$k_c = \frac{wd^3}{4L^3}E \qquad (7.6)$$

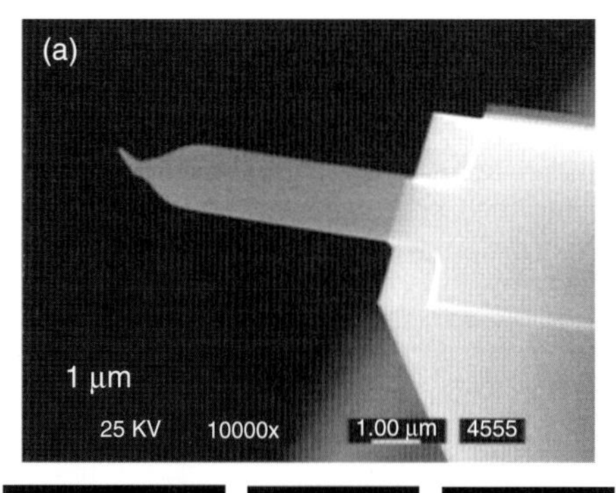

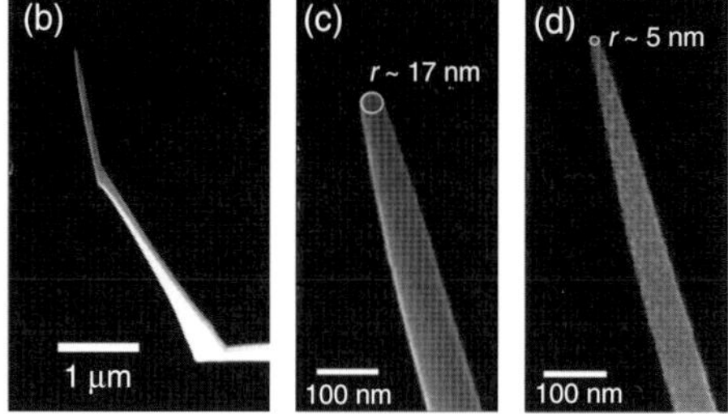

Figure 7.2 Electron micrographs of a small cantilever and EBD tips. (a) Small cantilever developed by Olympus; (b) sharp EBD tip grown on an original cantilever tip; (c) magnified view of an EBD tip before sharpening; and (d) magnified view of an EBD tip after sharpening by plasma etching in argon gas.

where E and ρ are the Young modulus and density of the material used, respectively. For given values of k_c and w and a given material, f_c increases in inverse proportion to $(Ld)^{1/2}$. Therefore, it is necessary to fabricate short and thin cantilevers to simultaneously achieve a high resonant frequency and a small spring constant [4, 12]. Collaborating with Olympus, we developed small cantilevers with $L = 6\ \mu m$, $w = 2\ \mu m$, and $d = 90$ nm using silicon nitride ($E = 1.46 \times 10^{11}$ N m^{-2} and $\rho = 3087$ kg m^{-3}) (Figure 7.2a). The resonant frequencies are 3.5 MHz in air and 1.2 MHz in water, and the spring constant and the quality factor in water are 0.1–0.2 and 2–3 N m^{-1}, respectively. Thus, a short response time of ~0.66 μs in water is achieved.

Short and thin cantilevers have an advantage of high detection sensitivity of their deflection, as the deflection detection sensitivity in the optical beam deflection method follows $\Delta\theta/\Delta z = 3/2L$ (Δz is the displacement and $\Delta\theta$ is the change in the angle of the cantilever free-end). They have an additional advantage that a large change in the resonant frequency, which results in a large phase change in tapping-mode AFM and phase modulation AFM [13–15], occurs by tip–sample interaction, as the resonant frequency change Δf_c follows $\Delta f_c \sim -0.5kf_c/k_c$ (k is the gradient of the force exerted between the tip and the sample). Moreover, the high resonant frequency results in a low thermal noise density because the thermal noise, which is determined only by the spring constant and the absolute temperature, spreads over a wide range of frequencies.

The tip of the small cantilevers that we developed is not sufficiently sharp (apex radius, ~15 nm). We thus attach an additional tip on the original tip by electron beam deposition (EBD) in phenol gas, and then sharpen the EBD tip by plasma etching in argon gas, which decreases the apex radius to 4–5 nm (Figure 7.2b–d).

7.3.2
Fast Amplitude Detector

We have developed two types of amplitude detectors. One is based on a peak-hold method and can output the amplitude signal every half cycle of cantilever oscillation [4]. The other is based on a Fourier method and can output the amplitude signal once every cycle of cantilever oscillation [7]. In the former type, the peak and bottom voltages of the cantilever oscillation signal are held and their difference is output as the peak-to-peak amplitude. The trigger signals for voltage holding are generated using either the cantilever oscillation signal itself (internal mode) or a phase-adjusted sinusoidal signal that is synchronized with the cantilever excitation signal (external mode). In the external mode, the detected amplitude is affected by the phase shift in the cantilever oscillation and is therefore more sensitive to the tip–sample interaction than the internal mode. In the peak-hold amplitude detector, only two point voltages in a cycle are used, hence the amplitude detection is prone to be affected by the thermal noise of cantilever oscillation. In the Fourier-method-based amplitude detector, the Fourier cosine and sine coefficients (a_1 and b_1) for the fundamental frequency are digitally calculated and $(a_1^2 + b_1^2)^{1/2}$ is output through a D/A board. The phase signal, $\arctan(b_1/a_1)$, is also output at every cycle of cantilever oscillation. The peak amplitude noise due to the thermal oscillation is three- to fourfold less than that of the peak-hold method [7].

7.3.3
High-Speed Scanner

Instead of the tube piezo scanners often used in conventional AFM, the scanner for high-speed AFM is constructed using stack piezoactuators and flexures monolithically fabricated within a metal base [16]. We employed the following mechanism: the y-scanner moves an xz-block connected to the base frame with two pairs of flexures, and in the xz-block, the x-scanner moves the z-scanner connected to the frame of the xz-block with two pairs of flexures (Figure 7.3). A sample stage consisting of a glass rod (1.5–2 mm in diameter) with a small mass is glued onto the top of the z-piezoactuator using nail enamel.

The rapid displacement of the z-piezoactuator exerts an impulsive force on the supporting base, which causes vibrations of the base and the surrounding structures and in turn of the z-piezoactuator itself. Therefore, the impulsive force has to be counteracted by applying a counterforce to the supporting base. There are two methods of applying this counterforce: (i) an additional z-piezoactuator is

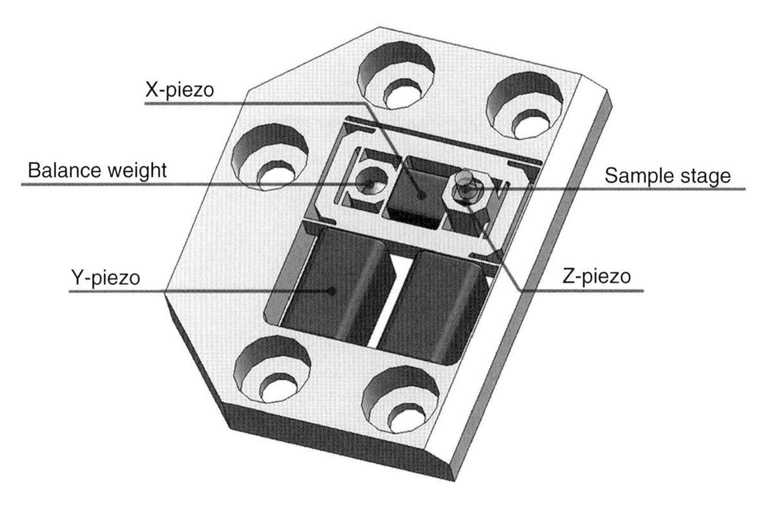

Figure 7.3 Sketch of high-speed scanner. In the z-scanner, one piezoactuator is held at four rims parallel to the displacement direction. The dimensions ($W \times L \times H$) of the z-piezoactuator are $2 \times 2 \times 2$ mm^3. The gaps in the scanner are filled with a soft elastomer for passive damping.

attached to the opposite side of the supporting base and the two z-piezoactuators are simultaneously displaced by the same length in the counter directions [4] and (ii) one z-piezoactuator is held so that its center of mass is stationary while the z-piezoactuator is displaced. The former type z-scanner is easy to make and its maximum displacement is approximately the same as that in free displacement. However, the resonant frequency becomes approximately half of that under free oscillation. The latter type z-scanner has an advantage that its resonant frequency is similar to that under free oscillation, although the maximum displacement becomes approximately half of that in the former type z-scanner. There may be several methods of holding a small piezoactuator so that the center of mass is kept stationary. In our recent high-speed scanner (Figure 7.3), a z-piezoactuator ($2 \times 2 \times 2$ mm^3) is held at its four rims parallel to the displacement direction, which does not deteriorate the maximum displacement. Using this method, a resonant frequency of ~500 kHz is achieved (Figure 7.4) [17].

The center of mass of the x-piezoactuator is also kept stationary by holding the x-piezoactuator using two pairs of identical flexures at the ends and by attaching a balance weight to the counter side. The mass of the balance weight is adjusted to be similar to the sum of the mass of the z-piezoactuator and the glass rod comprising the sample stage.

The above-mentioned designs for the z-scanner minimizes resonant vibrations arising from the structural mechanism to a level at which no further reduction of the vibrations by active damping is not necessary. The resonant vibrations only arise from the resonance of the z-piezoactuator itself. However, the resonant frequencies of the the x- and y-scanners are lower than those of the respective

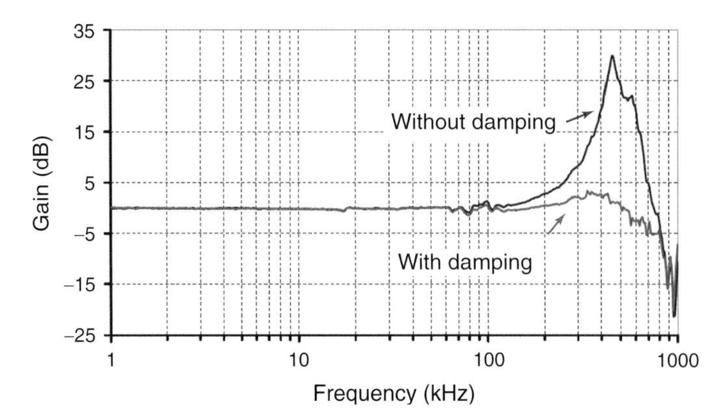

Figure 7.4 Frequency spectra of mechanical response of z-scanner and the effect of active damping. The gain spectra were measured for the z-scanner shown in Figure 7.3 with or without active Q-control damping. The Q-control gain could be increased until the resonant peak completely disappeared. However, this was not done because a lower quality factor increases the phase delay at frequencies $\lesssim$ 500 kHz, and moreover, the resonance at ~500 kHz would not have been excited by the feedback signal.

piezoactuators. The vibrations arising from the z-piezoactuator resonance and the structural resonances of the x- and y-scanners have to be actively damped.

7.3.4
Active Damping Techniques

Basically, there are two types of active damping method: feedback and feedforward. Since the z-scanner has to move quickly, only active damping by an analog circuit can be used. Moreover, it is difficult to construct an analog circuit characterized with the inverse transfer function of the z-scanner. Thus, only feedback-type active damping can be used. A feedback control method for active damping, the so-called Q-control, is known [18]. The z-scanner response is approximately described by a conventional equation of motion:

$$m\ddot{z} + \gamma\dot{z} + kz = f(t) \tag{7.7}$$

where m is the apparent mass, γ is the coefficient of friction, k is the spring constant, and $f(t)$ is the driving force. As the quality factor is reduced by increasing the frictional force, an additional driving force proportional to $-\dot{z}$ can suppress the resonant vibrations. To apply this method to the z-scanner, it is necessary to measure the displacement or speed of the z-scanner in real time, but this is difficult to achieve. To overcome this difficulty, we developed a new Q-control method in which a "mock z-scanner" consisting of an LRC circuit characterized with a transfer function similar to that of the real z-scanner is used [19] (Figure 7.5). The output signal from the mock z-scanner in response to the driving (feedback) signal is differentiated, inversely amplified with an appropriate gain, and then added to the driving signal. This simple method is reasonably effective in damping vibrations

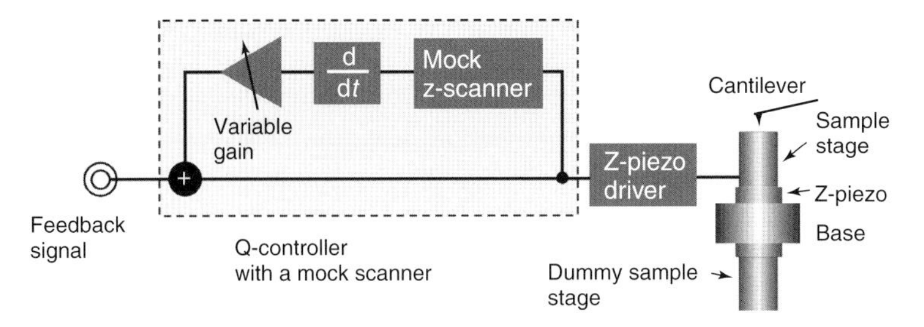

Figure 7.5 Feedback Q-control system for damping z-scanner vibrations. The mock z-scanner is constructed using an LRC circuit characterized with a transfer function similar to that of a real z-scanner. The block enclosed with the broken line indicates the Q-control circuit.

of the z-scanner and also in significantly increasing the response speed, which is inversely proportional to the quality factor.

When the z-scanner exhibits multiple resonant peaks, the simple Q-control method does not work effectively, particularly when a large second peak lies near the first resonant peak. When the resonant components are connected in series, we can use Q-controllers connected in series, each of which contains a mock z-scanner corresponding to one of the resonant elements of the z-scanner [7]. When resonant components with close resonant peaks are connected in parallel, active damping by Q-control is very difficult.

The signal waveform used to drive the x-scanner can be constructed using a computer and output via a D/A board. Therefore, digital feedforward control for active damping is easily applicable to the x-scanner even when the transfer function of the x-scanner is complicated. In addition, feedforward control can extend the bandwidth of the x-scanner beyond the first resonant frequency, provided the piezo driver has a sufficiently high bandwidth. Fourier transform of the periodic isosceles triangular signal is given by

$$F(\omega) = 2\pi X_0 \left[\frac{1}{2}\delta(\omega) - \frac{2}{\pi^2} \sum_{k=-\infty}^{+\infty} \frac{1}{k^2}\delta(\omega - k\omega_0) \right] \quad (k : \text{odd}) \tag{7.8}$$

where X_0 is the amplitude and ω_0 is the fundamental frequency. The signal $X(t)$ that drives the x-scanner, characterized with the transfer function $G(s)$, in the isosceles triangle waveform is given by the inverse Fourier transform of $F(\omega)/G(i\omega)$, which is expressed as

$$X(t) = \frac{X_0}{2} - \frac{4X_0}{\pi^2} \sum_{k=1}^{+\infty} \frac{1}{k^2} \frac{1}{G(ik\omega_0)} \cos(k\omega_0 t) \quad (k : \text{odd}) \tag{7.9}$$

Practically, the sum of the first ~10 terms in the series is a sufficient approximation to the ideal signal for driving the x-scanner. We can calculate Eq. (7.9) in advance to obtain numerical values of $X(t)$ and output them in succession from a computer through a D/A converter. This method is convenient and works well [7].

The y-scanner moves slowly while capturing an image and then moves back to the origin (starting point of imaging) after acquiring an image. Since the y-scanner moves the xz-block with a relatively large mass, its resonant frequency is low. Y-scanner vibrations, which may be excited when the y-scanner moves back to the origin, can be minimized using a smooth driving waveform. Even when the y-scanner is moved back to the origin at a relatively low speed, the time required is only a small fraction of the imaging time.

7.3.5
Suppression of Parachuting

During parachuting, the cantilever oscillation amplitude is constant at the free-oscillation amplitude A_0. Therefore, the error signal is saturated at $2A_0(1-r)$ irrespective of the distance between the tip end at the bottom of the swing and the sample surface. When the dimensionless set point r is very close to 1, this saturated error signal is very small, and hence, the feedback signal used to move the z-scanner is very small, resulting in prolonged parachuting. We cannot increase the feedback gain of the PID controller to shorten the parachuting period because a large gain would cause tip overshoot when scanning around the top area of a sample, inducing more frequent parachuting and instability of the feedback control.

In general, the cantilever peak-to-peak oscillation amplitude A_{p-p} is smaller than the peak-to-peak amplitude set point A_s when uphill regions of the sample are being scanned, whereas it is larger when downhill regions of the sample are being scanned. Therefore, the sign of $(A_{p-p} - A_s)$ is an indicator of which slope (uphill or downhill) is currently being scanned. If we can increase the feedback gain only when a downhill slope is being scanned, we will be able to shorten the parachuting period or completely avoid parachuting without causing tip overshoot.

We developed a new PID controller named "dynamic PID controller" whose gain parameters can be automatically altered depending on A_{p-p} [11]. A threshold level A_t is set at the set point (or slightly above the set point). When A_{p-p} exceeds A_t, the PID gains are increased. There are several methods of increasing the respective gains for PID. The simplest method is to add a pseudoerror signal to the true error signal. The broken lines in Figure 7.1 show the effect of dynamic PID control on the feedback bandwidth when the simplest method of gain alteration is used (compare the broken lines with the solid black lines in Figure 7.1). The feedback bandwidth becomes independent of the set point r as long as $r \lesssim 0.9$, indicating that no parachuting occurs when $r \lesssim 0.9$. To ensure that the dynamic PID control is effective even for conditions of a small A_0 and $r{\sim}0.9$, a low-noise amplitude signal and stable cantilever excitation are required. The former requirement can be satisfied using an amplitude detector based on the Fourier method.

The efficiency of cantilever excitation by a piezoactuator usually decreases with time owing to the temperature increase of the piezoactuator. When the cantilever free-oscillation amplitude A_0 decreases, A_{p-p} also decreases. The feedback system misinterprets this decrease in A_{p-p} as an excessively strong tip–sample interaction, and therefore, withdraws the sample stage from the cantilever. This withdrawal tends to completely dissociate the cantilever tip from the sample surface, particularly when r is close to 1. For example, when $A_0 = 1$ nm and $r = 0.9$, a decrease in A_0 by 0.1 nm results in complete dissociation. However, we cannot directly detect a change in A_0 while imaging. Before complete dissociation occurs, A_{p-p} is maintained constant at A_s by the feedback operation even when A_0 decreases, but the tip–sample interaction force decreases. Thus, we can detect a change in A_0 if a signal sensitive to the tip–sample interaction force is available.

When a cantilever tip oscillating with the lowest resonant frequency f_c taps a sample surface, higher harmonic oscillations ($2f_c, 3f_c, \ldots$) appear. The time-averaged amplitude of the second harmonic oscillation can be used as an indicator of drift in A_0 [20]. The power of the cantilever excitation is controlled using an I-controller with an integral time longer than the image acquisition time, so that the second harmonic amplitude is maintained constant without interfering the fast feedback control for imaging [19]. This drift compensator works effectively and thus enables stable low-invasive and high-speed imaging when operating together with the dynamic PID controller, even under the conditions of $A_0 = 1$ nm and $r = 0.95$.

The magnitude of the tapping force exerted on a sample by an oscillating cantilever tip is expressed as $k_c A_0 \sqrt{1-r^2} / Q_c$. When $A_0 = 1$–2 nm, $r = 0.9$, $k_c = 0.1$–0.2 N m^{-1}, and $Q_c = 2.5$, the tapping force is estimated to be 17–70 pN. Although this force appears to be very large for biological samples, note that the mechanical quantity that affects the sample is not the force itself but the impulse, that is, the product of the force and the time over which the force acts. In tapping-mode high-speed AFM, the acting time of the force is short ($<$100 ns), and therefore, the magnitude of the impulse is small.

7.3.6 Fast Phase Detector

As mentioned above, the resonant frequency of a cantilever is changed by the tip–sample interaction. When the cantilever excitation frequency is fixed at (or near) the cantilever resonant frequency as in the case of tapping-mode AFM and phase modulation-AFM, the change in the resonant frequency is accompanied by a change in the phase of the cantilever oscillation relative to the excitation signal. For conventional cantilevers, the change in resonant frequency is $\lesssim$100 Hz. However, it is ~1000 times larger for our small cantilevers because their f_c/k_c ratio is ~1000 times larger than that of conventional cantilevers. This large change results in a relatively large phase change even when the quality factor of the small cantilevers is small as in the case of a liquid environment.

A large phase change should be detectable without using a highly sensitive but slow detector such as a lock-in amplifier. As mentioned above, the Fourier-method-based amplitude detector can detect phase change. However, we developed another fast phase detector with higher functionality [21] by modifying the design given by Stark and Guckenberger [22]. A two-channel waveform generator produces sinusoidal and sawtooth waves with a given phase difference.

The sinusoidal wave is used to oscillate the cantilever. The cantilever oscillation signal is fed to the fast phase detector composed of a high-pass filter, a variable phase shifter, a zero-crossing comparator, a pulse generator, and a sample-and-hold circuit. A pulse signal is generated at either the rising or falling edge of the output signal of the zero-crossing comparator. The pulse signal acts as a trigger for the sample-and-hold circuit. When the trigger is generated, the amplitude of the voltage of the sawtooth signal is captured and retained by the sample-and-hold circuit. Thus, the sawtooth signal acts as a phase-voltage converter.

The fast phase detector can output a phase signal at every cycle of cantilever oscillation. Importantly, the phase detection timing can be varied by the variable phase shifter. This function is essential for obtaining an optimum-phase contrast image because the phase contrast markedly depends on the trigger timing [21]. Although not yet implemented, it is also possible to detect phase signals at multiple timings within an oscillation cycle, which enables the phase to be traced over time after the tip comes into contact with a sample surface.

7.4 Substrate Surfaces

The substrate surface on which a biological sample is placed is an important element in the dynamic AFM imaging of molecular processes. The substrate surface must not firmly bind sample molecules to ensure that their physiological functions are retained. However, extensively fast Brownian motion under weak sample–surface interaction makes imaging infeasible. Thus, the strength of the sample–surface association has to be controllable. When the biological process to be visualized contains dynamic interactions between two species of proteins, only one species can be immobilized to the substrate surface. Otherwise, two species of molecules can hardly interact with each other. Consequently, selective protein attachment to the substrate surface is required. Since AFM imaging is only possible from one direction, the surfaces of a sample in parallel to the imaging direction or those facing the substrate surface cannot be imaged. Therefore, the immobilization of a sample to the substrate surface in a desired orientation is sometimes required.

There is no versatile substrate surface applicable to a wide range of biological samples. Nevertheless, we here describe two types of surfaces that are useful for the dynamic imaging of some biological samples. Practical methods of preparing these surfaces and their application to biological samples are described elsewhere [23, 24].

7.4.1 Supported Planar Lipid Bilayers

Planar lipid bilayers are easily formed on an atomically flat mica surface by the deposition of small unilamellar vesicles onto the surface. A membrane surface with zwitterionic polar head groups such as phosphatidylcholine (PC) and phosphatidylethanolamine (PE) is resistant to the nonspecific binding of proteins. By including a lipid with a positively or negatively charged head group in the bilayer in an appropriate proportion, the surface can electrostatically bind biological molecules with an adequate affinity. Since lipids with functional groups at the polar head, such as biotin and Ni-NTA, are commercially available, we can prepare planar lipid bilayer surfaces for the specific and selective immobilization of samples.

7.4.1.1 Choice of Alkyl Chains

Except for the case of preparing two-dimensional (2D) crystals of proteins on a lipid bilayer surface, lipids with saturated alkyl chains such as dipalmitoylphosphatidylcholine (DPPC) are used to obtain low-fluidity bilayers. A bilayer that mainly consists of lipids with unsaturated alkyl chains such as 1,2-dioleoyl-sn-glycero-3-phosphocholine exhibits considerable fluidity at room temperature because of the weak chain–chain interaction. In fact, protein molecules immobilized on the surface move very rapidly and cannot be imaged even at an imaging rate of 30 frames per second.

7.4.1.2 Choice of Head Groups

For electrostatic immobilization, negatively charged head groups including phosphatidylserine (PS), phosphoric acid (PA), and phosphatidylglycerol (PG) and positively charged head groups such as trimethylammoniumpropane (TAP) and ethylphosphatidylcholine (EPC) can be used. Since the mica surface is negatively charged, positively charged lipid bilayers can be easily formed on the surface. On the other hand, according to our experience, negatively charged lipid bilayers cannot be formed on a mica surface using DPPA or DPPS (dipalmitoylphosphatidylserine). Using dipalmitoylphosphatidylglycerol (DPPG), negatively charged lipid bilayers can be formed over a wide area on a mica surface. This is because, unlike DPPA and DPPS, the negative charge of DPPG is not positioned at the distal end of the polar head, and hence, the surface of bilayers containing DPPG only weakly adsorbs positively charged proteins. Lipid bilayers containing DPPA or DPPS can be formed on positively charged bilayers [24]. Lipids having biotin or Ni-NTA at the polar head (such as biotin-DPPE and Ni-NTA-DOGS) can be used for selectively immobilizing biotinylated proteins or His-tag-conjugated proteins, respectively (Figure 7.6a). For the immobilization of biotinylated proteins, it is necessary to use streptavidin or NeutrAvidin as a mediator. When the mediators hamper the imaging of biotinylated proteins, we can use the surface of streptavidin 2D crystals.

7.4.2 Streptavidin 2D Crystal Surface

Two-dimensional crystals of streptavidin formed on biotin-containing lipid bilayers meet various requirements for substrate surfaces used in dynamic AFM imaging [23]. Streptavidin is composed of four identical subunits, each of which specifically

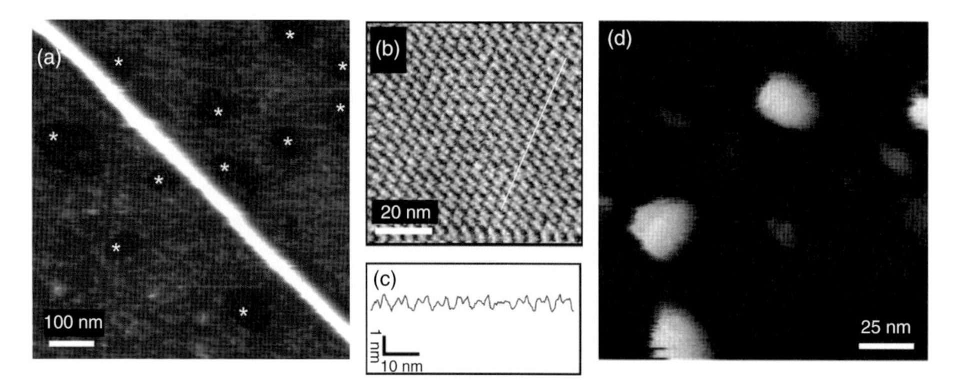

Figure 7.6 AFM images of streptavidin 2D crystal and protein samples placed on biotin-containing planar lipid bilayer or on streptavidin 2D crystal. (a) AFM image of microtubules immobilized on a biotin-containing planar lipid bilayer via streptavidin. The areas marked with asterisks represent the exposed mica surfaces, on some of which protofilaments are attached. The small particles that can be seen on the lipid bilayer are streptavidin molecules. (b) AFM image of type 3 crystal of streptavidin formed on a biotin-containing planar lipid bilayer surface. (c) Surface profile along the line indicated in the AFM image in (b). (d) AFM image of GroEL–GroES complexes formed in the presence of adenosine triphosphate (ADP). GroEL molecules biotinylated at the equatorial domain are immobilized onto a surface of streptavidin 2D crystal, while GroES molecules are free from the surface.

binds to one biotin molecule with strong affinity. In 2D crystals, two biotin binding sites face the solution and can therefore bind to biotinylated samples to be imaged [25, 26]. Since biotinylated Ni-NTA is commercially available, the surface can also bind to His-tag-conjugated recombinant proteins. Importantly, streptavidin is resistant to the nonspecific binding of many proteins, which safeguards the surface-bound proteins against dysfunction [27] and allows the selective surface attachment of one species of protein in a multicomponent system. In addition, it allows protein attachment in a controlled orientation through the biotin or His-tag conjugation sites in the protein.

On highly fluidic planar lipid layers containing a biotinylated lipid, streptavidin self-assembles into three distinct crystalline arrangements (types 1–3), depending on the crystallization conditions such as pH [28] and ionic strength [29]. The 2D crystals with each arrangement have *P*2 symmetry and the crystal forms remain unchanged for a long time (>12 h) on changing the buffer solution after crystallization. The properties of the 2D crystals including the stability, degree of order, surface roughness, and nonspecific binding have been well characterized [23]. Type 3 crystals, which are formed at pH 4.0 and a high ionic strength, have the smallest surface roughness (<1 nm) and the strongest intermolecular association (and are hence the most stable) (Figure 7.6b,c). It is known that binding of biotin to the crystal surface weakens the subunit–subunit interaction between neighboring streptavidin molecules. Therefore, when higher mechanical strength is required, the crystals are treated with glutaraldehyde. After the treatment, the crystal surfaces still retain sufficiently strong affinity for practical use with biotinylated proteins [23].

When the attachment of a protein molecule to the crystal surfaces is mediated through a single biotin–streptavidin bond, the protein molecule rotates around the bond. Less motility can be achieved using reactive dibiotin compounds. However, when a protein to be imaged contains domains connected by a flexible moiety and is biotinylated only at one domain, it exhibits excessive rapid Brownian motion, and therefore, its imaging is infeasible. This rapid motion can only be slowed by introducing biotin to the protein at more than two sites [24]. Thus, the streptavidin 2D crystals are excellent substrates for oligomerized proteins (Figure 7.6a,d), but their usefulness is limited to monomeric proteins with flexible polypeptide chains.

7.5 Imaging of Dynamic Molecular Processes

Thus far, high-speed AFM has been used in an attempt to directly visualize several dynamic processes of proteins including bacteriorhodopsin (bR) [8, 10], myosin V [30], dynein, streptavidin [9], calmodulin [23], actin filaments [23], chaperonin GroEL-GroES, p97 (one of the AAA proteins), and FACT (facilitates chromatin transcription) protein (one of the intrinsically disordered proteins) [31]. Here, we only describe imaging studies on bR. Several AFM movies can be seen at a web site [32].

7.5.1 Bacteriorhodopsin Crystal Edge

bR is a light-driven proton pump found in the purple membranes (PM) of *Halobacterium salinarum* [33]. PM is composed of bR trimers arranged in a hexagonal crystal lattice. The crystal is in dynamic equilibrium with bR molecules located in the noncrystalline area. In fact, successive AFM images of PM placed on a mica surface show dynamic events in which bR trimers bind to and dissociate from different sites at the border between the crystal and noncrystal areas (Figure 7.7a–c) [8]. Not only bR trimers but also bR dimers and monomers are observed to bind to and dissociate from the crystal edge but with a frequency much lower than that of the bR trimer, indicating that trimers are mostly preformed in the noncrystal area. Interestingly, the residence time of the newly bound bR trimer depends on the number of interaction sites involved (types I–III; roman numerals indicate the number of interaction bonds involved; Figure 7.7d). A histogram of the lifetime of type II bond shows a single exponential, from which the average lifetime τ_2 is estimated to be 0.19 s. The average lifetime is independent of the imaging rate used, indicating the negligible effect of the tip–sample contact on the trimer–trimer association. The average lifetime τ_3 of the type III bond is 0.85 s. The longer lifetime of type III bond than type II bond arises from the relationship $E_3 < E_2 < 0$, where E_2 and E_3 are the association energies responsible

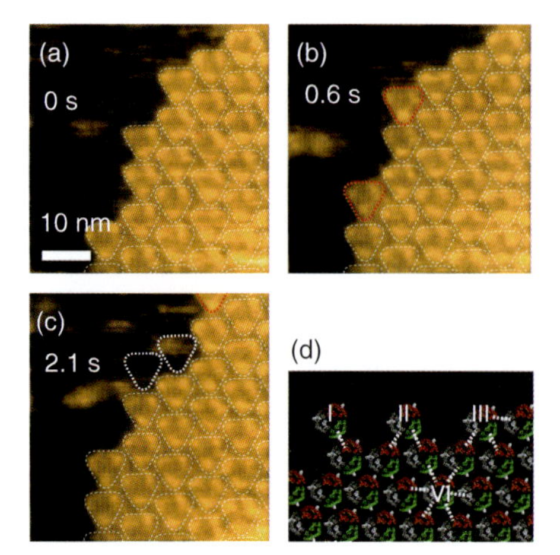

Figure 7.7 (a–c) High-speed AFM images of the border between crystal and noncrystal regions of purple membrane (cytoplasmic side). The bR molecules encircled by the red dotted lines indicate newly bound bR trimers. The white triangles indicate the previously bound trimers (at 2.1 s). Imaging rate, 3.3 frames per second; vertical brightness range, 3.8 nm. (d) Schematic representation of the binding of a trimer at the crystal edge (I, II, and III) and in the crystal interior (VI). Roman numerals indicate the number of interaction bonds (dotted lines) involving the W12 residue.

for type II and type III interactions, respectively. The average lifetime ratio, τ_2/τ_3, is given by

$$\frac{\tau_2}{\tau_3} = \exp\left(\frac{E_3 - E_2}{k_B T}\right) \tag{7.10}$$

where k_B is the Boltzmann constant and T is the temperature in kelvin. The energy difference $E_3 - E_2$ corresponds to the association energy of a single elementary bond. From the ratio $\tau_2/\tau_3 = 0.22$ and Eq. (7.10), this elementary association energy is estimated to be about $-1.5\ k_B T$, which corresponds to $-0.9\ \text{kcal mol}^{-1}$ at 300 K.

7.5.2 Photoactivation of Bacteriorhodopsin

Next, we show AFM images capturing the dynamic structural changes of bR on light illumination [10]. Although the photoactivated structural change in bR has been studied by electron microscopy and X-ray crystallography, no clear results have been obtained. We attempted to visualize the dynamic structural changes using high-speed AFM but could not do so even at an imaging rate of 10 ms per frame, as the photocycle is completed in 10 ms. Here, we used D96N bR mutant that has a long photocycle time (~10 s) but retains the ability of proton pumping. Figure 7.8 shows high-speed AFM images of the cytoplasmic surface of D96N under dark and illuminated conditions (532 nm green light, 0.5 μW). On light illumination, the center of mass of each bR molecule moves drastically outward by 0.7 nm and rotates counterclockwise around the trimer center by 7.4°. After

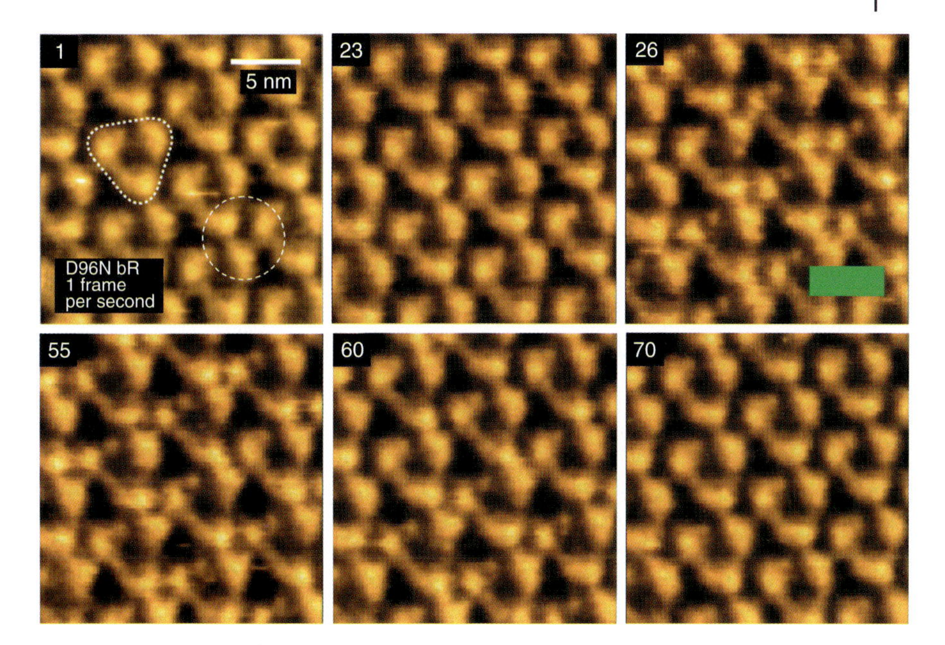

Figure 7.8 High-speed AFM images of cytoplasmic surface of D96N under dark and illuminated conditions. In frame 1, a bR trimer and a trefoil are highlighted by the white triangle and white circle, respectively. The green bar indicates the illumination of 532 nm green light with a power of 0.5 μW. The light was illuminated while capturing frames 25–54. The number of excited bR molecules decreased gradually after light was turned off.

the light is turned off, bR returns to the unphotolyzed state in a few seconds. The observed change is much larger than that suggested in previous studies. The overall position of each bR molecule does not change because of indiscernible alterations at the extracellular surface. Comparison of the AFM images of the cytoplasmic side with an atomic model of the unphotolyzed bR structure indicates that the outward movement of the center of mass is caused by the outward displacement of the E-F loop.

As a result of the outward displacement of the E-F loop, three nearest-neighbor bR monomers, each belonging to a different adjacent trimer, transiently assemble. The triad of nearest-neighbor monomers is designated *trefoil*. This transient assembly in a trefoil alters the decay kinetics of the activated state. When only one bR monomer in a trefoil is excited under weak illumination, it decays with a time constant of 7.3 s. When two or three bR monomers in a trefoil are excited under strong illumination, they decay with a short or long time constant (2 or 13 s). The decay time of each monomer depends on the order of its activation. The monomer that is activated latest among the activated monomers in the trefoil decays with the short time constant. On the other hand, the decay kinetics of the early activated monomers does not follow a single exponential, and the monomers decay with the long time constant on average. Thus, the monomer–monomer interaction in the trefoil engenders both positive and negative cooperative effects in the decay kinetics as the initial bR recovers. Consequently, the average rate of proton pumping by the activated bR is conserved independent of the light intensity. This is perhaps the reason why bR naturally exists in the form of a crystal in PM.

7.6 Future Prospects of High-Speed AFM

7.6.1 Imaging Rate and Low Invasiveness

At present, the maximum possible imaging rate is limited to 10–25 frames per second for a $250 \times 250\ nm^2$ scan range and 100 scan lines. Further increasing the imaging rate under a given set of imaging conditions should be possible but may be limited to a rate approximately twice as high as the present maximum rate. This is mainly because achieving a higher resonant frequency for small cantilevers is only possible by sacrificing the small spring constant (and hence sacrificing the low invasiveness). Thus, a breakthrough technique is required, such as a noncontact imaging technique compatible with high-speed imaging, to significantly increase the imaging rate. True noncontact imaging has already been achieved by scanning ion-conductance AFM [34, 35], but it appears to be impossible to increase the bandwidth of ion-conductance measurements up to 1 MHz. Although not yet attempted, it may be possible to use ultrasound for noncontact imaging. The wave front of high-frequency ultrasound ($>$10 MHz), which has propagated to the sample through the uniform sample stage, may trace the sample surface and may be detectable by a cantilever tip even when the tip is slightly apart from the sample surface. If noncontact imaging with high-bandwidth detection is possible, then we can use stiffer cantilevers with a resonant frequency much higher than $\sim$1 MHz.

7.6.2 High-Speed AFM Combined with Fluorescence Microscope

High-speed AFM is complementary to fluorescence microscopy in some respects, and therefore, high-speed AFM combined with fluorescence microscopy is expected to be very useful for biological studies. Conventional AFM has already been combined with fluorescence microscopy [36, 37], but the method used cannot be simply applied to sample-scan-type high-speed AFM because the optical view of the sample is blocked by the cantilever. The sample scan should be changed to a cantilever tip scan, and this change requires fast optical tracking of the scanning cantilever in the optical beam deflection detection system. This fast tracking may be accomplished by scanning a small focusing lens or using an acousto-optic modulator.

When simultaneous recording of optical and AFM images is performed, the strong light used for illumination in fluorescence microscopy may interfere with

AFM imaging through the luminescence of the cantilever tip and the photothermal bending of the cantilever. This problem can be solved by switching off the illuminating light during the trace scan and switching it on during the retrace scan [36].

Tip-scan-type high-speed AFM has an advantage of allowing the use of a wide sample stage (and hence large samples such as cells can be observed). In sample-scan-type high-speed AFM, the sample stage should be $\lesssim$2 mm in diameter to reduce the effect of its mass on the resonant frequency of the z-scanner and to reduce the effect of hydrodynamic pressure on the cantilever dynamics [38]. Even in tip-scan-type high-speed AFM, the effect of hydrodynamic pressure remains but can be minimized by using a large substrate with a projecting flat surface with a small diameter.

7.7 Conclusion

After a lengthy and extensive research carried out by a relatively small number of researchers, high-speed AFM is now applicable to directly visualizing dynamic biomolecular processes to gain more details of the functional mechanisms of biomolecules. It also can be applied to the studies of dynamic phenomena occurring at solid–liquid interfaces, such as cleaning by detergents, corrosion, catalytic reactions, and electrochemical reactions. Thus, this new microscopy will contribute to the progress in various fields. While applying high-speed AFM to a wide range of targets, we have to seek techniques that could break limitations existing in the present high-speed AFM, including the saturating feedback bandwidth and the limited maximum scan range and sample dimensions.

References

1. Huang, B., Bates, M., and Zhuang, X. (2009) Super-resolution fluorescence microscopy. *Annu. Rev. Biochem.*, **78**, 993–1016.
2. Binnig, G., Quate, C.F., and Gerber, C. (1986) Atomic force microscope. *Phys. Rev. Lett.*, **56**, 930–933.
3. Viani, M.B., Schäffer, T.E., Paloczi, G.T., Pietrasanta, L.I., Smith, B.L., Thompson, J.B., Richter, M., Rief, M., Gaub, H.E., Plaxco, K.W., Cleland, A.N., Hansma, H.G., and Hansma, P.K. (1999) Fast imaging and fast force spectroscopy of single biopolymers with a new atomic force microscope designed for small cantilevers. *Rev. Sci. Instrum.*, **70**, 4300–4303.
4. Ando, T., Kodera, N., Takai, E., Maruyama, D., Saito, K., and Toda, A. (2001) A high-speed atomic force microscope for studying biological macromolecules. *Proc. Natl. Acad. Sci. U.S.A.*, **98**, 12468–12472.
5. Ando, T., Uchihashi, T., Kodera, N., Yamamoto, D., Taniguchi, M., Miyagi, A., and Yamashita, H. (2007) Review: high-speed atomic force microscopy for observing dynamic biomolecular processes. *J. Mol. Recognit.*, **20**, 448–458.
6. Ando, T., Uchihashi, T., Kodera, N., Yamamoto, D., Taniguchi, M., Miyagi, A., and Yamashita, H. (2008) Invited review: high-speed AFM and

nano-visualization of biomolecular processes. *Pflügers Arch.-Eur. J. Physiol.*, **456**, 211–225.

7. Ando, T., Uchihashi, T., and Fukuma, T. (2008) High-speed atomic force microscopy for nano-visualization of dynamic biomolecular processes. *Prog. Surf. Sci.*, **83**, 337–437.
8. Yamashita, H., Voïtchovsky, K., Uchihashi, T., Antoranz Contera, S., Ryan, J.F., and Ando, T. (2009) Dynamics of bacteriorhodopsin 2D crystal observed by high-speed atomic force microscopy. *J. Struct. Biol.*, **167**, 153–158.
9. Yamamoto, D., Uchihashi, T., Kodera, N., and Ando, T. (2008) Anisotropic diffusion of point defects in two-dimensional crystal of streptavidin observed by high-speed atomic force microscopy, *Nanotechnology*, **19**, 384009 (9 pp).
10. Shibata, M., Yamashita, H., Uchihashi, T., Kandori, H., and Ando, T. (2010) High-speed atomic force microscopy shows dynamic molecular processes in photo-activated bacteriorhodopsin. *Nat. Nanotechnol.*, **5**, 208–212.
11. Kodera, N., Sakashita, M., and Ando, T. (2006) A dynamic PID controller for high-speed atomic force microscopy, *Rev. Sci. Instrum.*, **77**, 083704 (7 pp).
12. Walters, D.A., Cleveland, J.P., Thomson, N.H., and Hansma, P.K. (1993) Short cantilevers for atomic force microscopy. *Rev. Sci. Instrum.*, **67**, 3583–3590.
13. Sugawara, Y., Kobayashi, N., Kawakami, M., Li, Y.J., Naitoh, Y., and Kageshima, M. (2007) Elimination of nonlinear cantilever dynamics in phase modulation atomic force microscopy (PM-AFM) in constant amplitude mode, *Appl. Phys. Lett.*, **90**, 194104 (3 pp).
14. Kobayashi, N., Li, Y.J., Naitoh, Y., Kageshima, M., and Sugawara, Y. (2006) High-sensitivity force detection by phase-modulation atomic force microscopy, *Jpn. J. Appl. Phys.*, **45**, L793 (3 pp).
15. Fukuma, T., Kilpatrick, J.I., and Jarvis, S.P. (2006) Phase midulation atomic force microscope with true atomic resolution, *Rev. Sci. Instrum.*, **77**, 123703 (5 pp).
16. Ando, T., Uchihashi, T., Kodera, N., Miyagi, A., Nakakita, R., Yamashita, H., and Matada, K. (2005) High-speed AFM for studying the dynamic behavior of protein molecules at work. *Surf. Sci. Nanotechnol.*, **3**, 384–392.
17. Fukuma, T., Okazaki, Y., Kodera, N., Uchihashi, T., and Ando, T. (2008) High resonance frequency force microscope scanner using inertia balance support, *Appl. Phys. Lett.*, **92**, 243119 (3 pp).
18. Anczykowski, B., Cleveland, J.P., Krüger, D., Elings, V., and Fuchs, H. (1998) Analysis of the interaction mechanisms in dynamic mode SFM by means of experimental data and computer simulation. *Appl. Phys. A*, **66**, S885–S889.
19. Kodera, N., Yamashita, H., and Ando, T. (2005) Active damping of the scanner for high-speed atomic force microscopy, *Rev. Sci. Instrum.*, **76**, 053708 (5 pp).
20. Schiener, J., Witt, S., Stark, M., and Guckenberger, R. (2004) Stabilized atomic force microscopy imaging in liquids using second harmonic of cantilever motion for setpoint control. *Rev. Sci. Instrum.*, **75**, 2564–2568.
21. Uchihashi, T., Ando, T., and Yamashita, H. (2006) Fast phase imaging in liquids using a rapid scan atomic force microscope. *Appl. Phys. Lett.*, **89**, 213112 (3 pp).
22. Stark, M. and Guckenberger, R. (1999) Fast low-cost phase detection setup for tapping-mode atomic force microscopy. *Rev. Sci. Instrum.*, **70**, 3614–3619.
23. Yamamoto, D., Nagura, N., Omote, S., Taniguchi, M., and Ando, T. (2009) Streptavidin 2D crystal substrates for visualizing biomolecular processes by atomic force microscopy. *Biophys. J.*, **97**, 2358–2367.
24. Yamamoto, D., Uchihashi, T., Kodera, N., Yamashita, H., Nishikori, S., Ogura, T., Shibata, M., and Ando, T. (2010) High-speed atomic force microscopy techniques for observing dynamic biomolecular processes. *Methods Enzymol.*, **475**, 541–564.
25. Darst, S.A., Ahlers, M., Meller, P.H., Kubalek, E.W., Blankenburg, R., Ribi, H.O., Ringsdorf, H., and Kornberg, R.D. (1991) Two-dimensional crystals of streptavidin on biotinylated

lipid layers and their interactions with biotinylated macromolecules. *Biophys. J.*, **59**, 387–396.

26. Reviakine, I. and Brisson, A. (2001) Streptavidin 2D crystals on supported phospholipid bilayers: toward constructing anchored phospholipid bilayers. *Langmuir*, **17**, 8293–8299.
27. Heyes, C.D., Kobitski, A.Y., Amirgoulova, E.V., and Nienhaus, G.U. (2004) Biocompatible surfaces for specific tethering of individual protein molecules. *J. Phys. Chem. B*, **108**, 13387–13394.
28. Wang, S.W., Robertson, C.R., and Gast, A.P. (1999) Molecular arrangement in two-dimensional streptavidin crystals. *Langmuir*, **15**, 1541–1548.
29. Ratanabanangkoon, P. and Gast, A.P. (2003) Effect of ionic strength on two-dimensional streptavidin crystallization. *Langmuir*, **19**, 1794–1801.
30. Kodera, N., Yamamoto, D., Ishikawa, R., and Ando, T. (2010) Video imaging of walking myosin V by high-speed atomic force microscopy. *Nature*, **468**, 72–76.
31. Miyagi, A., Tsunaka, Y., Uchihashi, T., Mayanagi, K., Hirose, S., Morikawa, K., and Ando, T. (2008) Visualization of intrinsically disordered regions of proteins by high-speed atomic force microscopy. *Chem. Phys. Chem.*, **9**, 1859–1866.
32. Ando, T. *http://www.s.kanazawa-u.ac.jp/phys/biophys/index.htm* (accessed April 23 2010).
33. Lanyi, J.K. (2004) Bacteriorhodopsin. *Annu. Rev. Physiol.*, **66**, 665–688.
34. Hansma, P.K., Drake, B., Marti, O., Gould, S.A., and Prater, C.B. (1989) The scanning ion-conductance microscope. *Science*, **243**, 641–643.
35. Shevchuk, A.I., Frolenkov, G.I., Sanchez, D., James, P.S., Freedman, N., Lab, M.J., Jones, R., Klenerman, D., and Korchev, Y.E. (2006) Imaging proteins in membranes of living cells by high-resolution scanning ion conductance microscopy. *Angew. Chem. Int. Ed.*, **45**, 2212–2216.
36. Kassies, R., Van Der Werf, K.O., Lenferink, A., Hunter, C.N., Olsen, J.D., Subramaniam, V., and Otto, C. (2005) Combined AFM and confocal fluorescence microscope for applications in bio-nanotechnology. *J. Microsc.*, **217**, 109–116.
37. Peng, L., Stephens, B.J., Bonin, K., Cubicciotti, R., and Guthold, M. (2007) A combined atomic force/fluorescence microscopy technique to select aptamers in a single cycle from a small pool of random oligonucleotides. *Microsc. Res. Tech.*, **70**, 372–381.
38. Ando, T., Kodera, N., Maruyama, D., Takai, E., Saito, K., and Toda, A. (2002) A High-speed atomic force microscope for studying biological macromolecules in action. *Jpn. J. Appl. Phys.*, **41**, 4851–4856.

8
Integration of AFM with Optical Microscopy Techniques

Zhe Sun, Andreea Trache, Kenith Meissner, and Gerald A. Meininger

8.1
Introduction

Atomic force microscopy (AFM) was invented in the mid-1980s by Binnig *et al.* [1], and it was first introduced into the field of biology in the early 1990s [2]. Its ability to measure nanometer and subnanometer structures in an aqueous environment at ambient temperature has enabled multiple biological applications to evolve over the past two decades. As examples, AFM has been applied to obtain detailed topographical measurements of the cell surface [3], local cellular elasticity [4], protein–protein interactions [5], protein unfolding [6], protein molecular topography [2], and cell mechanical behavior [7]. New applications continue to be developed, particularly involving applications in which the AFM is integrated into optical microscope platforms. For example, AFM and total internal reflection fluorescence (TIRF) microscopy have been combined to examine the effect of force on the reorganization of focal adhesions in endothelial cells. The AFM was used to apply a compressing force to the cell surface, while cell adhesions at the bottom cell surface were monitored by TIRF microscopy to determine changes of focal adhesion area [8]. In addition to whole cell studies, there has also been interest in using AFM to obtain structural information of specific cell membrane domains. In these applications, cell membranes were isolated and labeled for specific proteins and/or domains using fluorescence probes and AFM was applied to map the membrane structures and dimensions. The AFM image and the fluorescence image of the same membrane were then matched to determine the specific components in the AFM-mapped structures [9, 10]. In addition, AFM technology itself is advancing to include higher spatial resolution [11] and higher scanning speeds [12]. This chapter briefly reviews several useful AFM techniques that have been successfully applied for studies of cell morphology, mechanical properties, dynamics, and adhesion. It also reviews some of the principles involved in configuring the AFM and optical microscope for fluorescence microscopy, TIRF, and Förster resonance energy transfer (FRET) [13] microscopy. We must acknowledge that there are numerous examples of studies and reviews describing applications of AFM, and we apologize that because of space limitations for this chapter, we were not able to thoroughly

Atomic Force Microscopy in Liquid: Biological Applications, First Edition.
Edited by Arturo M. Baró and Ronald G. Reifenberger.

credit the many scientists who have worked in this area. Several excellent reviews are available [14–17].

The AFM is a much more versatile tool for biological investigations than would otherwise be anticipated by investigators not entirely familiar with the capabilities of the instrument and the principles of its operation. As a scanning probe microscopy technique, the heart of the system is a probe (soft spring) with a tip (various configurations) that directly interacts with the biological sample (i.e., molecule, cell, or tissue). The probe can be selected with a variety of different spring constants and shapes and can be biologically functionalized, all of which can be tailored to specific applications. The following is a brief description of several common experimental approaches.

1) **Quantitative measurement of cell topography:** Much like rolling a wheel over the surface of a terrain, the AFM probe can be controlled to scan along the cell surface while tracking the cell height at each point (pixel). The AFM probe is brought into contact with the cell surface and operated so that the probe applies a constant force to the surface to maintain contact. When AFM probe moves from one location to an adjacent one on a cell surface, the force of contact is maintained constant by adjusting the height position of the probe in order to reproduce a fixed amount of deflection of the AFM probe. The continuous height adjustments of the probe are recorded to provide information on cell height. In applications in which there is a desire to minimize the shear forces, the AFM probe can be operated in tapping mode such that, while scanning across the cell, the tip oscillates above the cell surface and touches the cell only at the end of each oscillatory cycle. In either approach, the AFM can provide live cell topography that contains dimensional information of cell surface and near-membrane subsurface structures with high precision. In addition to obtaining the steady state topographical features of the cells, this technique has also been applied to track the changes of cell height, area, and cell volume induced by various treatments [18–21]. In many cases, the cortical cytoskeleton (composed of actin-containing stress fibers) is well illustrated in the AFM topographic images, mainly because of the higher rigidity of these structures compared with the surrounding cell membrane [22]. This feature of AFM topography has also been employed to determine the changes of cell cortical cytoskeletons [23, 24]. With new improvements in scanning speed, it will become possible in the near future to image cells at rates approaching the frame rate of a standard video camera (30 frames per second).
2) **Quantitative assessment of receptor–ligand adhesions on the cell surface:** AFM can accurately measure forces from the piconewton to the nanonewton scale, and consequently, its usefulness has been proved in recording the mechanical strength of receptor–ligand interactions on cell surface [25, 26], between two cells [27], and between a ligand and receptor-bound substrate [19, 28]. Measurement of receptor–ligand interactions by examination of unbinding forces has proved valuable for analyzing processes such as cell–matrix and cell–cell adhesions. Examples of the applications of this

AFM approach include determination of the integrin–fibronectin (FN) [29] binding force on the surface of osteoclast, fibroblast, vascular smooth muscle cells (VSMCs), and endothelial cells (EC) [19, 25, 30, 31]; evaluation of the dynamics of leukocytes binding with the endothelium [27]; examination of adhesive interactions between cancer cells and ECs [32]; mapping of the distribution of vitronectin receptor on ECs [33], examination of nicotinic receptors on neurons (C. G. Clark *et al.*, unpublished results) and assessment of cartilage adhesion properties [34]. These are but a few of the many examples of receptor–ligand relationships that can be studied with AFM. More recently, AFM measurements of adhesion have been extended into the time domain to examine changes in adhesion that occur with cell activation [35].

3) **Measurement of cell and tissue local stiffness/elasticity:** Cell and tissue stiffness/elasticity are measures of the compressional mechanical strength of local structures. It is a material property, generally dependent on the presence or absence of cytoskeletal structures and the existing tension levels or degree of activation of the cytoskeletal elements [36–38]. Using AFM, an increased cell stiffness has been observed in aorta VSMCs [39] and cardiomyocytes [40] isolated from aged animals, which could be associated with the stiffening of vessel and heart walls during aging. This technique has also been applied to measure the stiffness of osteoblast [41, 42] and to evaluate the pericyte contractility [43]. By measuring cell elasticity temporally, it is possible to track changes in cell activation and other mechanically dynamic events [44]. In whole tissue, it has been used to map the microelastic properties of articular cartilage tissue [45] and to determine the biomechanical properties of renal glomeruli [46]. Thus, force indentation measurements are proving very useful for providing information about cell/tissue mechanical properties and the dynamic nature of this variable.
4) **Measurement and application of mechanical forces to cells:** In addition to the ability of the AFM to measure the passive or static biomechanical properties of cells, it also has the ability to apply force to a particular site on a cell and to measure cell-generated forces. This approach has been used to investigate cellular mechanotransduction and to evaluate the active contractile process of cells. In VSMC, the AFM has been used to apply pulling force directly to single focal adhesion sites followed by measurement of the cellular mechanical response to the applied pull. When force was applied selectively through an integrin–FN focal adhesion site on the cell surface, the AFM probe was pulled downward by the cell in response to the pulling force [7]. These studies showed that the AFM can be mechanically coupled to cells to apply force and measure the mechanoresponsiveness of cells.

For each of the applications that are discussed above, it is important to recognize that a number of system parameters must be taken into consideration in configuring the AFM experiment. Special attention must be given to the selection of the appropriate AFM cantilever so that a compatible spring constant and tip size are specified for use. For example, probes with highly sharpened tips are suitable for mapping of molecular structures *in vitro*, but not necessarily for operation on a cell membrane because the probability of rupturing the cell membrane will be much higher. As such, when considering the tip size and profile, the size and properties of the sampled surface must be factored into the decision-making process. Similarly, the spring constant of the cantilever needs to be considered. Cantilevers with low spring constants ($<0.1\ \mathrm{N\,m^{-1}}$) are suitable for measurement of small forces (piconewton to nanonewton) and soft surfaces, as encountered during measurements of cell topography. However, low spring constants will be of limited usefulness when assessing very stiff materials or when applying higher forces ($>10\ \mathrm{nN}$) to a sample. In addition, if the tip is to be functionalized to interact with a biological sample, additional consideration needs to be given to the chemical functionalization process to optimize the usefulness of the AFM probe. In many adhesion studies, AFM probes have been functionalized by coating with biological molecules or by attaching cells. AFM probes are usually hydrophilic and can be easily coated with a variety of proteins, such as FN [30], antibodies, and vitronectin [33]. A variety of chemical modification strategies for the AFM probe tip are available [47, 48]. In some cases, it may be useful to modify the AFM probe tip by attaching a bead (2–5 μm) on the cantilever to increase the area of contact [49]. In order to assess the cell–matrix or cell–cell adhesion at a single cell level, live cells can also be attached to the AFM cantilever [27, 50]. Thus, the important message is that AFM probe needs to be selected with careful thought given to application.

8.1.1 Combining AFM with Fluorescence Microscopy

Integration of AFM with optical microscopy greatly enhances the capabilities of both techniques. From a practical point of view, coupling the AFM with an optical microscope facilitates selection of biological samples for studying and positioning the AFM probe over the selected samples. This is very critical for many AFM applications, especially for the studies of cultured cells or fluorescently transfected cells. The ability to simultaneously view the AFM probe and sample is also useful when it is necessary to carefully maneuver the AFM probe, as would be important for attaching a bead or cell to the AFM probe itself. More importantly, combining AFM with fluorescent microscopy enables simultaneous recording of the cell biomechanical events/properties with the cellular signaling events/biochemical processes, thus permitting correlations to be examined between the two data sets and cause–effect experiments to be designed. These types of experiments often yield new insight into how biochemical events (seen with the fluorescence microscope) are linked to biomechanical events (observed with or induced by AFM). As further examples of combing AFM with fluorescence microscopy, the following sections discuss different combination strategies for using the AFM with several types of advanced fluorescence microscopy techniques.

8.1.1.1 Epifluorescence Microscopy

For epi-fluorescence microscopy, fluorophores are utilized to label molecules of interest. When excited with specific wavelength (excitation light), the fluorophores emit light at higher wavelengths (emission). The emitted fluorescence can be visualized to determine the concentration, localization, and/or dynamics of the targeted molecule. An inverted fluorescence microscope provides the ideal platform that allows the AFM to be mounted on top of the microscope, usually with a specifically designed AFM-microscope coupling stage. Because the AFM is sensitive to noise and vibration, care must be taken to prevent the microscope and its components from transmitting mechanical vibrations to the AFM.

Several important optical components must be included in the optical fluorescence microscopy platform. These include

1) **Excitation light source:** For general application purposes, mercury or xenon lamps coupled with specific excitation filters can provide a wide range of excitation light source. An alternative is the use of a laser, which is commercially available in a wide range of wavelengths. In the selection of fluorophores, the emission wavelength of the probes used should be separable from the AFM laser diode wavelength in order to reduce overlap of its wavelength into the fluorescence image. Some AFM designs come with a piezo-driven x-y scanning stage, in which case, acquiring fluorescence images with the AFM operating in the scanning mode will be more complicated and will require either modification of the microscope or offline image alignment.
2) **Filter assembly:** Inverted fluorescence microscopes usually come with an optical cube to accommodate the excitation filter, dichroic mirror, and emission filter. The assembly of these filters is a common method for separating the fluorescence excitation signal from emission signal. The excitation filter allows only the wavelength of light matching the fluorophore excitation spectrum to pass through to the biological sample. The dichroic mirror reflects the excitation light to the sample, and also reflects any excitation light reflected by the substrate surface; on the other hand, it allows the fluorophore emission to pass through. The emission filter passes fluorescence only from the sample and blocks all other wavelengths. An illustration showing the light path through the filter assembly is shown in Figure 8.1a.
3) **Detection system:** The general choices for detecting fluorescence signals are CCD cameras for widefield illumination, or photon multiplier tube (PMT) for laser scanning microscopy.

8.1.2 Examples of Applications

8.1.2.1 Ca^{2+} Fluorescence Microscopy

Ca^{2+} is an important secondary messenger in mammalian cells, and one of its most important functions is to activate the cell's contractile apparatus involving interaction of actin and myosin. Not surprisingly, the activity of Ca^{2+} signaling is of great interest in the field of cell mechanotransduction. Several groups have combined AFM with calcium fluorescence imaging to directly test Ca^{2+} activity changes when applying mechanical stress on the apical surface of the cell using

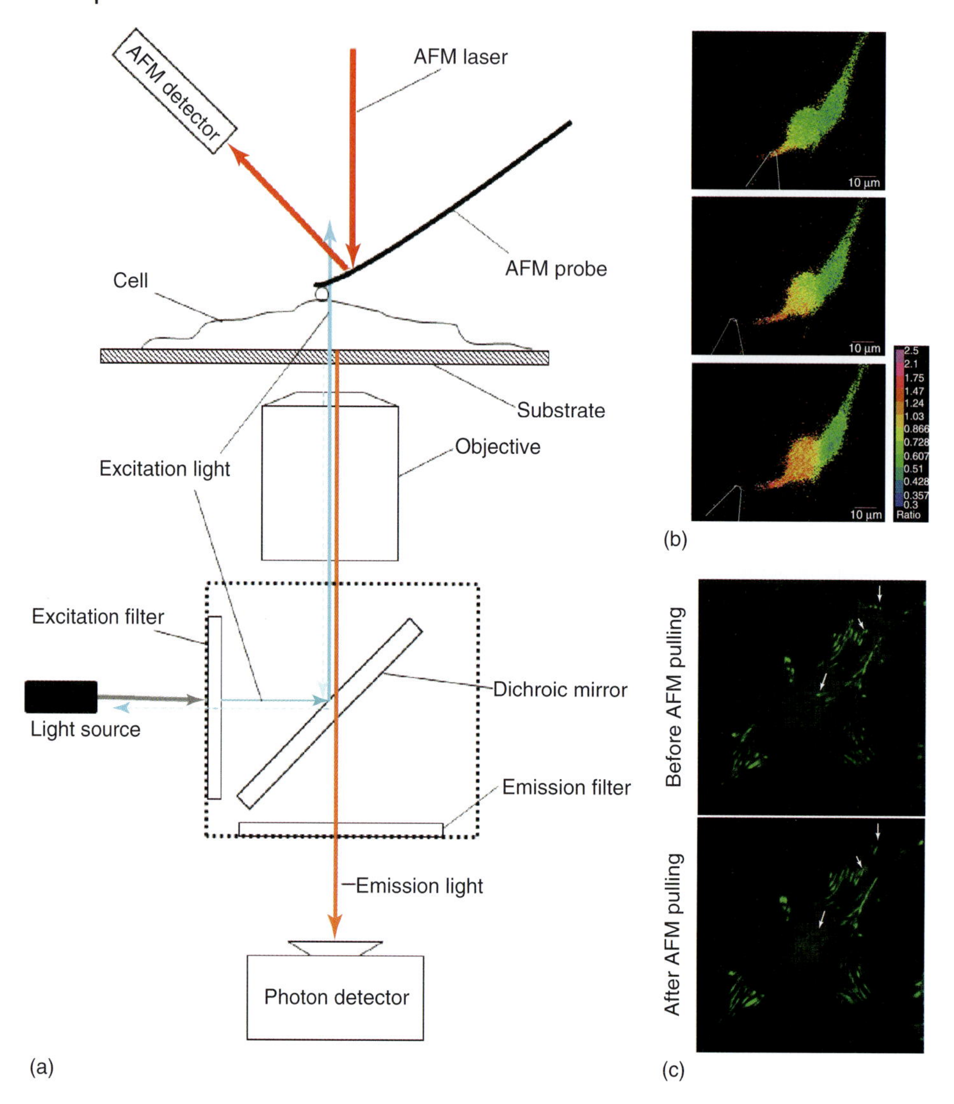

Figure 8.1 Combining AFM with fluorescence microscopy. (a) Diagram of the setup of combined AFM and epifluorescence microscopy. Blue lines indicate excitation light, red lines indicate the light of AFM laser, and the orange line indicates the emission of fluorophore. (b) AFM scanning with FN-coated-probe-induced calcium wave in VSMC. The calcium images were taken at 0.67 s intervals and are displayed in sequence from up to down. The AFM probe was coated with FN (0.5 mg mL^{-1}), and the position and shape of the probe is outlined by the dashed line; the calcium concentration (340/380 ratio) was pseudo-color indexed as represented by the color bar. (c) Application of pulling forces on the surface of VSMC induced reorganization of focal adhesions. GFP–integrin β_3 was transiently expressed in VSMC and used as a marker for focal adhesions. Mechanical force (~1500 pN) was applied through an FN-coated bead using AFM cantilever. Arrows indicate reorganization of focal adhesions after the pulling force was applied.

AFM. For example, Charras *et al.* [41] observed an increase in intracellular Ca^{2+} when applying indentation forces on cultured osteoblasts with AFM. A similar method was also employed by Huo *et al.* [51] to study the indentation-induced Ca^{2+} wave and its propagation across the bone cell network.

Our group has investigated the VSMC calcium changes induced by integrin–FN [29] interactions on VSMC surface. To detect intracellular Ca^{2+}, VSMCs were loaded with Fura-2*AM, and visualized using an Axiovert fluorescence microscope. To engage FN with cell surface integrins, the AFM probe was coated with human FN (0.5 $mg\,mL^{-1}$), and was scanned across the surface of the cell. For detection of Ca^{2+} concentration, the cells were excited with UV light of 340 and 380 nm wavelengths, and the dye emission was collected at 510 ± 40 nm. The fluorescence images were captured continuously (0.67 second per frame) during AFM scanning. As can be seen in Figure 8.1b, scanning of FN-coated probe triggered a calcium wave in the VSMC.

8.1.2.2 AFM – Epifluorescence Microscopy

Integrins are a family of transmembrane proteins that mechanically link cytoskeletal filaments with the extracellular matrix (ECM) and play a central role in the formation of focal adhesions in cultured cells. Abundant evidence has suggested that integrins play a critical role in mediating cell mechanotransduction. We examined whether externally applied mechanical forces affected focal adhesion remodeling in VSMC. To perform this experiment, VSMCs were transiently transfected to express integrin β_3–GFP fusion protein. The integrin β_3–GFP was diffusely distributed on the cell surface and was also localized and concentrated in focal adhesions at the basal surface of the cell (Figure 8.1c). An AFM probe was modified by attaching an FN-coated microsphere (5 μm) to the cantilever. The probe was brought into contact with the cell, and kept in contact for 20 min to allow formation of a strong focal adhesion. When AFM pulling forces (~1500 pN) were applied to the cell-attached microsphere on the AFM tip, integrin β_3-containing focal adhesions were observed to reorganize, presumably as part of a process to allow the cell to adapt to the externally applied force (Figure 8.1c).

Combining AFM and fluorescence microscopy techniques is becoming increasingly useful for a large number of applications in the biological field. From single molecule studies to the investigations of cells and tissues, new applications are being developed all the time.

8.2 Combining AFM with IRM and TIRF microscopy

8.2.1 Interference Reflection Microscopy

Interference reflection microscopy (IRM) is an optical method that can be used for visualizing the attachment sites of living cells (i.e., focal adhesions) in nonfluorescent samples. The optical effect is based on the contrast obtained from interference of light from the glass surface (0 phase shift on reflection) and the sample near the glass surface (180 phase shift on reflection). The light must be monochromatically polarized before entering glass or sample. Owing to interference, the cell–glass interface appears as various shades of gray, with the darkest areas indicating areas of closest contact of cell membrane with the coverslip [52, 53].

8.2.1.1 Optical Setup

Monochromatic light for the IRM technique is selected through a narrowband filter (e.g., 545 nm) from an epifluorescence lamp. The IRM illuminator is equipped with a field diaphragm, used to eliminate stray light from areas outside the field of interest, and an aperture diaphragm, used to control the illuminating numerical aperture at the sample. The incoming light is linearly polarized before being reflected into the objective (Figure 8.2a). This optical method is very sensitive to the reflections inside the objective, which may produce enough stray light to degrade the IRM image [54]. Therefore, the use of an Antiflex objective equipped with a plane-parallel 1/4 wave plate [55] will ensure that only the reflected light that comes from the sample is participating in the image formation [56]. This is realized by properly orienting the 1/4 wave plate with respect to the optical axis, such that the incident linearly polarized light will become circularly polarized after passing through the plate. After the light is reflected back from the sample, the light becomes linearly polarized again, but with a $\pi/2$ phase difference, passes through the analyzer, and then reaches the CCD camera.

8.2.2 Total Internal Reflection Fluorescence Microscopy

TIRF microscopy is an imaging method able to selectively excite fluorophores located in the immediate vicinity of the glass coverslip. Total internal reflection of light is an optical phenomenon that occurs at high- to low-index transition planes (e.g., glass–water interface), and it produces an evanescent field that decays exponentially with the distance from the glass–water interface, with a penetration depth of $\sim$100 nm. This method of surface illumination greatly improves the signal-to-noise ratio in fluorescence microscopy when observing events within the evanescent field.

TIRF microscopy has been extensively used to study organization of protein complexes in cultured cells, such as focal adhesions and endocytosis/exocytosis, and has also been used for single molecule visualization to capture protein dynamics and rare events associated with specific protein behavior [57, 58].

8.2.2.1 Optical Setup

The optical requirements for a TIRF microscope are very similar to the widefield fluorescence microscope with the addition of several specific components [59]. Many optical arrangements are available for TIRF. The use of high-numerical-aperture

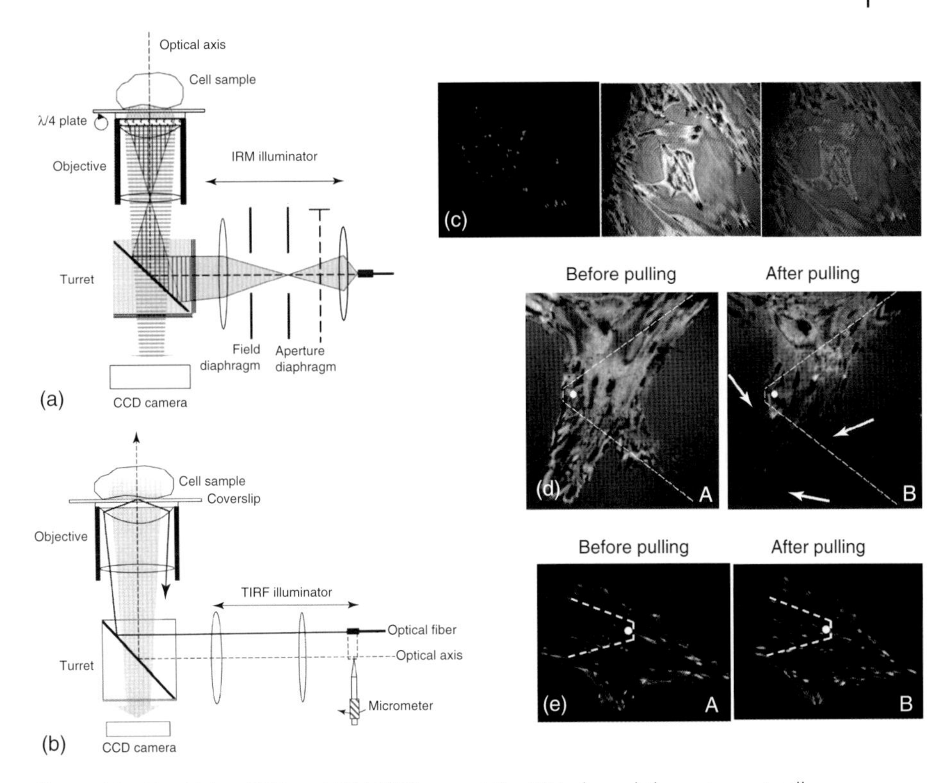

Figure 8.2 Combining AFM and IRM/TIRF. (a) IRM optical path. The monochromatic light passes through the aperture and field diaphragm; then the light is linearly polarized and directed into the objective. The light is circularly polarized by the 1/4 wave plate on its way to the sample. After reflection from the object, the light is again linearly polarized, passes the analyzer, and is further sent to a CCD camera, where the image is recorded. (b) TIRF microscopy optical path. To achieve the condition of total internal reflection at the sample level, it is necessary to position the optical fiber off axis. The reflected light from the sample plane is captured by the objective as a widefield image and is sent further to the CCD camera, where the image is recorded. (c) VSMC transiently transfected with GFP–vinculin was imaged by TIRF and IRM microscopy. TIRF microscopy offers a good contrast fluorescence image with dark background, and the IRM showed the regions at cell base membrane, which are in close contact with the glass substrate. Overlay of the two images illustrates the recruitment of GFP–vinculin to the focal adhesions. Image size is 100×74 μm. (d) The IRM images acquired on the same cell before (A) and after (B) AFM mechanical cell stimulation show differences in the cell shape and reorganization of focal adhesions (white arrows). The white dashed lines represent the position of the AFM tip above the cell. Image size is 82×90 μm. (e) Live VSMC transfected with vinculin–GFP and actin–mRFP was imaged by TIRF microscopy before (A) and (B) after the AFM stimulation. Focal adhesions and actin fibers at basal cell membrane reorganized following the mechanical stimulation. White dashed lines represent the AFM tip above the cell. Panels a,c and d were reproduced from reference [56].

objectives allows sample excitation in TIRF mode and collection of emitted light through the objective in widefield mode. The use of prisms for sample excitation separates the excitation and emission paths. Both configurations can be constructed in the laboratory for custom-built systems. A comprehensive presentation of different variations of TIRF systems can be found in [60].

Most of the commercial TIRF systems are based on high-numerical-aperture objectives combined with laser excitation delivered through optical fiber (Figure 8.2b). The ability to control the off-axis position of the optical fiber with respect to the optical axis of the microscope allows the phenomenon of total internal reflection to take place at the glass–water interface. There is a direct relationship between the off-axis position of the optical fiber and the point of focus of light at the back focal plane of the objective. By precisely positioning the optical fiber with a micrometer, one can control the angle of incidence at the glass–water interface in a reproducible manner. When the angle of incidence is higher than the critical angle

$$sin\,\theta_{\text{critical}} = \frac{n_2}{n_1}$$

where n_1 is the medium with high refractive index (i.e., glass), n_2 is medium with low refractive index (i.e., cell, water), an evanescent field $I(z)$ is produced. The evanescent wave has enough energy to excite the fluorophores only in the immediate vicinity of the glass coverslip. The evanescent field intensity $I(z)$ decreases exponentially with the distance z from the glass–water interface

$$I(z) = I_0 \exp\left(-\frac{z}{d}\right)$$

and has a depth of penetration

$$d = \frac{\lambda}{4\pi}\frac{1}{\sqrt{n_1^2\,sin\,\theta^2 - n_2^2}}$$

where λ is the wavelength of light and θ is the angle of incidence. The fluorescence is collected by the objective, and an image is formed on a CCD camera. Because the evanescent light excites only a thin section of the fluorescent sample (<100 nm), TIRF images have very low background, with virtually no out-of-focus fluorescence (Figure 8.2c). TIRF microscopy works well for long time-lapse live cell studies because of the low phototoxicity induced at the cell level.

The unique features of TIRF make this microscopy technique an important tool for biological studies, which is finding increasing application especially in the area of high-resolution microscopy [61].

8.2.2.2 Applications of Combined AFM–TIRF and AFM–IRM Microscopy

The TIRF–IRM combination with AFM in sequential or simultaneous measurements on live cells or single molecules has been employed for a wide range of applications: (i) micromechanical cell properties can be correlated with structural cellular changes [8], (ii) single molecule manipulation with AFM can be followed with TIRF [62–64], (iii) cross-correlated imaging makes it possible to analyze distribution of self-assembled myosin filaments [65], (iv) selected labeled molecules can be identified by fluorescence [66], and (v) spatial distribution of individual molecules can be analyzed in lipid bilayers [67, 68].

In VSMC mechanotransduction studies, mechanical stimulation applied at the apical cell surface using AFM induces significant changes in cell shape and focal adhesion reorganization that can be monitored in real time using TIRF or IRM. The mechanical stimulation is induced by a functionalized AFM probe (2 μm glass bead coated with FN), which is set at a chosen *x*-*y* coordinate on the cell surface [69]. After a strong focal adhesion is formed, the tip is moved upward in discrete steps. Following each controlled upward movement of the functionalized cantilever tip along the *z*-axis (i.e., pulling vertically away from the cell), the cell responds to the mechanical stimulation [56]. By comparing the IRM images of the same cell at the beginning (Figure 8.2d (A)) and at the end of the experiment (Figure 8.2d (B)), it is evident that the focal adhesions have reorganized, presumably to establish a new mechanical homeostatic state. Performing the same experiment in a live cell cotransfected with vinculin–GFP and actin–monomeric red fluorescent protein (mRFP) (Figure 8.2e), similar dramatic reorganization of focal adhesions is observed. Moreover, the rearrangement of actin cytoskeleton is also evident [70].

8.3 Combining AFM and FRET

8.3.1 FRET

FRET [13] is the nonradiative transfer of energy from a donor to an acceptor. Although fluorescence is usually used to measure the energy transfer efficiency, fluorescence by itself is not a part of the FRET mechanism [71, 72]. When a donor absorbs light, it moves from the ground state to a higher energy state, an excited state. The donor excitation decays to a metastable level of the excited state from where it can decay further to the ground state. If an acceptor is properly aligned and very close to the donor, the energy may be directly transferred to the acceptor via resonance energy transfer (RET), after which the acceptor undergoes a transition to an excited state and will eventually decay to its ground state emitting a photon.

The FRET process is the result of "long-range" dipole–dipole interactions between donor and acceptor and takes place when the two molecules are within ~10 nm, with the transfer rate varying with the sixth power of the separation distance between the donor and acceptor. In addition to the distance, efficient FRET requires that the donor and acceptor dipoles be properly oriented, there is significant overlap between the donor emission and acceptor excitation spectra, and the donor has a high quantum yield [13, 73].

The FRET process is characterized by the Förster distance, R_0, the distance at which half the excitation energy of the donor is transferred to the acceptor, while the remaining excitation is otherwise dissipated. Explicitly, the Förster distance can be calculated as

$$R_0 = 9.79 \times 10^3 \left(JK^2 n^{-4} Q_D\right)^{1/6} \quad (8.1)$$

where J quantifies the spectral overlap between the donor emission and the acceptor absorption, K is the dipole orientation factor, n is the index of refraction of the medium, and Q_D is the quantum yield of the donor. The efficiency of the energy transfer can be calculated as

$$E = \left[1 + \left(\frac{r}{R_0}\right)^6\right]^{-1} \quad (8.2)$$

where r is the distance between the donor and the acceptor. The Förster distance is generally of the order of 2–10 nm and limits the spatial extent of the FRET interaction. This approach provides high-resolution nanoscale sensing. In biological applications, FRET is most often used to monitor molecular interactions and conformational changes or to determine molecular distances [74, 75].

8.3.2
FRET and Near-Field Scanning Optical Microscopy (NSOM)

Given the myriad of scanning probe microscopy techniques, near-field scanning optical microscopy (NSOM) is the technique most readily associated with FRET. In this technique, light is generally delivered to the sample through an optical fiber that is tapered to yield a subwavelength aperture, usually of the order of 50 nm [76]. This aperture enables high-resolution imaging. However, the use of FRET from fluorophores on the NSOM probe has been shown to yield even higher resolution images by coating the probe with donor [77–79] or acceptor fluorophores [80–83]. The light can be used to excite a donor either on the probe or on the surface, while the FRET signal is collected from the donor and/or acceptor emission generally using epifluorescence or confocal microscopy. Detection down to a single FRET pair has been claimed using the FRET-NSOM technique [79]. However, FRET NSOM is technically difficult: there can be minimal resolution improvement because of the simultaneous participation of many FRET pairs, the distance from the surface has to be maintained within the Förster distance (~10 nm), there can be issues with photobleaching over long scan times due to low excitation/collection efficiency, and the probes can collect debris during scanning.

8.4
FRET-AFM

Unlike FRET–NSOM, the combination of FRET with AFM does not seek to further reduce spatial resolution in the imaging plane; AFM already has the capability to render images far below the diffraction limit. Rather, the nanoscale optical functionality added by FRET complements traditional AFM imaging, which is primarily based on force. Vickery and Dunn [83] demonstrated FRET transfer from surface-deposited donors to AFM-tip-mounted acceptors. Although the spatial resolution of the system was degraded because of the coating of AFM tip, the work demonstrated the high z-axial resolution fluorescence imaging enabled by FRET. The Banin group demonstrated FRET–AFM with a spatial resolution down to 30 nm using an AFM tip functionalized with a semiconductor nanocrystal [84].

As an AFM-mounted FN-coated bead is applied to the cell, focal adhesion complexes form between the cell and the bead. We sought to quantify the scale of integrin–ECM interactions during this process. Standard optical imaging techniques such as confocal or multiphoton microscopy do not possess the resolution to localize only those interactions occurring at the cell membrane, that is, the fluorescence signals of integrins that are bound to the bead, and the ones that are just adjacent to the beads will all be recorded. FRET offers a high-resolution method that can significantly improve the detection of integrin–FN interactions at the bead–cell interface. To implement this strategy, semiconductor nanocrystals were attached to the outside of the beads and cells were transiently transfected to express an αv-integrin–red fluorescent protein (RFP) fusion protein [85]. The QD-coated beads were coated with FN, were brought into contact with cell membrane using AFM, and induced the integrin-linked RFP binding to the QD-coated bead. In this system, the semiconductor nanocrystals (emission center wavelength 540 nm) served as the donor and the integrin-linked RFP served as the acceptor (excitation wavelength 561 nm) (Figure 8.3a,b). QD-bead was allowed to engage with cell surface for 20 min to allow formation of integrin–ECM bindings, and then a photobleaching method was used to measure FRET at the bottom surface of the bead (Figure 8.3c). As shown in Figure 8.3d,e, FRET efficiencies of the order of 15% were measured, from which a separation distance of ~7 nm was calculated. This measurement matches with the experimental conditions since the RFP was on the C-terminus of the αv integrin and thus on the inside of the cell membrane, which is approximately 4 nm thick. This result demonstrates the ability of FRET–AFM to enable nanoscale measurements on live cells and will hopefully inspire novel experimental investigations.

8.5 Sample Preparation and Experiment Setup

8.5.1 Cell Culture, Transfection, and Fura-Loading

VSMCs were cultured in Dulbecco's modified eagle medium (DMEM/F-12, Invitrogen Inc., Carlsbad, CA) supplemented with 10–20% fetal bovine serum (FBS, Invitrogen Inc.), L-glutamine (2 mM), sodium pyruvate (1 mM), penicillin (100 unit mL^{-1}), HEPES (10 mM), streptomycin (100 μg mL^{-1}), and amphotericin B (0.25 μg mL^{-1}) (all from Invitrogen). Cells were allowed to grow in a 50 mm tissue culture dishes with a #1 glass coverslip bottom (World Precision Instruments, Inc., Sarasota, FA), and in a humidified incubator (Heraeus Instruments, Inc.,

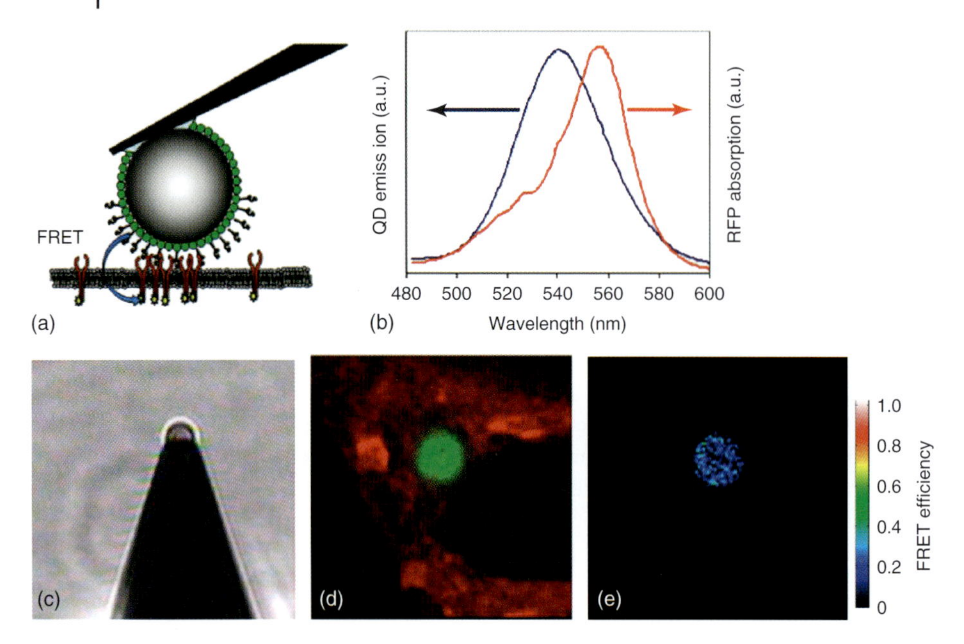

Figure 8.3 Illustration of the AFM–FRET setup. (a) Schematics of FRET setup showing the nanocrystal-coated microsphere and integrin αv–RFP at the cell membrane. The microsphere was mounted on an AFM probe and functionalized with FN. When brought into contact with cell surface by AFM, FN–integrin adhesions were formed, which brought QD close to integrin-tagged RFP to induce FRET. (b) Spectral overlap between the QD emission and the absorption of monomeric RFP. (c) Picture of a QD-microsphere mounted on an AFM cantilever. (d) Dual-channel confocal image of QD-microsphere (green) in contact with a HeLa cell expressing integrin αv-RFP (red). (e) Image of FRET efficiency between the QD-microsphere and the integrin-tagged RFP on the cell surface.

Newtown, CT) with 5% CO_2 at 37°C. To express GFP- and RFP-tagged proteins in VSMC, a Turbofectin transfection reagent (Origene, Inc) was used for transfecting the cells following the manufacturer's recommended protocol. Briefly, 2 ng DNA was mixed with 10 μL Turbofectin and 100 μL DMEM, and incubated for 25 min. The mixture was then added into the VSMC culture, and the expression of fluorescent proteins can usually be observed in 48 h.

8.5.2
Cantilever Preparation

As previously reported, the AFM cantilever was first decorated by mounting a glass microsphere (5 μm) to the tip of cantilever using a marine epoxy [86]. After the epoxy was cured, the microsphere was then further functionalized by incubating with FN (0.5 mg mL^{-1}) for 5 min, and was then washed five times with phosphate buffered saline (PBS). The FN coating was prepared fresh on the day of experiment.

8.5.3
Typical Experimental Procedure

Before the experiment, the cell culture media was changed to serum-free medium, and then a VSMC that was either loaded with Fura-2 or expressed fluorescence proteins was identified under the fluorescence microscope. An AFM probe with FN-coated microsphere was then manually aligned on top of the cell. AFM was set in contact mode operation, and the scanning size was set to 0 μm. In order to apply pulling forces onto the VSMC, the FN-coated microsphere was first allowed to form a strong focal adhesion on cell surface by lodging the microsphere on cell membrane for 20 min. Step pulling forces can be applied through the microsphere by manually adjusting the deflection set point to a smaller value. We typically observed a cantilever sensitivity of $\sim$160 nm V^{-1}; therefore, if the set point was adjusted by 0.5 V, it would correspond to a step increase of pulling force of $\sim$800 pN. The force application and the cellular mechanical response were recorded by AFM height images. The fluorescent images of the cells were collected before, during, and after the force application using the fluorescence microscope software. When analyzing the results, the AFM data and the cell fluorescence images can be correlated in temporal scale in order to determine the force-related effect on VSMC cell signaling.

References

1. Binnig, G., Quate, C.F., and Gerber, C. (1986) Atomic force microscope. *Phys. Rev. Lett.*, **56**, 930–933.
2. Hoh, J.H., Lal, R., John, S.A., Revel, J.P., and Arnsdorf, M.F. (1991) Atomic force microscopy and dissection of gap junctions. *Science*, **253**, 1405–1408.
3. Hoh, J.H. and Schoenenberger, C.A. (1994) Surface morphology and mechanical properties of MDCK monolayers by atomic force microscopy. *J. Cell Sci.*, **107** (Pt. 5), 1105–1114.
4. Radmacher, M., Fritz, M., Kacher, C.M., Cleveland, J.P., and Hansma, P.K. (1996) Measuring the viscoelastic properties of human platelets with the atomic force microscope. *Biophys. J.*, **70**, 556–567.
5. Moy, V.T., Florin, E.L., and Gaub, H.E. (1994) Intermolecular forces and energies between ligands and receptors. *Science*, **266**, 257–259.
6. Rief, M., Gautel, M., Schemmel, A., and Gaub, H.E. (1998) The mechanical stability of immunoglobulin and fibronectin III domains in the muscle protein titin measured by atomic force microscopy. *Biophys. J.*, **75**, 3008–3014.
7. Sun, Z., Martinez-Lemus, L.A., Hill, M.A., and Meininger, G.A. (2008) Extracellular matrix-specific focal adhesions in vascular smooth muscle produce mechanically active adhesion sites. *Am. J. Physiol. Cell Physiol.*, **295**, C268–C278.
8. Mathur, A.B., Truskey, G.A., and Reichert, W.M. (2000) Atomic force and total internal reflection fluorescence microscopy for the study of force transmission in endothelial cells. *Biophys. J.*, **78**, 1725–1735.
9. Frankel, D.J., Pfeiffer, J.R., Surviladze, Z. *et al.* (2006) Revealing the topography of cellular membrane domains by combined atomic force microscopy/fluorescence imaging. *Biophys. J.*, **90**, 2404–2413.
10. Franz, C.M. and Muller, D.J. (2005) Analyzing focal adhesion structure by atomic force microscopy. *J. Cell Sci.*, **118**, 5315–5323.

11. King, G.M., Carter, A.R., Churnside, A.B., Eberle, L.S., and Perkins, T.T. (2009) Ultrastable atomic force microscopy: atomic-scale stability and registration in ambient conditions. *Nano Lett.*, **9**, 1451.
12. Ando, T., Kodera, N., Takai, E., Maruyama, D., Saito, K., and Toda, A. (2001) A high-speed atomic force microscope for studying biological macromolecules. *Proc. Natl. Acad. Sci. U.S.A.*, **98**, 12468–12472.
13. Forster, T., (1965) *Delocalized excitation and excitation transfer*, in Modern Quantum Chemistry (ed. Q. Sinanoglu), Academic press, New york, pp. 93–137.
14. Allison, D.P., Mortensen, N.P., Sullivan, C.J., and Doktycz, M.J. (2010) Atomic force microscopy of biological samples. *Wiley Interdiscip. Rev. Nanomed. Nanobiotechnol.*, **2**, 618–634.
15. Lyubchenko, Y.L., Kim, B.H., Krasnoslobodtsev, A.V., and Yu, J. (2010) Nanoimaging for protein misfolding diseases. *Wiley Interdiscip. Rev. Nanomed. Nanobiotechnol.*, **2**, 526–543.
16. Muller, D.J., Helenius, J., Alsteens, D., and Dufrene, Y.F. (2009) Force probing surfaces of living cells to molecular resolution. *Nat. Chem. Biol.*, **5**, 383–390.
17. Puchner, E.M. and Gaub, H.E. (2009) Force and function: probing proteins with AFM-based force spectroscopy. *Curr. Opin. Struct. Biol.*, **19**, 605–614.
18. Schneider, S.W., Yano, Y., Sumpio, B.E. *et al.* (1997) Rapid aldosterone-induced cell volume increase of endothelial cells measured by the atomic force microscope. *Cell Biol. Int.*, **21**, 759–768.
19. Trache, A., Trzeciakowski, J.P., Gardiner, L. *et al.* (2005) Histamine effects on endothelial cell fibronectin interaction studied by atomic force microscopy. *Biophys. J.*, **89**, 2888–2898.
20. Mosbacher, J., Langer, M., Horber, J.K., and Sachs, F. (1998) Voltage-dependent membrane displacements measured by atomic force microscopy. *J. Gen. Physiol.*, **111**, 65–74.
21. Quist, A.P., Rhee, S.K., Lin, H., and Lal, R. (2000) Physiological role of gap-junctional hemichannels. Extracellular calcium-dependent isosmotic volume regulation. *J. Cell Biol.*, **148**, 1063–1074.
22. Hofmann, U.G., Rotsch, C., Parak, W.J., and Radmacher, M. (1997) Investigating the cytoskeleton of chicken cardiocytes with the atomic force microscope. *J. Struct. Biol.*, **119**, 84–91.
23. Rotsch, C. and Radmacher, M. (2000) Drug-induced changes of cytoskeletal structure and mechanics in fibroblasts: an atomic force microscopy study. *Biophys. J.*, **78**, 520–535.
24. Lu, L., Oswald, S.J., Ngu, H., and Yin, F.C. (2008) Mechanical properties of actin stress fibers in living cells. *Biophys. J.*, **95**, 6060–6071.
25. Lehenkari, P.P. and Horton, M.A. (1999) Single integrin molecule adhesion forces in intact cells measured by atomic force microscopy. *Biochem. Biophys. Res. Commun.*, **259**, 645–650.
26. Franz, C.M., Taubenberger, A., Puech, P.H., and Muller, D.J. (2007) Studying integrin-mediated cell adhesion at the single-molecule level using AFM force spectroscopy. *Sci. STKE*, **2007**, pl5.
27. Zhang, X., Wojcikiewicz, E.P., and Moy, V.T. (2006) Dynamic adhesion of T lymphocytes to endothelial cells revealed by atomic force microscopy. *Exp. Biol. Med. (Maywood)*, **231**, 1306–1312.
28. Baumgartner, W., Hinterdorfer, P. Ness, W. *et al.* (2000) Cadherin interaction probed by atomic force microscopy. *Proc. Natl. Acad. Sci. USA* **97**, 4005–4010.
29. Schaffner-Reckinger, E., Gouon, V., Melchior, C., Plancon, S., and Kieffer, N. (1998) Distinct involvement of beta3 integrin cytoplasmic domain tyrosine residues 747 and 759 in integrin-mediated cytoskeletal assembly and phosphotyrosine signaling. *J. Biol. Chem.*, **273**, 12623–12632.
30. Li, F., Redick, S.D., Erickson, H.P., and Moy, V.T. (2003) Force measurements of the alpha 5 beta integrin-fibronectin interaction. *Biophys. J.*, **84**, 1252–1262.
31. Sun, Z., Martinez-Lemus, L.A., Trache, A. *et al.* (2005) Mechanical properties of the interaction between fibronectin and alpha5beta1-integrin on vascular smooth muscle cells studied using atomic force

microscopy. *Am. J. Physiol. Heart Circ. Physiol.*, **289**, H2526–H2535.

32. Reeves, K.J., Sun, Z., Meininger, G.A., and Brown, N.J. (2009) Measurement of adhesion forces between prostate cancer cells and bone marrow endothelial cells using atomic force microscopy. Annual Fall Meeting of the Microcirculatory Society, Columbia, Missouri.
33. Kim, H., Arakawa, H., Osada, T., and Ikai, A. (2003) Quantification of cell adhesion force with AFM: distribution of vitronectin receptors on a living MC3T3-E1 cell. *Ultramicroscopy*, **97**, 359–363.
34. Chan, S.M., Neu, C.P., Duraine, G., Komvopoulos, K., and Reddi, A.H. (2010) Atomic force microscope investigation of the boundary-lubricant layer in articular cartilage. *Osteoarthritis Cartilage*, **18**, 956–963.
35. Hong, Z., Sun, Z., Li, Z., and Meininger, G.A. (2011) *The Effects of Angiotensin II on the Biomechanical Properties of Microvascular Smooth Muscle Cells: Changes in Adhesive Behavior to Extracellular Matrix and Cell Elasticity*, University of Missouri.
36. Mathur, A.B., Collinsworth, A.M., Reichert, W.M., Kraus, W.E., and Truskey, G.A. (2001) Endothelial, cardiac muscle and skeletal muscle exhibit different viscous and elastic properties as determined by atomic force microscopy. *J. Biomech.*, **34**, 1545–1553.
37. Hong, A., Heinz, E., and Antonik, M.D., (1998) Relative microelastic mapping of living cells by atomic force microscopy. *Biophys. j.*, **74**, 1564–1578.
38. Rotsch, C., Braet, F., Wisse, E., and Radmacher, M. (1997) AFM imaging and elasticity measurements on living rat liver macrophages. *Cell Biol. Int.*, **21**, 685–696.
39. Qiu, H., Zhu, Y., Sun, Z. *et al.* (2010) Short communication: vascular smooth muscle cell stiffness as a mechanism for increased aortic stiffness with aging. *Circ. Res.*, **107**, 615–619.
40. Lieber, S.C., Aubry, N., Pain, J., Diaz, G., Kim, S.J., and Vatner, S.F. (2004) Aging increases stiffness of cardiac myocytes measured by atomic force microscopy nanoindentation. *Am. J. Physiol. Heart Circ. Physiol.*, **287**, H645–H651.
41. Charras, G.T., Lehenkari, P.P., and Horton, M.A. (2001) Atomic force microscopy can be used to mechanically stimulate osteoblasts and evaluate cellular strain distributions. *Ultramicroscopy*, **86**, 85–95.
42. Takai, E., Costa, K.D., Shaheen, A., Hung, C.T., and Guo, X.E. (2005) Osteoblast elastic modulus measured by atomic force microscopy is substrate dependent. *Ann. Biomed. Eng.*, **33**, 963–971.
43. Kotecki, M., Zeiger, A.S., Van Vliet, K.J., and Herman, I.M. (2010) Calpain- and talin-dependent control of microvascular pericyte contractility and cellular stiffness. *Microvasc. Res.*, **80**, 339–348.
44. Azeloglu E.U., Costa K.D. (2009) Dynamic AFM elastography reveals phase dependent mechanical heterogeneity of beating cardiac myocytes. *IEEE Eng. Med. Biol. Soc.*, **2009**, 7180–7183.
45. Stolz, M., Raiteri, R., Daniels, A.U., VanLandingham, M.R., Baschong, W., and Aebi, U. (2004) Dynamic elastic modulus of porcine articular cartilage determined at two different levels of tissue organization by indentation-type atomic force microscopy. *Biophys. J.*, **86**, 3269–3283.
46. Wyss, H.M., Henderson, J.M., Byfield, F.J. *et al.* (2011) Biophysical properties of normal and diseased renal glomeruli. *Am. J. Physiol. Cell Physiol.*, **300**, C397–C405.
47. Hinterdorfer, P., Baumgartner, W., Gruber, H.J., Schilcher, K., and Schindler, H. (1996) Detection and localization of individual antibody-antigen recognition events by atomic force microscopy. *Proc. Natl. Acad. Sci. U.S.A.*, **93**, 3477–3481.
48. Raab, A., Han, W., Badt, D. *et al.* (1999) Antibody recognition imaging by force microscopy. *Nat. Biotechnol.*, **17**, 901–905.
49. Na, S., Trache, A., Trzeciakowski, J., Sun, Z., Meininger, G.A., and Humphrey, J.D. (2008) Time-dependent changes in smooth muscle cell stiffness and focal adhesion area in response to

cyclic equibiaxial stretch. *Ann. Biomed. Eng.*, **36**, 369–380.
50. Friedrichs, J., Helenius, J., and Muller, D.J. (2010) Quantifying cellular adhesion to extracellular matrix components by single-cell force spectroscopy. *Nat. Protoc.*, **5**, 1353–1361.
51. Huo, B., Lu, X.L., Costa, K.D., Xu, Q., and Guo, X.E. (2010) An ATP-dependent mechanism mediates intercellular calcium signaling in bone cell network under single cell nanoindentation. *Cell Calcium*, **47**, 234–241.
52. Bereiter-Hahn, J., Fox, C.H., and Thorell, B. (1979) Quantitative reflection contrast microscopy of living cells. *J. Cell Biol.*, **82**, 767–779.
53. Curtis, A.S. (1964) The mechanism of adhesion of cells to glass. A study by interference reflection microscopy. *J Cell Biol.*, **20**, 199–215.
54. Izzard, C.S. and Lochner, L.R. (1976) Cell-to-substrate contacts in living fibroblasts: an interference reflexion study with an evaluation of the technique. *J. Cell Sci.*, **21**, 129–159.
55. Ploem, J.S. (1975) in *Mononuclear Phagocytes in Immunity, Infection and Pathology* (ed. R. Van Furth), Blackwell Scientific Publication, London, pp. 405–421.
56. Trache, A. and Meininger, G.A. (2005) Atomic force-multi-optical imaging integrated microscope for monitoring molecular dynamics in live cells. *J. Biomed. Opt.*, **10**, 064023.
57. Schmoranzer, J., Goulian, M., Axelrod, D., and Simon, S.M. (2000) Imaging constitutive exocytosis with total internal reflection fluorescence microscopy. *J. Cell Biol.*, **149**, 23–32.
58. Toomre, D., Steyer, J.A., Keller, P., Almers, W., and Simons, K. (2000) Fusion of constitutive membrane traffic with the cell surface observed by evanescent wave microscopy. *J. Cell Biol.*, **149**, 33–40.
59. Trache, A. and Meininger, G.A. (2008) Total internal reflection fluorescence microscopy, in *Current Protocols in Microbiology* (eds R. Coico, T. Kowalik, J.M.Quarles, B. Stevenson, and R.K. Taylor), John Wiley & Sons, Inc., pp. 2.1–2.22.
60. Axelrod, D. (2008) Chapter 07: total internal reflection fluorescence microscopy. *Methods Cell Biol.*, **89**, 169–221.
61. Betzig, E., Patterson, G.H., Sougrat, R. *et al.* (2006) Imaging intracellular fluorescent proteins at nanometer resolution. *Science*, **313**, 1642–1645.
62. Gumpp, H., Stahl, S.W., Strackharn, M., Puchner, E.M., and Gaub, H.E. (2009) Ultrastable combined atomic force and total internal reflection fluorescence microscope [corrected]. *Rev. Sci. Instrum.*, **80**, 063704.
63. Hugel, T., Holland, N.B., Cattani, A., Moroder, L., Seitz, M., and Gaub, H.E. (2002) Single-molecule optomechanical cycle. *Science*, **296**, 1103–1106.
64. Kellermayer, M.S., Karsai, A., Kengyel, A. *et al.* (2006) Spatially and temporally synchronized atomic force and total internal reflection fluorescence microscopy for imaging and manipulating cells and biomolecules. *Biophys. J.*, **91**, 2665–2677.
65. Brown, A.E., Hategan, A., Safer, D., Goldman, Y.E., and Discher, D.E. (2009) Cross-correlated TIRF/AFM reveals asymmetric distribution of force-generating heads along self-assembled, "synthetic" myosin filaments. *Biophys. J.*, **96**, 1952–1960.
66. Yamada, T., Afrin, R., Arakawa, H., and Ikai, A. (2004) High sensitivity detection of protein molecules picked up on a probe of atomic force microscope based on the fluorescence detection by a total internal reflection fluorescence microscope. *FEBS Lett.*, **569**, 59–64.
67. Ira Zou, S., Ramirez, D.M. *et al.* (2009) Enzymatic generation of ceramide induces membrane restructuring: correlated AFM and fluorescence imaging of supported bilayers. *J. Struct. Biol.*, **168**, 78–89.
68. Shaw, J.E., Epand, R.F., Sinnathamby, K. *et al.* (2006) Tracking peptide-membrane interactions: insights from in situ coupled confocal-atomic force microscopy imaging of NAP-22 peptide insertion and assembly. *J. Struct. Biol.*, **155**, 458–469.
69. Trache, A. and Meininger, G.A. (2008) Atomic force microscopy, in *Current*

Protocols in Microbiology, John Wiley & Sons, Inc., pp. 2.1–2.17.

70. Trache, A. and Lim, S.M. (2009) Integrated microscopy for real-time imaging of mechanotransduction studies in live cells. *J. Biomed. Opt.*, **14**, 034024.
71. Clegg, R.M. (1992) Fluorescence resonance energy transfer and nucleic acids, *Methods Enzymol.* **211**, 353–388.
72. Clegg, R.M. (1996) Fluorescence resonance energy transfer, in. (eds. X.F. Wang and B. Herman), *John Wiley & Sons Inc.*, pp. 179–251.
73. Periasamy, A. (2001) Fluorescence resonance energy transfer microscopy: a mini review. *J. Biomed. Opt.* **6**, 287–291.
74. Day, R.N., Periasamy, A., and Schaufele, F. (2001) Fluorescence resonance energy transfer microscopy of localized protein interactions in the living cell nucleus. *Methods*, **25**, 4–18.
75. Erickson, M.G., Liang, H., Mori, M.X., and Yue, D.T. (2003) FRET two-hybrid mapping reveals function and location of L-type Ca2+ channel CaM preassociation. *Neuron*, **39**, 97–107.
76. Pohl, D.W., Denk, W., and Lanz, M. (1984) Optical stethoscopy: image recording with resolution $\lambda/20$. *Appl. Phys. Lett.*, **44**, 651–653.
77. Ebenstein, Y., Mokari, T., and Banin, U. (2004) Quantum-dot-functionalized scanning probes for fluorescence-energy-transfer-based microscopy. *J. Phys. Chem. B*, **108**, 93–99.
78. Muller, F., Gotzinger, S., Gaponik, N., Weller, H., Mlynek, J., and Benson, O. (2004) Investigation of energy transfer between CdTe nanocrystals on polysterene beads and dye molecules for FRET-SNOM applications. *J. Phys. Chem. B*, **108**, 14527–14534.
79. Sekatskii, S.K., Dietler, G., and Letokhov, V.S. (2008) Single molecule fluorescence resonance energy transfer scanning near-field optical microscopy. *Chem. Phys. Lett.*, **452**, 220–225.
80. Shubeita, G.T., Sekatskii, S.K., Chergui, M., Dietler, G., and Letokhov, V.S. (1999) Investigation of nanolocal fluorescence energy transfer for scanning probe microscopy. *Appl. Phys. Lett.*, **74**, 3453–3455.
81. Shubeita, G.T., Sekatskii, S.K., Dietler, G., Potapova, I., Mews, A., and Basche, T. (2003) Scanning near-field optical microscopy using semiconductor nanocrystals as a local fluorescence and fluorescence resonance energy transfer source. *J. Microsc.*, **210**, 274–278.
82. Vickery, S.A. and Dunn, R.C. (1999) Scanning near-field fluorescence resonance energy transfer microscopy. *Biophys.J.*, **76**, 1812–1818.
83. Vickery, S.A. and Dunn, R.C. (2001) Combining AFM and FRET for high resolution fluorescence microscopy. *J. Microsc.*, **202**, 408–412.
84. Ebenstein, Y., Yoskovitz, E., Costi, R., Aharoni, A., and Banin, U. (2006) Interaction of scanning probes with semiconductor nanocrystals; physical mechanism and basis for near-field optical imaging. *J. Phys. Chem. A*, **110**, 8297–8303.
85. Sun, Z., Juriani, A., Meininger, G.A., and Meissner, K.E. (2009) Probing cell surface interactions using atomic force microscope cantilevers functionalized for quantum dot-enabled Forster resonance energy transfer. *J. Biomed. Opt.*, **14**, 040502.
86. Sun, Z. and Meininger, G.A. (2011) Atomic force microscope-enabled studies of integrin-extracellular matrix interactions in vascular smooth muscle and endothelial cells. *Methods Mol. Biol.*, **736**, 411–424.

Part II
Biological Applications

Atomic Force Microscopy in Liquid: Biological Applications, First Edition.
Edited by Arturo M. Baró and Ronald G. Reifenberger.

9
AFM Imaging in Liquid of DNA and Protein–DNA Complexes

Yuri L. Lyubchenko

9.1
Overview: the Study of DNA at Nanoscale Resolution

DNA is one of the extensively studied biological molecules. The structure–function relationship of DNA is perfectly explained by its double helical structure assembled from four complementary bases and is schematically shown in Figure 9.1. The stability of the double helix assures a safe storage of the genetic information. At the same time, the double helix can be unwound while the complementary base pairing provides a mechanism for precise copying of the DNA sequence during replication and transcription into RNA. Both processes are controlled and executed by specialized proteins. DNA is a long linear polymer whose structural properties and interactions with proteins can be observed directly by nanoimaging techniques. Historically, it was electron microscopy (EM) that was widely used to study the various properties of DNA. EM imaging was critical in proving the nucleosomal organization of DNA within the cell, illustrating the beads-on-a-string model of chromatin for unfolded chromatin and elucidating the higher-order structure of chromatin (see [1]). Importantly, the Nobel prize winning discovery of RNA splicing was made in the EM study of hybrids of adenovirus DNA with mRNA [2, 3]. EM also enabled the imaging of DNA melting, providing a full understanding of this fundamental property of the double-stranded DNA molecule and resulting in the theory predicting the melting process of long DNA molecules [4–6].

The advent of the atomic force microscopy (AFM) nanoimaging technique heralded new prospects to molecular biology for applications to study DNA and protein–DNA complexes. A gentle sample preparation methodology for AFM minimizes the sample preservation concern typically associated with EM studies. Moreover, AFM is capable of imaging samples in fully hydrated states and can do so by scanning aqueous solutions. Therefore, in addition to nanoscale static structural data of DNA and protein–DNA complexes, AFM visualizes the dynamics of conformational transitions of DNA and its various stages. This chapter describes the power of AFM imaging in liquid by a few examples, including the study of the conformational dynamics of DNA and a few types of protein–DNA complexes.

Atomic Force Microscopy in Liquid: Biological Applications, First Edition.
Edited by Arturo M. Baró and Ronald G. Reifenberger.

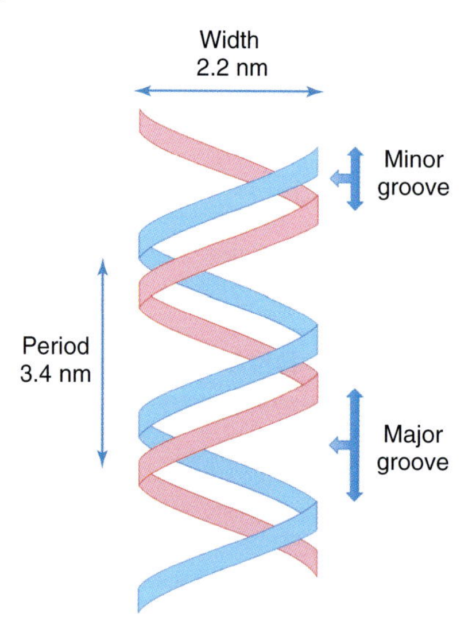

Figure 9.1 Scheme for B-form DNA double helix. Two complementary strands are shown as ribbons of different colors. Major characteristics of the double helix are indicated in the figure.

9.2
Sample Preparation for AFM Imaging of DNA and Protein–DNA Complexes

Sample preparation is a key step for any imaging, and AFM is no exception. The progress in reliable imaging of DNA was made in the early 1990s when there were two initially proposed methods for DNA sample preparation for AFM [7–10]. In both methods, mica was used as the AFM substrate primarily due to its very smooth surface topography. The mica surface is almost atomically flat over microns of area. However, the surface is negatively charged and cannot bind negatively charged DNA molecules. The methods mentioned above and described below overcome this problem.

In the first method [7, 8], the mica surface is treated with Mg^{2+} to increase the affinity of the negatively charged mica surface to DNA. As a result, the DNA molecules were held in place strongly enough to permit reliable imaging by AFM. Later studies showed that pretreatment of mica is not required, but Mg^{2+} cations must be present in the buffer solution [11]. Instead of Mg^{2+} cations, other divalent cations can also be used [12]. Ni^{2+} cations are the second most widely used cations in AFM studies of DNA and protein–DNA complexes. The correlation between the DNA-binding activity of the cation and its hydrated radii was found in [13], suggesting that divalent cations bridge the negatively charged DNA backbone at the mica surface. A later paper [14] elaborated on this idea and proposed a mechanism for DNA binding at the mica surface. In addition to DNA, the cation-assisted procedure was applied to imaging of various protein–DNA complexes (reviewed in [15]).

An alternative method [9] utilizes the pretreatment methodology of mica and other silica surfaces with aminopropyltriethoxy silane (APTES). After APTES covalently binds to mica, amino groups are brought to the surface, resulting in aminopropyl mica. The methodology developed in [16], termed *AP-mica*, enables the preparation of a very smooth functionalized mica. The AP-mica surface remains positively charged even at alkali pH values ($pK_a = 10.4$), and therefore, AP-mica is capable of binding to negatively charged DNA in the pH range of stable DNA duplexes. Divalent cations are not required in this method. In addition, AP-mica retains its DNA-binding activity for a few weeks after the preparation [16]. These features of AP-mica were critical in studies of local and global conformational transitions of DNA and properties of different protein–DNA complexes, including chromatin (reviewed in [15, 16]). Images of supercoiled plasmid DNA (~6 kb) and reconstituted chromatin produced by this method are shown in Figure 9.2a,b, respectively. The images of supercoiled DNA have a clear plectonemic morphology that is perfectly in line with EM data and numerous studies in solution [17]. The cation-assisted methods failed to obtain such morphology of supercoiled DNA, making this methodology problematic for studies of supercoiled DNA.

Another mica functionalization technique similar to the AP-mica procedure was developed, where aminopropyl silatrane is used to yield APS-mica [18]. This functionalized mica surface has properties very similar to AP-mica, but APS-mica is better suited for the imaging of DNA and protein–DNA complexes in aqueous solutions due to a smoother surface of the APS-mica in water compared to AP-mica [19, 20].

In addition to these major methods, a number of other methods of mica surface preparation have been developed. In [21], the procedure for DNA immobilization on mica in the presence of monovalent cations was proposed; however, a rather long incubation period was required for the DNA immobilization. The method utilizing the spreading of DNA onto a carbon-coated mica substrate mediated by the denaturation of cytochrome *c* at an air–water interface was developed in [22] and

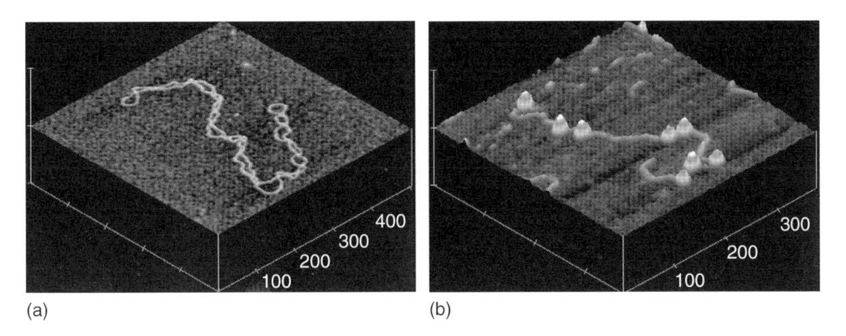

Figure 9.2 AFM images of supercoiled (a) pSA509 plasmid DNA (3.7 kb) and (b) reconstructed chromatin. The samples were deposited onto an AP-mica surface, rinsed with water, and argon-dried. The images were acquired in air with a NanoScope III microscope operated in the tapping mode. (Reprinted from Ref. [15]; Copyright ©2004, Springer.)

applied to imaging of DNA with AFM. The same group later proposed an approach in which mica was coated with a cationic lipid bilayer [23]. The coating of mica with polylysine (PL) yielded a good substrate suitable for imaging of DNA [24], although supercoiled DNA appeared as molecules with a loosely twisted morphology or a completely relaxed shape. The procedure includes immobilization of PL during 30 s, followed by rinse-drying steps and immediate deposition of the sample. This substrate was successfully used for imaging of reconstituted chromatin after drying of the sample [25]. Mica can also be coated with spermidine, and this procedure was used for imaging the dynamics of DNA and reconstituted chromatin with time-lapse AFM in aqueous solutions [26].

9.3 AFM of DNA in Aqueous Solutions

Imaging of DNA and protein–DNA complexes in aqueous solutions is the only way to characterize these molecules in fully hydrated states. Importantly, the capability of AFM to operate in aqueous environments opens prospects for direct imaging of DNA dynamics, including global and local structural transitions, as well as interactions of DNA with proteins. Therefore, attempts to image DNA in fully hydrated states, including imaging in aqueous solutions, were made just after the sample preparation methods were developed. The use of the AP-mica procedure made it possible for the first time to image DNA in aqueous solutions [27]. The images were stable during multiple continuous scans over the same molecules. Importantly, the lengths of fully hydrated DNA molecules were very close to the anticipated lengths for B-form DNA. Another important observation from these studies was the enhanced contrast of DNA compared to the low height for dried DNA samples. Importantly, these early images were obtained with the use of contact AFM mode.

The use of the cation-assisted method for DNA imaging was not that successful. Although images of DNA in aqueous solutions were obtained, pretreatment of the sample was required [28]. The samples deposited on mica have to be dehydrated with propanol. In addition, the use of electron-beam-deposited tips rather than regular commercial silicon nitride tips was required [27]. The measured lengths of DNA molecules in propanol and in aqueous solutions corresponded to the B-DNA geometry (Figure 9.1). However, rescanning of DNA in propanol led to shortening of the DNA length that was explained by the transition of B-DNA to the dehydrated A-DNA conformation. Instability of the images and even cutting of DNA with the tip was reported in this study. Implementation of tapping-mode AFM decreased the tip–sample interaction, enabling reliable imaging in aqueous solutions using the cation-assisted technique [29].

9.3.1 Elevated Resolution in Aqueous Solutions

The resolution of AFM depends on the tip sharpness, and this parameter is one of the resolution-limiting factors in AFM imaging. An additional problem with imaging in air, in terms of the resolution, is the capillary effect [30]. Upon approaching the AFM tip to the surface, a water bridge is formed, complicating the high-resolution imaging because of strong capillary forces [31]. The use of an oscillating tip can break the capillary force, and this methodology dramatically facilitated imaging of biological samples that are typically hydrophilic [32]. The capillary effect is fully eliminated if the imaging is performed in aqueous solution, enabling an increase in the resolution. This capability was demonstrated in [23], in which densely and uniformly packed arrays of DNA molecules and the cationic bilayer were scanned in an aqueous solution. The measured width of the DNA on the AFM images was close to the 2.2 nm width of the DNA double helix (Figure 9.1). In addition, the lateral periodicity along the DNA filaments was resolved and the lateral modulation was in agreement with the known pitch of the double helix (3.4 ± 0.4 nm, Figure 9.1). Given the fact that imaging in aqueous solutions preserves biological samples, the ability to provide high-resolution imaging is another attractive factor for imaging of DNA and protein–DNA complexes in liquids. High-resolution imaging was achieved in [33], where supercoiled DNA molecules were imaged with the use of the AP-mica methodology. Figure 9.3 shows the scan of a mini plasmid DNA (688 bp); the measurements of the DNA width and height were performed along the line. Because the mobility of molecules contributes to widening of the DNA images, the width measurements were made for nonstreaky parts of plasmid DNA in Figure 9.3. The width of DNA molecules determined from different cross sections of images using the procedure, illustrated in Figure 9.3, is 2.8 ± 0.8 nm, which is very close to the 2.2 nm crystallographic width of B-DNA. In this comparison, we need to keep in mind the inevitable widening of DNA because of the finite size of the AFM probes [30]. It should be noted that the image in Figure 9.3 was obtained with a regular tip, suggesting that these tips were terminated with sharp spikes (asperities) providing the necessary high-resolution topographs. This assumption was tested in [34], in which direct cryo-AFM analysis of regular AFM tips was performed.

9.3.2
Segmental Mobility of DNA

Time-lapse imaging was used for studies of global and local dynamics of DNA. The global dynamics of DNA primarily reflects the segmental mobility of DNA defined by the persistence length. The segmental dynamics of DNA was illustrated in a series of studies in which DNA mobility on the surface was imaged directly by time-lapse AFM imaging. Figure 9.4a–c shows three consecutive images of mini plasmid DNA (688 bp) taken with 4.5 min intervals by scanning in Tris-EDTA (TE) buffer (pH 7.5) using the AP-mica methodology [33]. The initial image (a) is smooth, but streaky sections on the second image (b) indicate the movement of segments of the molecule that substantially change in shape on image (c). Combined segmental mobility and overall dynamics of large supercoiled DNA is illustrated in images d–f of the same figure. This set of frames shows the evolution of the shape of the DNA molecule as a function of time. The two elongated, thick

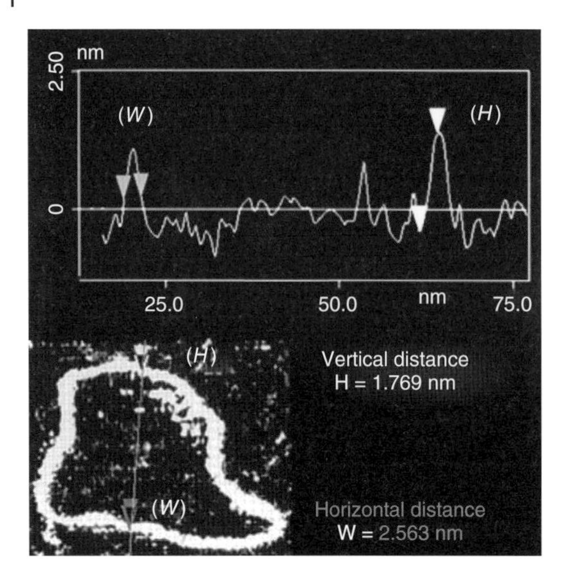

Figure 9.3 High-resolution image of mini plasmid DNA. The graph above the image shows the cross-sectional data for width (*W*) and height (*H*) along the line in the image. The measurements were done for the part of the molecule that was stable during scanning. For imaging in liquid, 1 × 1 cm pieces of AP-mica were mounted on the sample stage on the microscope (NanoScope III, Digital Instruments, Santa Barbara, CA), and 25 μl of DNA solution was injected into the gap between the glass tip holder and the AP-mica. (Reprinted from Ref. [33]; Copyright ©1997, The National Academy of Sciences of the United States.)

regions of close DNA–DNA contacts are indicated with arrows in Figure 9.4d–f. The DNA strands in these regions move apart slightly, leading to the formation of small loops.

Segmental dynamics of DNA was studied systematically in [35]. Figure 9.5a shows three selected AFM images of linearized pBR322 DNA molecules (about 1.5 μm in length; two molecules on the images) acquired during 4 h of continuous observations. Traces of these molecules assembled from 104 frames are superimposed and are shown as bundles in (b) and (c). The data graphically illustrate the time-dependent segmental dynamics of DNA. The measurements of the local flexibility of DNA were performed as illustrated in (d). The measured variability of the site-specific DNA flexure was in good concordance with the predictions of the sequence-specific DNA curvature.

The advent of high-speed atomic force microscopy (HS-AFM) [36, 37] made it possible to extend the ability of AFM to detect molecular motion at a subsecond timescale. The DNA fragment (491 bp, 167 nm) in 10 mM HEPES buffer containing 5 mM Mg^{2+} cations was imaged with a HS-AFM instrument (RIBM Co., Ltd). The data acquisition rate was 3 frames per second. One frame is shown in Figure 9.6a. The shape of the DNA molecule is consistent with images obtained with traditional AFM. The DNA path on this and other images was traced and superimposed by placing the top left ends of the molecules at one point. Figure 9.6b shows a bundle

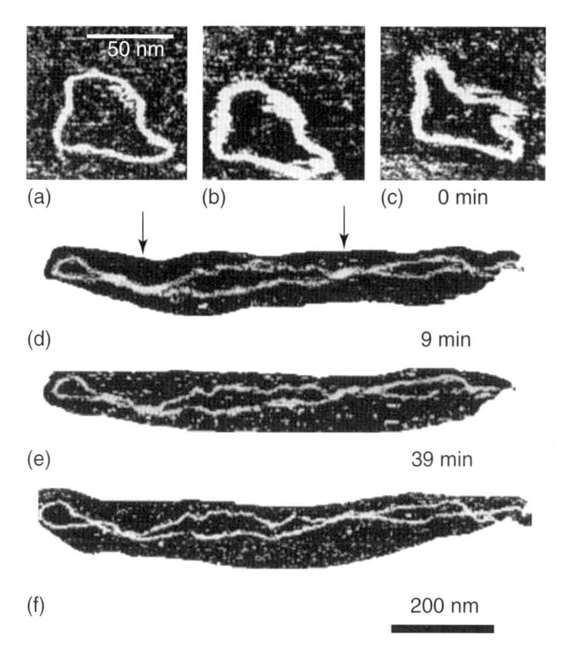

Figure 9.4 DNA motion during scanning. (a–c) AFM images of mini plasmid DNA (688 bp) obtained by time-lapse scanning in Tris-EDTA buffer over a 300 × 300 nm area. The images in (a–c) were taken with an interval of 4.5 min. (d–f) AFM images of an individual DNA molecule, demonstrating segmental movement of the molecule at the solid–liquid interface. The images were taken in TE buffer with 80 mM NaCl. The images in (e) and (f) were taken at 9 and 39 min, respectively, after the image in (d) was recorded. The regions of close DNA–DNA contacts are indicated with arrows. See Figure 9.2 for experimental details. (Reprinted from Ref. [33]; Copyright ©1997, The National Academy of Sciences of the United States.)

of traces of the same molecule taken over 20 frames. These images demonstrate that segmental dynamics at the surface–liquid interface occurs at the millisecond timescale rather than the minute scale, which was reported previously. Such a high segmental mobility of DNA suggests that DNA is not in tight contact with the surface and is capable of interacting with proteins and other ligands that can be monitored with the time-lapse technique.

9.4
AFM Imaging of Alternative DNA Conformations

9.4.1
Cruciforms in DNA

Although the double helical structure of DNA with crystallographic B-form geometry shown in Figure 9.1 is the predominant conformation of DNA, depending on the

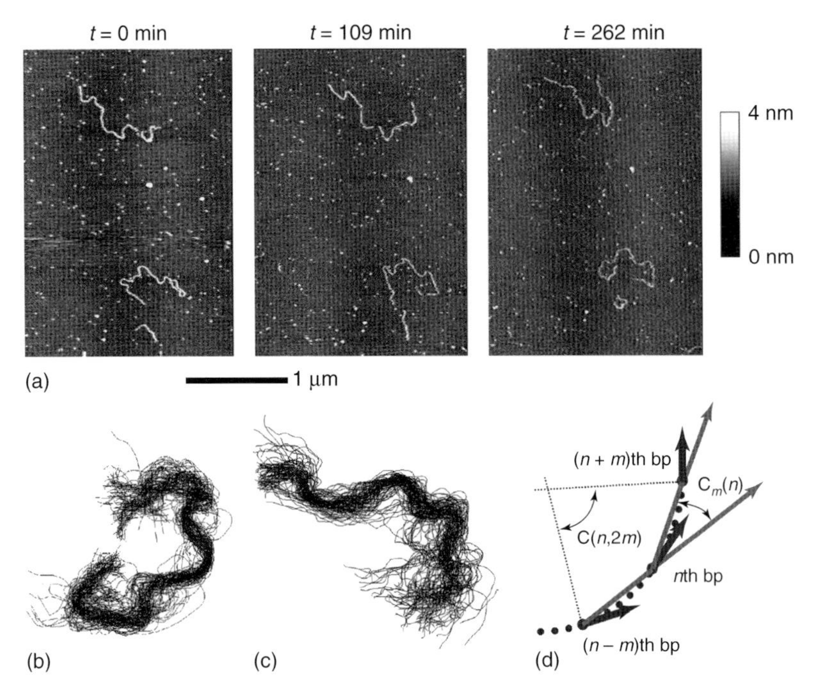

Figure 9.5 Segmental dynamics of DNA. (a) Three samples of AFM images out of a 4 h time-sequence observation of two linearized pBR322 DNA molecules. Each molecule is 1.5 μm long. The data were acquired with tapping-mode AFM with the use of the cation-assisted sample preparation method. pBR322 plasmid molecules were deposited on a disc of freshly cleaved ruby mica (Mica New York Corp., NY) for ~1 min from a 1 $\mu g\,ml^{-1}$ DNA solution that contained 4 mM HEPES and 1 mM $MgCl_2$ (pH 6.8–7.4). (b,c) Superimposition of 107 and 94 traces derived from AFM images of the two linearized pBR322 DNAs. The images were obtained at time intervals of about 2 min. (d) Schematic representation of the segmental curvature. (Reprinted from Ref. [35]; Copyright © 1998, with permission from Elsevier.)

sequence and environmental conditions, DNA can adopt different geometries. Conformational dynamics is a fundamental property of DNA, enabling it to accomplish various genetic functions. A set of non-B-conformations of DNA primarily includes left-handed Z-DNA, cruciforms, intramolecular triple helices, and unwound regions, the biological role of which is widely discussed [38–41]. Various biochemical and gel electrophoresis techniques were instrumental in detecting alternative conformations [38], but no direct data on the structure of local conformations within a long DNA molecule were available.

Inverted repeating sequences were discovered in DNA as soon as genome sequences became available. The biological significance of the DNA regions containing inverted repeats was widely discussed owing to the ability of these regions to form alternative structures, such as cruciforms (reviewed in [38]). However, their direct imaging in naturally existing DNA molecules was revealed only with the use of AFM [42, 43]. AFM with the AP-mica procedure was used for

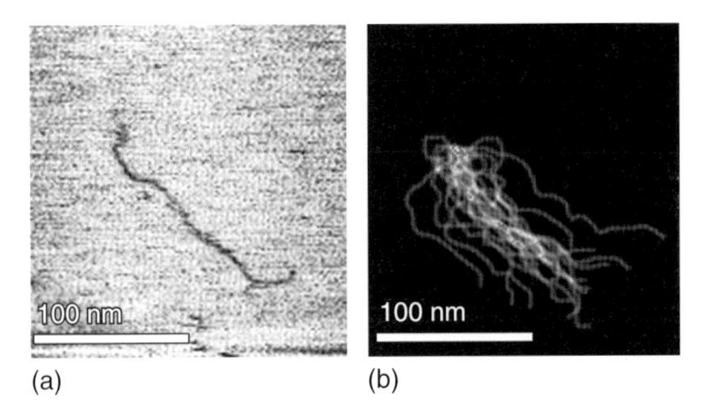

Figure 9.6 (a) Single frame of HS-AFM image obtained at 3 frames per second. (b) DNA strands were traced in 20 successive frames, and the traces were overlaid by superimposing the top ends of the molecules.

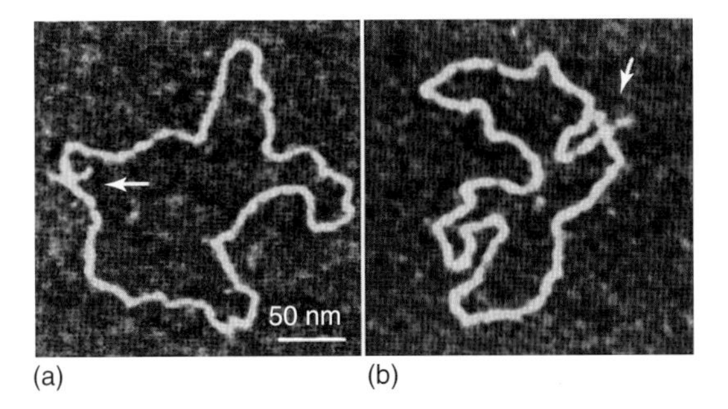

Figure 9.7 (a,b) AFM images of pUC8F14C plasmid DNA obtained by imaging in air. The cruciforms are indicated with arrows. The plasmid DNA had a low supercoiling density ($Lk - Lk_0$, from −7 to −13) leading to circular DNA molecules with a few or no supercoiling turns after the formation of the cruciform. The sample was prepared using the AP-mica methodology. See paper [42] for details. (Reprinted from Ref. [42]; Copyright © 1998, with permission from Elsevier.)

imaging of the cruciform structure in plasmid DNA containing a 106 bp inverted repeat with an expected length of 53 bp for the arms [42]. The AFM images of two DNA molecules with cruciforms for the sample prepared by deposition from a low salt solution (TE buffer) are shown in Figure 9.7. The cruciforms appear as clear-cut extrusions indicated on these images with arrows. The sizes of the arms for extended extrusions are 15–20 nm, which are in full agreement with the expected length of the hairpins containing 53 bp (18 nm for B-helix DNA geometry). The new finding revealed by these studies with AFM was that cruciforms adopt primarily two different conformations, an unfolded conformation with approximately a 180° angle between the hairpin arms (a) and a folded conformation with an acute angle between the arms, ~60° on image in (b). These data suggest that the cruciform is a dynamic structure, and time-lapse AFM was applied to directly visualize this process.

The images assembled in three columns in Figure 9.8 are the frames from a series of scans over the same area with three DNA molecules containing cruciforms

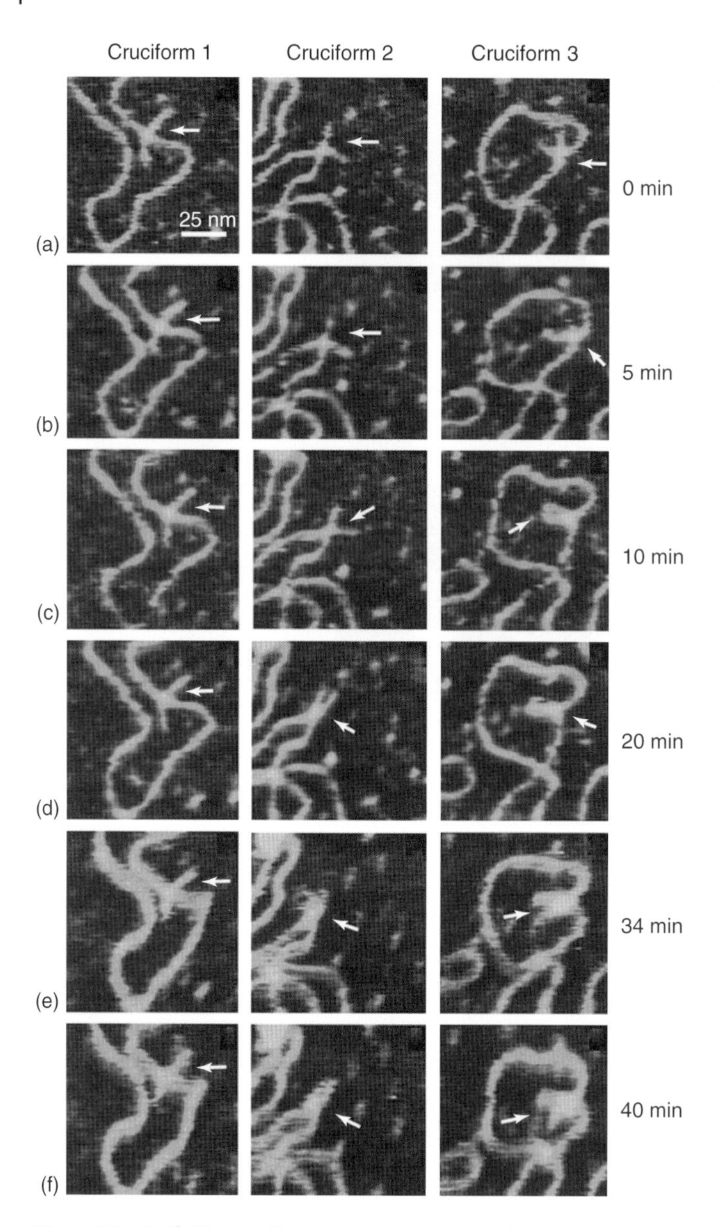

Figure 9.8 (a–f) The cruciforms in motion: consecutive scans of the fragments of molecules with three different initial geometries of cruciforms zoomed around only one cruciform; the time is indicated on the right. (Reprinted from Ref. [42]; Copyright © 1998, with permission from Elsevier.)

of different geometries [42]. Cruciform 1 is in an extended conformation, which generally remained unchanged during all observation times. The angles between the hairpins and the main DNA strands were measured. The measurements for the opposite (vertical) acute angles, 76°(±11) and 79°(±10), indicate that the hairpin arms of extended cruciforms have a very low mobility. The hairpin arms of cruciform 2 were initially quite far away from each other (with an angle of about 100°), but then they moved (frame b) and finally adopted a parallel orientation (frame c). The behavior of cruciform 3 was quite different. Initially, the angle between the hairpins was about 60°, then the arms became parallel (frame b) and moved apart again at the end of the experiment (frame c). Thus, the X-type structure of cruciforms is very mobile, which explains the large variability of the angle values analyzed in [42]. A relatively low mobility of the main DNA strands of cruciforms in experiments in liquid can be explained by their stronger binding to the surface compared with the short hairpin arms.

Unlike the X-type conformation, the extended cruciform (cruciform 1) geometry is less dynamic. The major difference between the unfolded and folded conformations of the cruciform is the shape of the DNA around the cruciform. The cruciform in the X-type conformation induces a kink between the main DNA helices, while the helices are smooth with no kink in the unfolded conformation. These observations led to the hypothesis that the conformational transition of the cruciform functioned as a molecular switch for conformational dynamics of supercoiled DNA [43]. The interplay between the local DNA structures and the global DNA conformations (topology) plays a significant role in the function of DNA. The cruciforms formed by inverted repeats in supercoiled DNA adopt different conformations, forming alternative DNA structures that govern various DNA functions. This conformational transition can have a strong effect on global DNA structure and dynamics, playing the role of a molecular switch and influencing the search of homologous regions within the supercoiled DNA molecule or in a topological domain [43].

Detailed studies of the conformational transitions of the cruciform with the use of time-lapse AFM in liquid in various conditions were performed in [44]. The results obtained in this paper provided a number of important characteristics to the structure and dynamics of the cruciform, depending on the DNA supercoiling and ionic conditions. While the conformation of synthetic junctions is defined by ionic conditions, primarily the presence of divalent cations, DNA supercoiling is the most critical factor in determining the cruciform conformation. Therefore, the cruciform of DNA with high supercoiling density adopts a predominately folded conformation even at low ionic strength and the absence of Mg^{2+} cations; however, these cations further shift the equilibrium between the folded and unfolded conformations toward the folded state. In topoisomers with low superhelical density, the population of DNA with folded conformations is 50–80%, depending on the ionic strength of the buffer and the type of cation added. Overall, time-lapse AFM studies led to the conclusion that DNA supercoiling is a key factor in determining the conformational dynamics of the cruciform structure.

9.4.2 Intramolecular Triple Helices

An intramolecular DNA triplex, or H-DNA, formed by homopurine-homopyrimidine (Pu·Py) tracts is another biologically important alternative DNA structure [45]. Structural studies of intramolecular triplexes are instrumental in understanding the mechanism of how H-DNA is involved in genetic processes. The challenge in imaging H-DNA is the requirement of low pH for a stable existence of the structure. This requirement was met in a study [46] in which the APS-mica procedure was used. A high-resolution AFM image of the pUC19 plasmid with a 46 bp long purine–pyrimidine repeat prepared at acidic pH is shown in Figure 9.9. A distinct feature of the molecules prepared at acidic pH is the formation of a clear kink with a short protrusion indicated by an arrow. The formation of a sharp kink is fully consistent with the model of the intramolecular DNA triplex [45] inserted in Figure 9.9, although various conformations of H-DNA in plasmid DNA were observed [46] pointing to the dynamic feature of this structure in supercoiled DNA.

Time-lapse AFM imaging was performed to visualize the dissociation process of H-DNA with the use of the APS-mica methodology [47]. The plasmid DNA containing the H-DNA tract (Figure 9.9) was stable at pH 5, but changes occurred after raising the pH of the buffer to the neutral condition, illustrated by a subset of images in Figure 9.10. The initial image of DNA (a) was indistinguishable from the image taken at acidic pH, with the H-DNA protrusion indicated with an arrow. The image in (b) shows that the protrusion becomes smaller, eventually resulting in a complete unfolding of the H-DNA, as shown in (c). A complete set of 21 consecutive images [47] demonstrated that the local conformational transition induced the change in the global shape of the DNA plasmid. This finding graphically illustrates the link between global and local structural transitions in supercoiled

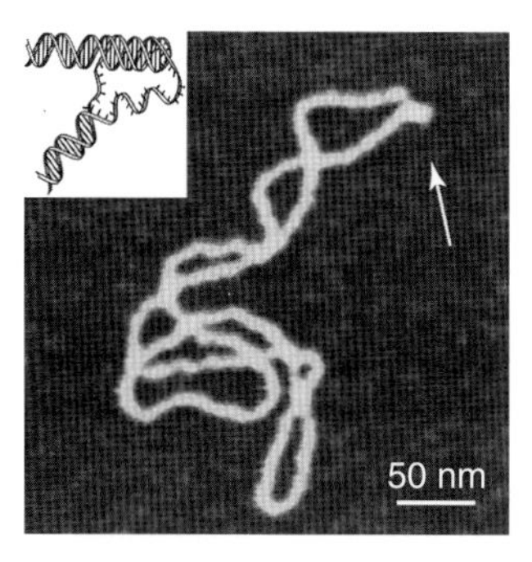

Figure 9.9 AFM image of plasmid pCW2966 containing a 46 bp mirror repeat forming H-DNA at acidic pH. H-DNA regions are indicated with an arrow. The DNA sample was incubated in 50 mM sodium acetate buffer at pH 5.0 and then deposited onto APS-mica in the same buffer. The images for plasmid-containing H-DNA were acquired in air with a NanoScope III microscope, operated in tapping mode. A sharp kink at the base of a thick protrusion is indicated with an arrow. Inset shows the model for H-DNA.

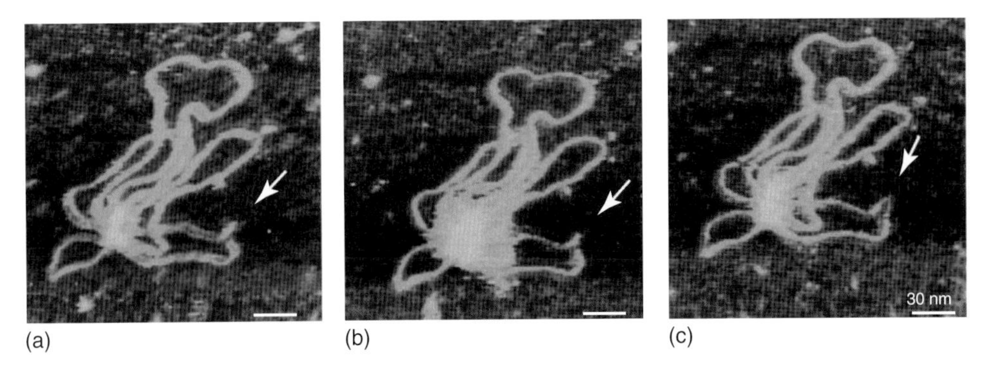

Figure 9.10 (a–c) Time-lapse AFM imaging of H-DNA dissociation after replacing the acidic buffer with a neutral one. The position of the protrusion that stably existed at acidic buffer (H-DNA) is indicated with arrows. (Reprinted from Ref. [47]; Copyright © 2002, Cambridge University Press.)

DNA. Detailed analysis of the time-lapse data [47] showed that the H-DNA region undergoes a series of structural changes after the disappearance of the protrusion, suggesting that the H- to B-form transition is not a simple two-state conformational transition. There are a number of conformational substates, including unwound regions, that are capable of being visualized with time-lapse AFM.

9.4.3
Four-Way DNA Junctions and DNA Recombination

Four-way DNA junctions or Holliday junctions (HJs) were suggested in 1964 by Robin Holliday as central intermediates in homologous and site-specific recombination [48]. Two models explaining the mechanism of branch migration were suggested. According to the early Alberts-Meselson-Sigal (AMS) model [49, 50], exchanging DNA strands around the junction remain parallel during branch migration. Kinetic studies of branch migration [51] suggest an alternative model in which the junction adopts an extended conformation. The model was tested by direct imaging of the branch migration process using a special design of HJ suitable for direct observation of branch migration with AFM [52]. Figure 9.11 illustrates the results of such an approach. (a) shows HJ in the folded conformation, which, according to the AMS model, is the geometry needed for branch migration. (b) shows one of the images in the middle of the observation period, and (c) shows the final event – dissociation of homologous DNA strands. The data were analyzed [52], and the results of the length measurements are presented as plots in Figure 9.11e. They show changes in lengths of the arms during the entire observation period. There is a substantial change in the lengths of the arms starting in frame 7 after the transition from the cis to trans conformation occurred. Short arms after image 9 start growing and slightly fluctuate in length before the abrupt dissociation of the junction occurs (Figure 9.11d). The lengths of the long arms change in an opposite fashion, as one would expect for coordinated arms changing lengths during branch

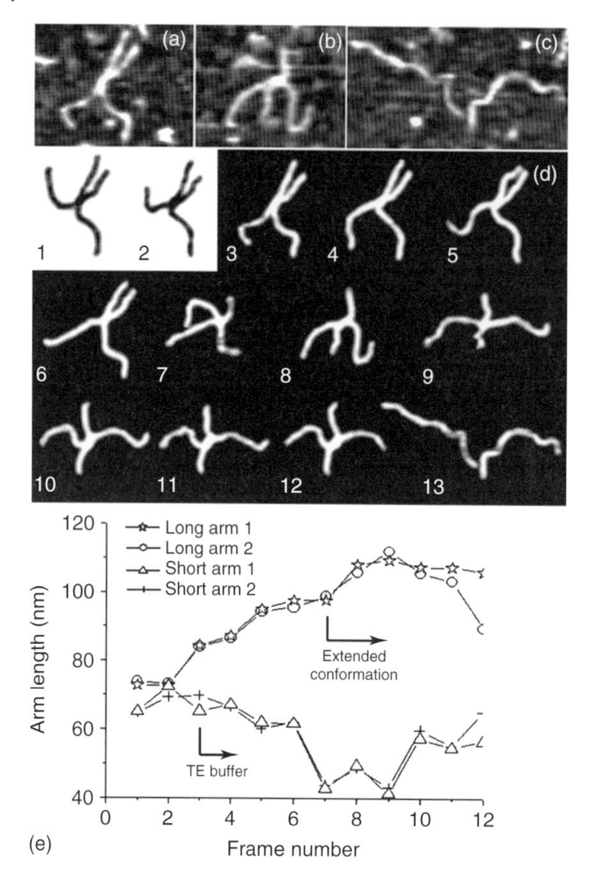

Figure 9.11 Time-lapse AFM images illustrating the dynamics and strand exchange of another molecule. (a–c) Images of the same molecule at different times. (d) A set of 13 images illustrating the dynamics of the same Holliday junction. The capturing times for each image are as follows – TNM buffer containing 10 mM Tris–HCl, pH 7.9, 50 mM NaCl, 10 mM MgCl2 (black molecules on a white background): 1, 0 min and 2, 31 min, and TE buffer (the molecules in white on a black background): 3, 37 min; 4, 40 min; 5, 48.5 min; 6, 60.0 min; 7, 64.5 min; 8, 65.7 min; 9, 70.0 min; 10, 71.5 min; 11, 72.8 min; 12, 74.3 min; and 13, 75.6 min. (e) The measured arm lengths for each molecule in (d). (Reprinted from Ref. [52]; Copyright © 2003, by the American Society for Biochemistry and Molecular Biology.)

migration. The strand dissociation also occurs abruptly (images 12 and 13), which is in line with other observations in the paper [52]. The complete sets of images for both experiments assembled as animated files can be viewed in the supplementary materials of the paper [52]. Overall, the time-lapse AFM observations directly and unambiguously support the model for branch migration in which the extended conformation of the HJ, rather than folded ones, is required for branch migration. Note that unfolding of the HJ is required for protein-mediated branch migration. In

Escherichia coli, a special protein, RuvA, accomplishes this unfolding, as confirmed by AFM imaging [43].

9.5 Dynamics of Protein–DNA Interactions

9.5.1 Site-Specific Protein–DNA Complexes

The time-lapse AFM technique was successfully applied to studies of protein–DNA complexes of various types. An early study [53] revealed the assembly process of RNA polymerase (RNAP)–DNA complexes in a series of time-lapse images. Later, the same group visualized the sliding of the polymerase along DNA [54], concluding that diffusion of RNAP along DNA constitutes a mechanism for accelerated promoter location. Further studies [55] showed that in addition to sliding, an enzyme could jump from one site to another. The results are illustrated in Figure 9.11, which shows a time-lapse sequence of a transcribing RNAP complex imaged in transcription buffer with AFM. A stalled elongation complex imaged before the addition of nucleoside triphosphates to the liquid cell is shown in Figure 9.12a,c–f,h–j. On nucleotide injection, RNAP began to thread the DNA template in a processive and unidirectional manner (Figure 9.12b,g) toward the shorter arm of the DNA, consistent with the orientation of the promoter in the template. After transcription, RNAP usually released the DNA. The cation-assisted procedure was used in these studies.

A similar immobilization approach enabled the observation of the interaction of DNA with photolyase by AFM [56]. Nonspecific photolyase–DNA complexes were visualized, showing association, dissociation, and movement of photolyase over the DNA. The final movement suggests a sliding mechanism by which photolyase can scan DNA for damaged sites.

Detailed studies of the interaction of p53 protein with DNA were described in [57] (Figure 9.13). Initially, the protein is located close to the left end of the DNA (i′). The DNA begins moving and flips from one side of the protein to the other. It is suggested that the flipping was caused by a local dissociation of the DNA from the surface with subsequent movement over the top of the protein, as the protein does not change its position and morphology. The flipping process also suggests that relatively strong non-sequence-specific interactions between p53 and DNA take place. The protein dissociated from the DNA (l′), but bound the same flank of DNA later (o′), followed by a sliding event (about 100 nm sliding distance). These dynamics were analyzed by the length measurements, and according to this analysis, the nanoscale dynamics of the complex (DNA flipping, protein sliding, dissociation, and association) occurs at the minute timescale. This observation is in agreement with an early theoretical model in which sliding and three-dimensional diffusions were proposed as modes for the site-search mechanism by a protein [58, 59].

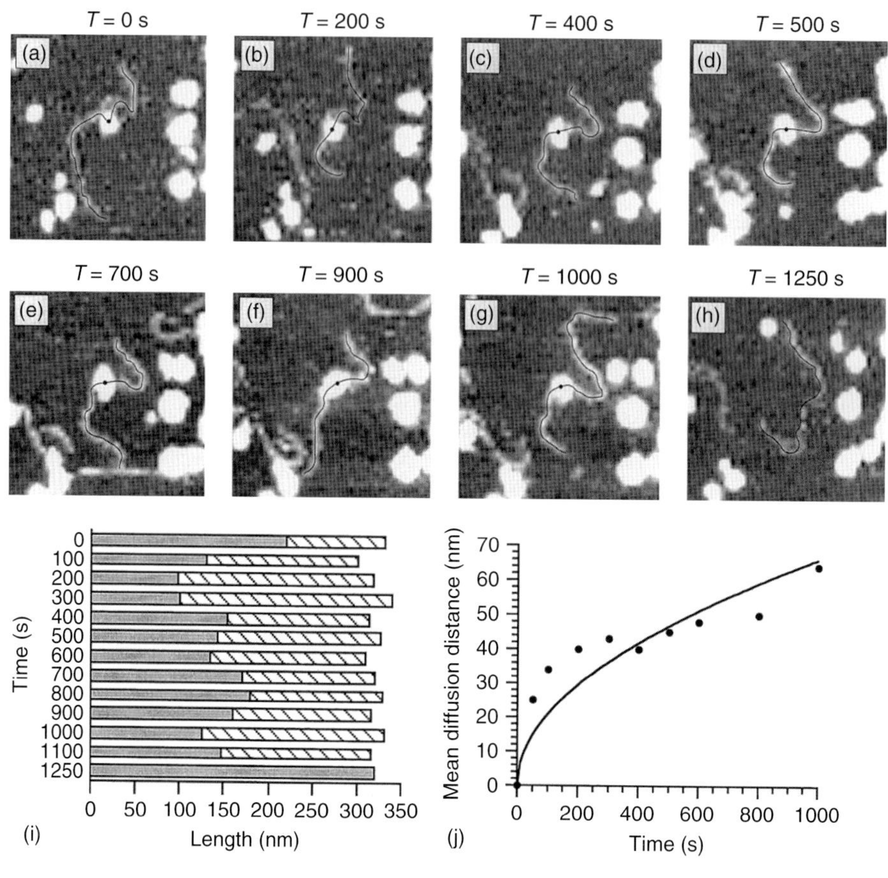

Figure 9.12 Time-lapse AFM images of a transcribing RNAP. (a) Image of a stalled elongation complex (A70), as judged from the length of the DNA arms, before the injection of NTPs into the liquid cell. (b–h) Selected images after transcription were resumed by injecting 5 mM NTPs in transcription buffer through the flow cell. During transcription, which proceeds toward the shorter arm, the RNAP remained stably bound to the surface, while the DNA was pulled through the protein. The DNA contour is traced with a thin line, and the center of the RNAP is marked with a dot. Image size is 350 × 350 nm. (i) Bar plot representation of the length of the two DNA arms (*hatched* and *solid bars*) in 13 sequential images. (j) Plot of mean diffusion distance D_l versus time t to illustrate square root dependence typical of diffusion-controlled motion. (Reprinted from Ref. [55]; Copyright © 1999, with permission from Elsevier.)

In addition to protein recognition of one site on DNA, there is a large family of proteins that need two specific sites for functionality. The incorporation of two binding sites is a complication for theoretical study, and two-site binding also narrows the assortment of compatible experimental approaches. Recent studies showed that AFM is a method of choice for characterizing protein–DNA systems of this type [19, 60, 61]. Time-lapse HS-AFM in liquid [37] has been recently applied to interactions of the ATP-dependent type III restriction enzyme EcoP15I with

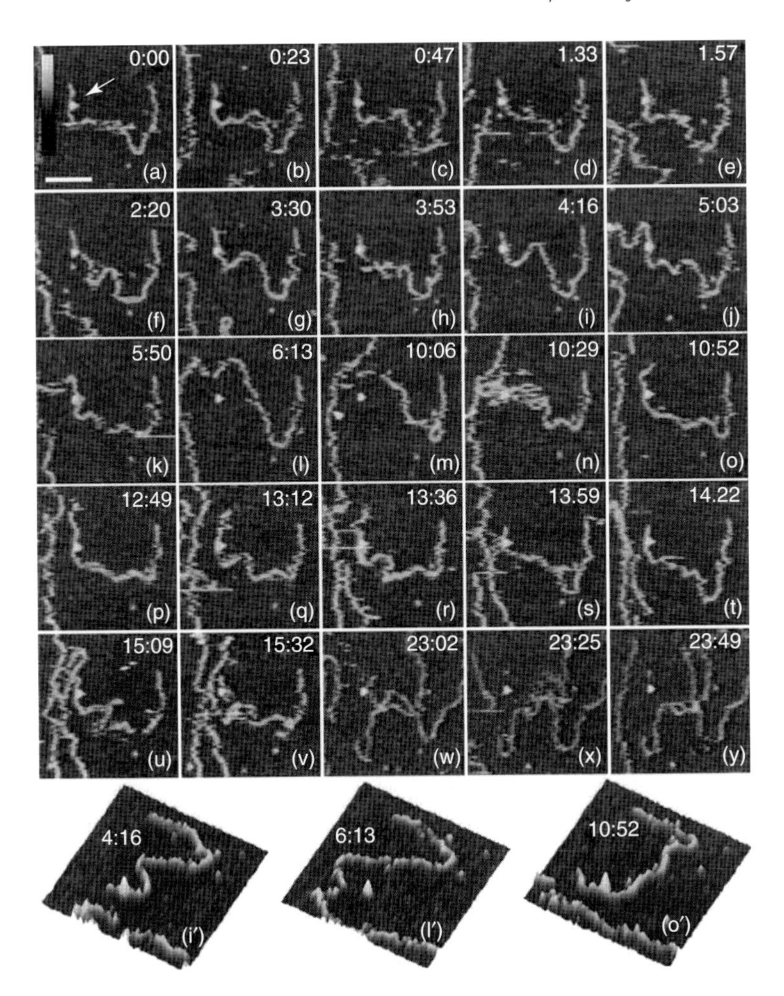

Figure 9.13 (a–y) Time-lapse AFM images of p53–DNA interactions. An arrow in (a) points to the protein bound to DNA. (i′, l′, o′) Three frames of the sequence are displayed as 3D surface plots. All images were cut out of a sequence recorded with an original scan size of 1.6×1.6 μm^2, 256×256 pixels at a rate of 23.3 s per image. Image size: 620×620 nm^2. Time unit: minutes, seconds. (Reprinted from Ref. [57]; Copyright © 2001, with permission from Elsevier.)

DNA [62]. This restriction enzyme needs to interact with two recognition sites, separated by up to 3500 bp before it can cleave DNA. Dynamics of the EcoP15I-DNA system was imaged in real time at scan rates of 1–3 frames per second [37]. The study showed that EcoP15I translocated over the DNA for long distances at a rate of 79 ± 33 bp s^{-1}, resulting in accumulated supercoiling of DNA in an ATP-dependent manner. It was concluded from this study that EcoP15I used two distinct mechanisms to communicate between two recognition sites: diffusive DNA loop formation and ATPase-driven translocation of the intervening DNA contour.

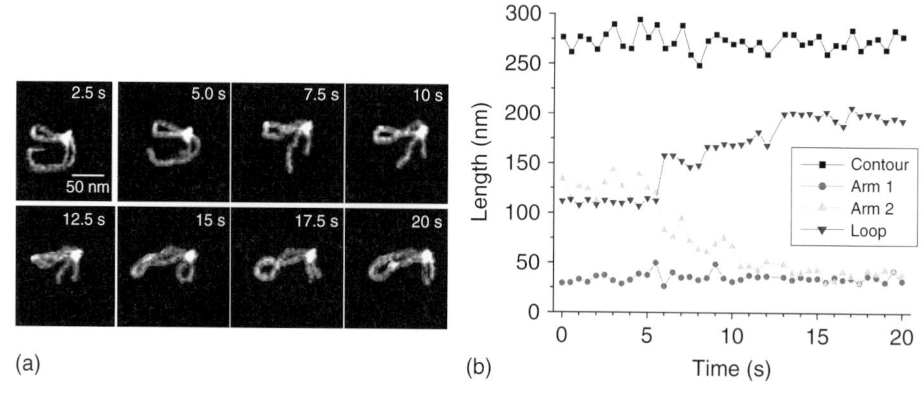

Figure 9.14 High-speed AFM analysis of EcoRII translocation. (a) Individual frames for every 2.5 s are shown. (b) The change in lengths of the entire DNA, arms, and loops over a period of 20 s, measured in 0.5 s intervals. As the long arm gradually gets shorter, the loop length gradually increases. The contour lengths of the entire molecule and the short arm have consistent length over the entire timescale. The sliding over the length of 300 bp occurs within 10 s. (Reprinted from Ref. [63]; Copyright © 2009, with permission from the American Chemical Society.)

Another multisite system, the EcoRII–DNA complex, has been studied using the same HS-AFM approach [63]. Similar to the one-site DNA-binding proteins, EcoRII shows multiple binding-unbinding events and translocations along the DNA. Measurements of the diffusion coefficients revealed 50 times faster 2D mobility compared with the linear diffusion. However, most striking was the observation of the dynamics of the system in which the protein occupies two specific sites on the DNA. Figure 9.14a displays eight frames of the event spanning 20 s and imaged at a rate of 2 frames per second in which the initial complex with a short loop (2.5 s) undergoes sliding dynamics (5–12.5 s), ending up with the formation of the specific complex (15 s) involving the third specific site on DNA, identified by the formation of a stable large loop. Figure 9.14b shows the results of measurements of the loop sizes and the arm lengths over time. The contour length of the entire DNA throughout the imaging period remains constant, whereas other parameters change. Short arm 1 also changes in length, staying at a value close to the expected 34 nm for specific binding 100 bp from the end, which shows that the protein remains specifically bound at this recognition site. On the contrary, at about 6 s, another arm decreases in size in parallel with the increase in the loop size. After 10 s, the protein stops translocating and binds to another specific DNA site. This is validated by the length measurements, which show that arm 2 stops at the length of 34 nm, expected for binding 100 bp from the end where the third specific binding site is located. The change in the loop length occurs over a period of about 10 s, covering a distance of about 300 bp (102 nm) until it stops at another recognition site. This means that the protein in this complex moves at a rate of about 30 bp s^{-1} (10.2 nm s^{-1}).

Two important conclusions emerge from this paper. First, multisite binding proteins, in addition to facilitated diffusion mechanisms, utilize a novel loop mechanism in which a protein bound to one DNA site searches for another site by threading the DNA filament through the complex. This threading process continues until the enzyme finds the second recognition site, on which it forms a stable synaptic complex. Second, threading of DNA is not accompanied by DNA rotating, as it was anticipated based on the helical geometry of DNA, suggesting that the sliding process occurs by a linear movement of the protein along the DNA surface with no rotation. These are entirely novel findings never reported before. Importantly, only time-lapse AFM imaging is capable of detecting such phenomena. Moreover, the process is quick, taking <10 s, which corresponds to a rate of about 30 bp s^{-1}(10.2 nm s^{-1}); therefore, HS-AFM was required for direct visualization of these rapid events.

9.5.2
Chromatin Dynamics Time-Lapse AFM

Nucleosome dynamics is a key property of chromatin, providing access to DNA wrapped around the histone core. The recent single-molecule studies using fluorescence microscopy and spectroscopy techniques detected a transient dissociation of DNA from nucleosomes in the absence of remodeling proteins [64–67], but they did not provide the answer to the spatial range of such dynamics. Time-lapse AFM was instrumental in the direct observation of the dynamics of nucleosomes assembled on a 353 bp DNA substrate containing a 147 bp nucleosome positioning sequence [19, 68]. With such a design, mapping of the nucleosome position relative to the central 147 bp region over the observation period was performed and the number of DNA turns around the histone core was measured.

Figure 9.15a shows the images and the data for a pathway characterized by two-step changes of the nucleosome. Initially, the nucleosome slightly changes its conformation (frames 1 and 2) and then unwraps on frame 3. It retains its geometry over three frames in a row (frames 3–5). Then, between frames 5 and 6, it loosens and unwraps again on frame 6, stays unchanged on frame 7, and finally undergoes full dissociation on frame 8. This dataset also shows that nucleosome dynamics is accompanied by the elongation of the both DNA arms and a decrease in the size of the nucleosome particle, as well as a change in the interarm angle. The AFM images were analyzed quantitatively, and the values of the arm lengths and particle volume yielded the nucleosome morphology (in number of DNA turns), plotted in Figure 9.15b,c according to the frame number. The nucleosome particle initially has $\sim$2.5 turns (frame 1), remains unchanged at frame 2, releases $\sim$0.75 turn during frames 3 and 4, stays with 1.75 turns between frames 4 and 5, and then unwraps for 1.2 turns between frames 5 and 6, followed by histone core dissociation on frame 8.

Overall, the time-lapse AFM study showed that local DNA dynamics could lead to unwrapping of DNA regions as large as dozens of base pairs. Eventually, a complete unfolding of nucleosomes occurs and the histone core dissociates

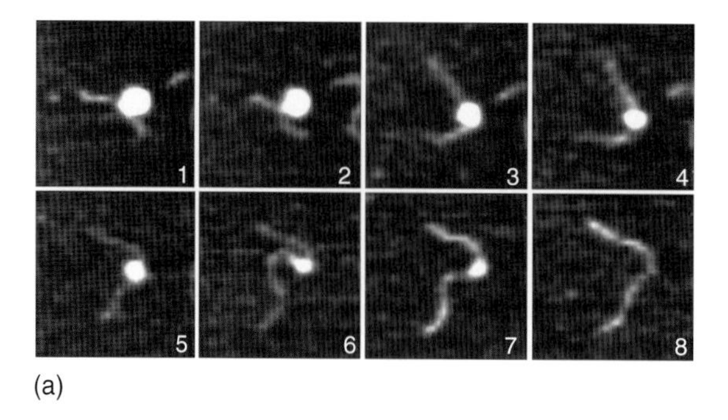

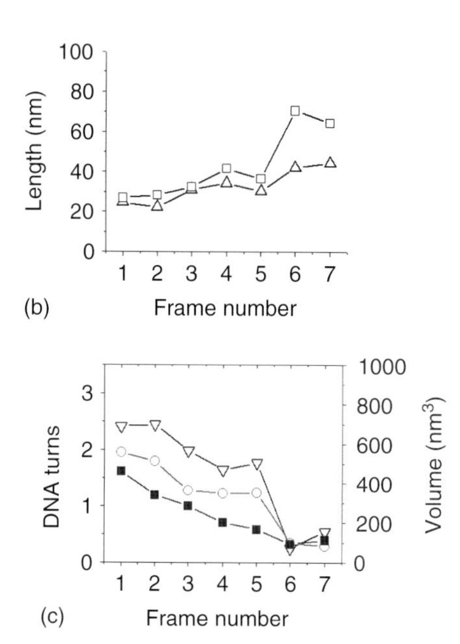

Figure 9.15 Two-step unwrapping process. (a) Consecutive AFM images of nucleosomes with a two-step unwrapping process taken during continuous scanning in the buffer. Each frame is 200 nm. (b) Dependence of arm lengths on the frame number. (c) The dependence of the number of DNA turns around the core calculated from arm lengths (line with inverted triangles) and from the angle between DNA arms (line with circles) on the frame number. The dependence of nucleosome volume on the frame number is shown in the line with squares (plot against right *y*-axis). Each frame takes about 170 s to scan. (Reprinted from Ref. [68]; Copyright (2009), with permission from the American Chemical Society.)

from the complex. These studies demonstrate that nucleosome unwrapping could occur in the absence of ATP-dependent chromatin remodeling proteins. They also suggest that the inherent dynamics of nucleosomes can contribute to the chromatin unwrapping process. It was also hypothesized [68] that electrostatic interactions of the nucleosome with remodeling proteins modulate the stability of the nucleosome, facilitating the ATP-driven unwrapping process performed by remodeling proteins.

Additional information on nucleosome dynamics was provided by Suzuki *et al.* [26], in which the molecular dynamics of polynucleosomal arrays in solution were analyzed by HS-AFM. The time-lapse data at the subsecond scanning rate made it possible to image nucleosome dissociation, as described above, as well as sliding of the nucleosome core particle. These dynamics appear as fluctuations within ~50 nm along the DNA strand. The time-lapse data also revealed two pathways for nucleosome dissociation. In addition to the dissociation of the histone octamers, a sequential dissociation of tetramers was observed.

9.6 DNA Condensation

DNA is a stiff polymer, characterized by a persistence length as large as ~50 nm [69]. In solution, DNA molecules as large as the lambda phage genome (~16 μm) [27] have a random coil shape with the size of ~10^3 nm. At the same time, the phage particle is considerably smaller in size, ~55 nm, suggesting the existence of a special mechanism enabling the DNA to pack so tightly. The problem associated with such tight packing is a high DNA charge. DNA is a linear polymer with a very high linear charge density; therefore, neutralization of the inevitable electrostatic repulsion is required for DNA packing. It was shown by numerous studies (reviewed in [70]) that condensation is facilitated in the presence of multivalent cations. In the presence of multivalent cations, DNA can be condensed into particles of very small sizes and different morphologies, including spheres, rods, and toroids [70].

AFM studies demonstrated that DNA condensation could occur during interactions with surfaces of different properties. The use of cationic bilayers in [23] led to the arrangement of plasmid DNA into tightly assembled arrays. Condensation of DNA into particles of different shapes, including toroids, was observed with AFM in [9], in which the mica surface was treated with APTES in organic solvents. Later, studies which used various silanes for the treatment of mica were performed [71]. Selecting the type of functional groups on the surface, DNA, and the salt concentrations controlled the yield of DNA aggregates and the types of aggregate morphologies. These studies shed a new light on the interactions that are involved in the DNA condensation processes *in vivo*. In addition, the finding that DNA can be condensed with cationic liposomes [70] highlighted this process for those interested in developing efficient gene delivery systems for eukaryotic cells. It was shown in [72] that polyamidoamine dendrimers (G4) were capable of efficiently condensing plasmid DNA. Time-lapse AFM was instrumental in the study of structure and dynamics of aggregates prepared at different concentrations of dendrimers. G4–DNA complexes were dynamic in nature, showing a degree of mobility. The formation of the G4–DNA complexes decreased the susceptibility of DNA to nucleases, an important characteristic in the development of complexes resistant to the nuclease attack within cells. This protection is related to the structural morphology of the formed complex, which is itself shown to be dependent on the G4/DNA ratios and the time allowed for complex formation.

9.7 Conclusions

AFM, with its capability to image biological samples in a fully hydrated state, is coming of age. The conformational dynamics of DNA in physiologically relevant supercoiling states can be assessed with AFM. Recent experimental studies and advanced theoretical analyses suggest that DNA dynamics, rather than static DNA structure, plays a critical role in the control of gene expression, DNA replication, recombination, and repair. Local and global DNA conformations are in a dynamic equilibrium. Competing transitions between different local structures accompanied by dynamic changes of global DNA conformations may have a profound role in various DNA functions. The ability to use time-lapse AFM to image DNA in liquid in complex with various proteins, including ligands, enables direct imaging of critical genetic processes, such as transcription, replication, and recombination. Recent advances in HS-AFM have enabled the technique to become available to a broader biomedical community. At the current stage of development, HS-AFM is capable of imaging molecular dynamics with the rate approaching video optical microscopy. However, the unique capability to provide the dynamics at the nanoscale range opens novel prospects for nanoimaging. Moreover, the current HS-AFM design operates with the oscillation amplitude in the range of a nanometer, thereby dramatically minimizing the potential deformation effect of the tip. This technological advance is an important step for further improvement of the instrument, enabling AFM to operate in the least invasive mode (attractive tip–sample interaction regime). This will almost eliminate the tip-induced modification of the sample. The tip geometry is the major resolution-limiting factor for AFM. Recent advances in the tip manufacturing process provided the AFM practitioners with probes as sharp as a few nanometers. At the same time, there is room for improvement in instrumentation, such as high speed for large samples and sample preparation techniques. Finally, the technological advances of AFM analysis of DNA and protein–DNA complexes can be extended to other biomolecular systems and thus will provide structural biologists pursuing single-molecule approaches with novel, powerful nanoimaging tools.

Acknowledgments

I am grateful to Luda Shlyakhtenko for her valuable comments and critical reading of the manuscript, as well as the current and former members of the group for their contribution to works incorporated into the manuscript. The work was supported by grants from NIH (GM 62235), NSF **(NSF-EPSCOR EPS-0701892, PHY-061590)**, DOE **(DE-FG02-08ER64579)**,NATO **(CBN.NR.NRSFP 983204)**, and Nebraska Research Initiative (NRI).

References

1. Watson, J.D., Baker, T.A., Bell, S.P., Gann, A., Levine, M. *et al.* (2008) *Molecular Biology of the Gene*, Pearson Education, Inc., San Francisco, CA.
2. Chow, L.T., Gelinas, R.E., Broker, T.R., and Roberts, R.J. (1977) An amazing sequence arrangement at the 5' ends of adenovirus 2 messenger RNA. *Cell*, **12**, 1–8.
3. Berget, S.M., Moore, C., and Sharp, P.A. (1977) Spliced segments at the 5' terminus of adenovirus 2 late mRNA. *Proc. Natl. Acad. Sci. U.S.A.*, **74**, 3171–3175.
4. Borovik, A.S., Kalambet, Y.A., Lyubchenko, Y.L., Shitov, V.T., and Golovanov, E.I. (1980) Equilibrium melting of plasmid ColE1 DNA: electron-microscopic visualization. *Nucleic Acids Res.*, **8**, 4165–4184.
5. Lyubchenko, Y.L., Kalambet, Y.A., Lyamichev, V.I., and Borovik, A.S. (1982) A comparison of experimental and theoretical melting maps for replicative form of phi X174 DNA. *Nucleic Acids Res.*, **10**, 1867–1876.
6. Lyubchenko, Y.L., Vologodskii, A.V., and Frank-Kamenetskii, M.D. (1978) Direct comparison of theoretical and experimental melting profiles for RF II phiX174 DNA. *Nature*, **271**, 28–31.
7. Bustamante, C., Vesenka, J., Tang, C.L., Rees, W., Guthold, M. *et al.* (1992) Circular DNA molecules imaged in air by scanning force microscopy. *Biochemistry*, **31**, 22–26.
8. Vesenka, J., Guthold, M., Tang, C.L., Keller, D., Delaine, E. *et al.* (1992) Substrate preparation for reliable imaging of DNA molecules with the scanning force microscope. *Ultramicroscopy*, **42–44**, 1243–1249.
9. Lyubchenko, Y.L., Gall, A.A., Shlyakhtenko, L.S., Harrington, R.E., Jacobs, B.L. *et al.* (1992) Atomic force microscopy imaging of double stranded DNA and RNA. *J. Biomol. Struct. Dyn.*, **10**, 589–606.
10. Lyubchenko, Y.L., Jacobs, B.L., and Lindsay, S.M. (1992) Atomic force microscopy of reovirus dsRNA: a routine technique for length measurements. *Nucleic Acids Res.*, **20**, 3983–3986.
11. Bustamante, C. and Rivetti, C. (1996) Visualizing protein-nucleic acid interactions on a large scale with the scanning force microscope. *Annu. Rev. Biophys. Biomol. Struct.*, **25**, 395–429.
12. Thundat, T., Allison, D.P., Warmack, R.J., Brown, G.M., Jacobson, K.B. *et al.* (1992) Atomic force microscopy of DNA on mica and chemically modified mica. *Scanning Microsc.*, **6**, 911–918.
13. Hansma, H.G. and Laney, D.E. (1996) DNA binding to mica correlates with cationic radius: assay by atomic force microscopy. *Biophys. J.*, **70**, 1933–1939.
14. Pastre, D., Pietrement, O., Fusil, S., Landousy, F., Jeusset, J. *et al.* (2003) Adsorption of DNA to mica mediated by divalent counterions: a theoretical and experimental study. *Biophys. J.*, **85**, 2507–2518.
15. Lyubchenko, Y.L. (2004) DNA structure and dynamics: an atomic force microscopy study. *Cell Biochem. Biophys.*, **41**, 75–98.
16. Lyubchenko, Y.L., Gall, A.A., and Shlyakhtenko, L.S. (2001) in *DNA-Protein Interactions; Principles and Protocols*, Methods in Molecular Biology (ed. T. Moss,) Humana Press, Totowa, NJ, pp. 569–578.
17. Rybenkov, V.V., Vologodskii, A.V., and Cozzarelli, N.R. (1997) The effect of ionic conditions on the conformations of supercoiled DNA. I. Sedimentation analysis. *J. Mol. Biol.*, **267**, 299–311.
18. Shlyakhtenko, L.S., Gall, A.A., Filonov, A., Cerovac, Z., Lushnikov, A. *et al.* (2003) Silatrane-based surface chemistry for immobilization of DNA, protein-DNA complexes and other biological materials. *Ultramicroscopy*, **97**, 279–287.
19. Lyubchenko, Y.L. and Shlyakhtenko, L.S. (2009) AFM for analysis of structure and dynamics of DNA and protein-DNA complexes. *Methods*, **47**, 206–213.
20. Lyubchenko, Y.L., Shlyakhtenko, L.S., and Gall, A.A. (2009) Atomic force microscopy imaging and probing of DNA, proteins, and protein DNA complexes: silatrane surface chemistry. *Methods Mol. Biol.*, **543**, 337–351.

21. Ellis, J.S., Abdelhady, H.G., Allen, S., Davies, M.C., Roberts, C.J. *et al.* (2004) Direct atomic force microscopy observations of monovalent ion induced binding of DNA to mica. *J. Microsc.*, **215**, 297–301.
22. Yang, J., Takeyasu, K., and Shao, Z. (1992) Atomic force microscopy of DNA molecules. *FEBS Lett.*, **301**, 173–176.
23. Mou, J., Czajkowsky, D.M., Zhang, Y., and Shao, Z. (1995) High-resolution atomic-force microscopy of DNA: the pitch of the double helix. *FEBS Lett.*, **371**, 279–282.
24. Bussiek, M., Mucke, N., and Langowski, J. (2003) Polylysine-coated mica can be used to observe systematic changes in the supercoiled DNA conformation by scanning force microscopy in solution. *Nucleic Acids Res.*, **31**, e137.
25. Bussiek, M., Toth, K., Brun, N., and Langowski, J. (2005) DNA-loop formation on nucleosomes shown by in situ scanning force microscopy of supercoiled DNA. *J. Mol. Biol.*, **345**, 695–706.
26. Suzuki, Y., Higuchi, Y., Hizume, K., Yokokawa, M., Yoshimura, S.H. *et al.* (2010) Molecular dynamics of DNA and nucleosomes in solution studied by fast-scanning atomic force microscopy. *Ultramicroscopy*.
27. Lyubchenko, Y., Shlyakhtenko, L., Harrington, R., Oden, P., and Lindsay, S. (1993) Atomic force microscopy of long DNA: imaging in air and under water. *Proc. Natl. Acad. Sci. U.S.A.*, **90**, 2137–2140.
28. Hansma, H.G., Bezanilla, M., Zenhausern, F., Adrian, M., and Sinsheimer, R.L. (1993) Atomic force microscopy of DNA in aqueous solutions. *Nucleic Acids Res.*, **21**, 505–512.
29. Hansma, H.G., Laney, D.E., Bezanilla, M., Sinsheimer, R.L., and Hansma, P.K. (1995) Applications for atomic force microscopy of DNA. *Biophys. J.*, **68**, 1672–1677.
30. Lyubchenko, Y.L. (2011) Preparation of DNA and nucleoprotein samples for AFM imaging. *Micron*, **42**, 196–206.
31. Freund, J., Halbritter, J., and Horber, J.K. (1999) How dry are dried samples? Water adsorption measured by STM. *Microsc. Res. Tech.*, **44**, 327–338.
32. Zhong, Q., Inniss, D., Kjoller, K., and Elings, V.B. (1993) Fractured polymer/silica fiber surface studied by tapping mode atomic force microscopy. *Surf. Sci. Lett.*, **290**, L688–L692.
33. Lyubchenko, Y.L. and Shlyakhtenko, L.S. (1997) Visualization of supercoiled DNA with atomic force microscopy in situ. *Proc. Natl. Acad. Sci. U.S.A.*, **94**, 496–501.
34. Sheng, S., Czajkowsky, D.M., and Shao, Z. (1999) AFM tips: how sharp are they? *J. Microsc.*, **196**, 1–5.
35. Scipioni, A., Zuccheri, G., Anselmi, C., Bergia, A., Samori, B. *et al.* (2002) Sequence-dependent DNA dynamics by scanning force microscopy time-resolved imaging. *Chem. Biol.*, **9**, 1315–1321.
36. Ando, T., Kodera, N., Naito, Y., Kinoshita, T., Furuta, K. *et al.* (2003) A high-speed atomic force microscope for studying biological macromolecules in action. *Chemphyschem*, **4**, 1196–1202.
37. Ando, T., Uchihashi, T., Kodera, N., Yamamoto, D., Taniguchi, M. *et al.* (2007) High-speed atomic force microscopy for observing dynamic biomolecular processes. *J. Mol. Recognit.*, **20**, 448–458.
38. Sinden, R.R. (1994) *DNA Structure and Function*, Academic Press, San Diego, CA.
39. Hatfield, G.W. and Benham, C.J. (2002) DNA topology-mediated control of global gene expression in Escherichia coli. *Annu. Rev. Genet.*, **36**, 175–203.
40. Sinden, R.R., Potaman, V.N., Oussatcheva, E.A., Pearson, C.E., Lyubchenko, Y.L. *et al.* (2002) Triplet repeat DNA structures and human genetic disease: dynamic mutations from dynamic DNA. *J. Biosci.*, **1** (Suppl. 27), 53–65.
41. Wells, R.D. (2009) Discovery of the role of non-B DNA structures in mutagenesis and human genomic disorders. *J. Biol. Chem.*, **284**, 8997–9009.
42. Shlyakhtenko, L.S., Potaman, V.N., Sinden, R.R., and Lyubchenko, Y.L. (1998) Structure and dynamics of supercoil-stabilized DNA cruciforms. *J. Mol. Biol.*, **280**, 61–72.

43. Shlyakhtenko, L.S., Hsieh, P., Grigoriev, M., Potaman, V.N., Sinden, R.R. *et al.* (2000) A cruciform structural transition provides a molecular switch for chromosome structure and dynamics. *J. Mol. Biol.*, **296**, 1169–1173.
44. Mikheikin, A.L., Lushnikov, A.Y., and Lyubchenko, Y.L. (2006) Effect of DNA supercoiling on the geometry of holliday junctions. *Biochemistry*, **45**, 12998–13006.
45. Soyfer, V.N. and Potaman, V.N. (1996) *Triple-Helical Nucleic Acids*, Springer, New York.
46. Tiner, W.J. Sr., Potaman, V.N., Sinden, R.R., and Lyubchenko, Y.L. (2001) The structure of intramolecular triplex DNA: atomic force microscopy study. *J. Mol. Biol.*, **314**, 353–357.
47. Lyubchenko, Y.L., Shlyakhtenko, L.S., Potaman, V.P., and Sinden, R.R. (2002) Global and local DNA structure and dynamics. Single molecule studies with AFM. *Microsc. Microanal.*, **8**, 170–171.
48. Holliday, R. (1964) A mechanism for gene conversion in fungi. *Genet. Res. Camb.*, **5**, 282–304.
49. Meselson, M. (1972) Formation of hybrid DNA by rotary diffusion during genetic recombination. *J. Mol. Biol.*, **71**, 795–798.
50. Sigal, N. and Alberts, B. (1972) Genetic recombination: the nature of a crossed strand-exchange between two homologous DNA molecules. *J. Mol. Biol.*, **71**, 789–793.
51. Panyutin, I.G. and Hsieh, P. (1994) The kinetics of spontaneous DNA branch migration. *Proc. Natl. Acad. Sci. U.S.A.*, **91**, 2021–2025.
52. Lushnikov, A.Y., Bogdanov, A., and Lyubchenko, Y.L. (2003) DNA recombination: holliday junctions dynamics and branch migration. *J. Biol. Chem.*, **278**, 43130–43134.
53. Guthold, M., Bezanilla, M., Erie, D.A., Jenkins, B., Hansma, H.G. *et al.* (1994) Following the assembly of RNA polymerase-DNA complexes in aqueous solutions with the scanning force microscope. *Proc. Natl. Acad. Sci. U.S.A.*, **91**, 12927–12931.
54. Kasas, S., Thomson, N.H., Smith, B.L., Hansma, H.G., Zhu, X. *et al.* (1997) Escherichia coli RNA polymerase activity observed using atomic force microscopy. *Biochemistry*, **36**, 461–468.
55. Guthold, M., Zhu, X., Rivetti, C., Yang, G., Thomson, N.H. *et al.* (1999) Direct observation of one-dimensional diffusion and transcription by Escherichia coli RNA polymerase. *Biophys. J.*, **77**, 2284–2294.
56. van Noort, S.J., van der Werf, K.O., Eker, A.P., Wyman, C., de Grooth, B.G. *et al.* (1998) Direct visualization of dynamic protein-DNA interactions with a dedicated atomic force microscope. *Biophys. J.*, **74**, 2840–2849.
57. Jiao, Y., Cherny, D.I., Heim, G., Jovin, T.M., and Schaffer, T.E. (2001) Dynamic interactions of p53 with DNA in solution by time-lapse atomic force microscopy. *J. Mol. Biol.*, **314**, 233–243.
58. Berg, O.G., Winter, R.B., and von Hippel, P.H. (1981) Diffusion-driven mechanisms of protein translocation on nucleic acids. 1. Models and theory. *Biochemistry*, **20**, 6929–6948.
59. Winter, R.B., Berg, O.G., and von Hippel, P.H. (1981) Diffusion-driven mechanisms of protein translocation on nucleic acids. 3. The Escherichia coli lac repressor–operator interaction: kinetic measurements and conclusions. *Biochemistry*, **20**, 6961–6977.
60. Lushnikov, A.Y., Potaman, V.N., and Lyubchenko, Y.L. (2006) Site-specific labeling of supercoiled DNA. *Nucleic Acids Res.*, **34**, e111; 111–117.
61. Lushnikov, A.Y., Potaman, V.N., Oussatcheva, E.A., Sinden, R.R., and Lyubchenko, Y.L. (2006) DNA strand arrangement within the SfiI-DNA complex: atomic force microscopy analysis. *Biochemistry*, **45**, 152–158.
62. Crampton, N., Yokokawa, M., Dryden, D.T., Edwardson, J.M., Rao, D.N. *et al.* (2007) Fast-scan atomic force microscopy reveals that the type III restriction enzyme EcoP15I is capable of DNA translocation and looping. *Proc. Natl. Acad. Sci. U.S.A.*, **104**, 12755–12760.
63. Gilmore, J.L., Suzuki, Y., Tamulaitis, G., Siksnys, V., Takeyasu, K. *et al.* (2009) Single-molecule dynamics of the DNA-EcoRII protein complexes

revealed with high-speed atomic force microscopy. *Biochemistry*, **48**, 10492–10498.

64. Bucceri, A., Kapitza, K., and Thoma, F. (2006) Rapid accessibility of nucleosomal DNA in yeast on a second time scale. *EMBO J.*, **25**, 3123–3132.

65. Koopmans, W.J., Brehm, A., Logie, C., Schmidt, T., and van Noort, J. (2007) Single-pair FRET microscopy reveals mononucleosome dynamics. *J. Fluoresc*, **17**, 785–795.

66. Li, G., Levitus, M., Bustamante, C., and Widom, J. (2005) Rapid spontaneous accessibility of nucleosomal DNA. *Nat. Struct. Mol. Biol.*, **12**, 46–53.

67. Tims, H.S. and Widom, J. (2007) Stopped-flow fluorescence resonance energy transfer for analysis of nucleosome dynamics. *Methods*, **41**, 296–303.

68. Shlyakhtenko, L.S., Lushnikov, A.Y., and Lyubchenko, Y.L. (2009) Dynamics of nucleosomes revealed by time-lapse atomic force microscopy. *Biochemistry*, **48**, 7842–7848.

69. Livshits, M.A., Amosova, O.A., and Lyubchenko, Y.L. (1990) Flexibility difference between double-stranded RNA and DNA as revealed by gel electrophoresis. *J. Biomol. Struct. Dyn.*, **7**, 1237–1249.

70. Bloomfield, V.A. (1996) DNA condensation. *Curr. Opin. Struct. Biol.*, **6**, 334–341.

71. Fang, Y. and Hoh, J.H. (1998) Surface-directed DNA condensation in the absence of soluble multivalent cations. *Nucleic Acids Res.*, **26**, 588–593.

72. Abdelhady, H.G., Allen, S., Davies, M.C., Roberts, C.J., Tendler, S.J. *et al.* (2003) Direct real-time molecular scale visualisation of the degradation of condensed DNA complexes exposed to DNase I. *Nucleic Acids Res.*, **31**, 4001–4005.

10
Stability of Lipid Bilayers as Model Membranes: Atomic Force Microscopy and Spectroscopy Approach

Lorena Redondo-Morata, Marina Inés Giannotti, and Fausto Sanz

10.1
Biological Membranes

10.1.1
Cell Membrane

Cells can be thermodynamically defined as open systems in constant exchange of mass, energy, and information with the environment. To this end, the cell membrane is a key structure that defines the confines of the cell and some of its internal compartments. Besides, it is not only essential as a structural part of the cell but also provides a support matrix for many types of proteins that are inserted on it [1]. In the early 1970s, Singer and Nicolson proposed the fluid mosaic model [2], depicturing cell membranes as two-dimensional liquids where all lipid and protein molecules diffuse easily (Figure 10.1a). Concerning the composition, besides all the protein and carbohydrates, lipids are the main components in molar fraction of the cell membrane [3]. Lipids are a broad family that covers many different chemical structures (some of them are exemplified in Figure 10.1b): sphingolipids (ceramides (CERs), sphingomyelin, gangliosides, and sphingosines); sterols (cholesterol (Chol) and vitamins); or phospholipids.

10.1.2
Supported Lipid Bilayers

Lipid bilayers are among the most important self-assembled structures in nature. In particular, phospholipid bilayers resemble cell membranes in key aspects; for instance, they retain two-dimensional fluidity and suppose an excellent environment for the insertion of membrane proteins. Therefore, model bilayer systems are very manageable platforms to investigate the biological processes that occur at the cellular or subcellular level. Such model membranes have proved to be appropriate for studies of cell signaling [6], pathogen attack [7], and ligand–receptor interactions [8], among others. The first time a planar bilayer system was reported to be used for research purposes was in the 1960s by Mueller *et al.* [9], consisting of

Atomic Force Microscopy in Liquid: Biological Applications, First Edition.
Edited by Arturo M. Baró and Ronald G. Reifenberger.

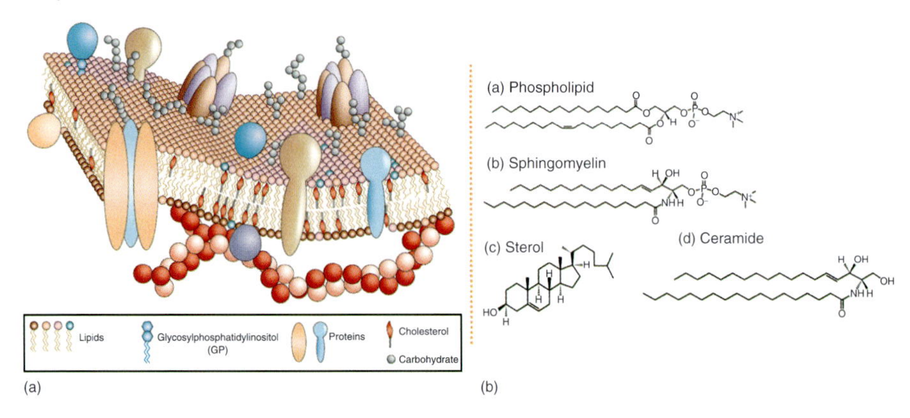

Figure 10.1 (a) The fluid mosaic model of the cell membrane. Like a mosaic, the cell membrane is a complex structure made up of many different parts, such as proteins, phospholipids, and cholesterol. The relative amounts of these components vary from membrane to membrane, and the types of lipids in membranes can also vary [4]. (b) Chemical structure of selected examples of common membrane lipids;(A) phospholipid molecule (1-palmitoyl-2-oleoyl-*sn*-glycero-3-phosphocholine, POPC), (B) sphingomyelin, (C) sterol (cholesterol), and (D) ceramide (*N*-octadecanoyl-D-*erythro*-sphingosine) ((a) Reprinted with permission from [5] Copyright 2004 Nature Publishing Group.)

two solution chambers with a 1 mm hole painted by lipid molecules, the so-called black lipid membranes [10]. Since then, several supported systems have emerged in order to develop more sophisticated approaches for both biophysical studies and sensor design such as solid supported lipid bilayers (SLBs), self-assembled monolayer–monolayer systems, polymer-cushioned phospholipid bilayers, arrays of supported lipid phospholipid membranes, and bilayer-coated microfluidics. For review about solid SLBs, see [11].

SLBs are robust and stable systems suitable to be studied by atomic force microscopy (AFM). The most used preparation methods are Langmuir–Blodgett (LB) and the liposome rupture methods which are schematized in Figure 10.2. The LB technique consists of a lipid monolayer disposed in a water–air interface, with lateral mobile barriers by which the monolayer is compressed [12]. Then, the phospholipid monolayer may be transferred to a solid substrate at a controlled surface pressure and speed (when the substrate is disposed perpendicular respect to the monolayer they are called Langmuir-Blodget, and when it is parallel to the monolayer, Langmuir–Schaefer) [13]. To obtain a lipid bilayer, a second transference can be made on top of the previous monolayer; if it is of different composition, it would form an asymmetric SLB [14, 15]. On the other hand, the liposome rupture method is a simple and very popular method to prepare SLBs [16]. The method consists of the deposition of a suspension of small unilamellar vesicles (SUVs) onto the substrate, in order for the SUVs to adsorb from the bulk onto the surface. Despite the mechanism not being fully understood, it is generally acknowledged that the SUVs in the substrate may fuse with one another, deform, flatten, and finally rupture to form a continuous SLB [17]. The final structure of the

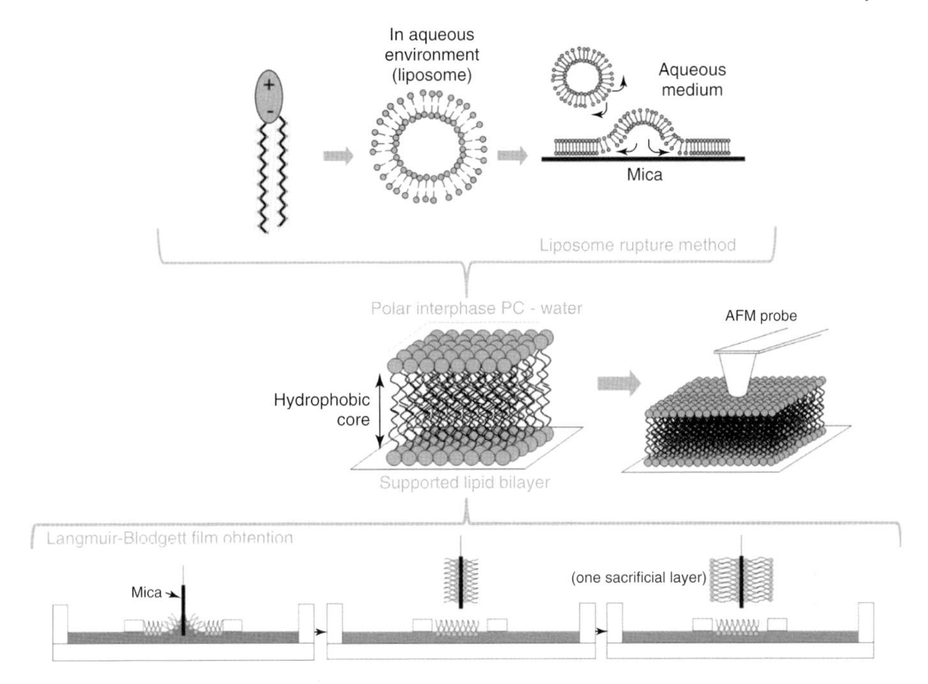

Figure 10.2 Supported lipid bilayers (SLBs) onto hydrophilic surfaces have been extensively used to mimic the biophysical properties of biological membranes. The upper schematic diagram shows the formation of SLBs via the rupture liposome method. In the lower diagram, it can be seen an SLB formation by means of Langmuir–Blodgett technique, performing two consecutive extractions of a lipid monolayer.

SLB has been reported to be strongly influenced by factors such as SUVs size, lipid composition and concentration, pH, temperature, ionic strength, the presence of divalent cations (especially Ca^{2+} and Mg^{2+}), surface roughness, and surface charge density [18]. There are also other alternative methods to prepare SLBs, although not as popular as the two herein described, for instance, the hydration of spin-coated films [19].

Suitable substrates for SLBs formation should be clean, smooth, and hydrophilic. Thus, mica substrate is the most popularized one [20], although other substrates such as fused silica [21, 22], borosilicate glass [21, 23], oxidized silicon [21], or thin films such as indium-tin oxide [24] can be used.

There are a myriad of experimental techniques useful to study the properties of supported and nonsupported lipid membranes, including nuclear magnetic resonance [25], neutron reflectivity [26], X-ray reflection [27] and diffraction [28] methods, fluorescence microscopy [29], Brewster angle microscopy [30], ellipsometry [31], fluorescence recovery after photobleaching [32], X-ray photoelectron spectroscopy [33], and time-of-light secondary ion mass spectrometry [34]. Among them, AFM in liquid environment sprung up as a well-established technique for imaging the lateral organization of SLBs. The main advantage with respect to other techniques is that the structure of biological samples, such as cell or lipid membranes, can be

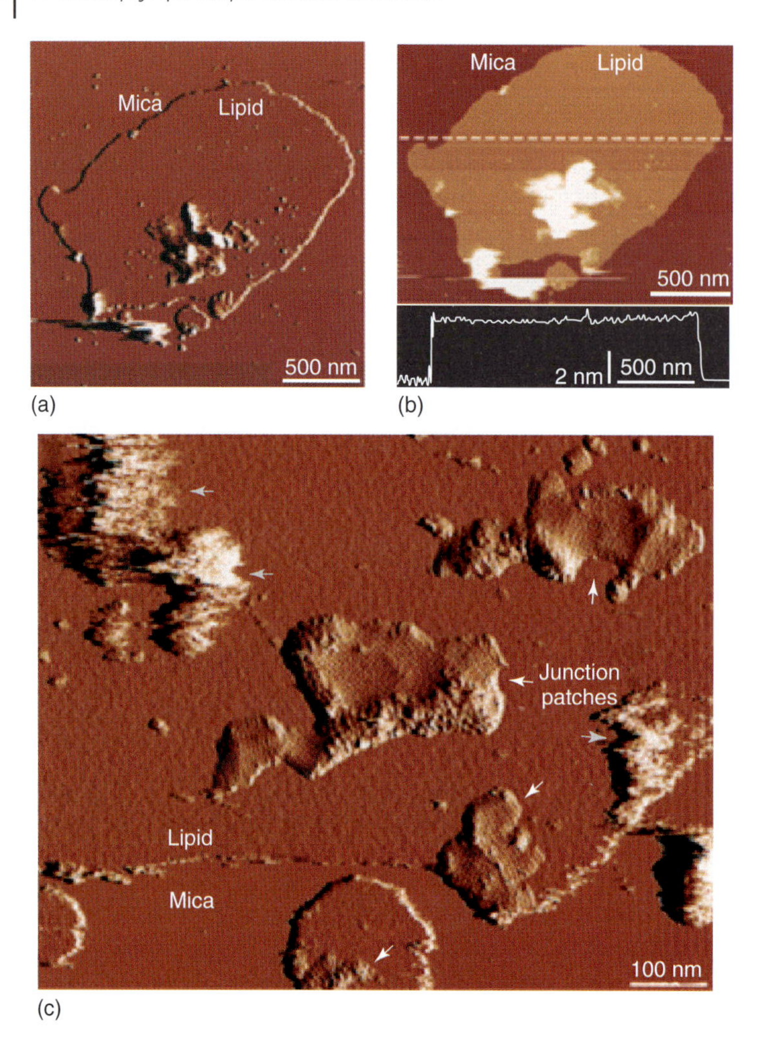

Figure 10.3 AFM imaging allows SLBs observations. (a) Deflection and (b) height image of a native membrane patch isolated from the lens core. The height profile along the dashed line in (b) is shown in the lower panel and shows that the thickness of the bilayer is 46 Å. (c) Removal of vesicular structures that were attached to the membrane patches (black arrows) with the AFM tip showed the presence of corrugated patches that protruded from the lipid bilayer (white arrows). The patches contained aquaporin arrays. (Reprinted by permission from Macmillan Publishers Ltd: EMBO Reports [37], Copyright 2007.)

explored not only in liquid media but also in real time with (sub)nanometer resolution [35]. Figure 10.3 illustrates an example of high-resolution AFM imaging used to study the native membranes isolated from a lens where patches of the highly functionalized membrane containing arrays of aquaporin can be appreciated. For a review about the most outstanding analysis in SLBs using AFM imaging, see [36].

10.2
Mechanical Characterization of Lipid Membranes

Lipid bilayers are closely related to biomembrane function, not only because of their structural role under a complex combination of forces [38] but also because of their contribution to the function of several membrane proteins [39]. Specifically, the chemical composition of lipid bilayers has shown to take part in the gating process of ion and mechanosensitive channels [40]. The concrete mechanochemical change of the membrane is also significantly related to the correct unfolding [41] and aggregation processes [42] of membrane proteins. Therefore, there has been an increasing interest in understanding the interplay between the lateral lipid organization and the overall membrane function.

The mechanical properties of lipid bilayers have been assessed by means of different techniques. Probably, the most remarkable one is the micropipette aspiration technique [43], which has provided a great deal of information regarding quantitative values for membrane elastic moduli for shear and bending and also for interbilayer friction [44]. However, this technique is limited to the use of giant vesicles, thus providing a microscopic insight on bilayer mechanical stability. At the nanometer scale, the surface force apparatus (SFA) has provided valuable information regarding lipid bilayer adhesion, fusion, and healing as well as interaction forces between lipid bilayers [45].

Within the last decades, AFM-based force spectroscopy (FS) has emerged as an essential tool to quantitatively characterize the mechanical properties of lipid membranes, with the advantage of high spatial range sensitivity and versatility, as well as the possibility of locating and probing confined areas of membranes under environmentally controlled conditions.

10.2.1
Breakthrough Force as a Molecular Fingerprint

FS with AFM has proven to be a suitable technique to test the mechanical properties of a wide variety of systems at the nanoscale. Among the most outstanding, it is worth to mention the indentation of hard materials during the approach of the AFM tip to the surface [46] and the stretching of discrete macromolecular structures such as polysaccharides [47], proteins [48], and DNA [49], while the tip retracts away from the surface. The métier of this technique is to directly measure the interaction forces when a small number of molecules are taking part. In the case of lipid membranes, FS is especially valuable in terms of spatial accuracy and force resolution. Classically, FS experiments are conducted under constant velocity conditions, that is, the deflection of the cantilever is measured while the AFM tip is approaching and retracting from the surface. When the spring constant of the cantilever is known, the measured deflection is converted into force. To deepen in this scenario, a schematic representation of a typical AFM–FS force curve for the indentation of an SLB is shown in Figure 10.4.

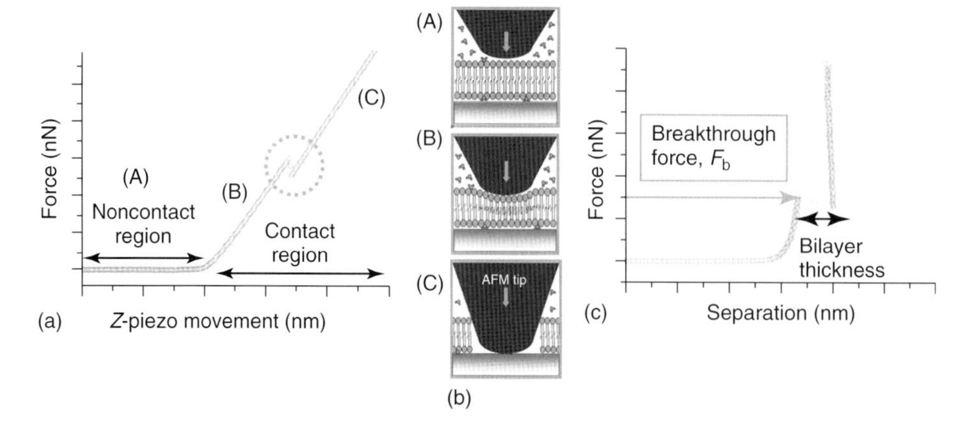

Figure 10.4 Probing the mechanical stability of a lipid bilayer with AFM–FS. (a) Typical plot of the vertical force versus the vertical piezo movement toward the surface. Three different parts can be distinguished in the curve associated to the progression described in the cartoons(b): (A) tip and sample are not yet in mechanical contact; (B) the tip elastically deforms the lipid bilayer; and (C) the tip ruptures the sample, thus becoming in contact with the mica substrate. (c) Force is represented versus separation, where the breakthrough force, F_b, the force at which the bilayer ruptures, is highlighted. With this kind of plot, bilayer thickness can be measured.

Experimentally, the lipid bilayer spread onto the mica substrate is identified by means of AFM imaging. Then, a set of force curves are performed in the center of the bilayer domains. When the cantilever tip is away from the bilayer sample, there is no mechanical interaction between the tip and the bilayer (Figure 10.4a and 10.4b (A)). On further approaching the cantilever tip to the bilayer sample, they become in mechanical contact. As the tip approaches the sample, the short-range interactions between the tip and sample start arising. These interaction forces have different origins, being the most remarkable Derjaguin–Landau–Verwey–Overbeek (DLVO) forces (corresponding to the charge of the so-called diffuse layer and van der Waals forces), hydration forces, and steric forces [50]. The origin and magnitude of these forces are decidedly dependent on the bilayer chemistry, the AFM tip chemistry, and the physicochemical properties of the liquid environment. Commonly, the combination of these forces gives rise to a few hundred piconewtons. Then, the bilayer is elastically deformed by the AFM probe (Figure 10.4a and 10.4b (B)) until the tip ruptures the membrane and gets in contact with the substrate (Figure 10.4a and 10.4b (C)). As depicted in Figure 10.4a and 10.4c, a discontinuity can be observed on the approaching force curve; this is interpreted as the penetration of the AFM tip through the SLB [51]. The vertical force at which this discontinuity happens is the maximum force that the bilayer is able to withstand before breaking and is the so-called breakthrough force (F_b). This discontinuity measures circa 4 nm in separation, which correlates well with the height of a lipid bilayer. Such breakthrough event typically occurs at several nanonewtons of force. Noticeably, these force measurements are highly reproducible in sequential experiments; even for different tips and samples, the distribution of F_b values is consistent, and the error is almost entirely derived from the calibration of the spring constant of the cantilever (when equipartition theorem is used, error in calibration is around 10–15% [52]). Analogous to the force required to indent hard materials as single crystals [53], the dehybridization force of DNA [54], the chair-boat transition force in polysaccharides [55], or the unfolding force of single proteins [56], the breakthrough force during indentation of lipid bilayers clearly fingerprints the (nano)mechanical stability of these systems.

As the force required to puncture a lipid membrane is in the nanonewton regime, one should accurately decide which is the spring constant of the cantilever necessary to properly match the stiffness of the probed sample, just to have the appropriate sensitivity. For instance, indenting hard substrates require high-spring constant cantilevers (k_S *ca.* 300 N m^{-1}), while uncoiling a single polysaccharide molecule with force requires very soft cantilevers (k_S *ca.* 0.06 N m^{-1}).

This kind of breakthrough events have been detected during the (nano)mechanical study of a wide variety of thin films. It was first reported in the case of a surfactant bilayer [57], but it has also been observed probing the ordering mechanisms of supported monolayers [58] and characterization of squeezed liquid films [59]. In the case of the lipid bilayers, the first report of a breakthrough event was performed by Dufrene *et al.* [15, 60] when analyzing DSPE (1,2-distearoyl-*sn*-glycero-3-phosphoethanolamine), DGDG (digalactosyl diglyceride), and *DOPE* (1,2-dioleoyl-*sn*-glycero-3-phosphoethanolamine) bilayer systems. Since then, F_b value can be considered as the fingerprint of the bilayer stability under the experimental conditions in which the measurements are performed. So far, different variables were reported to affect the mechanical stability of lipid bilayers and, as a consequence, the F_b value, such as temperature, lipid chemistry, pH, ionic strength, and electrolyte composition [61–63]. The influence of these variables on F_b is discussed in the following sections.

10.2.2 AFM Tip-Lipid Bilayer Interaction

As stated above, when the cantilever starts feeling the bilayer surface, interaction forces between the tip and the sample start being measurable, mainly, electrostatic and van der Waals forces. If the bilayer and the tip are oppositely charged, a jump to contact occurs (cantilever deflects toward the surface); otherwise, if they have the same sign of electric charge, repulsion can be observed in the force curve (cantilever deflects away from surface). As stated above, when the tip enters in mechanical contact with the bilayer, a combination of hydration and steric forces govern the tip–sample interaction. Then, the tip elastically deforms the membrane until the aforementioned breakthrough event. Further compression provokes the hard contact of the AFM tip with the substrate. The retracting part of the force–distance curve might show hysteresis, as, in some cases, the AFM tip gets adhered to the surface, depending on the chemistry of both the tip and SLB. Once the adhesion force is overcome, the cantilever detaches from the surface and retracts to its equilibrium position.

Focusing on the interaction forces arising between the tip and the lipid surface, one of the most relevant are the electrostatic ones [64, 65]. Most of the lipid membranes are negatively charged, including the zwitterionic ones, which expose the negative charge to the interface as it has been shown by zeta potential measurements [61, 66]. Furthermore, the tips most regularly used to perform FS experiments are made of silicon nitride, which is also known to be slightly negatively charged at physiological pH [62]. In this respect, the DLVO theory, which takes both electrostatic and van der Waals forces into account, has been useful to quantify the repulsive force between a negatively charged lipid bilayer and a negatively charged AFM tip, showing interactions up to tens of piconewtons [61, 66, 67], a value that is expected to be small due to the low surface charge density of the silicon nitride tip [68]. If the chemistry of both the tip and the bilayer is modified, these electrostatic contributions can be experimentally assessed. Chemical titration of lipid bilayers by FS has shown how, when the tip is carboxyl functionalized, the pH can be experimentally changed to tune the electrostatic interaction [69]. If the pH is higher than the corresponding pK_a, both the tip and lipid bilayer are negatively charged; therefore, a long-range electrostatic repulsion can be observed, which can reach up to hundreds of piconewtons. On the other hand, hydration and steric forces are the most remarkable short-range interactions. Hydration forces have been reported to decay exponentially with distance, expanding up to *ca.* 5 nN. Indeed, AFM imaging at the water–lipid interface visualizes individual hydration layers, demonstrating that the intrinsic hydration layers of the lipid bilayers are stable enough [70]. Besides, when two opposite lipid bilayers are considered, the force–distance repulsion is supposed to correlate with the force required to remove water molecules of the interface when both surfaces come into contact [71].

In addition to the above-mentioned long-range electrostatic (DLVO) and short-range hydration–steric interaction forces, a mechanical contribution concerning the elastic deformation of the membrane before the breakthrough point should be considered to fully comprehend the force–distance curve. Relating to the elastic deformation of the sample by the AFM tip, the first model proposed to explain the experimental data was the Hertz model [31], although some authors showed the failure of this model to fit the experimental curve [61, 72]. Instead, a simple elastic model built up by parallel springs seems to reproduce the results better [53, 61].

A theoretical model that relates the force of the film rupture to microscopic properties of the bilayer has been described by Butt *et al.* [73]. According to this model, the rupture of the membrane is an activated process, with an associated energy barrier that follows the Arrhenius law, a fact supported by the exponential dependence of the F_b to the tip velocity [74]. The distribution of forces necessary to create a hole under the AFM tip is closely related to the line tension (Γ), which represents the free energy associated with the unsaturated bonds of the molecules at the edge of the hole and with the effective spreading pressure (S), which is used to quantify the tendency of the film to spread into the gap between the tip and the substrate. This model has some limitations since, for instance, it simplifies the molecular nature of the membrane, as it is considered a fluid in the lateral

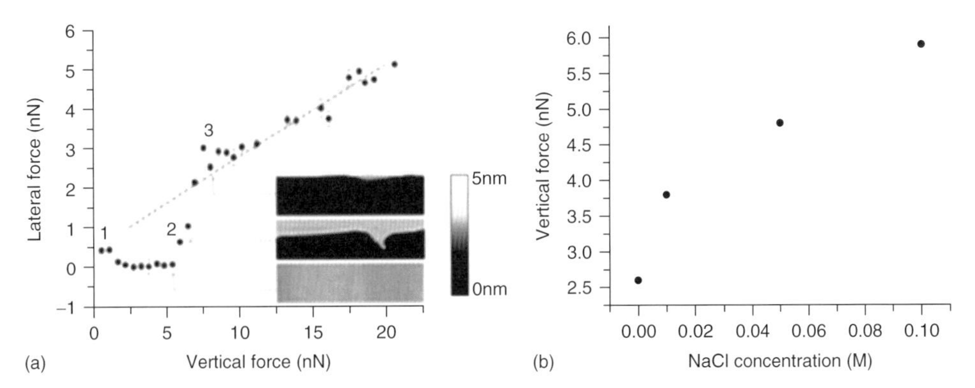

Figure 10.5 (a) Lateral force versus vertical force curves on DMPC bilayers in an aqueous environment. NaCl concentration is 0.1 M. Graphs is accompanied by bilayer topographic image obtained simultaneously with the friction data. The error bars represent the confidence intervals (at 95%, number of samples 256). A reference lateral force versus vertical force curve obtained on mica is included in the panel (dashed line). Curves were divided into three regions, the beginning of each of which is labeled 1, 2, and 3. The first region shows an extremely low friction force because of the repulsive electrostatic interactions due to DMPC polar heads and the tip. During the second region, the tip creates defects or begins to break the bilayer and the lateral force increases steeply. In the third region, the tip contacts the mica beneath the bilayer. (b) Vertical forces at which the second region begins as a function of NaCl concentration. (Adapted from [78] with permission, Copyright 2005 American Chemical Society.)

dimension; however, it has been shown to provide realistic quantitative numbers to explain the film indentation on supported substrates under a wide variety of conditions [67, 75, 76].

Finally, it is noteworthy to highlight the relation between force curves and friction (lateral force). By means of lateral force microscopy technique, it is possible to study tribological properties of the lipid bilayers. Trunfio-Sfarghiu *et al.* [77] combined an AFM–FS mode with a tribometer to correlate the mechanical resistance and the friction forces of a DOPC (1,2-dioleoyl-*sn*-glycero-3-phosphocholine) bilayer. They showed how the friction coefficient is related to the F_b value, since for higher the F_b values, the friction coefficients were lower and more stable. In addition, Oncins *et al.* [78] studied the structural changes caused by Na^+ on a DMPC (1,2-myristoyl-*sn*-glycero-3-phosphocholine) bilayer. They observed three different friction regimes as the vertical force exerted by the AFM probe on the SLB increased, as it is shown in Figure 10.5.

10.2.3
Effect of Chemical Composition on the Mechanical Stability of Lipid Bilayers

Using AFM–FS technique, experimental work can be focused on the precise measurement of the F_b value on a variety of chemically different lipid bilayers as a straightforward experimental way to gain an insight in the relationship between membrane conformation and its corresponding (nano)mechanical properties.

In a recent work, Sanz and collaborators [67] employed FS on chemically distinct SLBs to quantitatively explore the molecular determinants that provide a mechanical stability to the supported membrane. Changing the chemical composition of the hydrophilic head systematically, the length of the hydrophobic tail, the number of chain unsaturations and the Chol concentration, insights into the mechanical stability of each bilayer can be gained. The published results demonstrate that the overall mechanical stability of the lipid bilayer results from a complex mechanochemical balance, where the chemical composition of both the phospholipid headgroup and tail has a crucial effect. By systematically probing a set of chemically different SLBs, they first showed that both the phospholipid headgroup and tail have a decisive effect on their mechanical properties. Concerning the hydrocarbon chain length, the mechanical stability of the probed SLBs linearly increases by 3.3 nN on the introduction of each additional $-CH_2-$ in the chain, as it is shown in Figure 10.6.This increase originates from a most favorable packing between neighboring hydrophobic chains. Moreover, it has been reported that long hydrocarbon tails interact better with each other than shorter ones, increasing the van der Waals interactions between neighboring molecules. In particular, on the introduction of an extra $-CH_2-$ in the chain, the enthalpy of the system increases *ca.* 2 $kJ\,mol^{-1}$ [79].

Regarding the mechanical effect of the hydrocarbon tail unsaturations, the membrane mechanical stability as a function of the number of unsaturations and molecular branching in the chemical structure of the apolar tails is summarized in Figure 10.7. The introduction of a –*cis* double bond in the chain supposes a bend in the molecular structure. The resulting molecular tilt reduces the effective packing between molecules, that is, the lateral interactions, leading to a lower melting temperature [79]. This explanation correlated well with the FS observations and leads to conclude that, as the number of chain unsaturations increases, the mechanical resistance of the bilayer is greatly reduced. Indeed, this is of common occurrence in nature where bacteria regulate the fluidity of their membrane by changing the number of double bonds and the length of the phospholipid hydrocarbon chains.

Finally, FS observations demonstrated that mechanical stability of lipid bilayers exhibits a significant dependence on the phospholipid headgroup as well. Altogether, the work highlights the compelling effect of subtle variations in the chemical structure of phospholipid molecules on the membrane response when exposed to mechanical forces.

10.2.4
Effect of Ionic Strength on the Mechanical Stability of Lipid Bilayers

The role of the physicochemical environment is also susceptible to be studied by means of FS. It is known that ionic strength increases the electrostatic interaction between phospholipid headgroups due to a charge screening effect, and this fact leads to a higher proximity of the hydrophobic phospholipid tails and the consequent increase of van der Waals interactions [64, 80–84]. A pioneer study

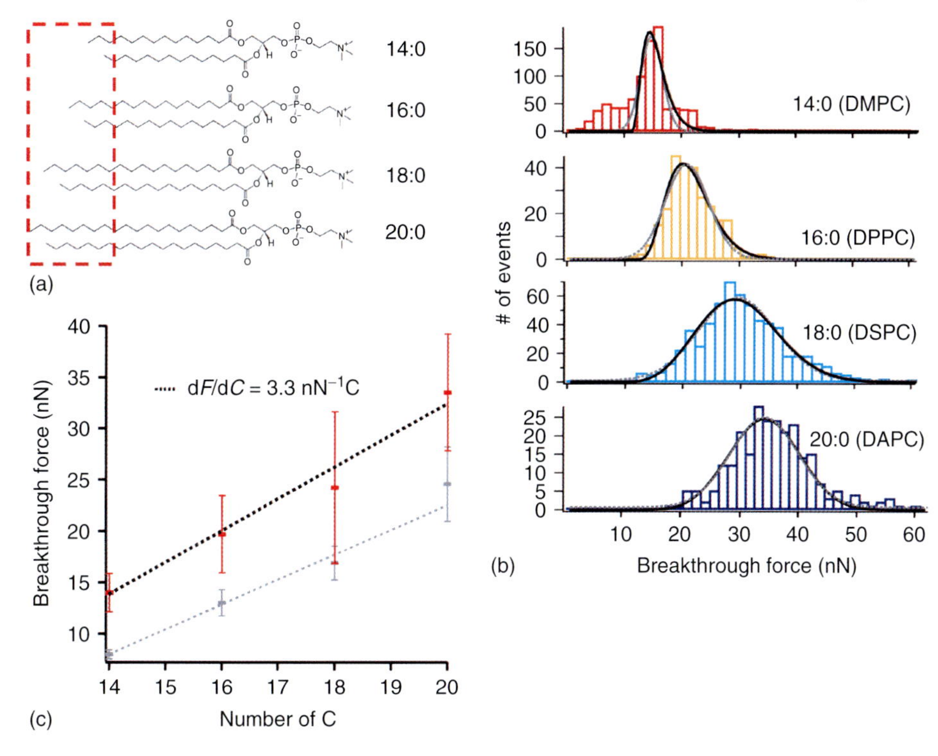

Figure 10.6 Mechanical resistance of the lipid bilayer is highly dependent on the length of the phospholipid tail. (a) Chemical structures of the tested PC phospholipids. The red square highlights the difference in the chain length. (b) Distribution of F_b values for the probed PC phospholipids with varying lengths. Gaussian fits to the data (dotte gray lines) yield a mean F_b value of 13.95 (1.87 nN, $n = 889$), DMPC (red); 19.66 (3.78 nN, $n = 266$), DPPC (1,2-dipalmitoyl-*sn*-glycero-3-phosphocholine, orange); 24.2 (7.4 nN, $n = 1796$), DSPC (1,2-distearoyl-*sn*-glycero-3-phosphocholine, blue); 33.5 (5.7 nN, $n = 263$), and DAPC (1, 2-diarachidonoyl-*sn*-glycero-3-phosphocholine, dark blue). Solid black lines represent the fitting to the continuum nucleation model. (c) Plot of the measured average F_b value as a function of the number of carbons present in each phospholipid chain (red data points). Linear fit to the data (dotted line) yields a slope of 3.3 nN/-CH_2- moiety. Gray data points stand for the average F_b value for the probed phospholipids under 150 mM NaCl but in the absence of Mg^{2+}. (Reprinted with permission from [67], copyright 2010 American Chemical Society.)

about the effect of ionic strength in the mechanical stability of lipids was performed by Garcia-Manyes *et al.* [61], showing that high ionic strength induces not only a better and faster deposition of phosphatidylcholine (PC) bilayers onto mica [62] but also an increase of the F_b value, on both for Na^+ and Mg^{2+} addition. In this work, DPPC (1,2-dipalmitoyl-*sn*-glycero-3-phosphocholine), DMPC, DOPC, and DLPC (1,2-dilauroyl-*sn*-glycero-3-phosphocholine) model membranes as well as natural *Escherichia coli* membranes, were investigated to probe the effect of ionic strength on the mechanical stability of lipid bilayers. As shown in Figure 10.8, the force required to puncture the lipid bilayer in all cases is significantly inferior at

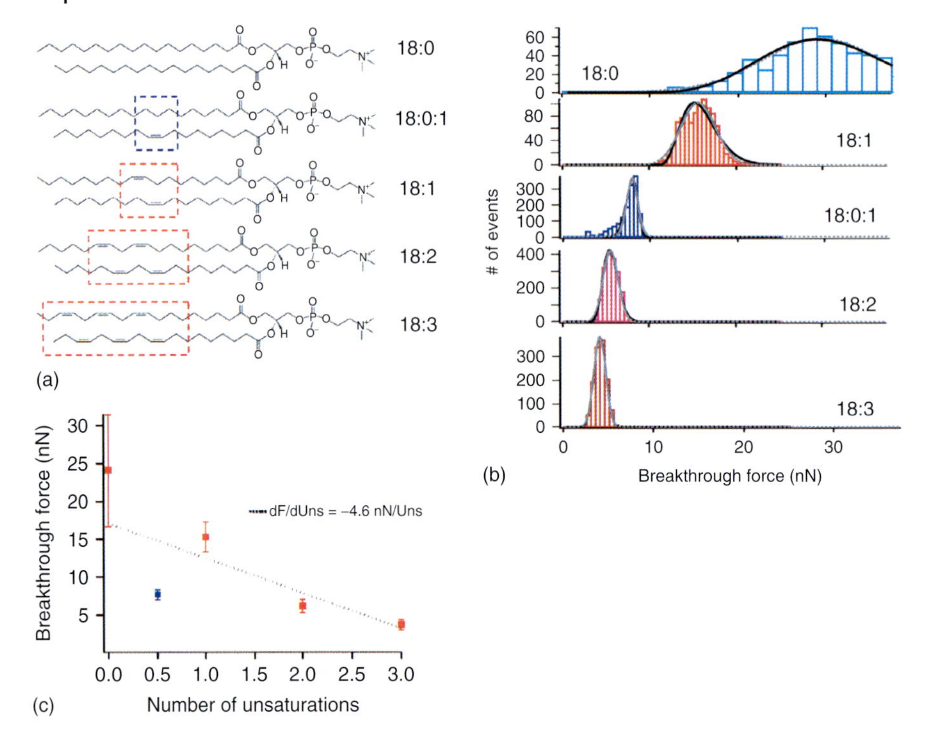

Figure 10.7 Mechanical resistance of the lipid bilayer depends on the number of unsaturations present in the hydrophobic tail. (a) Chemical structures of the probed phospholipids. Red squares mark the number and position of the symmetric unsaturations within the phospholipid structure. Blue gray square highlights the position of the asymmetric unsaturation in the case of 18 : 0–18 : 1 PC. (b) Distribution of F_b values for the probed phospholipids, where Gaussian fits to the data (dotted gray lines), yield a mean F_b value of 24.2 ± 7.4 nN, $n = 1796$, DSPC (18 : 0, blue); 15.4 ± 2 nN, $n = 1011$ (18 : 1, red); 6.26 ± 0.87 nN, $n = 1802$ (18 : 2, pink); 3.78 ± 0.66 nN, $n = 1270$ (18 : 3, orange); and 7.7 ± 0.66 nN, $n = 1509$ (18 : 0–18 : 1, drak blue). Fits to the continuum nucleation model are represented as solid black lines. (c) Plot of the measured F_b values as a function of the number of symmetric unsaturations present in each phospholipid chain. In the case of the asymmetrically unsaturated phospholipid (18 : 0–18 : 1, blue data point), the unsaturation is represented as 0.5. Linear fit (dotted line) to the data corresponding to lipids exhibiting symmetric unsaturations (red data points) yields a slope of −4.6 nN/symmetric unsaturation. The blue data point (which is not included in the fit to the mechanical stability data for the bilayers exhibiting symmetric unsaturations) falls below the measured trend, highlighting the reduced mechanical stability promoted by one asymmetric unsaturation. (Reprinted with permission from [67], copyright 2010 American Chemical Society.)

low-ionic strength conditions. Furthermore, on the addition of divalent cations, such as Mg^{2+}, the mechanical stability of SLBs is further increased.

It is well established that electrostatic interactions have a pivotal role in the structural and dynamic properties of lipid bilayers [61, 62, 85]. This is the case of Na^{+} or Ca^{2+}, which strongly interact with the carbonyl oxygen groups of PC polar

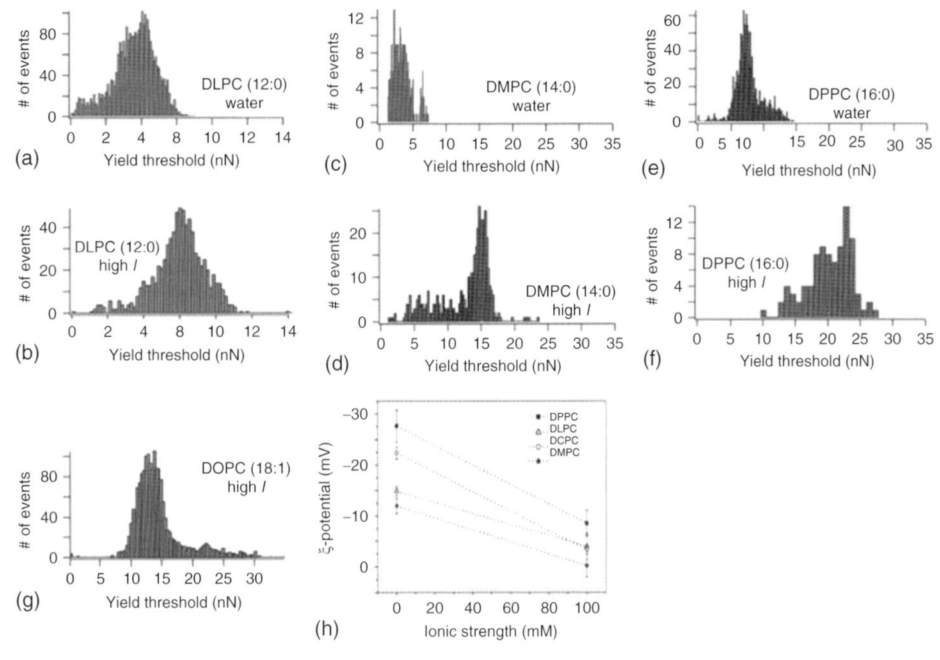

Figure 10.8 (a,b) Histograms of the yield threshold value for DLPC in water and 150 mM NaCl + 20 mM $MgCl_2$, respectively, where (a) $F_b = 3.78 \pm 0.04$ nN ($n = 2656$) and (b) $F_b = 6.12 \pm 0.08$ nN ($n = 794$). (c,d) Histograms of the yield threshold value for DMPC in water and 150 mM NaCl + 20 mM $MgCl_2$, respectively, where (c) $F_b = 2.76 \pm 0.11$ nN ($n = 396$) and (d) $F_b = 14.93 \pm 0.09$ nN ($n = 427$). (e,f) Histograms of the yield threshold value for DPPC in water and 150 mM NaCl + 20 mM $MgCl_2$, respectively, where (e) $F_b = 7.02 \pm 0.06$ nN ($n = 1272$) and (f) $F_b = 20.31 \pm 0.29$ nN ($n = 785$). (g) Histogram of the yield threshold force value for DOPC (18 : 1), where $F_b = 14.4 \pm 0.10$ nN ($n = 1691$). All results correspond to a Gaussian fitting of the data shown in the histogram. Results are presented as the Gaussian center. (h) ζ-Potential (zeta potential) values of the PC liposomes as the ionic strength increases. Every point in the graph is the average of 15 independent measurements. Error bars stand for standard deviation value within the measurement. Lines are a guide to the eye. (Reproduced from [61] with permission. Copyright 2005 Elsevier.)

moieties while changing the orientation of the phospholipids and their molecular packing [83, 86, 87]. Interestingly, this increment in the lipid order also influences the brittleness of the lipid bilayer. In a study of the friction properties of PC bilayers, Oncins *et al.* [78] concluded that the cohesion of the film and the F_b value are reduced in the absence of NaCl.

10.2.5
Effect of Different Cations on the Mechanical Stability of Lipid Bilayers

It is clear then that the specific chemical environment strongly modifies both the structural and thermodynamic behavior of SLBs. Regarding the influence of the electrolyte solution on the membrane, "specific" effects for different ions of the same charge are found that, in many cases, follow a trend as a function of the ion size [84, 88]. Detailed information about the location of ions with respect to the phospholipid membranes can be obtained from molecular dynamics (MD) simulations [80, 87, 89–91], which show that ion binding modifies the area per lipid, lipid ordering, orientation of the lipid head dipole, and the charge distribution along the system, among others. One of the most studied ions, partly because of its biological relevance, is K^+ that has been an object of controversy in literature. MD simulations showed that Na^+ ions have a strong effect on PC bilayers, increasing the lateral interactions between phospholipid molecules. Although the binding of K^+ was found to be much weaker, mainly because of the larger size of K^+ compared with Na^+ [90], these results were not conclusive because of the uncertainty in the estimation of the strength of the different force fields [81, 91]. In addition, in some simulations, it was pointed out that the size of K^+ ion could have been exaggerated [81]. Despite the current computational resources, the dynamics simulations of ions within a lipid bilayer cannot be as long as they should in order to fully understand the interaction between the ion and the surrounding phospholipids. Nevertheless, differences in the residence times and binding behavior between different ion species were observed [90–92].

FS spectroscopy has shown to be a very suitable technique to shed light in this particular topic. In a recent work, Redondo-Morata *et al.* systematically probed SLBs of distinct composition immersed in electrolytes composed of a variety of monovalent and divalent metal cations, providing information which clearly demonstrates that there is an independent and important contribution of each ion to the gross mechanical resistance of the lipid membranes [66]. As illustrated in Figure 10.9, it is unambiguous that Na^+ and K^+ have a great contribution to the mechanical stability of the membranes, being high F_b values representative of the tightest phospholipid packing, while Li^+ and Cs^+ do not show this dramatic effect. Accordingly, the penetration of the cations on the polar moiety of the membrane and the consequent decrease in the intermolecular distances is controlled not only by the cations size, charge density, and hydration but also by the initial distance between the SLB phospholipids. Moreover, it has been experimentally proved that K^+ has a contribution to the (nano)mechanical stability of SLBs. In the case of alkaline earth cations, while Sr^{2+} shows poor interaction with the membranes, Ca^{2+} and Mg^{2+} exhibit a great influence in SLBs, giving rise to higher F_b values, specifically Ca^{2+} in the case of DPPC (solid-like phase) and Mg^{2+} in the case of DLPC (liquid-like phase).

As a result, there are several mechanisms that contribute to determine the specific adsorption of ions to SLBs. The nature of the metallic cation, size, and charge density directly condition the penetration of these ions into the polar region of PC membranes. Besides, the intermolecular distance of phospholipids in the bilayer contributes to a preferential ionic adsorption according to the real size of the cations in solution.

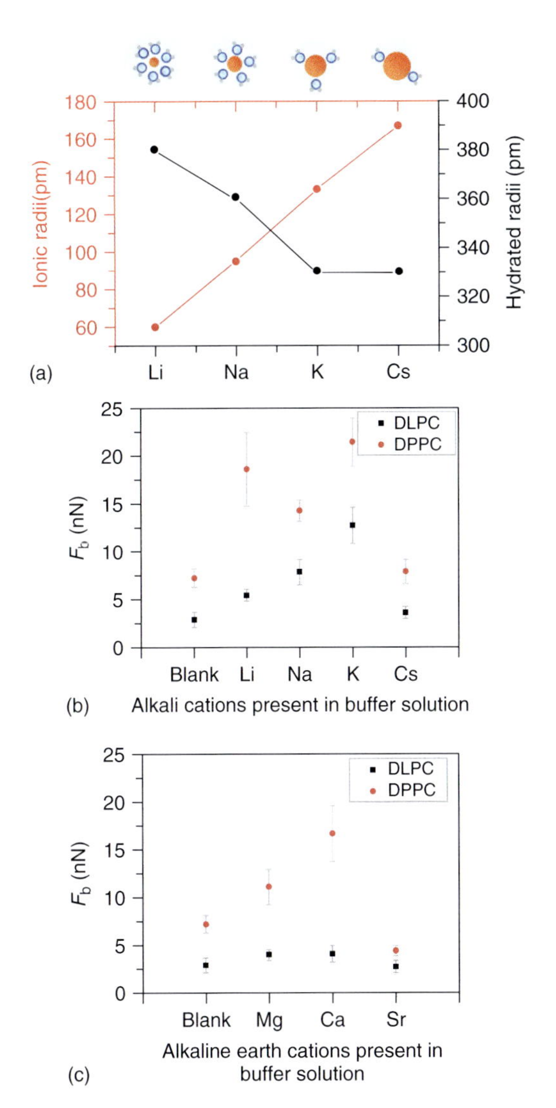

Figure 10.9 (a) Representation of the ionic radii and the hydrated radii for different alkaline cations. (b) F_b values versus alkali cation composition for all the obtained force plots on DPPC and DLPC bilayers at 150 mM. Error bars stand for standard deviation. (c) F_b values versus alkaline earth cation composition for DPPC and DLPC bilayers at 50 mM. Error bars stand for standard deviation.

10.2.6
Effect of Temperature on the Mechanical Stability of Lipid Bilayers

The complex structural dynamics of SLBs are governed by temperature-dependent parameters, and they can exist in several lamellar phases as a function of temperature; that is, gel phase (L_β), liquid crystalline or fluid phase (L_α), and ripple phase

(P_β) [93]. Often, phase behavior is dominated by individual molecules in a cooperative way with important variations in their lateral interactions, in close relationship with the (nano)mechanical properties of the bilayer. Temperature-controlled AFM has been established as a suitable tool to analyze both the topographical and mechanical evolution of biologically relevant issues that are temperature dependent at the nanometer scale.

In this framework, Leonenko *et al.* investigated the phase transitions in DPPC and DOPC mica-supported bilayers by means of AFM with temperature control, and they identified a broad $L_\beta - L_\alpha$ transition for DPPC [72]. For this system, F_b decreases as the temperature increases, and beyond the main transition, DPPC shows similar instability in the force plot as fluid-phase DOPC bilayers. Further studies by Sanz and coworkers related the evolution of the force required to penetrate SLBs with an AFM tip with the structural phase transition observed on AFM contact imaging [63]. They generalized that the main phase transition ($L_\beta - L_\alpha$) is not sharp but occurs in a range of temperatures of *ca.* $10 - 13^\circ$C (ΔT_M, Figure 10.10a) and that the solid-like phase shows a much higher resistance on breakthrough than the fluid-like phase, as evidenced for DMPC and DPPC-SLBs (Figure 10.10b). Furthermore, it was verified that the temperature has a strong effect on the (nano)mechanics of the solidlike phase but has a less impact on the yield threshold value of the liquidlike phase. The temperature range in which phase transition occurs in SLBs has been proposed to be due to the individual melting of the lipid leaflets in two subsequent temperature transitions, as demonstrated by Keller *et al.* [94] on DPPC-SLBs. The low-temperature transition was proposed to be related to the melting of the leaflet far from the substrate surface and the second transition to the phase transition of the leaflet in contact with the mica. The detection of a single phase transition event observed by Oncins *et al.* when studying topographic and mechanical evolution with temperature in single DPPC monolayers, strengthened this interpretation [95]. In a related context, 1-palmitoyl-2-oleoyl-*sn*-glycero-3-phosphoethanolamine (POPE) was described to undergo two distinct phase transitions, the first one involving the transition from the gel to the L_α phase and the second one involving the formation of intermediate structures in the transition from L_α to the inverted hexagonal phase exhibiting higher mechanical stability [96].

10.2.7
The Case of Phase-Segregated Lipid Bilayers

Following the results reviewed herein, it is undeniable that AFM stands as a remarkable tool to both topographically and (nano)mechanically characterize more complex processes, such as phase separation in bilayers composed of a mixture of different lipids. As previously reviewed, F_b is a characteristic of the chemical structure and molecular organization of a particular bilayer at specific environmental conditions; thus, in multicomponent systems, the F_b can be directly associated to the bilayer composition both in homogeneous multicomponent systems or phase-segregated domains. As a direct example, in a recent report

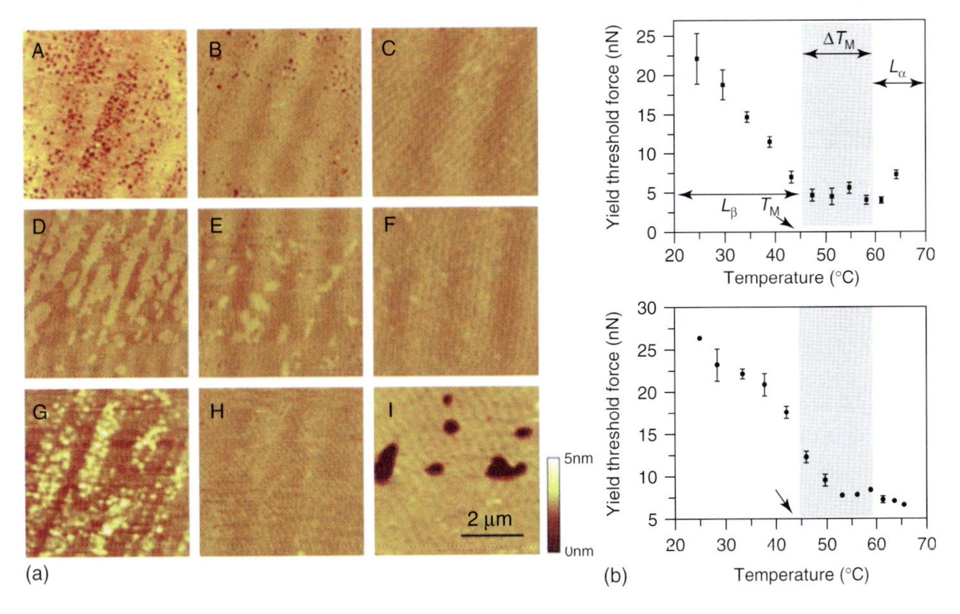

Figure 10.10 Effect of temperature on a DPPC-supported bilayer. (a) AFM contact mode images showing the main phase transition on heating the sample (A) 24.5°C, (B) 37.7°C, (C) 38.7°C, (D) 44.8°C, (E) 48.3°C, (F) 51.4°C, (G) 52.9°C, (H) 59.4°C, and cooling back to (I) 34.2°C. (b) Yield threshold force dependence with temperature for two independent experiments (different tip and different sample). All measurements were performed under high ionic strength solution. Each point in the graph corresponds to the center value of a Gaussian fitting to the obtained histogram. Error bars stand for SD of the Gaussian fitting to the yield threshold force histograms. T_M stands for the main transition temperature obtained from DSC measurements. Dark areas stand for the temperature range (ΔT_M) in which phase transitions are observed in supported planar bilayers through AFM images (a). (Adapted from [63] with permission. Copyright 2005 Elsevier.)

on an SLB composed of a POPE:1-palmitoyl-2-oleoyl-*sn*-glycero-3-phosphoglycerol (POPG) mixture, AFM images revealed the existence of two separated phases, the higher showing a region with protruding subdomains [97]. FS was applied to clarify the nature of each phase. The values of F_b, adhesion force, and height extracted from the force curves were assigned to the corresponding gel (L_β) (the mechanically stable domains) and fluid (L_a) phase (the mechanically labile phase). Recently, the group of Sanz demonstrated by using AFM–FS that the incorporation of sterols such as Chol and ergosterol to the SLBs affects their mechanical stability linearly not only for systems in the liquid phase (DLPC) but also for phospholipids present in the gel phase (DPPC), contrary to previous beliefs [67].

AFM force mapping is the collection of an array of force curves with record of their spatial position. If a force map is collected following the acquisition of a topographical image in a phase-segregated SLB, a direct correlation of the (nano)mechanical properties and the different domains in the bilayer can be done. This is exemplified in Figure 10.11 for a DPPC/Chol SLB. When 20 mol% of Chol is incorporated into

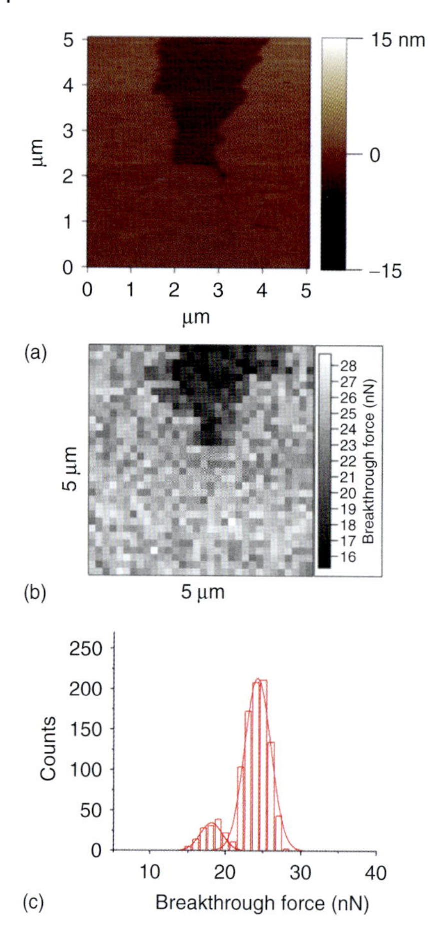

Figure 10.11 AFM topographical image of a DPPC/Chol (80 : 20) mixture (a), the corresponding $32 \times 32 F_b$ map (b), and the histograms of the F_b (c). F_b average values for the two distributions correspond to 18.0 ± 1.3 and 24.0 ± 1.6 nN. Measurements performed in 10 mM HEPES, 150 mM NaCl, 20 mM $MgCl_2$, pH 7.4, at 27°C.

a DPPC bilayer, the system phase segregates into domains of different composition and can be detected by AFM imaging and assigned to a gel-like phase and a Chol-enriched liquid-ordered-like phase (Figure 10.11a). FS of such system reveals different mechanical stability for each phase, and a bimodal distribution of F_b is obtained (Figure 10.11c). When the F_b are plotted in a 2D map, it permits the straight structure and mechanics correlation in phase-segregated systems (Figure 10.11b) (J. Relat-Goberna, M.I. Giannotti, and F. Sanz, unpublished results).

Recently, increasing evidence about the role of lateral organization in biological membranes has been presented. In particular, great interest has been centered in specific microdomains enriched in sphingolipids and sterols called *membrane rafts*

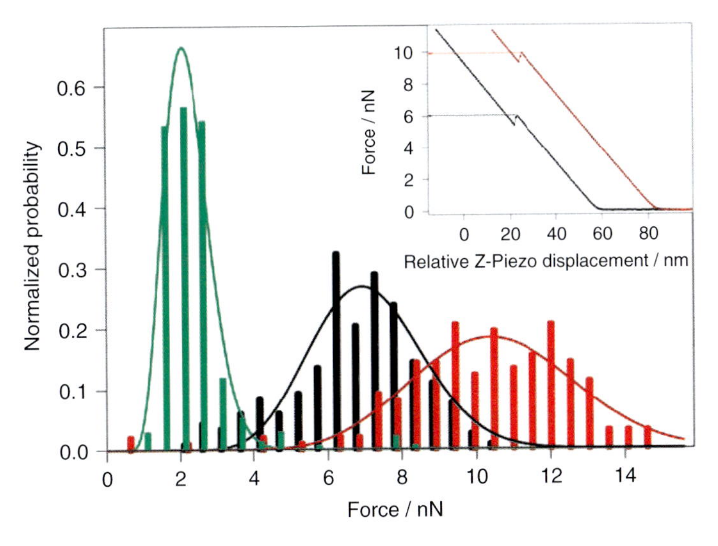

Figure 10.12 Normalized histograms of the force needed to pierce the bilayer with the AFM tip (threshold yield), measured for different samples at 25°C. Histogram bars refer to a piercing experiment performed on a DOPC-supported bilayer (red) and on the L_o (black) and L_d (green) phases in a DOPC/SM/Cholbilayer. The solid lines refer to the fit of the histograms to the continuum nucleation model. Inset: typical force curves for L_o (red) and L_d (black) phases. (Reproduced from [75] with permission. Copyright 2006 Wiley-VCH.)

[98]. These microdomains are described as a liquid-ordered (L_o) phase, structurally and dynamically different from the rest of the membrane, a liquid-disordered (L_d) phase [99]. Many studies have been performed on ternary lipid mixtures of PC, sphingomyelin (SM), and Chol as model membranes to mimic the rafts in cells, since these mixtures lead to coexisting L_o andL_d domains [100]. Schwille and coworkers have extensively analyzed the raft domain organization in DOPC/SM/Chol model membranes by means of AFM in combination with fluorescence correlation spectroscopy (FCS [75, 101, 102] and by performing FS experiments on the different phases they have shown that the average force needed to pierce the L_o phase of the DOPC/SM/Chol bilayer is about 4 nN higher than the one needed to pierce the L_d phase (Figure 10.12). Additionally, Schwille *et al.* [76] demonstrated via AFM–FS measurements how a peptide derived from Bax, a protein regulator of physiological cell death and involved in the formation of partially lipidic pores, affects the mechanical properties of the bilayers. They observed by AFM and confocal microscopy that the pore-forming peptide Bax-α5 affects the morphological organization of domain-exhibiting DOPC/SM/Chol membranes, and FS experiments revealed that the force needed to punch through lipid bilayers in the presence of Bax-α5 decreased, contributing to elucidate the mechanism of pore formation in lipid bilayers. In a parallel study, Zou and coworkers have explored the influence of different Chol contents on the morphology and (nano)mechanical stability of phase-segregated DOPC/SM/Chol lipid bilayers by means of AFM imaging and force mapping

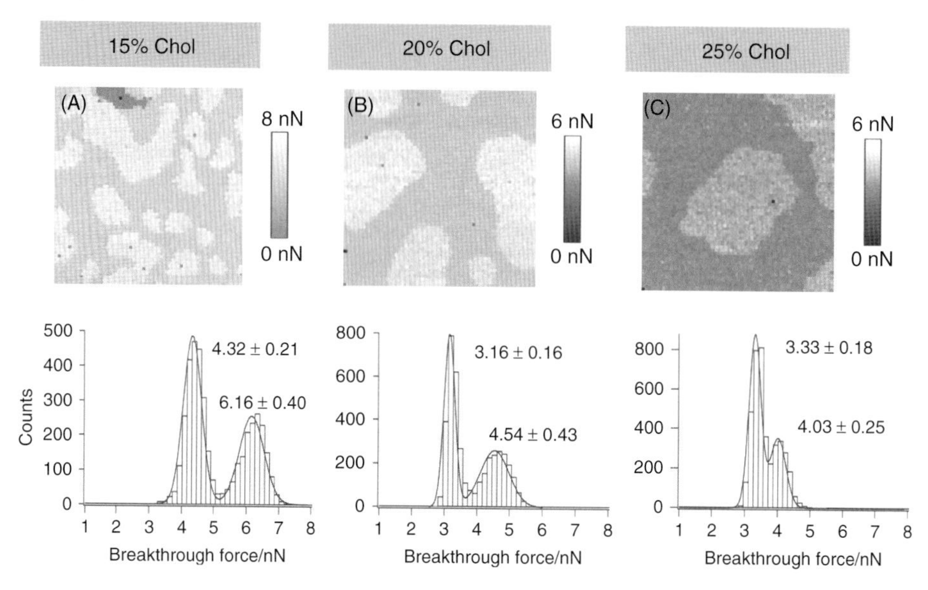

Figure 10.13 Representative breakthrough force maps and corresponding histograms of DOPC/SM/Chol bilayers. Breakthrough force maps were reconstructed from 64 × 64-pixel force-mapping experiments and the corresponding histograms on bilayers with 15(A), 20(B), and 25(C)% Chol, at loading rate of 2000 nm s^{-1}. All histograms consist of 4096 force curves, and the numbers in the histogram represent mean ± SD breakthrough force values from Gaussian fits. All breakthrough force maps are 3 × 3 μm^2. (Adapted from [103] with permission. Copyright 2010 Elsevier.)

[103]. They observed consistently higher F_b in the SM/Chol-enriched L_o domains than in the DOPC-enriched L_d phase and that the (nano)mechanical stability of both coexisting phases decreases with increasing Chol content (Figure 10.13). By following the dependence of the F_b with the loading rate in both phases, they calculated the activation energies (ΔE_a) of bilayer rupture at zero applied force. ΔE_a values obtained were in the range 75–125 kJ mol^{-1} for the various Chol contents and did not show a dependence on the different lipid phases. CER incorporation on coexisting fluid phase and ordered domains in phase-separated binary and ternary lipid mixtures has been shown to affect the lipid spatial organization in domains, with the appearance of a CER-enriched gel-like phase and the displacement of Chol from rafts [101, 104]. In order to directly probe and quantify the (nano)mechanical stability and rigidity of the CER-enriched platforms, Zou *et al.* [105] performed AFM topographical images and force mapping that allowed the determination of the F_b and elastic modulus of the different phases. They confirmed the expulsion of Chol from sphingolipid/Chol-enriched domains as a result of CER incorporation and observed an increased mechanical stability attributed to the influence of CER in the lipid organization and packing behavior.

Beyond doubt, AFM force mapping has proven to be a very powerful tool to complement other biophysical techniques currently used in studying multicomponent lipid bilayers toward understanding of natural membranes.

10.3
Future Perspectives

In the future, SLBs will continue to play a pivotal role in the biophysical investigation and in the technology field concerning the development of nanodevices and sensors. Here, synthetic lipid membranes are presented as simple one-component or multicomponent models. Nevertheless, up-to-date investigations reveal that extension of these studies to more biological relevant membranes has a promising future. For instance, among the ongoing studies on more sophisticated systems, it is worth mentioning the analysis of asymmetric bilayers [106], the insertion of membrane proteins [107], or mixtures with different sterols [108] or surfactants [109]. The ultimate goal is to fully understand the high mechanochemical complexity of biological membranes.

Besides, the interaction between the bilayer and the supporting substrate still remains unclear, and different configurations or experimental systems may help to shed light on the particular effect of the substrate to the membrane (nano)mechanics. For example, the effect of the substrate corrugation [110] or porosity [111], or the use of tethered lipid membranes [112], lipid membranes on a polymer cushion [113], or pore-spanning lipid bilayers [114], shows potential perspective to study how (nano)mechanical properties of the bilayers are affected by the substrate. It has been reported lately that the interleaflet coupling is highly dependent on the substrate and the physical parameters of the measuring system [115]; an interesting issue that remains in need of further investigation.

The experimental approach provided by AFM–FS technique represents an ideal platform to bridge the gap with MD simulations, which expose a variety of molecular mechanisms that are not accessible with mesoscopic experimental techniques [116]. In order to deepen in the physical perspective, the rupture of the lipid bilayer by the AFM tip has been modeled as a barrier-limited two-state process. Nonetheless, literature reveals evidence that more than a single physicochemical determinant is involved in the breakthrough process. Thus, it is of great interest to experimentally assess the energy landscape of the molecular interactions between phospholipid neighboring molecules. For such purposes, further development of both dynamic force spectroscopy, that is, indenting the lipid bilayers at different loading rates, and force-clamp spectroscopy, where constant force is applied during the experiment, is underway.

References

1. Dowhan, W. (1997) *Ann. Rev. Biochem.*, **66**, 199–232.
2. Singer, S.J. and Nicolson, G.L. (1972) *Science*, **175**, 720–731.
3. van Meer, G., Voelker, D.R., and Feigenson, G.W. (2008) *Nat. Rev. Mol. Cell Biol.*, **9**, 112–124.
4. Edidin, M. (2003) *Nat. Rev. Mol. Cell Biol.*, **4**, 414–418.
5. Pietzsch, J. (2004) *Horizon Symposia: Living Frontier*, Nature Publishing Group, pp. 1–4.
6. (a) Qi, S.Y., Groves, J.T., and Chakraborty, A.K. (2001) *Proc. Natl.*

Acad. Sci. U.S.A., **98**, 6548–6553; (b) Kasahara, K. and Sanai, Y. (2001) *Trends Glycosci. Glycotechnol.*, **13**, 587–594; (c) Stoddart, A., Dykstra, M.L., Brown, B.K., Song, W.X., Pierce, S.K., and Brodsky, F.M. (2002) *Immunity*, **17**, 451–462; (d) Heldin, C.H. (1995) *Cell*, **80**, 213–223.

7. (a) Ono, A. and Freed, E.O. (2001) *Proc. Natl. Acad. Sci. U.S.A.*, **98**, 13925–13930; (b) Xu, L., Frederik, P., Pirollo, K.F., Tang, W.H., Rait, A., Xiang, L.M., Huang, W.Q., Cruz, I., Yin, Y.Z., and Chang, E.H. (2002) *Hum. Gene Ther.*, **13**, 469–481.
8. (a) Yang, T.L., Baryshnikova, O.K., Mao, H.B., Holden, M.A., and Cremer, P.S. (2003) *J. Am. Chem. Soc.*, **125**, 4779–4784; (b) Plant, A.L., Brighamburke, M., Petrella, E.C., and Oshannessy, D.J. (1995) *Anal. Biochem.*, **226**, 342–348.
9. Mueller, P., Rudin, D.O., Tien, H.T., and Wescott, W.C. (1962) *Nature*, **194**, 979–980.
10. Ottovaleitmannova, A. and Tien, H.T. (1992) *Prog. Surf. Sci.*, **41**, 337–445.
11. Castellana, E.T. and Cremer, P.S. (2006) *Surf. Sci. Rep.*, **61**, 429–444.
12. Talham, D.R., Yamamoto, T., and Meisel, M.W. (2008) *J. Phys.: Condens. Matter*, **20**., 13pp.
13. Vie, V., Van Mau, N., Lesniewska, E., Goudonnet, J.P., Heitz, F., and Le Grimellec, C. (1998) *Langmuir*, **14**, 4574–4583.
14. Rinia, H.A., Demel, R.A., van der Eerden, J., and de Kruijff, B. (1999) *Biophys. J.*, **77**, 1683–1693.
15. Dufrene, Y.F., Barger, W.R., Green, J.B.D., and Lee, G.U. (1997) *Langmuir*, **13**, 4779–4784.
16. (a) Brian, A.A. and McConnell, H.M. (1984) *Proc. Natl. Acad. Sci. U.S.A: Biol. Sci.*, **81**, 6159–6163; (b) Jass, J., Tjarnhage, T., and Puu, G. (2000) *Biophys. J.*, **79**, 3153–3163; (c) Richter, R.P. (2006) *Langmuir*, **22**, 3497–3505.
17. Mingeot-Leclercq, M.P., Deleu, M., Brasseur, R., and Dufrene, Y.F. (2008) *Nat. Protoc.*, **3**, 1654–1659.
18. (a) Reimhult, E., Hook, F., and Kasemo, B. (2003) *Langmuir*, **19**, 1681–1691; (b) Stelzle, M., Weissmuller, G., and Sackmann, E. (1993) *J. Phys. Chem.*, **97**, 2974–2981.
19. (a) Mennicke, U. and Salditt, T. (2002) *Langmuir*, **18**, 8172–8177; (b) Simonsen, A.C. and Bagatolli, L.A. (2004) *Langmuir*, **20**, 9720–9728; (c) Pompeo, G., Girasole, M., Cricenti, A., Cattaruzza, F., Flamini, A., Prosperi, T., Generosi, J., and Castellano, A.C. (2005) *Biochim. Biophys. Acta: Biomembr.*, **1712**, 29–36.
20. (a) Zasadzinski, J.A.N., Helm, C.A., Longo, M.L., Weisenhorn, A.L., Gould, S.A.C., and Hansma, P.K. (1991) *Biophys. J.*, **59**, 755–760; (b) Egawa, H. and Furusawa, K. (1999) *Langmuir*, **15**, 1660–1666.
21. Tamm, L.K. and McConnell, H.M. (1985) *Biophys. J.*, **47**, 105–113.
22. Lagerholm, B.C., Starr, T.E., Volovyk, Z.N., and Thompson, N.L. (2000) *Biochemistry*, **39**, 2042–2051.
23. Cremer, P.S., Groves, J.T., Ulman, N., and Boxer, S.G. (1998) *Biophys. J.*, **74**, A311–A311.
24. (a) Yang, J.L. and Kleijn, J.M. (1999) *Biophys. J.*, **76**, 323–332; (b) Gritsch, S., Nollert, P., Jahnig, F., and Sackmann, E. (1998) *Langmuir*, **14**, 3118–3125.
25. (a) Grelard, A., Couvreux, A., Loudet, C., and Dufourc, E.J. (2009) in *Methods in Molecular Biology*, vol. 462 (eds R. Woscholski, B. Larijani, and C.A. Rosser), Human Press, Totowa, pp. 111–133; (b) Bechinger, B. (1999) *Biochim. Biophys. Acta: Biomembr.*, **1462**, 157–183.
26. Reinl, H., Brumm, T., and Bayerl, T.M. (1992) *Biophys. J.*, **61**, 1025–1035.
27. Kago, K., Matsuoka, H., Yoshitome, R., Yamaoka, H., Ijiro, K., and Shimomura, M. (1999) *Langmuir*, **15**, 5193–5196.
28. Mohwald, H. (1988) *Thin Solid Films*, **159**, 1–15.
29. Slotte, J.P. (1995) *Biochim. Biophys. Acta: Lipids Lipid Metab.*, **1259**, 180–186.
30. Honig, D. and Mobius, D. (1991) *J. Phys. Chem.*, **95**, 4590–4592.
31. Dufrene, Y.F., Boland, T., Schneider, J.W., Barger, W.R., and Lee, G.U. (1998) *Faraday Discuss.*, **111**, 79–94.

32. Schram, V., Lin, H.N., and Thompson, T.E. (1996) *Biophys. J.*, **71**, 1811–1822.
33. Deleu, M., Paquot, M., Jacques, P., Thonart, P., Adriaensen, Y., and Dufrene, Y.F. (1999) *Biophys. J.*, **77**, 2304–2310.
34. Linton, R., Guarisco, V., Lee, J.J., Hagenhoff, B., and Benninghoven, A. (1992) *Thin Solid Films*, **210**, 565–570.
35. (a) Mueller, D.J. and Dufrene, Y.F. (2008) *Nat. Nanotechnol.*, **3**, 261–269; (b) Dufrene, Y.F. (2008) *Nat. Rev. Microbiol.*, **6**, 674–680.
36. El Kirat, K., Morandat, S., and Dufrene, Y.F. (2010) *Biochim. Biophys. Acta: Biomembr.*, **1798**, 750–765.
37. Buzhynskyy, N., Hite, R.K., Walz, T., and Scheuring, S. (2007) *EMBO Rep.*, **8**, 51–55.
38. Vogel, V. and Sheetz, M. (2006) *Nat. Rev. Mol. Cell Biol.*, **7**, 265–275.
39. (a) Phillips, R., Ursell, T., Wiggins, P., and Sens, P. (2009) *Nature*, **459**, 379–385; (b) Ursell, T., Kondev, J., Reeves, D., Wiggins, P.A., and Phillips, R. (2008) The role of lipid bilayer mechanics in mechanosensation, in *mechanosensitivity in Cells and Tissues 1: Mechanosensitive Ion Channels*, (eds A. Kamkin and I. Kiseleva) Springer-Verlag.
40. (a) Schmidt, D., Jiang, Q.-X., and MacKinnon, R. (2006) *Nature*, **444**, 775–779; (b) Schmidt, D. and MacKinnon, R. (2008) *Proc. Natl. Acad. Sci. U.S.A.*, **105**, 19276–19281; (c) Oliver, D., Lien, C.C., Soom, M., Baukrowitz, T., Jonas, P., and Fakler, B. (2004) *Science*, **304**, 265–270; (d) Ramu, Y., Xu, Y., and Lu, Z. (2006) *Nature*, **442**, 696–699; (e) Perozo, E., Kloda, A., Cortes, D.M., and Martinac, B. (2002) *Nat. Struct. Biol.*, **9**, 696–703; (f) Suchyna, T.M., Tape, S.E., Koeppe, R.E., Andersen, O.S., Sachs, F., and Gottlieb, P.A. (2004) *Nature*, **430**, 235–240.
41. Hong, H.D. and Tamm, L.K. (2004) *Proc. Natl. Acad. Sci. U.S.A.*, **101**, 4065–4070.
42. Reynwar, B.J., Illya, G., Harmandaris, V.A., Mueller, M.M., Kremer, K., and Deserno, M. (2007) *Nature*, **447**, 461–464.
43. (a) Evans, E. and Rawicz, W. (1990) *Phys. Rev. Lett.*, **64**, 2094–2097; (b) Rawicz, W., Olbrich, K.C., McIntosh, T., Needham, D., and Evans, E. (2000) *Biophys. J.*, **79**, 328–339.
44. (a) Heinrich, V. and Rawicz, W. (2005) *Langmuir*, **21**, 1962–1971; (b) Evans, E., Heinrich, V., Ludwig, F., and Rawicz, W. (2003) *Biophys. J.*, **85**, 2342–2350.
45. (a) Benz, M., Gutsmann, T., Chen, N.H., Tadmor, R., and Israelachvili, J. (2004) *Biophys. J.*, **86**, 870–879; (b) Helm, C.A., Israelachvili, J.N., and McGuiggan, P.M. (1989) *Science*, **246**, 919–922; (c) Marra, J. and Israelachvili, J. (1985) *Biochemistry*, **24**, 4608–4618.
46. Corcoran, S.G., Colton, R.J., Lilleodden, E.T., and Gerberich, W.W. (1997) *Phys. Rev. B*, **55**, 16057–16060.
47. Giannotti, M.I. and Vancso, G.J. (2007) *ChemPhysChem*, **8**, 2290–2307.
48. Fisher, T.E., Oberhauser, A.F., Carrion-Vazquez, M., Marszalek, P.E., and Fernandez, J.M. (1999) *Trends Biochem. Sci.*, **24**, 379–384.
49. Lee, G.U., Chrisey, L.A., and Colton, R.J. (1994) *Science*, **266**, 771–773.
50. Butt, H.J., Cappella, B., and Kappl, M. (2005) *Surf. Sci. Rep.*, **59**, 1–152.
51. Franz, V., Loi, S., Muller, H., Bamberg, E., and Butt, H.H. (2002) *Colloids Surf. B: Biointerfaces*, **23**, 191–200.
52. Proksch, R., Schaffer, T.E., Cleveland, J.P., Callahan, R.C., and Viani, M.B. (2004) *Nanotechnology*, **15**, 1344–1350.
53. Fraxedas, J., Garcia-Manyes, S., Gorostiza, P., and Sanz, F. (2002) *Proc. Natl. Acad. Sci. U.S.A.*, **99**, 5228–5232.
54. (a) Krautbauer, R., Clausen-Schaumann, H., and Gaub, H.E. (2000) *Angew. Chem. Int. Ed.*, **39**, 3912–3915; (b) Clausen-Schaumann, H., Rief, M., Tolksdorf, C., and Gaub, H.E. (2000) *Biophys. J.*, **78**, 1997–2007.
55. Marszalek, P.E., Oberhauser, A.F., Pang, Y.P., and Fernandez, J.M. (1998) *Nature*, **396**, 661–664.
56. Carrion-Vazquez, M., Oberhauser, A.F., Fowler, S.B., Marszalek, P.E., Broedel, S.E., Clarke, J., and Fernandez, J.M. (1999) *Biophys. J.*, **76**, A173–A173.
57. Ducker, W.A. and Clarke, D.R. (1994) *Colloids Surf. A: Physicochem. Eng. Aspects*, **93**, 275–292.

58. (a) Oncins, G., Vericat, C., and Sanz, F. (2008) *J. Chem. Phys.*, **128**, 044701; (b) Torrent-Burgues, J., Pla, M., Escriche, L., Casabo, J., Errachid, A., and Sanz, F. (2006) *J. Colloid Interface Sci.*, **301**, 585–593; (c) Oncins, G., Torrent-Burgues, J., and Sanz, F. (2008) *J. Phys. Chem. C*, **112**, 1967–1974; (d) Torrent-Burgues, J., Oncins, G., and Sanz, F. (2008) *Colloids Surf. A: Physicochem. Eng. Aspects*, **321**, 70–75.
59. (a) Hofbauer, W., Ho, R.J., Hairulnizam, R., Gosvami, N.N., and O'Shea, S.J. (2009) *Phys. Rev. B*, **80**, 134104-1–134104-5; (b) O'Shea, S.J., Gosvami, N.N., Lim, L.T.W., and Hofbauer, W. (2010) *Jpn. J. Appl. Phys.*, **49**, 08LA01-1–08LA01-9.
60. Dufrene, Y.F., Boland, T., Schneider, J.W., Barger, W.R., and Lee, G.U. (1998) *Faraday Discuss*, **11**, 79–94.
61. Garcia-Manyes, S., Oncins, G., and Sanz, F. (2005) *Biophys. J.*, **89**, 1812–1826.
62. Garcia-Manyes, S., Oncins, G., and Sanz, F. (2006) *Electrochim. Acta*, **51**, 5029–5036.
63. Garcia-Manyes, S., Oncins, G., and Sanz, F. (2005) *Biophys. J.*, **89**, 4261–4274.
64. Butt, H.J. (1992) *Nanotechnology*, **3**, doi:10.1088/0957-4484/1083/1082/1003
65. Butt, H.J. (1991) *Biophys. J.*, **60**, 777–785.
66. Redondo, L., Oncins, G., and Sanz, F. (2010) *Biophys. J.*, **98**, 626a–627a.
67. Garcia-Manyes, S., Redondo-Morata, L., Oncins, G., and Sanz, F. (2010) *J. Am. Chem. Soc.*, **132**, 12874–12886.
68. (a) Zhmud, B.V., Sonnefeld, J., and Bergstrom, L. (1999) *Colloids Surf. A: Physicochem. Eng. Aspects*, **158**, 327–341; (b) Yin, X.H. and Drelich, J. (2008) *Langmuir*, **24**, 8013–8020; (c) Butt, H.J. (1991) *Biophys. J.*, **60**, 1438–1444.
69. Garcia-Manyes, S., Gorostiza, P., and Sanz, F. (2006) *Anal. Chem.*, **78**, 61–70.
70. (a) Fukuma, T., Higgins, M.J., and Jarvis, S.P. (2007) *Biophys. J.*, **92**, 3603–3609; (b) Higgins, M.J., Polcik, M., Fukuma, T., Sader, J.E., Nakayama, Y., and Jarvis, S.P. (2006) *Biophys. J.*, **91**, 2532–2542.
71. Israelachvili, J.N. (1992) *Intermolecular and Surface Forces*, 2nd edn, Academic Press, London.
72. Leonenko, Z.V., Finot, E., Ma, H., Dahms, T.E.S., and Cramb, D.T. (2004) *Biophys. J.*, **86**, 3783–3793.
73. (a) Butt, H.J. and Franz, V. (2002) *Phys. Rev. E*, **66**, 031601-1–031601-9. (b) Loi, S., Sun, G., Franz, V., and Butt, H.-J. (2002) *Phys. Rev. E. Stat. Nonlinear, Soft. Matter Phys.*, **66**, 031602-1–031602-7.
74. Sisquella, X., de Pourcq, K., Alguacil, J., Robles, J., Sanz, F., Anselmetti, D., Imperial, S., and Fernandez-Busquets, X. (2010) *FASEB J.*, **24**, 4203–4217.
75. Chiantia, S., Ries, J., Kahya, N., and Schwille, P. (2006) *ChemPhysChem*, **7**, 2409–2418.
76. Garcia-Saez, A.J., Chiantia, S., Salgado, J., and Schwille, P. (2007) *Biophys. J.*, **93**, 103–112.
77. Trunfio-Sfarghiu, A.-M., Berthier, Y., Meurisse, M.-H., and Rieu, J.-P. (2008) *Langmuir*, **24**, 8765–8771.
78. Oncins, G., Garcia-Manyes, S., and Sanz, F. (2005) *Langmuir*, **21**, 7373–7379.
79. Chiu, S.W., Jakobsson, E., Mashl, R.J., and Scott, H.L. (2002) *Biophys. J.*, **83**, 1842–1853.
80. (a) Gurtovenko, A.A. (2005) *J. Chem. Phys.*, **122**, 244902-1–244902-10; (b) Petrache, H.I., Tristram-Nagle, S., Harries, D., Kucerka, N., Nagle, J.F., and Parsegian, V.A. (2006) *J. Lipid Res.*, **47**, 302–309.
81. Gurtovenko, A.A. and Vattulainen, I. (2008) *J. Phys. Chem. B*, **112**, 1953–1962.
82. (a) Kandasamy, S.K. and Larson, R.G. (2006) *Biochim. Biophys. Acta: Biomembr.*, **1758**, 1274–1284; (b) Petrache, H.I., Kimchi, I., Harries, D., Tristram-Nagle, S., Podgornik, R., and Parsegian, V.A. (2004) *Biophys. J.*, **86**, 379A–379A.
83. (a) Vacha, R., Siu, S.W.I., Petrov, M., Bockmann, R.A., Barucha-Kraszewska, J., Jurkiewicz, P., Hof, M., Berkowitz, M.L., and Jungwirth, P. (2009) *J. Phys. Chem. A*, **113**, 7235–7243;

(b) Vacha, R., Jurkiewicz, P., Petrov, M., Berkowitz, M.L., Bockmann, R.A., Barucha-Kraszewska, J., Hof, M., and Jungwirth, P. (2010) *J. Phys. Chem. B*, **114**, 9504–9509; (c) Pandit, S.A., Bostick, D., and Berkowitz, M.L. (2003) *Biophys. J.*, **84**, 3743–3750.

84. (a) Aroti, A., Leontidis, E., Dubois, M., and Zemb, T. (2007) *Biophys. J.*, **93**, 1580–1590; (b) Leontidis, E., Aroti, A., Belloni, L., Dubois, M., and Zemb, T. (2007) *Biophys. J.*, **93**, 1591–1607; (c) Petrache, H.I., Zemb, T., Belloni, L., and Parsegian, V.A. (2006) *Proc. Natl. Acad. Sci. U.S.A.*, **103**, 7982–7987.

85. (a) Pabst, G., Hodzic, A., Strancar, J., Danner, S., Rappolt, M., and Laggner, P. (2007) *Biophys. J.*, **93**, 2688–2696; (b) Pedersen, U.R., Leidy, C., Westh, P., and Peters, G.H. (2006) *Biochim. Biophys. Acta: Biomembr.*, **1758**, 573–582; (c) Porasso, R.D. and Cascales, J.J.L. (2009) *Colloids Surf. B: Biointerfaces*, **73**, 42–50; (d) Sinn, C.G., Antonietti, M., and Dimova, R. (2006) *Colloids Surf. A: Physicochem. Eng. Aspects*, **282**, 410–419; (e) Vernier, P.T., Ziegler, M.J., and Dimova, R. (2009) *Langmuir*, **25**, 1020–1027.

86. Bockmann, R.A., Hac, A., Heimburg, T., and Grubmuller, H. (2003) *Biophys. J.*, **85**, 1647–1655.

87. Bockmann, R.A. and Grubmuller, H. (2004) *Biophys. J.*, **86**, 370A–370A.

88. (a) Garcia-Celma, J.J., Hatahet, L., Kunz, W., and Fendler, K. (2007) *Langmuir*, **23**, 10074–10080; (b) Leontidis, E., Aroti, A., and Belloni, L. (2009) *J. Phys. Chem. B*, **113**, 1447–1459; (c) Leontidis, E. and Aroti, A. (2009) *J. Phys. Chem. B*, **113**, 1460–1467.

89. (a) Gurtovenko, A.A., Miettinen, M., Karttunen, M., and Vattulainen, I. (2005) *J. Phys. Chem. B*, **109**, 21126–21134; (b) Pandit, S., Bostick, D., and Berkowitz, M. (2004) *Biophys. J.*, **86**, 368A–368A; (c) Yi, M., Nymeyer, H., and Zhou, H.X. (2008) *Phys. Rev. Lett.*, **101**, 038103-1–038103-4.

90. Cordomi, A., Edholm, O., and Perez, J.J. (2008) *J. Phys. Chem. B*, **112**, 1397–1408.

91. Cordomi, A., Edholm, O., and Perez, J.J. (2009) *J. Chem. Theory Comput.*, **5**, 2125–2134.

92. Miettinen, M.S., Gurtovenko, A.A., Vattulainen, I., and Karttunen, M. (2009) *J. Phys. Chem. B*, **113**, 9226–9234.

93. Nagle, J.F. and Tristram-Nagle, S. (2000) *Biochim. Biophys. Acta: Rev. Biomembr.*, **1469**, 159–195.

94. Keller, D., Larsen, N.B., Moller, I.M., and Mouritsen, O.G. (2005) *Phys. Rev. Lett.*, **94**, 025701-1–025701-4.

95. Oncins, G., Picas, L., Hernandez-Borrell, J., Garcia-Manyes, S., and Sanz, F. (2007) *Biophys. J.*, **93**, 2713–2725.

96. Picas, L., Montero, M.T., Morros, A., Oncins, G., and Hernandez-Borrell, J. (2008) *J. Phys. Chem. B*, **112**, 10181–10187.

97. Picas, L., Montero, M.T., Morros, A., Cabanas, M.E., Seantier, B., Milhiet, P.-E., and Hernandez-Borrell, J. (2009) *J. Phys. Chem. B*, **113**, 4648–4655.

98. Simons, K. and Ikonen, E. (1997) *Nature*, **387**, 569–572.

99. (a) McMullen, T.P.W., Lewis, R., and McElhaney, R.N. (2004) *Curr. Opin. Colloid Interface Sci.*, **8**, 459–468; (b) Harder, T. and Simons, K. (1997) *Curr. Opin. Cell Biol.*, **9**, 534–542.

100. Dietrich, C., Bagatolli, L.A., Volovyk, Z.N., Thompson, N.L., Levi, M., Jacobson, K., and Gratton, E. (2001) *Biophys. J.*, **80**, 1417–1428.

101. (a) Chiantia, S., Kahya, N., Ries, J., and Schwille, P. (2006) *Biophys. J.*, **90**, 4500–4508; (b) Chiantia, S., Kahya, N., and Schwille, P. (2007) *Langmuir*, **23**, 7659–7665.

102. Chiantia, S., Ries, J., Chwastek, G., Carrer, D., Li, Z., Bittman, R., and Schwille, P. (2008) *Biochim. Biophys. Acta: Biomembr.*, **1778**, 1356–1364.

103. Sullan, R.M.A., Li, J.K., Hao, C.C., Walker, G.C., and Zou, S. (2010) *Biophys. J.*, **99**, 507–516.

104. Ira, Z.S., Ramirez, D.M., Vanderlip, S., Ogilvie, W., Jakubek, Z.J., and Johnston, L.J. (2009) *J. Struct. Biol.*, **168**, 78–89.

105. (a) Sullan, R.M.A., Li, J.K., and Zou, S. (2009) *Langmuir*, **25**, 7471–7477;

(b) Sullan, R.M.A., Li, J.K., and Zou, S. (2009) *Langmuir*, **25**, 12874–12877; (c) Zou, S. and Johnston, L.J. (2010) *Curr. Opin. Colloid Interface Sci.*, **15**, 489–498.
106. Manno, S., Takakuwa, Y., and Mohandas, N. (2002) *Proc. Natl. Acad. Sci. U.S.A.*, **99**, 1943–1948.
107. Domenech, O., Redondo, L., Picas, L., Morros, A., Montero, M.T., and Hernandez-Borrell, J. (2007) *J. Mol. Recognit.*, **20**, 546–553.
108. Tierney, K.J., Block, D.E., and Longo, M.L. (2005) *Biophys. J.*, **89**, 2481–2493.
109. (a) Finot, E., Leonenko, Y., Moores, B., Eng, L., Amrein, M., and Leonenko, Z. (2010) *Langmuir*, **26**, 1929–1935; (b) Leonenko, Z., Finot, E., Vassiliev, V., and Amrein, M. (2006) *Ultramicroscopy*, **106**, 687–694; (c) Custers, J.P.A., Kelemen, P., van den Broeke, L.J.P., Stuart, M.A.C., and Keurentjes, J.T.F. (2005) *J. Am. Chem. Soc.*, **127**, 1594–1595.
110. Goksu, E.I., Nellis, B.A., Lin, W.-C., Satcher, J.H. Jr., Groves, J.T., Risbud, S.H., and Longo, M.L. (2009) *Langmuir*, **25**, 3713–3717.
111. Nussio, M.R., Oncins, G., Ridelis, I., Szili, E., Shapter, J.G., Sanz, F., and Voelcker, N.H. (2009) *J. Phys. Chem. B.*, **113**, 10339–10347.
112. Koeper, I. (2007) *Mol. Biosyst.*, **3**, 651–657.
113. Smith, H.L., Jablin, M.S., Vidyasagar, A., Saiz, J., Watkins, E., Toomey, R., Hurd, A.J., and Majewski, J. (2009) *Phys. Rev. Lett.*, **102**, 228102-1–228102-4.
114. Mey, I., Stephan, M., Schmitt, E.K., Mueller, M.M., Ben Amar, M., Steinem, C., and Janshoff, A. (2009) *J. Am. Chem. Soc.*, **131**, 7031–7039.
115. Seeger, H.M., Marino, G., Alessandrini, A., and Facci, P. (2009) *Biophys. J.*, **97**, 1067–1076.
116. (a) Rog, T., Pasenkiewicz-Gierula, M., Vattulainen, I., and Karttunen, M. (2009) *Biochim. Biophys. Acta: Biomembr.*, **1788**, 97–121; (b) Berkowitz, M.L., Bostick, D.L., and Pandit, S. (2006) *Chem. Rev.*, **106**, 1527–1539; (c) Berkowitz, M.L. (2009) *Biochim. Biophys. Acta: Biomembr.*, **1788**, 86–96.

11
Single-Molecule Atomic Force Microscopy of Cellular Sensors

Jürgen J. Heinisch and Yves F. Dufrêne

11.1
Introduction

11.1.1
Mechanosensors in Living Cells

Life is defined by its ability to reproduce and to interact with its environment. Cells, which form the basic units of life from unicellular microbes to those embedded in the tissues of multicellular organisms, share these properties. Their reaction to external cues is frequently mediated by more or less complex signal transduction pathways, which detect environmental changes of either physical or chemical nature at the cell surface and trigger the appropriate intracellular responses. These pathways frequently alter the transcriptional program in the nucleus and consequently the cells proteome (see [1–4] for some reviews on random examples from bacteria, fungi, insects, and mammals, respectively).

Most of these signal transduction pathways start with sensor proteins, which span the plasma membrane and either bind specific ligands (hormones, adhesion molecules, etc.) or can detect mechanical forces and other changes in the physical environment. The detection of mechanical forces may have been one of the first sensing mechanisms that appeared in evolution [5]. Accordingly, mechanosensing and mechanotransduction, which allow the cells to convert these mechanical forces into biochemical signals, play key roles in regulating processes such as cell adhesion, cell growth, cell differentiation, cell shape, and cell death [6–8]. In the processes studied so far, mechanical forces can alter the protein conformation and its biological activity by various mechanisms: (i) the force-induced exposure of otherwise cryptic peptide sequences (such as in the cell adhesion proteins fibronectin and integrin [6–9]), (ii) the opening of mechanosensitive ion channels (such as the bacterial mechanosensitive K^+ channel, MscL [10]), (iii) the strengthening of receptor–ligand interactions by tensile mechanical force (catch bonds such as those occurring in the *Escherichia coli* fimbrial adhesive protein FimH [11, 12]), and (iv) the binding of cytoskeletal proteins (such as vinculin) to mechanically stretched cytoplasmic proteins (such as talin rod molecules) [13], to name just a few examples.

Atomic Force Microscopy in Liquid: Biological Applications, First Edition.
Edited by Arturo M. Baró and Ronald G. Reifenberger.

Although single-molecule techniques [14, 15] have increased our understanding of the molecular basis of mechanosensing and mechanotransduction phenomena to some extent, the exact modes by which mechanical stimuli are translated into biochemical signals within the cell remain often unclear.

11.1.2
Yeast Cell Wall Integrity Sensors: a Valuable Model for Mechanosensing

Compared to the examples from mammalian physiology given above, unicellular microbes, such as the baker's yeast *Saccharomyces cerevisiae*, are frequently exposed to more drastic environmental changes, for instance, rapid shifts in temperature or acidity, in medium osmolarity, or in the availability of nutrients. Another important difference to mammalian cells is the presence of a rigid cell wall, which is essential for survival and determines the cell shape [16]. Therefore, the fungal cell wall and the interference with its biosynthesis constitute an ideal target for the development of specific antibiotics [17]. This point is not trivial since pathogenic fungi are also eukaryotes and thus share the basic cellular organization with their host organisms (infected humans, livestock, and plants). In *S. cerevisiae*, the cell wall consists primarily of glucans and mannoproteins, with some minor contribution from chitin, which is mostly localized to the bud neck in dividing cells and to the bud scar after cytokinesis [18, 19].

During the normal yeast life cycle, the cell wall has to be constantly remodeled to allow for bud and cell growth or for the formation of mating projections. In addition, extracellular stresses such as compounds affecting the cell wall or the yeast plasma membrane, as well as a temperature rise or a shift to low osmolarity conditions, trigger the local enforcement of the protective wall by directed synthesis of cell wall material (polysaccharides and wall proteins) at the sites of lesions. In all these cases, the requirement for cell wall biosynthesis and remodeling is signaled by the so-called CWI (cell wall integrity) pathway [16, 20]). The major components of this signal transduction cascade are outlined in Figure 11.1. In brief, sensors at the cell surface, which are described in more detail below, receive the stress signal and transmit it to the activation of the small G-protein Rho1 by its nucleotide exchange factor Rom2. In its active GTP-bound state, Rho1 interacts with and presumably activates the sole yeast protein kinase C [21]. This in turn triggers a conserved MAPK (for mitogen-activated protein kinase) cascade, which ultimately leads to the phosphorylation of transcription factors (Rlm1 and the SBF complex) that then trigger the appropriate transcriptional response in the nucleus. Genes for cell wall biosynthetic enzymes and cell wall proteins are thus expressed and provide the basis for the described local enforcements of the cell wall structures.

The only components potentially amenable to live cell atomic force microscopy (AFM) analyses in the CWI pathway are the sensor molecules that traverse the plasma membrane. Five sensors (Wsc1, Wsc2, Wsc3, Mid2, and Mtl1) have been identified in *S. cerevisiae* as putative upstream components of the signaling cascade. They are characterized by some general structural features as summarized in Figure 11.2. These include a relatively short C-terminal cytoplasmic tail, which

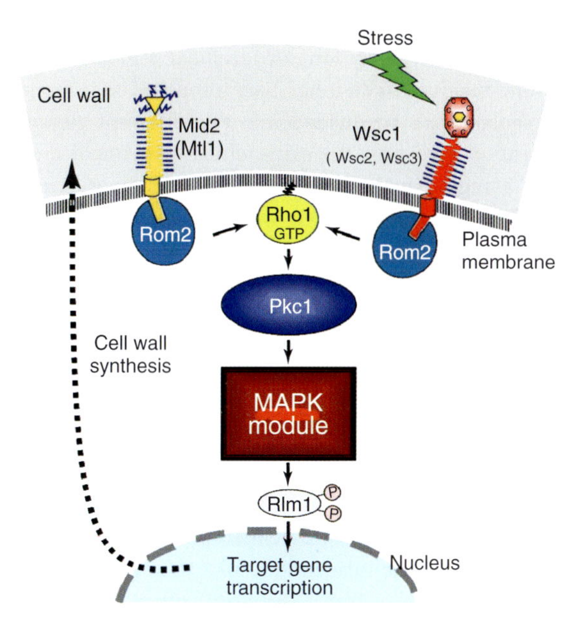

Figure 11.1 Schematic representation of cell wall integrity (CWI) signaling in the yeast *S. cerevisiae*. Plasma membrane spanning sensors reach out into the cell wall (gray) and form a nanospring (wavy lines), presumably contacting the wall with their head groups (red hexagon, Wsc1; yellow triangle, Mid2). Blue lines attached to the sensors designate glycosyl chains. See text for details on the downstream signaling cascade from the GEF Rom2 to the transcription factor Rlm1. More details on the CWI pathway are reviewed in [16, 22].

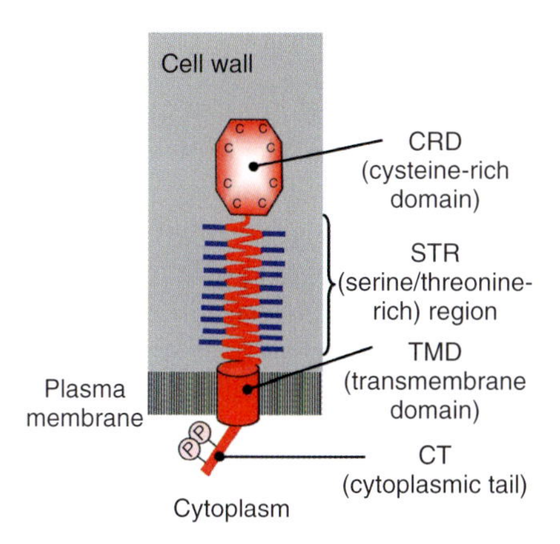

Figure 11.2 Domain structure of Wsc-type sensors of the yeast CWI signaling pathway. An aminoterminal CRD is presumably anchored to cell wall polysaccharides and connected to the single membrane-spanning domain (TMD) via a highly mannosylated, serine/threonine-rich (STR) region, which behaves like a nanospring [24]. The cytoplasmic tail (CT) can be phosphorylated and connects the sensor to the downstream signaling cascade [22].

connects the sensors to the downstream signaling components; a single putative transmembrane domain (TMD); and a relatively long extracellular region rich in glycosylated serine and threonine residues, which has been proposed to assume a rodlike structure [23]. The sensors can be divided into two different classes on the basis of the kinds of head groups near the extracellular N-terminal end: the Wsc-type sensors contain a region with eight conserved cysteines (CRD (cysteine-rich domain), also called *Wsc-domain* [20]), whereas Mid2 carries an N-glycosylated asparagine residue, which is also conserved in Mtl1. The head groups are believed to contact the polysaccharides within the cell wall and thus serve as a second anchor for the presumed mechanosensors in addition to the TMD. Calculating from the primary sequence of their extracellular regions, the different sensors within each subgroup penetrate the cell wall to different depths, which could determine their specificities [22].

In this chapter, we summarize recent progress made in understanding the mechanics, clustering, and functions of yeast sensors, owing to the integration of modern tools of molecular genetics (protein design in live cells), with the powerful set of AFM techniques (live cell imaging, single-molecule manipulation). While we focus here on the yeast *S. cerevisiae*, the unicellular model eukaryote, our findings should be useful to understand the behavior and function of other sensors as well, and we anticipate that the technology described here would be applicable to virtually all cell types.

11.2 Methods

11.2.1 Atomic Force Microscopy of Live Cells

AFM is emerging as one of the most powerful single-cell and single-molecule tools in microbiology, as is evident from the continuous increase in papers published in the field (for reviews, see [25–28]). AFM analysis of single cells requires the firm attachment of the cells onto a surface, which is far from being a simple task. Unlike animal cells, microbes have a well-defined shape and usually have no tendency to spread on surfaces under experimental conditions. As a result, the contact area between a cell and a support is very small, often leading to cell detachment by the scanning tip. Therefore, several approaches have been developed to promote cell attachment [29]. A first approach is to attach negatively charged cells onto glass (or mica) supports functionalized with positively charged macromolecules, such as poly-L-lysine or polyethylenimine. This method has been successfully applied to lactic acid bacteria, gram-negative bacteria, and diatoms, providing novel information on their surface structure and elasticity. Cells can also be immobilized onto gelatin-coated supports. To achieve this, thin freshly cleaved mica disks are vertically dipped into gelatin solution and air dried overnight by allowing them to stand on the edge of a paper towel. An aliquot of concentrated cell suspension is spread on the support, allowed to stand for 10 min, and rinsed in deionized water. Another convenient method for cell immobilization is the use of porous polymer membranes. For this purpose, a concentrated cell suspension is gently sucked through a porous membrane with pore sizes similar to the size of the cells to be investigated. This procedure offers two advantages – it is fairly simple and does not involve a macromolecular support, thus preventing the risk of cell surface or tip contamination. However, the success of this method strongly depends on the cell geometry and on the cell surface properties. In particular, the method is generally not suited for rod-shaped bacteria.

Figure 11.3a shows a topographic image of a living *S. cerevisiae* cell trapped into a porous membrane. Using this approach, high-resolution images of the cell surface were obtained [24], revealing a very smooth surface that was consistent with the presence of the glucan/mannoprotein outer layer. By contrast, the surface of *Aspergillus fumigatus* spores is covered by a crystalline layer of regularly arranged rodlets (Figure 11.3b) [30]. These structures are composed of hydrophobins, a family of small, moderately hydrophobic proteins of biological importance, since they enhance spore dispersion by air currents and mediate the adherence to host cells. Notably, dramatic changes in the cell surface structure were observed after germination, that is, the rodlet layer changed into a layer of amorphous material, presumably caused by the underlying polysaccharides. Hence, AFM is particularly well suited to track cell wall remodeling associated with growth, thus complementing structural information obtained by optical and electron microscopy [30, 31].

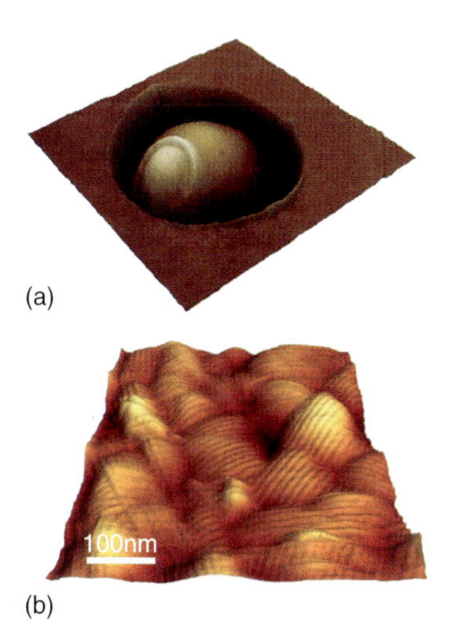

Figure 11.3 High-resolution imaging of living microbial cells. (a) AFM image of a *S. cerevisiae* yeast cell protruding from a porous polymer support. The cell shows a circular bud scar left after detachment of the daughter cell. (b) Higher-resolution AFM image of the surface of an *Aspergillus fumigatus* spore documenting the presence of ordered nanoscale rodlets. (Images reprinted with permission from Ref. [32].)

11.2.2
AFM Detection of Single Sensors

Biomolecular recognition by AFM is classically performed by single-molecule force spectroscopy (SMFS) [28, 33], in which the cantilever deflection is recorded as a function of the vertical displacement of the scanner, that is, as the sample is pushed toward the tip and retracted. This results in a cantilever deflection versus scanner displacement curve, which can be transformed into a force–distance curve using appropriate corrections. Force curves can be recorded at multiple locations of the *xy*-plane to yield spatially resolved force maps. The characteristic adhesion (or unbinding) force between tip and sample observed during retraction is then used to map receptor sites.

An important prerequisite for successful molecular recognition experiments is to functionalize the AFM tip with appropriate ligands (or receptors) [33]. The forces that immobilize the molecules have to be stronger than the intermolecular force being studied, and the attached biomolecules should have enough mobility so that they can freely interact with complementary molecules. It is also recommended to minimize the contribution of nonspecific adhesion to the measured forces and to attach the biomolecules at a low surface density to ensure single-molecule detection. A good strategy to covalently anchor proteins on tips is to use a polyethylene glycol (PEG) cross-linker that provides motional freedom and prevents denaturation. Tips are first modified with amino groups, further reacted with PEG linkers carrying benzaldehyde functions that are then directly attached to proteins through their lysine residues. Biomolecules may also be attached via self-assembled monolayers of alkanethiols on gold tips.

The site-directed nitrilotriacetic acid (NTA)-polyhistidine (His_n) system has proved to be particularly well suited for single-molecule detection [34, 35]. Surprisingly, the method can be readily applied to live cells, for example, for fishing recombinant histidine-tagged proteins expressed on the cell surface using AFM tips coated with NTA groups [24, 36]. The NTA-His_6 binding chemistry, well known for affinity purification of recombinant proteins, involves the formation of a hexagonal complex between the tetradental ligand NTA and divalent metal ions such as Ni^{2+}. Since NTA occupies four of the six coordination sites of Ni^{2+}, the two remaining sites are accessible to other Lewis bases, for example, the histidines of tagged proteins. To gain insight into the force of the NTA-His interaction, we recorded force–distance curves between AFM tips modified with Ni^{2+}-NTA-terminated alkanethiols and solid supports functionalized with His_6-Gly-Cys peptides (Figure 11.4a) [35]. Consistent with earlier work [34], the adhesion force histogram showed three maxima at rupture forces of 153 ± 57, 316 ± 50, and 468 ± 44 pN attributed to monovalent and multivalent interactions between a single His_6 moiety and one, two, and three NTA groups, respectively (Figure 11.4b). The plot of adhesion force versus log of the loading rate revealed a linear regime, from which a kinetic off-rate constant of dissociation was deduced. The obtained value was in the range of that estimated for the multivalent interaction involving two NTAs, using fluorescence measurements, and may account for an

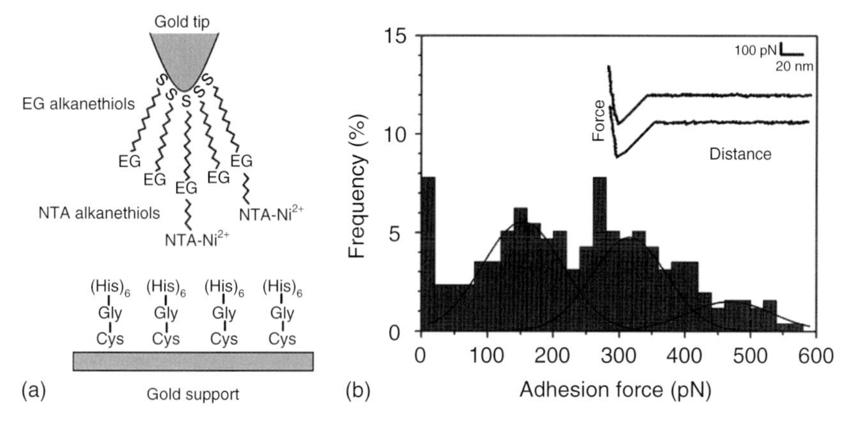

Figure 11.4 Strength of the NTA-His$_6$ bond. (a) Surface chemistry used to modify AFM tips with Ni^{2+}-NTA-terminated alkanethiols and solid supports with His$_6$-Gly-Cys peptides. EG, ethylene glycol. (b) Force histogram and typical force curve obtained between the NTA and His$_6$ surfaces. (Reprinted with permission from Ref. [35].)

increased binding stability of the NTA-His bond. This study demonstrates that the NTA-His system is a well-suited platform for the oriented detection (or attachment) of proteins in SMFS studies.

11.2.3
Bringing Yeast Sensors to the Surface

Since AFM is a surface technique, how can we use it to manipulate sensors that are embedded within the cell wall? A simple calculation indeed reveals that the native Wsc1 sensor supplied with a His-tag should extend no more than 86 nm above the plasma membrane. As the cell wall is approximately 110 nm thick, native sensors cannot reach the outermost cell surface [37]. To solve this problem, we added an additional 186 amino acids (which correspond to a length of 67 nm) with a His-tagged extension of the Ser/Thr-rich (STR) region of another CWI sensor, Mid2. This amounts to a total external length of 153 nm and renders the chimeric sensor accessible to an Ni^{2+}-NTA tip. Owing to this novel procedure, it is now possible to force probe by AFM not only on natural surface proteins but also those usually located well beneath the surface within the cell wall.

As indicated above, the STR region of the CWI sensors has been postulated to assume a rodlike structure and was indeed shown in AFM force measurements to mechanistically function as a nanospring [24]. We therefore reasoned that it may penetrate the cell wall like a needle and that adding a similar STR region from another sensor would allow this structure to punch through the entire cell wall and reach the surface. For experimental reasons concerning genetic stability, we chose the sequence of the STR region from the Mid2 sensor for this purpose [36]. It should be noted that the two STR regions from Wsc1 and Mid2 are separated by the

CRD of the original Wsc1 and that the Mid2 part was also equipped with a His-tag to allow for detection by AFM. The chimeric construct was obtained by fusion of the respective coding DNA sequences using the yeast *in vivo* recombination technique and by subsequently expressing the construct in a yeast strain lacking the endogenous *WSC1* gene. It is of utmost biological relevance that the chimeric construct was fully able to function like the wild-type Wsc1 sensor in phenotypic analyses under stress conditions since this suggests that the genetic manipulations did not create any physiological artifacts [36].

Although we demonstrated the merits of this technique as a proof of principle only for the yeast Wsc1 sensor, it clearly could be applied to any surface protein that at least in part assumes a linear structure and penetrates the cell wall of other fungi, bacteria, or plants. The ease and speed of genetic manipulations in *S. cerevisiae*, owing to its high capacity for homologous recombination, are in favor of using this system at least to obtain the relevant chimeric constructs for any organism. Apart from AFM nanomechanical measurements, the chimeric Wsc1/Mid2 sensors described above can also serve other purposes. Thus, we constructed a molecular ruler set suitable to determine cell wall thickness under different growth and stress conditions *in vivo* [38]. Moreover, since the chimeric sensor does reach the cell surface, any peptide sequence introduced at its N-terminal end will be exposed to the environment. Thus, appropriate derivatives could serve as a yeast surface display system for antigen presentation to trigger immunological responses and produce specific antisera, an application currently restricted to the use of the yeast agglutinins [39].

11.3
Probing Single Yeast Sensors in Live Cells

11.3.1
Measuring Sensor Spring Properties

How do cellular proteins respond to mechanical stimuli? Clarification of this issue is central to our understanding of many cellular protein functions, including mechanosensors. SMFS has allowed researchers to probe the mechanical properties of water-soluble and membrane proteins [6, 7, 14, 15, 40–43]. Yet, studying the molecular elasticity of cellular proteins embedded in their native environment remains very challenging.

Using the above strategy (SMFS combined with genetic manipulations), we detected and stretched single Wsc1 sensors on living yeast cells, revealing that they behave like nanosprings capable of resisting high mechanical force and of responding to cell surface stress (Figure 11.5) [24, 36]. For this purpose, *S. cerevisiae* cells expressing His-tagged elongated Wsc1 proteins were immobilized on a polycarbonate membrane (Figure 11.3a), and single sensors were detected by scanning the cell surface with an Ni^{2+}-NTA tip (Figure 11.5a). Adhesion force data revealed the localization of individual molecules, thus confirming that they reached the cell surface. In contrast, His-tagged Wsc1 sensors that were not elongated could generally not be detected at the yeast surface, except in the very localized region of the bud scar.

Force–extension curves (Figure 11.5a), obtained at a stretching rate of 10 000 pN s^{-1}, displayed a first regime where the molecule was stretched at nearly constant force. Presumably, this region corresponds to the straightening of the extracellular polypeptide chain and/or to stretching of the plasma membrane. The constant force region was followed by a linear region where force was directly proportional to extension, thus characteristic of a Hookean spring. Hence, there are two linear springs in series, one being the AFM cantilever (k_c) and the other, the sensor spring constant (k_s). Using the slope (s) of the linear portion of the raw deflection versus piezo displacement curves and the equation $k_s = (k_c \cdot s)/(1 - s)$, the spring constant of Wsc1 was found to be 4.6 ± 0.4 pN nm^{-1} at a stretching rate of 10 000 pN s^{-1}. This value is very close to that reported for ankyrin repeats, as measured by AFM [44]. The linear elasticity of ankyrin was suggested to modulate the activity of ankyrin-associated transporters in response to mechanical strain and/or to generate tension in the plane of the membrane bilayer. For the proteins tested so far, such a nanospring behavior is very unusual in nature and is in sharp contrast to the properties of modular proteins. Stretching modular proteins leads to nonlinear force peaks that are well described by the worm-like-chain model, where each peak reflects the force-induced unfolding of secondary structures [40, 41].

When stretching Wsc1 proteins on different locations and on different cells, we found that their stretching length increased linearly with the adhesion force, indicating that different proteins subjected to different forces have the same linear elasticity. We therefore suggest that the observed force-induced lengthening reflects straightening of Wsc1 secondary or tertiary structure elements with coil-like shape, and possibly of the whole protein-cell-wall complex as well. Consistent with this view, we never observed nonlinear force peaks reflecting the unfolding of secondary α-helical or β-sheet structures, even for pulling forces of up to 250 pN.

Glycosylation of the extracellular STR region of Wsc1 is believed to be responsible for its stiff, extended conformation. We therefore analyzed two different mutants altered in the extracellular STR region (i.e., a genomic *pmt4* deletion that leads to a reduced number of glycosylations due to the defect in the encoded mannosyltransferase, and a strain producing a Wsc1 derivative with an insertion of a stretch of nonglycosylated glycines into the STR region). We found that the mechanical properties of these sensor variants were dramatically altered and did no longer exhibit a linear Hookean spring behavior (Figure 11.5b). Force peaks could not be fitted with a WLC model but were well described with an exponential fit, $F = F_0 \exp(X/X_0)$, where F_0 is the initial force and X_0 is the characteristic length constant. This shows that glycosylation of the sensor adds to the stiffness of the extracellular STR region and is required for its spring properties.

Does the sensor stiffness respond to stressing conditions? Lowering the salt concentration or increasing the incubation temperature of the yeast cells resulted in a substantial reduction of the sensor spring constant, indicating that the mechanical properties of Wsc1 are influenced by cell surface stress (Figure 11.5c).

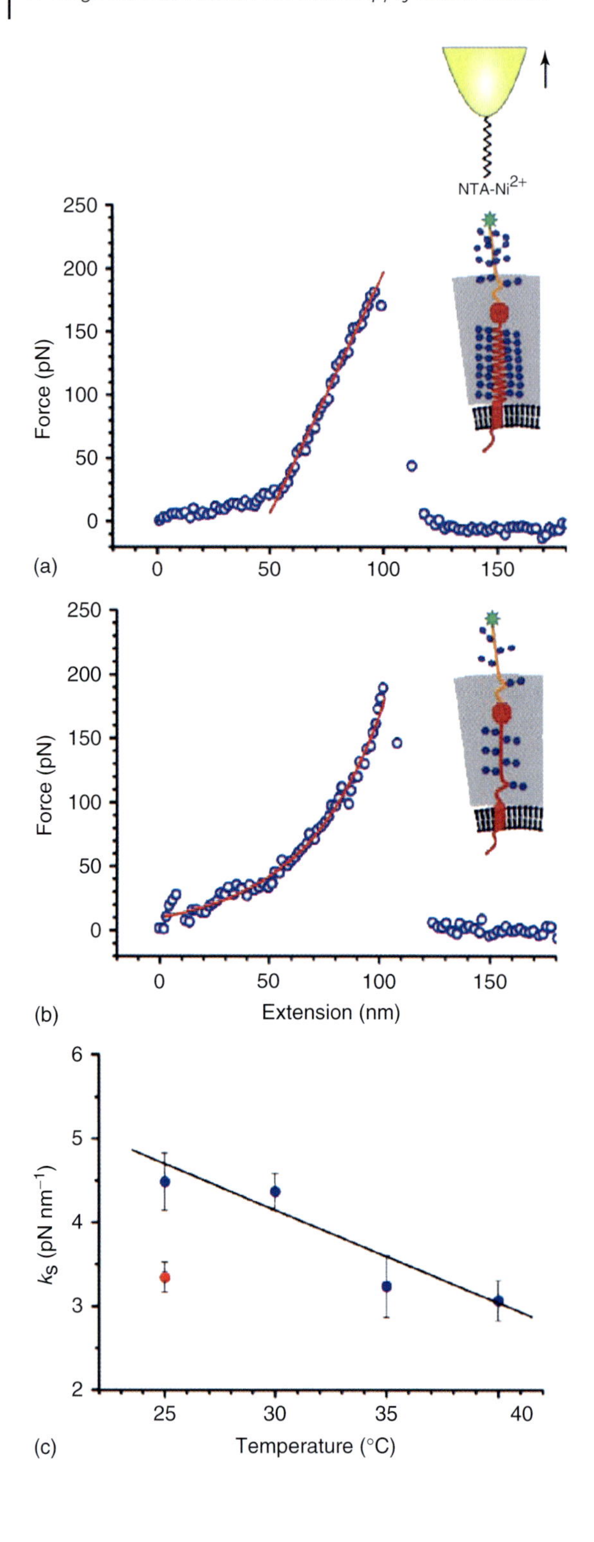
NTA-Ni2+
250
200
150
100
50
0
Force (pN)
(a)
0
50
100
150
Force (pN)
Extension (nm)
(b)
6
5
4
3
2
kS (pN nm−1)
25
30
35
40
Temperature (°C)
(c)

The reasons for such sensor softening may be direct or indirect. Indeed, stressing by hypoosmotic shock or heat shock may alter the elasticity and tension of the cell wall and/or plasma membrane, which would alter the apparent protein stiffness. On the other hand, temperature and turgor pressure could also directly affect the mechanical properties of the protein. For instance, the reduction in the molecular spring constant of filamin on increasing the temperature [45] has been suggested to result from a shift in the nature of the interactions responsible for mechanical stability from hydrogen bonds to hydrophobic interactions. In summary, the single-molecule experiments described above open new avenues for investigating how proteins respond to forces in living cells – primarily those of yeast and bacteria, but also of mammals – and how mechanosensing events proceed *in vivo*.

11.3.2
Imaging Sensor Clustering

Understanding how membrane proteins assemble to form membrane micro- and nanodomains is another important question in cell biology. In higher eukaryotes, lateral microdomains (or "lipid rafts") enriched in sphingolipids and sterols have been speculated to favor segregation of specific membrane proteins like receptors or GPI-anchored proteins [46–48]. If they indeed exist, which is still a matter of controversial debate, microdomains in higher eukaryotes are 20–100 nm in size and very transient, making their direct visualization in live cells very challenging. In contrast, stable microdomains have been observed within the plasma membrane of *S. cerevisiae* harboring specific GFP-labeled marker proteins [49]. These microdomains are distributed into the so-called MCPs (membrane compartment with Pma1), which form a networklike structure defined by their constituent proton ATPase Pma1, and MCCs (membrane compartment with Can1), which display 300 nm patches and house a number of proton symporters, as well as a component of the eisosomes [50].

This raises the question whether yeast sensors form clusters as well. Fluorescence microscopy has shown that Wsc1-GFP proteins form membrane patches [51], but the structure–function relationships of this clustering remained mysterious. To address this problem, we determined the distribution of single Wsc1 sensors

Figure 11.5 AFM shows that Wsc1 behaves like a nanospring. (a) Representative force curve obtained on stretching a single Wsc1 molecule. As shown in the figure, *S. cerevisiae* cells expressing Wsc1 sensors with an extended His-tag were probed using Ni^{2+}-NTA tips. Clearly visible are two extension regimes, reflecting elongation at nearly zero force, followed by a Hookean spring behavior (red line). (b) Representative force curve obtained on stretching single Wsc1 sensors in a *pmt4* deletion mutant in which mannosylation of the extracellular Ser/Thr-rich region is substantially reduced. The force increases are no longer linear but are well described with an exponential fit (red curve). (c) Comparison of the Wsc1 spring constant values measured in native conditions (buffer at $T = 25\,^{\circ}C$, left blue dot) and under stressing conditions, that is, either deionized water (red dot) or $>25\,^{\circ}C$ (three blue dots on the right). (Reprinted with permission from Ref. [24].)

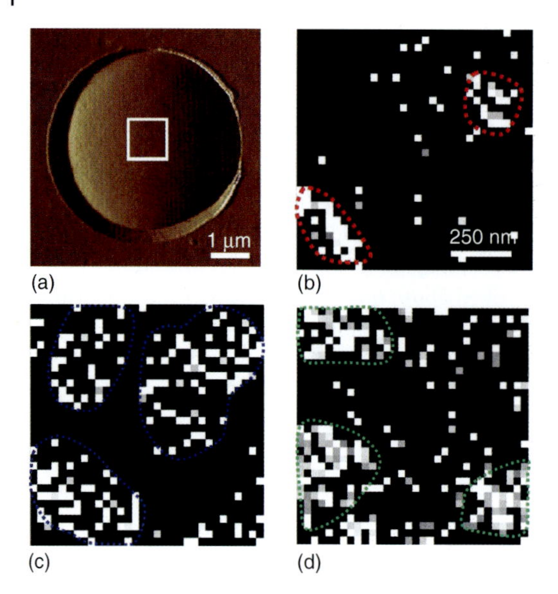

Figure 11.6 Clustering of Wsc1 is stimulated under stressing conditions. (a) AFM deflection image of a yeast cell trapped into a porous polymer membrane, recorded in buffer solution at 25 °C. The cells express elongated, fully functional Wsc1-Mid2 hybrid sensor bearing a His-tag. (b) Representative adhesion force map obtained by scanning a 1 µm × 1 µm area on the cell surface with an Ni^{2+}-NTA tip. The heterogeneous distribution of the bright pixels, which represent the detection of single sensors, clearly documents the formation of nanoscale clusters (highlighted by dotted red lines). (c,d) Adhesion force maps (1 µm × 1 µm) recorded with an Ni^{2+}-NTA tip either in buffer solution at 37 °C (c, heat shock) or in deionized water at 25 °C (d, hypoosmotic shock). Stressing conditions strongly enhance Wsc1 clustering (clusters are highlighted by dotted blue and green lines, respectively). (Reprinted with permission from Ref. [52].)

in living *S. cerevisiae* cells by SMFS [52]. Adhesion maps of 1 µm × 1 µm size were recorded on cells expressing the above-mentioned His-tagged elongated sensors using Ni^{2+}-NTA tips (Figure 11.6). Many sensor molecules appeared to cluster in areas of ~0.04 μm^2 and an equivalent diameter of 230 nm. Notably, this size of Wsc1 clusters matches the range of the 300 nm large patches reported for MCC marker proteins in *S. cerevisiae* and is clearly larger than the putative size of rafts in higher eukaryotes.

To our surprise, both the total amount of wild-type Wsc1 sensors and their tendency to cluster increased when cells where stressed by either heat or a medium with low osmolarity (Figure 11.6c,d). This coincides with the observation that stress conditions activate the CWI pathway and suggests that clustering is a stress-responsive process that is intimately connected to signaling. We propose that clustering is a means developed by yeast to locally concentrate Wsc1 sensors and their interacting downstream components. This process ultimately enhances the signal strength and the corresponding cellular response. A similar mechanism has been suggested for bacterial chemoreceptors, which on stimulation are rearranged to form clusters of variable sizes (up to 250 nm) [53].

In the same set of experiments, we also provided hints as to the molecular mechanism that governs the clustering of Wsc1. As described above, all Wsc-type sensors bear a CRD with eight conserved cysteine residues essential for sensor function. We therefore mapped the distribution of Wsc1 in CRD mutants, that is, mutants in which some of these cysteines were replaced by alanine residues, leading to nonfunctional sensors (Figure 11.7). Such mutants displayed adhesion frequencies, adhesion values, and spring behaviors similar to those of the wild-type protein. Thus, the sensors' mechanical properties are exclusively determined by the STR region as stated above and are not influenced by the CRD. However, adhesion maps of the CRD mutants demonstrated major differences in the lateral sensor

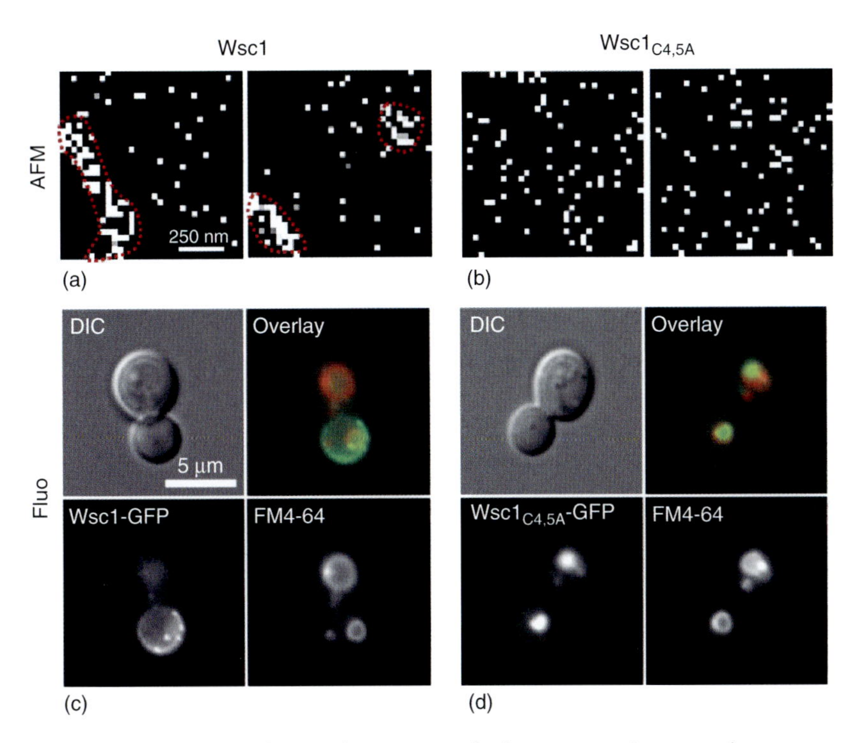

Figure 11.7 The conserved CRD of Wsc1 is essential for clustering. (a,b) Adhesion force maps (1 μm × 1 μm) recorded on the surface of cells expressing wild-type Wsc1 ((a) fully functional Wsc1 with an extended His-tag) and mutant $Wsc1_{C4,5A}$ ((b) altered in cysteine residues), with an Ni^{2+}-NTA tip in buffer solution at 25 °C. While the mutant showed a surface density similar to that of the wild-type protein, the sensors were evenly distributed and did no longer cluster. (c,d) Fluorescence microscopy shows vacuolar accumulation of nonclustering sensors. A representative number of cells expressing GFP fusions of either the wild-type Wsc1 or the $Wsc1_{C4,5A}$ construct were examined by differential interference contrast (DIC) microscopy and fluorescence microscopy. Wsc1-GFP signals are shown in green. The vacuolar membrane was stained with FM4-64 and is shown in red. (Reprinted with permission from Ref. [52].)

organization at the surface as compared to wild-type cells (Figure 11.7a,b): mutant sensors appeared to be evenly distributed rather than clustered.

To confirm these findings, fluorescence microscopy was performed using Wsc1-GFP proteins to visualize the distribution of the sensors within the plasma membrane of growing yeast cells (Figure 11.7c,d). In wild-type cells, Wsc1-GFP fusions are dynamically distributed depending on the cell cycle, with the signal concentrating at the emerging bud, then appearing within the cell and to some extent in the vacuole, before concentrating again at the bud neck during cytokinesis [54]. The protein appears to form spots within the plasma membrane, which where greatly diminished in number when CRD mutants were investigated. In such mutants, the GFP signals were generally much stronger within the vacuoles as compared to the wild-type cells (Figure 11.7d). Taken together, these data demonstrate the importance of the CRD in sensor clustering and turnover. Therefore, besides the proposed role of carbohydrate binding – which is based purely on amino acid sequence similarities to a domain in fungal exonucleases – we suggest an additional role for CRD in mediating protein–protein interactions.

In conclusion, we propose that in yeast, like in higher eukaryotes, sensor function is coupled to a localized enrichment of sensors/receptors in membrane patches. We expect our newly observed Wsc1 nanoclusters, for which we propose the term Wsc1 "sensosome" (Figure 11.8), to promote the concomitant accumulation of downstream signaling components of the CWI signaling pathway at the inner leaflet of the plasma membrane. Regarding the mechanics of the signal reception, stretching of either the cell wall or the plasma membrane could alter the conformation of CRD, thereby exposing interfaces to promote intermolecular protein–protein interactions. These interactions would trigger the association of further sensor molecules within the plasma membrane. In addition to these changes in CRD, all sensor molecules will also assume a conformation of their cytoplasmic domains, which makes them competent for interaction with the downstream CWI signaling components.

11.3.3 Using Sensors as Molecular Rulers

How thick are microbial cell walls and how does wall thickness adjust to external agents or mutations? Today, this question remains largely unanswered owing to the lack of suitable probing techniques in live cells. Thin-section transmission electron microscopy (TEM) is currently the only method available to determine cell wall thickness [55]. Using the combination of Wsc1 sensors and single-molecule AFM, we recently presented a new method for measuring cell wall thickness in living *S. cerevisiae* cells [38]. The idea relies on the expression of His-tagged sensors of increasing lengths in yeast and their subsequent specific detection at the cell surface using a modified AFM tip. Different sensors with a calculated extracellular length increasing from 90 to 154 nm were designed. These sensor lengths were approximated by assuming a fully linear peptide chain and a peptide bond length

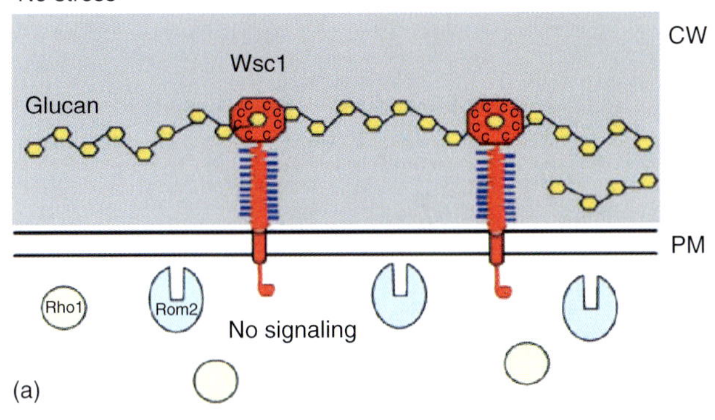

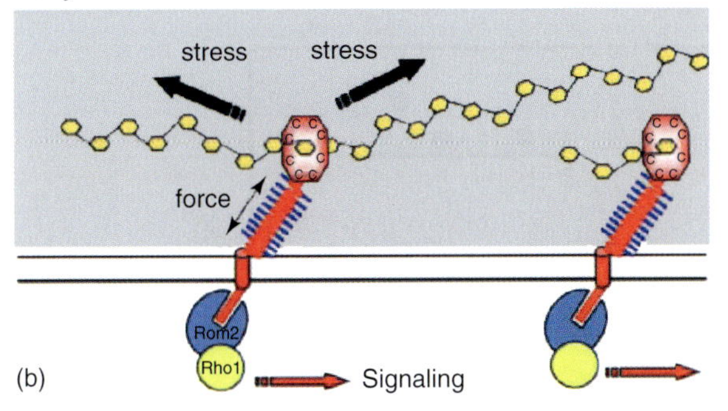

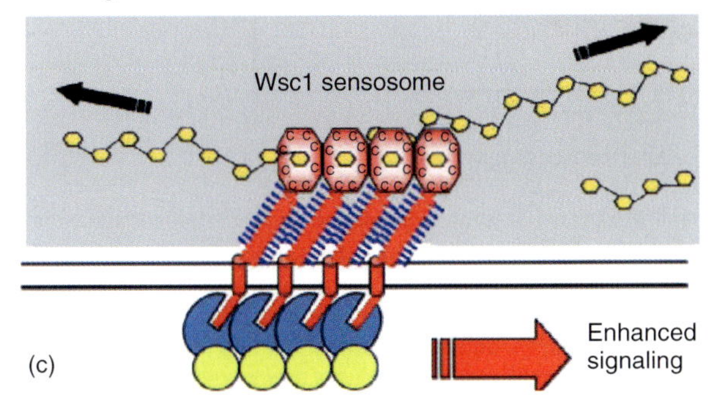

Figure 11.8 AFM unravels the Wsc1 "sensosome." The schematic drawing presents our current working model of sensor function and signal enhancement by clustering. Sensor domains and downstream signaling components are drawn as in Figures 11.1 and 11.2. In the absence of stress (a) the sensors do not cluster, they are in loose contact with the cell wall (CW) by their CRD, and their cytoplasmic tail (CT) does not interact with the downstream signaling components. Under stress conditions, the sensors are stretched by mechanical forces (b), leading to a conformational change in the CT and a recruitment of the interaction partners at the cytoplasmic face of the plasma membrane (PM). Protein–protein interactions of the deformed CRDs induce sensor clustering, including the downstream components, thus forming a "sensosome" with enhanced signaling capacity (c). (Reprinted with permission from Ref. [52].)

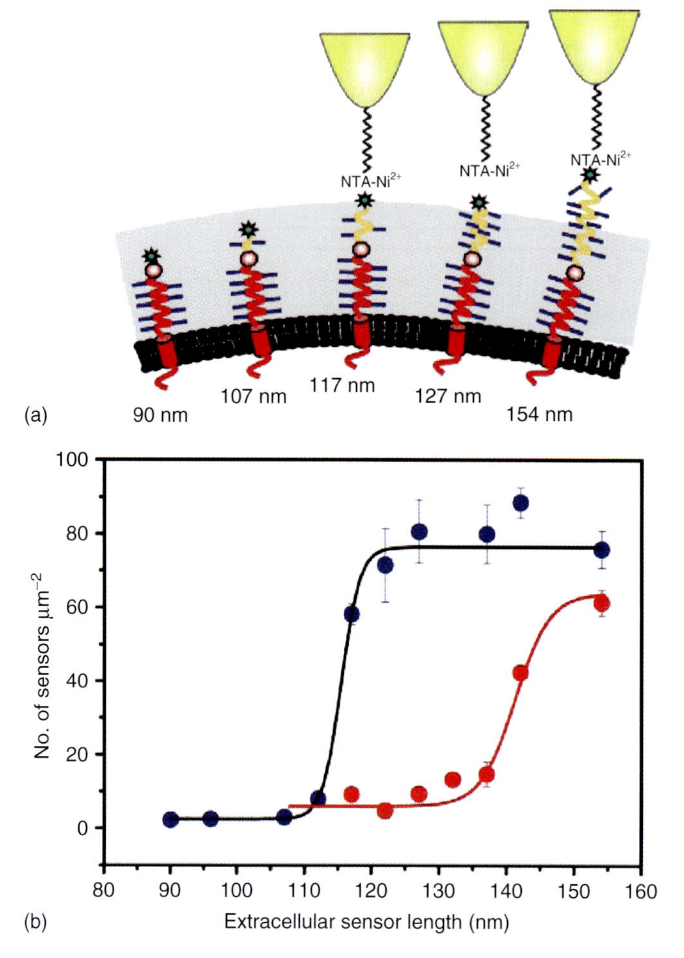

Figure 11.9 Measuring cell wall thickness using Wsc1 molecular rulers. (a) Use of molecular rulers to measure cell wall thickness. His-tagged and modified Wsc1 membrane sensors of various lengths were detected using AFM tips functionalized with Ni^{2+}-NTA groups. Only the sensors that were long enough to reach the surface were detected by the tip. (b) Sensograms, that is, variation of the amount of detected sensors as a function of extracellular sensor length, obtained for wild-type cells (in blue) and for mutant cells with thicker cell walls (in red). (Reprinted with permission from Ref. [38].)

of 0.36 nm. The shortest sensor to be detected under different growth conditions should thus provide an estimate of the cell wall thickness.

We established the method using wild-type cells expressing Wsc1 sensors of increasing extracellular lengths (Figure 11.9). Force curves recorded across the cell surface using an Ni^{2+}-NTA tip showed specific adhesion events reflecting the detection of single sensors. The sensogram obtained by plotting the variation of the amount of detected sensors as a function of sensor length showed a sharp increase around 115 nm. Hence, shorter sensors (<110 nm) were never detected, indicating that they do not reach the surface, while longer ones (>120 nm) were detected with a rather constant value of 80 sensors μm^{-2}. From the derivative of this curve, we estimated the cell wall thickness to be 115 nm, thus slightly larger than the average TEM value of 105 nm [37]. Presumably, the thinner cell wall measured by TEM could be an artifact caused by the invasive sample preparation procedure (e.g., dehydration).

We then showed the ability of our method to measure changes in cell wall thickness, resulting from mutations or biochemical treatments. We first analyzed

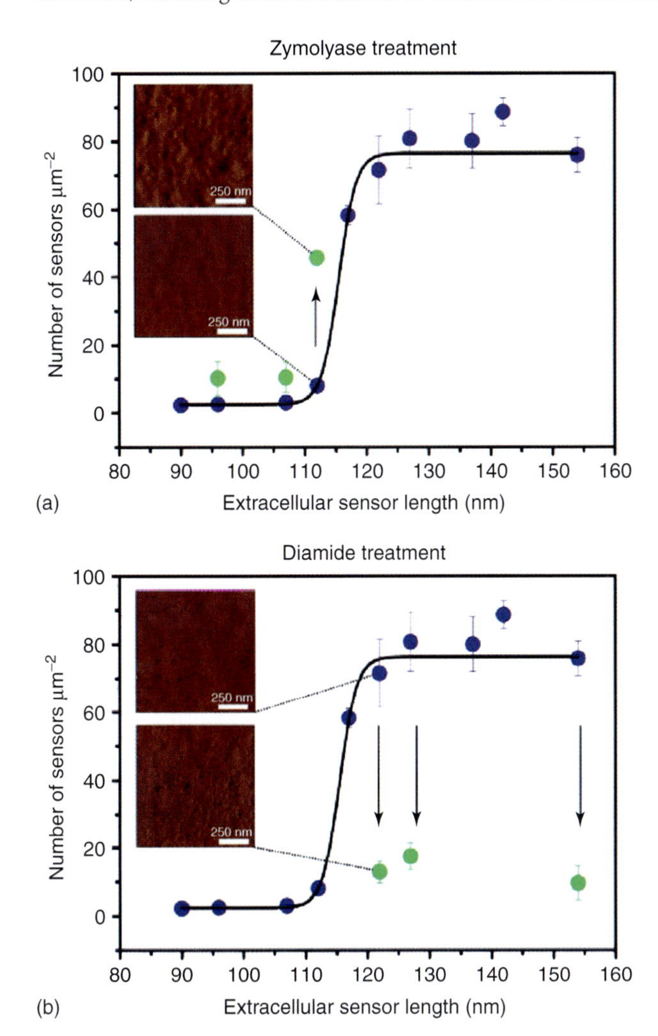

Figure 11.10 Sensograms reveal changes in cell wall thickness resulting from (bio)chemical treatments. Sensograms obtained for native wild-type cells (a and b, blue) and for wild-type cells after Zymolyase treatment (a, green) and after diamide treatment (b, green). Insets: high-resolution topographic images of the cells documenting a rougher surface after treatment. (Reprinted with permission from Ref. [38].)

mutants in which Wsc1 endocytosis was abolished and for which thicker cell walls are expected due to more pronounced cell wall synthesis (Figure 11.9b). Indeed, only the two longest sensors could be detected at the surface of such mutants, with a computed *in vivo* cell wall thickness of ~140 nm. Zymolyase is an enzyme mixture that degrades cell wall polysaccharides and is thus expected to rapidly produce thinner cell walls on addition to yeast cells. Figure 11.10a shows that sensors with a length of 112 nm could still be detected in large amount on Zymolyase-treated cells, thus reflecting their thinner cell wall. Only cells with very short sensors (<112 nm) showed a lower number of detectable molecules (10 sensors μm^{-2}), indicating that the enzymatic digestion essentially affected a 10 nm thick outer layer and that the resulting cell wall displayed a higher degree of heterogeneity. Consistent with this, AFM topographic images suggested that treated cells were rougher than native ones. Diamide is known to cause oxidative stress, which also results in the strengthening of the yeast cell wall. Accordingly, Figure 11.10b shows that even the longer sensors (up to 154 nm) could hardly be detected at the surface of diamide-treated cells, with a computed increase in cell wall thickness of more than 40 nm. This treatment, too, produced a general heterogeneity in cell wall thickness, presumably because of the oxidative stress exerted on the cell wall proteins, which resulted in the detection of a lower number of sensors of various lengths and a generally rougher surface than the one of untreated controls (Figure 11.10b).

11.4
Conclusions

The experiments surveyed in this chapter demonstrate that the combination of single-molecule AFM with genetic manipulations is a powerful tool in cell biology. This integrated platform can be applied for measuring the mechanical properties of single sensors in live cells and for imaging the sensor localization, thus revealing whether they are isolated or form clusters. Our AFM methodology is unique, in that it can force probe not only surface-associated proteins but also proteins that are embedded within the cell wall and are thus well beneath the surface. We anticipate that this novel approach will enable a paradigm shift in cell biology, so that pertinent questions can be addressed, such as understanding the mechanical response of single sensors and receptors and how these proteins distribute across the cell surface [32]. In the future, an exciting technological challenge will be the combination of the described single-molecule analyses with advanced light microscopy, thus enabling cell surface constituents to be simultaneously identified by light microscopy and force probed by AFM. Also, the use of more sophisticated AFM experiments, in which the approach/retraction speeds and contact time are varied, should provide a novel insight into the adhesive and mechanical properties of cell surface proteins, as well as into their possible recruitment and clustering.

The results discussed here dramatically increase our understanding of the biophysical properties of yeast sensors, with the following major findings. First, Wsc1 behaves like a linear nanospring, with the ability to resist high mechanical force and to respond to cell surface stress. Second, analysis of STR mutants supports the important role of glycosylation at the extracellular STR region for the nanomechanical response. Third, the formation of Wsc1 nanoclusters is mediated by the CRD and is enhanced under stress conditions that activate the CWI signaling pathway. Fourth, the correlation of the clustering and signaling capacity of Wsc1 suggests that sensor function is coupled to a localized enrichment of sensors within membrane patches. While we focused here on a CWI sensor from the yeast *S. cerevisiae* as a prototype of unicellular eukaryote, our findings should be useful to understand the behavior and function of signal receptors throughout the biological kingdom.

Moreover, our single-molecular ruler strategy provides a powerful tool for measuring the thickness of microbial cell walls under varying environmental conditions and in different mutants. The method can thus be employed to address a number of exciting biological and medical questions. Compared to fluorescence methods traditionally used in cell biology, this AFM approach is still more sensitive (single-molecule detection) and surface specific (no signal from the underlying layers).

Acknowledgments

Work in the J. J. H. team was funded by the Deutsche Forschungsgemeinschaft (DFG) within the framework of the SFBs 431 and 944. Work in the Y. F. D. team was supported by the National Foundation for Scientific Research (FNRS); the Université catholique de Louvain (Fonds Spéciaux de Recherche); the Région wallonne; the Federal Office for Scientific, Technical and Cultural Affairs (Interuniversity Poles of Attraction Programme); and the Research Department of the Communauté française de Belgique (Concerted Research Action). Y.F.D. is the Senior Research Associate at the FNRS.

References

1. Buelow, D.R. and Raivio, T.L. (2010) Three (and more) component regulatory systems – auxiliary regulators of bacterial histidine kinases. *Mol. Microbiol.*, **75**, 547–566.
2. Hohmann, S. (2002) Osmotic stress signaling and osmoadaptation in yeasts. *Microbiol. Mol. Biol. Rev.*, **66**, 300–372.
3. Silbering, A.F. and Benton, R. (2010) Ionotropic and metabotropic mechanisms in chemoreception: 'chance or design'? *EMBO Rep.*, **11**, 173–179.
4. McIntosh, B.E., Hogenesch, J.B., and Bradfield, C.A. (2010) Mammalian Per-Arnt-Sim proteins in environmental adaptation. *Annu. Rev. Physiol.*, **72**, 625–645.
5. Kee, Y.S. and Robinson, D.N. (2008) Motor proteins: myosin mechanosensors. *Curr. Biol.*, **18**, R860–R862.
6. Vogel, V. and Sheetz, M. (2006) Local force and geometry sensing regulate cell functions. *Nat. Rev. Mol. Cell Biol.*, **7**, 265–275.
7. Brown, A.E.X. and Discher, D.E. (2009) Conformational changes and signaling in cell and matrix physics. *Curr. Biol.*, **19**, R781–R789.

8. Schwartz, M.A. (2009) The force is with us. *Science*, **323**, 588–589.
9. Friedland, J.C., Lee, M.H., and Boettiger, D. (2009) Mechanically activated integrin switch controls $\alpha 5\beta 1$ function. *Science*, **323**, 642–644.
10. Li, Y., Wray, R., Eaton, C., and Blount, P. (2009) An open-pore structure of the mechanosensitive channel MscL derived by determining transmembrane domain interactions upon gating. *FASEB J.*, **23**, 2197–2204.
11. Sokurenko, E.V., Vogel, V., and Thomas, W.E. (2008) Catch-bond mechanism of force-enhanced adhesion: counterintuitive, elusive, but... widespread? *Cell Host Microbe*, **4**, 314–323.
12. Yakovenko, O., Sharma, S., Forero, M., Tchesnokova, V., Aprikian, P., Kidd, B., Mach, A., Vogel, V., Sokurenko, E., and Thomas, W.E. (2008) FimH forms catch bonds that are enhanced by mechanical force due to allosteric regulation. *J. Biol. Chem.*, **283**, 11596–11605.
13. del Rio, A., Perez-Jimenez, R., Liu, R., Roca-Cusachs, P., Fernandez, J.M., and Sheetz, M.P. (2009) Stretching single talin rod molecules activates vinculin binding. *Science*, **323**, 638–641.
14. Bustamante, C., Macosko, J.C., and Wuite, G.J.L. (2000) Grabbing the cat by the tail: manipulating molecules one by one. *Nat. Rev. Mol. Cell Biol.*, **1**, 130–136.
15. Sotomayor, M. and Schulten, K. (2007) Single-molecule experiments *in vitro* and *in silico*. *Science*, **316**, 1144–1148.
16. Levin, D.E. (2005) Cell wall integrity signaling in *Saccharomyces cerevisiae*. *Microbiol. Mol. Biol. Rev.*, **69**, 262–291.
17. Heinisch, J.J. (2005) Baker's yeast as a tool for the development of antifungal kinase inhibitors – targeting protein kinase C and the cell integrity pathway. *Biochim. Biophys. Acta*, **1754**, 171–182.
18. Klis, F.M., Boorsma, A., and De Groot, P.W. (2006) Cell wall construction in *Saccharomyces cerevisiae*. *Yeast*, **23**, 185–202.
19. Lesage, G. and Bussey, H. (2006) Cell wall assembly in *Saccharomyces cerevisiae*. *Microbiol. Mol. Biol. Rev.*, **70**, 317–343.
20. Heinisch, J.J. and Dufrene, Y.F. (2010) Is there anyone out there? Single molecule atomic force microscopy meets yeast genetics to study sensor functions. *Integr. Biol.*, **2**, 408–415.
21. Schmitz, H.P. and Heinisch, J.J. (2003) Evolution, biochemistry and genetics of protein kinase C in fungi. *Curr. Genet.*, **43**, 245–254.
22. Rodicio, R. and Heinisch, J.J. (2010) Together we are strong: cell wall integrity sensors in yeasts. *Yeast*, **27**, 531–540.
23. Philip, B. and Levin, D.E. (2001) Wsc1 and Mid2 are cell surface sensors for cell wall integrity signaling that act through Rom2, a guanine nucleotide exchange factor for Rho1. *Mol. Cell Biol.*, **21**, 271–280.
24. Dupres, V., Alsteens, D., Wilk, S., Hansen, B., Heinisch, J.J., and Dufrêne, Y.F. (2009) The yeast Wsc1 cell surface sensor behaves like a nanospring *in vivo*. *Nat. Chem. Biol.*, **5**, 857–862.
25. Dufrêne, Y.F. (2004) Using nanotechniques to explore microbial surfaces. *Nat. Rev. Microbiol.*, **2**, 451–460.
26. Dufrêne, Y.F. (2008) Towards nanomicrobiology using atomic force microscopy. *Nat. Rev. Microbiol.*, **6**, 674–680.
27. Müller, D.J. and Dufrêne, Y.F. (2008) Atomic force microscopy as a multifunctional molecular toolbox in nanobiotechnology. *Nat. Nanotechnol.*, **3**, 261–269.
28. Müller, D.J., Helenius, J., Alsteens, D., and Dufrêne, Y.F. (2009) Force probing surfaces of living cells to molecular resolution. *Nat. Chem. Biol.*, **5**, 383–390.
29. Dufrêne, Y.F. (2008) Atomic force microscopy and chemical force microscopy of microbial cells. *Nat. Protoc.*, **3**, 1132–1138.
30. Dague, E., Alsteens, D., Latgé, J.P., and Dufrêne, Y.F. (2008) High-resolution cell surface dynamics of germinating *Aspergillus fumigatus* conidia. *Biophys. J.*, **94**, 656–660.
31. Plomp, M., Leighton, T.J., Wheeler, K.E., Hill, H.D., and Malkin, A.J. (2007) *In vitro* high-resolution structural dynamics of single germinating bacterial

spores. *Proc. Natl. Acad. Sci. U.S.A.*, **104**, 9644–9649.

32. Dupres, V., Alsteens, D., Andre, G., and Dufrêne, Y.F. (2010) Microbial nanoscopy: a closer look at microbial cell surfaces. *Trends Microbiol.*, **18**, 397–405.
33. Hinterdorfer, P. and Dufrêne, Y.F. (2006) Detection and localization of single molecular recognition events using atomic force microscopy. *Nat. Methods*, **3**, 347–355.
34. Kienberger, F., Kada, G., Gruber, H.J., Pastushenko, V.P., Riener, C., Trieb, M., Knaus, H.G., Schindler, H., and Hinterdorfer, P. (2000) Recognition force spectroscopy studies of the NTA-His6 bond. *Single Mol.*, **1**, 59–65.
35. Verbelen, C., Gruber, H.J., and Dufrêne, Y.F. (2007) The NTA-His6 bond is strong enough for AFM single-molecular recognition studies. *J. Mol. Recognit.*, **20**, 490–494.
36. Heinisch, J.J., Dupres, V., Alsteens, D., and Dufrene, Y.F. (2010) Measurement of the mechanical behavior of yeast membrane sensors using single-molecule atomic force microscopy. *Nat. Protoc.*, **5**, 670–677.
37. Backhaus, K., Heilmann, C.J., Sorgo, A.G., Purschke, G., de Koster, C.G., Klis, F.M., and Heinisch, J.J. (2010) A systematic study of the cell wall composition of *Kluyveromyces lactis*. *Yeast*, **27**, 647–660.
38. Dupres, V., Dufrene, Y.F., and Heinisch, J.J. (2010) Measuring cell wall thickness in living yeast cells using single molecular rulers. *ACS Nano*, **4**, 5498–5504.
39. Shibasaki, S., Maeda, H., and Ueda, M. (2009) Molecular display technology using yeast--arming technology. *Anal. Sci.*, **25**, 41–49.
40. Rief, M., Gautel, M., Oesterhelt, F., Fernandez, J.M., and Gaub, H.E. (1997) Reversible unfolding of individual titin immunoglobulin domains by AFM. *Science*, **276**, 1109–1112.
41. Oberhauser, A.F., Marszalek, P.E., Erickson, H.P., and Fernandez, J.M. (1998) The molecular elasticity of the extracellular matrix protein tenascin. *Nature*, **393**, 181–185.
42. Oesterhelt, F., Oesterhelt, D., Pfeiffer, M., Engel, A., Gaub, H.E., and Müller, D.J. (2000) Unfolding pathways of individual bacteriorhodopsins. *Science*, **288**, 143–146.
43. Sapra, K.T., Damaghi, M., Köster, S., Yildiz, O., Kühlbrandt, W., and Muller, D.J. (2009) One βHairpin after the other: exploring mechanical unfolding pathways of the transmembrane β-barrel protein OmpG. *Angew. Chem. Int. Ed.*, **44**, 8306–8308.
44. Lee, G., Abdi, K., Jiang, Y., Michaely, P., Bennett, V., and Marszalek, P.E. (2006) Nanospring behaviour of ankyrin repeats. *Nature*, **440**, 246–249.
45. Schlierf, M. and Rief, M. (2005) Temperature softening of a protein in single-molecule experiments. *J. Mol. Biol.*, **354**, 497–503.
46. Simons, K. and Ikonen, E. (1997) Functional rafts in cell membranes. *Nature*, **387**, 569–572.
47. Jacobson, K. and Dietrich, C. (1999) Looking at lipid rafts? *Trends Cell Biol.*, **9**, 87–91.
48. Lingwood, D. and Simons, K. (2010) Lipid rafts as a membrane-organizing principle. *Science*, **327**, 46–50.
49. Grossmann, G., Opekarova, M., Malinsky, J., Weig-Meckl, I., and Tanner, W. (2007) Membrane potential governs lateral segregation of plasma membrane proteins and lipids in yeast. *EMBO J.*, **26**, 1–8.
50. Strádalová, V., Stahlschmidt, W., Grossmann, G., Blazíková, M., Rachel, R., Tanner, W., and Malinsky, J. (2009) Furrow-like invaginations of the yeast plasma membrane correspond to membrane compartment of Can1. *J. Cell Sci.*, **122**, 2887–2894.
51. Straede, A. and Heinisch, J.J. (2007) Functional analyses of the extra- and intracellular domains of the yeast cell wall integrity sensors Mid2 and Wsc1. *FEBS Lett.*, **581**, 4495–4500.
52. Heinisch, J.J., Dupres, V., Wilk, S., Jendretzki, A., and Dufrene, Y.F. (2010) Single-molecule atomic force microscopy reveals clustering of the yeast plasma-membrane sensor Wsc1. *PLoS One*, **5**, e11104.

53. Keymer, J.E., Endres, R.G., Skoge, M., Meir, Y., and Wingreen, N.S. (2006) Chemosensing in *Escherichia coli*: two regimes of two-state receptors. *Proc. Natl. Acad. Sci. U.S.A.*, **103**, 1786–1791.
54. Wilk, S., Wittland, J., Thywissen, A., Schmitz, H.P., and Heinisch, J.J. (2010) A block of endocytosis of the yeast cell wall integrity sensors Wsc1 and Wsc2 results in a reduced fitness *in vivo*. *Mol. Genet. Genomics*, **284**, 217–229.
55. Osumi, M. (1998) The ultrastructure of yeast: cell wall structure and formation. *Micron*, **29**, 207–233.

12
AFM-Based Single-Cell Force Spectroscopy

Clemens M. Franz and Anna Taubenberger

12.1
Introduction

Adhesive interactions of cells with the extracellular matrix (ECM) or with other cells maintain the integrity of tissues and are indispensable for proper tissue function. In addition, cell adhesion governs a wide range of additional processes, such as intercellular communication, inflammation, tumor progression, and cell migration. Specific cell adhesion is mediated by different cell adhesion molecules (CAMs). Most CAMs are transmembrane proteins containing ligand-binding sites in their extracellular domain and interaction sites for intracellular binding partners and the cytoskeleton in their cytoplasmic domain [1]. Better understanding of the adhesion forces transmitted by CAMs is critical for explaining the complex interplay of cells and ECM in tissues and has been a main objective of biomedical research.

Cell adhesion strength to proteins of the ECM is generally measured by the ability of cells to remain attached when exposed to a detachment force. A popular method to study cell adhesion is washing assays [2]. In these assays, cells are first seeded onto an adhesive surface. After an attachment period, the nonadhered cells are washed off by flushing a buffer solution over the surface, and the remaining cells are counted. These relatively simple bulk assays have provided a surprising wealth of adhesion information and identified a number of key adhesion receptors [3]. Since a large a number of cells are analyzed in each experiment, statistically relevant data is provided within a short time frame. On the other hand, washing assays suffer from poor reproducibility and provide only qualitative adhesion information averaged over the entire cell population. Small differences in cell adhesion that are of potential biological significance are difficult to detect. For instance, adhesive subpopulations arising from different functional states of individual cells cannot be identified. In addition, short contact times are difficult to control, while the shear forces of the buffer stream may be too weak to dislodge tightly adhering cells after longer attachment periods.

Quantitative analysis of cell adhesion therefore requires single-cell techniques. Furthermore, because cell adhesion is usually mediated by several bonds, assays are

Atomic Force Microscopy in Liquid: Biological Applications, First Edition.
Edited by Arturo M. Baró and Ronald G. Reifenberger.

necessary that can distinguish between single and multiple receptor adhesion [4]. Several techniques that measure adhesion forces of single cells have been developed over the past years, including the biomembrane force probe (BFP) [5], laser [6] and magnetic tweezers [7], and atomic force microscopy (AFM)-based single-cell force spectroscopy (SCFS) [8, 9], and all of these methods have been applied to study single-molecule bond rupture events in living cells. Of these techniques, the BFP and optical trapping offer the highest force sensitivity (~0.1–0.01 nN), making them suitable to study single-molecule adhesion events. However, the maximum forces that can be detected or applied with these techniques (~1000 pN) are usually below the detachment forces required to remove well-adhering cells. In contrast, AFM-based methods allow a wider force range to be measured (~10 pN–100 nN), spanning roughly four orders of magnitude. Thus, compared with other SCFS techniques, AFM-based SCFS offers the most versatile force range and can therefore be used to address a broader range of biological questions [10]. In particular, the ability to measure forces with high resolution over a wide range makes AFM a valuable tool to study cellular adhesion forces across dimensions from the single-molecule level to overall cell (total cell adhesion mediated by a large number of receptors) [11, 12].

In SCFS, a living cell is attached to a functionalized AFM cantilever and then lowered onto a substrate or another cell with defined force. During subsequent cantilever withdrawal, the force required to separate the cell from its binding partner is measured by monitoring the cantilever deflection. Alternatively, a ligand-functionalized cantilever can be approached onto an immobilized cell (Figure 12.1). Recording the cantilever force as a function of its vertical position during cell retraction generates a force–distance (F–D) curve from which cell adhesion forces can be determined.

By varying the contact time interval, the build up of cell adhesion forces can be monitored. Generally, receptor-mediated adhesion is initiated on the cell membrane by the engagement of individual receptors with their respective ligand [13, 14]. The number of receptor–ligand pairs may then grow by increasing the cell–substrate contact area and by receptor diffusion within that zone. Subsequent adhesion strengthening occurs through receptor clustering and linkage to the

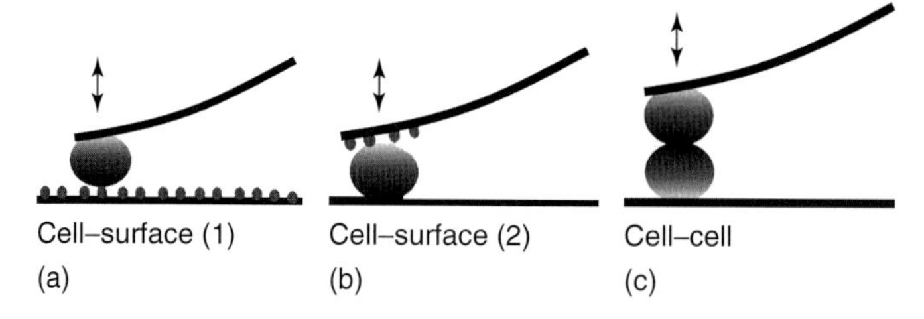

Figure 12.1 Different SCFS experimental setups to measure cell interactions with adhesive substrates. (a) A single cell is immobilized on the AFM cantilever and approached to an adhesive substrate. (b) A cell is immobilized on a surface, and a ligand-coated cantilever is pushed onto of the cell. (c) To quantify cell–cell adhesion, a cell attached is pushed onto a second cell immobilized on a surface.

cytoskeleton. Studying the kinetics of cell adhesion formation requires that the contact time can be precisely controlled. The piezo-driven positioning system of AFM provides exquisite temporal and spatial control over the cell–substrate interaction. By varying the interaction time, progression from single-receptor to cooperative-receptor binding can be monitored on the force level and correlated with the establishment of overall adhesion [15, 16].

To address the contribution of individual cell surface receptors to cell adhesion, SCFS can be performed with single-molecule sensitivity. By probing a specific receptor-ligand bond over a range of loading rates, a dynamic force spectrum (DFS) can be generated, from which bond parameters, such as the bond strength, the bond lifetime, or the width of the potential barrier, can be determined. Thus, SCFS can provide detailed information about the dissociation mechanism of specific receptor–ligand bonds. However, single-molecule experiments in the complex context of a living cell require rigorous controls ensuring molecular specificity of the measures interaction.

AFM–SCFS probes receptors presented on the surface of living cells. Considering that intracellular binding partners and associated membrane proteins or linkage of adhesion receptors to the cytoskeleton may modulate the mechanics of the adhesive bridge, studying receptor-mediated adhesion in the environment of a living cell provides an important advantage over *in vitro* systems that use purified proteins. Furthermore, the effects of inhibitors or stimulators of cell adhesion can be monitored directly in the cellular context.

Examples of receptor–ligand pairs that have been studied using single-molecule SCFS include $\alpha_5\beta_1$-fibronectin [17–19], $\alpha_4\beta_1$-VCAM-1 [20, 21], $\alpha_2\beta_1$-collagen type I [16], $\alpha_L\beta_2$-ICAM-I [20, 22], and $\alpha_L\beta_2$-ICAM-2 [23]. In addition, different cadherin (VE- and N-cadherins) [24, 25] and selectin (E-, L-, P-selectins) [26–28] interactions have been analyzed on the single-molecule level. Furthermore, the effect of activating molecules on integrin-mediated adhesion has been investigated. For instance, it was shown that 12-O-tetradecanoylphorbol-13-acetate (TPA), an activator of integrin binding, increases integrin avidity and strengthens the linkage between integrin and cytoskeleton [29], while treatment with inhibitors of actomyosin contractility decreases integrin binding strength [16].

A further advantage of AFM-based SCFS is the comparative ease of use and the commercial availability of the required instrumentation. SCFS can be performed using a standard AFM setup featuring a fluid cell and an incorporated light microscopy unit. A temperature-controlled chamber set to 37 °C should be used when working with mammalian cells. In addition, as different cell types tend to deform strongly under load along the pulling direction, complete cell–substrate or cell–cell separation may require an extended z-range. AFMs featuring a pulling range of 100 μm are commercially available. By combining SCFS with optical microscopy, cells can be positioned to assess cellular interactions at a given location on a functionalized surface, tissue, or another cell [11].

In this chapter, we provide an overview of AFM-based SCFS techniques to study cell adhesion both with single-molecule force resolution and at the single-cell scale. We provide detailed description of different cantilever functionalization protocols and cell attachment procedures. We illustrate F–D curve recording and analysis and demonstrate how specific bond parameters can be extracted from dynamic force spectra. Finally, we provide a short outlook on current and future developments of AFM-based SCFS.

12.2 Cantilever Choice

Soft cantilevers with spring constants below 0.1 N m^{-1} commonly used for contact mode imaging in fluid are useful for SCFS. While these cantilevers are sensitive enough to measure single-molecule unbinding forces, their detection range also covers cell-scale forces building up during contact times of up to 20 min. However, when longer adhesion times are tested, cell adhesion forces may quickly exceed the maximum force detectable with these cantilevers. In this case, the analysis of single-molecule deadhesion events is usually irrelevant and stiffer cantilevers can be used [30]. In principle, standard tip-carrying cantilevers are compatible with SCFS. However, tipless cantilevers have become commercially available, and they have several advantages in SCFS. For instance, the lack of the tip on theses cantilevers allows precise positioning of the cell at the cantilever apex, ensuring that the cantilever bending force exerted by the cell is applied over the entire length of the cantilever. In addition, the AFM tip may protrude over well-spread or small cells, such as T cells, thereby preventing cell–substrate contact. Finally, the tip may puncture the cell during substrate contact if attachment occurs in the tip vicinity.

12.3 Cantilever Functionalization

Reproducible force measurements require stable attachment of the probed cell to the AFM cantilever. Importantly, cell attachment to the cantilever has to be strong compared to the adhesive interactions occurring with the substrate in order to prevent cell detachment from the cantilever during force measurements. Cell attachment to the cantilever is augmented by functionalizing the cantilever with an adhesive coating. Coating with a lectin, such as concanavalin A (ConA) [31] or wheat germ agglutinin (WGA) [11], is a gentle method to immobilize a wide range of cell types via sugar residues exposed by glycosylated cell surface proteins. When only low adhesion forces (<5 nN) are expected, it is often sufficient to functionalize the cantilever by simple ConA physisorption. Higher forces may require a more elaborate multistep functionalization protocol [12, 31]. In this case, the lectin–sugar bridge easily withstands cell adhesion forces of up to 50 nN, typically built up over the first 20 min or so of cell–substrate contact. Because of the ubiquity of cell surface glycosylation, lectin functionalization immobilizes a wide range of cell types, including fibroblast, endothelial, or epithelial cells. However, lectin-mediated cell attachment may not be suitable for cells of the immune system, such as T cells, which may become immunologically activated on receptor clustering on the cell surface induced by the lectin coating. In this case, receptor-independent cell immobilization, such as using mussel adhesive [32], may be preferable. Alternatively, strong cantilever–cell coupling can be achieved by attaching biotinylated cells to streptavidin-functionalized cantilevers [4, 25] or by chemically cross-linking ECM proteins onto the cantilever [30].

12.4 Cantilever Calibration

True spring constants of cantilevers usually deviate from their nominal values, making it necessary to calibrate each cantilever before being used in SCFS experiments. Furthermore, the sensitivity of the system (the factor that relates the voltage change measured by the photodiode to the cantilever deflection) is influenced by cantilever geometry and mounting and also needs to be determined every time the cantilever is mounted into the holder. For sensitivity determination, an F–D curve is recorded with a pulling range of several micrometers and intermediate pulling speeds ($2.5\,\mu\mathrm{m\,s^{-1}}$) on a stiff substrate. If the adhesion substrate itself is sufficiently stiff, for instance, when ECM-coated glass slides are used, the F–D curve can be recorded *in situ* on the adhesion substrate. Softer substrates, such as flexible ECM gels, are not suitable, and an alternative hard surface must be used for calibration. While the cantilever is in contact with the surface, cantilever deflection and vertical piezo movement are equal and the sensitivity can be obtained from a fit of the linear part of the F–D curve. For subsequent spring constant determination, the cantilever is raised above the surface, allowing it to oscillate freely. Spring constant determination is most frequently performed using the thermal noise method [33]. This method measures the thermal fluctuations of cantilever deflection linked to the spring constant by the function

$$\frac{1}{2}k_{\mathrm{B}}T = \frac{1}{2}k\Delta x^2 \tag{12.1}$$

where k_{B} is the Boltzmann constant, T is the absolute temperature, k is the cantilever spring constant, and Δx is the fluctuation of the cantilever deflection. Many commercial AFMs contain built-in calibration routines based on the thermal noise method. While the thermal noise method may yield calibration errors of up to 15%, its ease of use makes it the preferred method for SCFS.

12.5 Cell Attachment to the AFM Cantilever

There are three general procedures of cell attachment to the cantilever for SCFS measurements: (i) attaching a suspended cell to a functionalized cantilever immediately before force measurement, (ii) attaching an adherent, well-spread cell to an ECM-functionalized cantilever, and (iii) cultivating cells directly on the cantilever for several hours before measurement.

To attach a suspended cell to the cantilever, a small volume of a cell suspension (typically containing $\sim 10^3$ cells) is added to the sample chamber. A single candidate cell is then selected, the cantilever is positioned over the cell, and briefly (1–10 s) and gently ($F < 1$ nN) pushed onto the cell. On cantilever retraction, the cell usually remains on the cantilever. After a rest phase of several minutes, during which the cell is able to form a strong bond with the cantilever, force measurements can be performed.

Cells growing in suspension can be added into the sample chamber directly. Adherent cells, on the other hand, have to be removed from the tissue culture substrate first, for instance, by short treatment with EDTA, trypsin, or other proteases, potentially degrading adhesion receptors on the cell surface. Nevertheless, attaching a suspended cell immediately before SCFS measurements has several advantages. Cells usually remain roughly spherical in shape, leading to geometrically and mechanically reproducible cell–substrate interactions and similar cell–substrate contact areas over a number of force cycles. This procedure is also comparatively fast and can be completed in 10 min. The coupling between cantilever and cell is usually strong enough to withstand cellular adhesion forces formed within the first minutes of substrate contact and can therefore be used to test the formation of early adhesive contacts, including single-molecule binding events. On the other hand, this coupling method cannot sustain high cell adhesion forces normally formed after longer cell–substrate attachment times (several hours). Furthermore, maintaining constant contact conditions (force, temperature, pH, etc.) for extended periods is difficult and puts a high demand on microscope usage time. Taken together, long-contact-time adhesion measurements by SCFS require different cell attachment approaches.

To test higher adhesion forces, such as those that occur in well-adhered cells, a cantilever coated with an ECM component, such as fibronectin or laminin, can be positioned on top of an adhered cell for ~5–15 min, enabling the cell to attach to the cantilever firmly [30, 34]. The ECM component can either be adsorbed to the cantilever, or for even stronger coupling, covalently attached to the cantilever using glutaraldehyde. However, in this case, extensive washes of the functionalized cantilever or chemically quenching reactions should be performed to ensure that the living cells are not exposed to unreacted, cytotoxic aldehyde groups. With this procedure, adhesion forces in excess of 100 nN have been measured, and cells can be detached from the substrate even after several hours of incubation. As a drawback, for a considerable time during cell attachment to the cantilever, the cell receives two different adhesive cues (basally from the adhesive substrate and apically from the functionalized cantilever), which may induce signaling events in the cell and change cell adhesion behavior. Interestingly, the cellular response to an adhesive substrate can be influenced by the type of ECM molecule used to coat the cantilever, indicating cross-talk between adhesion receptors across the cells [35].

ECM-functionalized cantilever can also be used to probe adhesion of immobile tissue culture, preserving tissue integrity [36].

Finally, cells can be cultured directly on the cantilever for several hours before adhesion measurements [4]. While this procedure provides extremely robust cell attachment, usually more than one cell attaches to the cantilever, complicating the performance of single-cell experiments. In addition, extensive cell spreading on the cantilever may cause mechanically ill-defined cell–substrate contact.

12.6 Recording a Force–Distance Curve

A force cycle is initiated by approaching the cantilever with an attached cell to the adhesive surface using the piezo-driven positioning system of the AFM until a preset contact force is reached. The cantilever is then kept stationary for a defined contact period. Subsequently, the cell is withdrawn at a constant retraction velocity. During retraction, bonds formed between the cell and substrate during the contact interval will break sequentially until the cell and substrate are completely separated. During approach and retraction, the force exerted on the cantilever, which is proportional to the cantilever deflection, is recorded as a function of the cantilever distance from the surface in an F–D curve. The sequential rupture of individual adhesive contacts formed during the contact phase will leave a characteristic adhesion signature in the retract curve (Figure 12.2).

During retraction, cells will deform considerably under force, so that long piezo travel distances (up to 100 μm) are required for complete cell–substrate separation. Piezo hysteresis (nonlinear movement) may have a significant influence on the resulting long F–D curves. This effect can be offset by coupling the piezo height extension with a height sensor measuring the actual piezo extensions. In a feedback system ("closed-loop" mood), the piezo extension can then be adjusted according

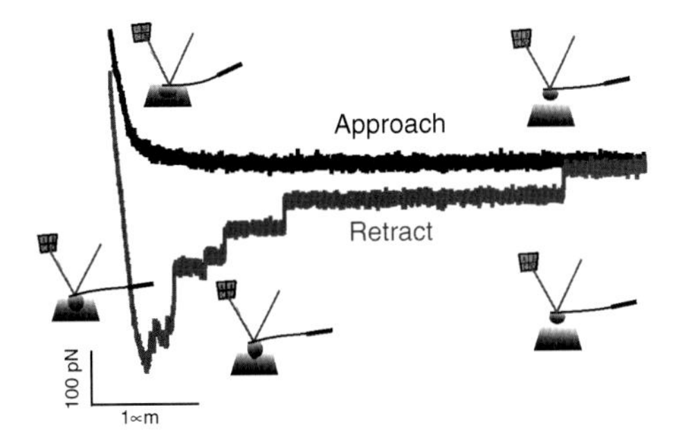

Figure 12.2 Recording a force–distance curve. Schematic representation of the F–D cycle. The sequential rupture of individual adhesive contacts formed during the contact phase will leave a characteristic adhesion signature in the retract curve.

to the measured height, yielding unskewed piezo extension and linear force curves. In addition, the "closed-loop" also compensates for piezo "creep" (piezo position overshoot after reaching the force set point), preventing an unwanted force increase during extended cell–substrate contact times.

A drawback of the "closed-loop" mode is somewhat increased noise levels in the force signal. Single-molecule experiments usually require the highest possible force sensitivity but only relatively short piezo travel distances and contact times. Because piezo hysteresis can be neglected during short pulling distances and creep during short contact times, single-molecule experiments are better run in the "open-loop" mode to reduce signal noise and to achieve maximal force resolution.

Implementation of a "closed-loop" feedback system provides additional leeway in controlling the force during cell–substrate contact. In SCFS experiments, usually, a moderate contact force (<2 nN) is used, which gives the cell the opportunity to establish contacts with the adhesive surface rather than pushing it onto the surface with high force. Nevertheless, even small contact forces induce noticeable cell deformation. The initial cell deformation after reaching the preset contact force then increases further because of viscoelastic relaxation of the cell. As a result, the effective force applied to the cell may drop by 30–90% within the first seconds of substrate contact. In "constant force" mode, the z-piezo feedback loop can counteract this force drop so that constant force is maintained throughout the entire measurement period. Although the "constant force" mode establishes defined and constant force conditions, it may lead to excessive cell spreading as the cell is evading the cantilever pushing force. Nevertheless, "constant force" settings are necessary when the effect of contact force on cell adhesion is tested. In contrast, in "constant height" mode, the cantilever (z-piezo position) is arrested after the preset contact force is reached. In this case, the effective contact force will initially drop and later fluctuate, as the cells progress through alternating pulling and pushing cycles. However, usually an equilibrium state between cantilever pushing force and cellular shape changes is reached, so that relatively stable interaction conditions are established, making "constant height" the preferred mode for long contact time measurements. By recording the acting cantilever force throughout the F–D cycle, regular force cycle progression can be monitored. Figure 12.3 depicts a complete "constant height" mode force run, including the approach, contact, and retraction phases.

Between force cycles, the cell should be allowed to recover, and the recovery time should not be shorter than the cell–substrate contact time. The number of force curves that can be recorded with a single cell critically depends on the chosen contact time. For short contact times (<30 s), more than 20 F–D curves can easily be recorded per cell. However, when longer contact times (~10–20 min) are tested, the cell may not completely recover after the force cycle and start displaying abnormal adhesion patterns. In this case, the number of performed F–D cycles should be reduced. Occasionally, cells build up very high forces, in which case the cell may detach from the cantilever during retraction.

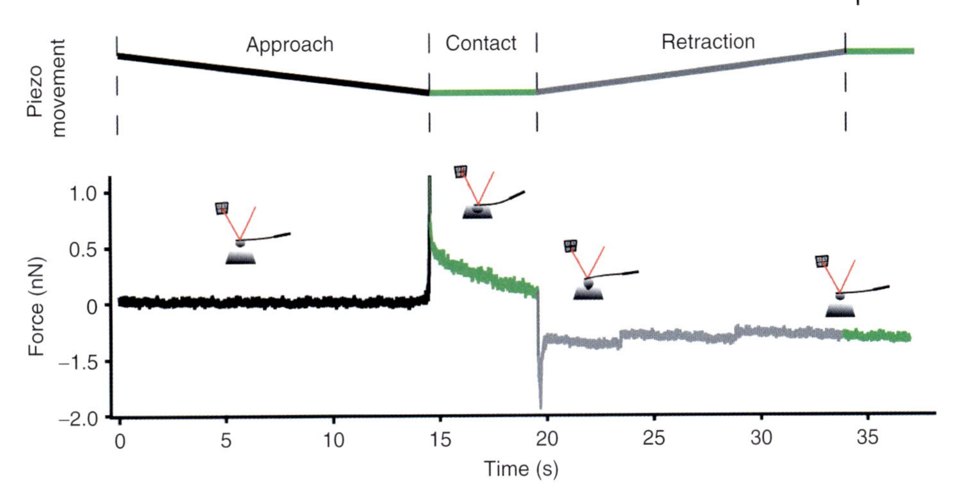

Figure 12.3 Monitoring the force signal during an F–D cycle. Piezo movement (top) and force signal (bottom) versus time during an F–D cycle. The piezo extends until a preset force set-point is reached (black). In "constant height" mode, the piezo position is arrested during contact (green). Owing to its viscous properties, the cell relaxes and the force on the cantilever decays during the first seconds of contact. In the retraction curve (gray), the cantilever deflects downward to counteract adhesion forces between cell and substrate. The baseline force level is reached when all linkages between cell and substrate are broken.

12.7 Processing F–D Curves

Unprocessed F–D curves occasionally have tilted baselines and offsets between trace and retrace baselines. Trace and retrace baselines tilted in opposite direction are typical for thermal drift (Figure 12.4a). Thermal drift occurs because of the cantilever's sensitivity to temperature gradients and is particularly pronounced at the beginning of SCFS experiments (before the establishment of a thermal equilibrium) and when cantilevers with asymmetric coatings of chromium and gold are used. As these metal layers have different thermal expansion coefficients, temperature changes induce cantilever bending. Thermal drift is easily corrected in F–D curves by subtracting a line fit to the baselines form the force data (Figure 12.4b).

As the cantilever moves through the medium, it experiences viscous drag, which results in deflection of the cantilever in the opposite direction of its movement (Figure 12.5a). Hydrodynamic drag forces increase with the viscosity of the medium, the proximity of the cantilever from the surface, and linearly with the speed of the cantilever (Figure 12.5b). At low pulling speeds, the hydrodynamic drag is below the uncertainty of the force measurements and therefore negligible. However, at high pulling speeds ($>10\,\mu m\,s^{-1}$), the hydrodynamic drag force may reach the magnitude of a single-molecule binding strength ($\sim$50 pN). Consequently, rupture forces will be underestimated unless corrected [37, 38]. The drag force (F_h) acting

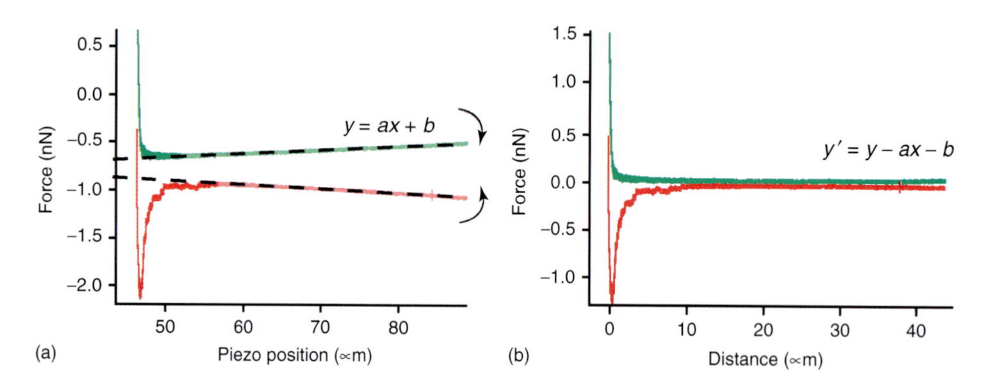

Figure 12.4 Correction of F–D curves for thermal drift. (a) Tilted and offset baselines indicate thermal drift. (b) A line is fitted to the baselines and subtracted from the force data.

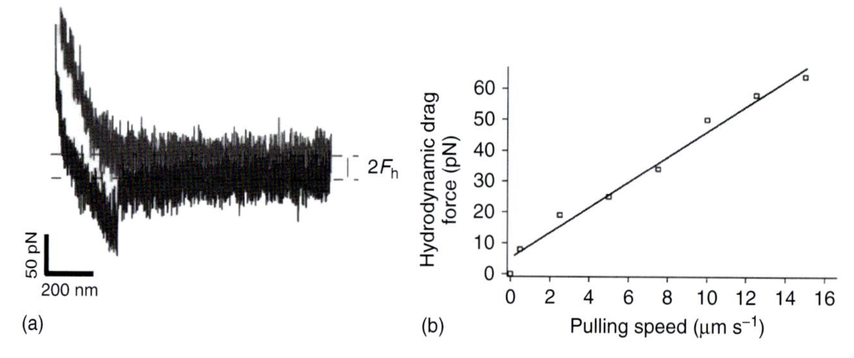

Figure 12.5 Hydrodynamic drag forces versus pulling speed. (a) The approach and retract baselines are vertically shifted because of hydrodynamic drag (F_h) acting on the cantilever while it moves through the medium. The difference between the baselines corresponds to the double drag force because the drag force acts during the approach and the retraction in the opposite direction. (b) Linear relationship of hydrodynamic drag force and pulling speed. At high pulling speeds (>10 μm s^{-1}), a correcting drag force may have to be added to the rupture force to avoid underestimating the rupture force [37, 38]. (Figure adopted from Ref. [39].)

on the cantilever can be calculated as follows:

$$F_h = \frac{6\pi \eta a_{\text{eff}}^2}{h + d_{\text{eff}}} \cdot v_{\text{tip}} \tag{12.2}$$

The coefficients a_{eff} (effective cantilever size) and d_{eff} (effective cantilever thickness) depend on the geometry of the cantilever and can be determined by moving the cantilever freely through the medium while measuring the hydrodynamic drag in relation to the distance (h) from the surface. The coefficient v_{tip} denotes the tip velocity.

12.8
Quantifying Overall Cell Adhesion by SCFS

Retraction F–D curves usually show a characteristic and complex pattern, including a maximum force peak (F_D) and several smaller force steps. During cell removal, different specific (as well as unspecific adhesion) structures rupture simultaneously and leave a superimposed trace in the retraction force curve, complicating a meaningful force curve interpretation. Typically, retraction F–D curves can be separated into different phases (Figure 12.6).

As the cantilever is retracting, the cell cortex deforms and the force acting on individual adhesive contacts formed during the cell–substrate contact phase

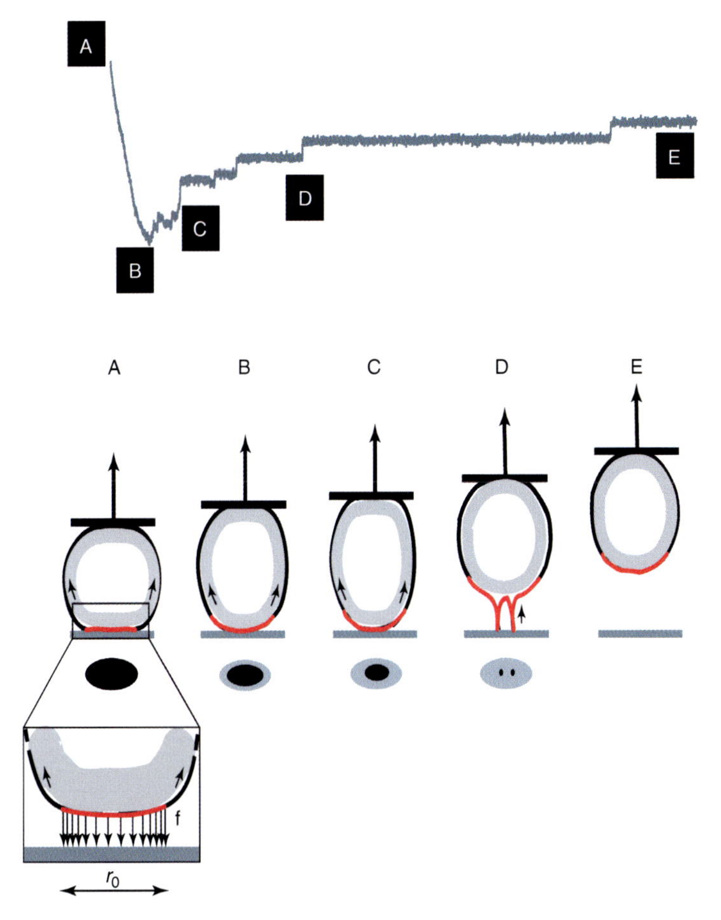

Figure 12.6 Schematic representation of the cell detachment process. The cell detachment process is separated into different phases. (A) The cell is in contact with the substrate. In the contact zone adhesive interactions occur. (B,C) During cell detachment established bonds rupture and the contact zone shrinks. When the cell body is separated from the surface, solely membrane nanotubes link cell and substrate (D) until the cell is fully detached from the substrate (E).

is increasing. The largest adhesion force that is recorded, the detachment force F_D, represents the maximum strength of cell–substrate binding (Figure 12.7). The maximal detachment force is frequently used as a measure of cell adhesion. Alternatively, the deadhesion work (the area enclosed by the F–D curve) provides a measure of cell adhesion strength, although this value is also highly influenced by the elastic deformation of the cell during detachment and nanotube extraction from the membrane.

After the maximal force is overcome, the force decreases quickly because the applied force load is shared by fewer receptors in the contact zone and the probability that these bonds resist rupture decreases [40]. The successive rupture of adhesive bridges during cell detachment introduces a series of discrete force steps into the F–D curve. Two different classes of force steps can be distinguished: force jumps, or "j" events, are typically preceded by a nonlinear increase in force loading, and tethers, or "t" events, are preceded by a force plateau (Figure 12.7).

Commonly, "j" events are considered to represent the unbinding of discrete adhesive units under load. These adhesive units may be constituted by a single or by a small group of receptors linked to the cytoskeleton. Stretching of the stiff receptor-membrane-cytoskeleton linker leads to a non–linear force increase before bond rupture [41]. The magnitude of the force steps reflects the stochastic survival of this receptor–ligand complex under an increasing force load [40, 41].

In contrast, in "t" events, adhesive units are pulled away from the cell cortex at the tip of a membrane tether. Tethers are cytoskeleton-free membrane tubes from a large membrane reservoir [42]. The force required to extend a tether depends on the lipid composition of the cellular membrane and on the mechanical properties of the of the cell cortex, but is independent of its length [43]. The lifetime of a membrane tether therefore depends on the receptor–ligand interaction at its tip, whereas the force required to maintain and extend a tether, which depends mainly on the properties of the cell membrane, is not [9, 44, 45]. Thus, "j" and "t" events denote mechanistically differing unbinding events, and they must be distinguished in the force curve analysis. In addition to the slope of the force curve

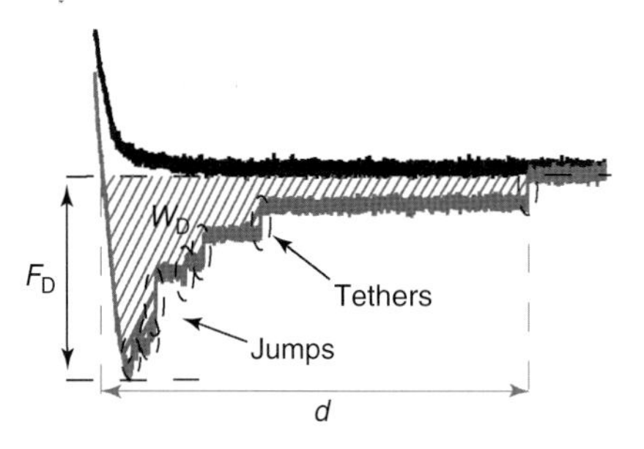

Figure 12.7 Schematic representation of the F–D curve. From the F–D curve, different parameters can be determined: F_D, maximal force required to detach the cell; W_D, adhesion work; d, distance required for separation; and discrete force steps (jumps and tethers).

preceding rupture, "j" and "t" events can be distinguished by the position within the F–D curve at which they occur. Tethers are easily extended to several tens of micrometers [12, 20] and preferentially rupture only in the final phase of cell detachment, when the cell body is not in contact with the substrate anymore [9].

"J" events represent receptor–ligand complexes unbinding under force, and they are therefore suitable for analysis in DFS experiments. The increase of "j" event rupture steps also provides information about progressive receptor clustering with increasing cell–substrate contact time. In contrast, tethers do not rupture during force loading and are excluded from DFS analysis. Likewise, the force step of "t" provides no information about the number of receptors in the adhesive bridge. However, tethers may be released on single receptor–ligand dissociation at the tether tip. In this case, the tether length can be used to calculate the lifetime of the receptor–ligand interaction under force [44]. An overview contrasting "j" and "t" unbinding events is provided in Figure 12.8.

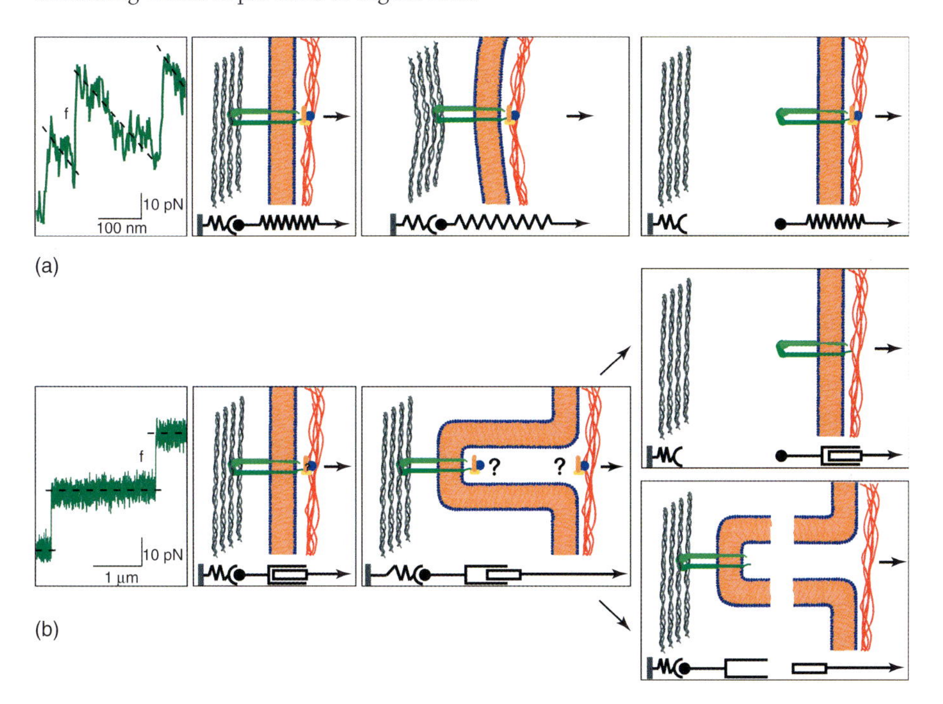

Figure 12.8 Sketch illustrating "j" and "t" events. (a) "j"-Events: a receptor anchored to the cytoskeleton binds to a ligand in the ECM (here collagen). On pulling on the cell during cantilever retraction, the receptor-membrane-cytoskeleton linker is stretched and the force on the cantilever increases. On bond rupture, the force on the cantilever rapidly decreases. (b) "t"-Event: a receptor that is not anchored to the cytoskeleton (alternatively, anchorage is disrupted during pulling) is extracted with a membrane nanotube ("tether") from the cell body. The force on the cantilever remains constant during tether extraction. When the receptor–ligand bond is released, the force on the cantilever decreases staircaselike (upper sketch). Alternatively, the nanotube may fail (sketched below) or the receptor might be pulled out of the membrane.

12.9
SFCS with Single-Molecule Resolution

SCFS experiments can be performed with single-molecule sensitivity, providing a unique way to measure the strength of a single receptor–ligand bond on the cell membrane. In contrast to single-molecule experiments using purified proteins, in SCFS, receptors are investigated in their physiological surrounding, ensuring that the receptors are oriented properly and presented within their native lipid bilayer environment. Furthermore, intracellular interactions partners are still able to bind to and regulate receptor function. In contrast, recombinant receptors are often purified in truncated forms and miss intracellular domains.

It is crucial for single-molecule experiments that the molecular identity of the probed adhesive interaction is known. Cells possess a great diversity of surface proteins, and adhesion receptors often exist in tens of thousands in number on the cell surface. Further complexity is added by the fact that some adhesion receptors recognize several different ECM components. In turn, many ECM components are recognized by more than one type of receptor. Ensuring that unbinding of precisely one bond of a specific receptor–ligand pair is probed requires rigorous controls. First, the contribution of nonspecific (either receptor mediated or not) has to be minimized or eliminated. This can be achieved by offering the cell a homogeneous matrix formed from a single ECM component. If an appropriate cell system is chosen so that the cells express only one type of receptor to this particular ECM component, adhesion events will be limited to one class of receptor interactions. For instance, Chinese hamster ovary (CHO) cells express no endogenous collagen receptors but can be engineered to express $\alpha_2\beta_1$.integrin as their sole collagen receptor [16, 46]. Nevertheless, the specificity of the interaction must be conclusively proven, for example, by inhibiting the probed receptor–ligand interaction with blocking antibodies or with an excess of ligand mimetic peptides. A number of CAMs, including several integrins and cadherin receptors, require divalent cations for ligand binding, and removing these extracellular divalent cations by chelation efficiently suppresses adhesion. Furthermore, receptor knock-out or knock-down systems may also provide stringent controls. Under these conditions, SCFS can be applied to elucidate the contribution of individual CAMs on the cell surface, regardless of the heterogeneity of the cell surface and the complex molecular mechanisms determining overall cell adhesion.

After the molecular identity of the adhesive interaction is established, binding has to be limited to a single ligand–receptor pair, which can be achieved by short contact times (<400 ms) and low contact forces (300 pN). Under these conditions, receptor–ligand bonds form only in a small subset of total cycles. The probability of bond formation follows Poisson statistics [47], and the probability that a single unbinding event occurs in repeated F–D curves is given by [48]

$$P\left(N_{\rm b} = \frac{1}{nb} > 0\right) = \frac{\lambda}{\exp(\lambda) - 1} \tag{12.3}$$

where $N_{\rm b}$ is the number of bonds and γ is the frequency of binding events. If less than 30% of all force cycles yield F–D curves containing a single rupture event,

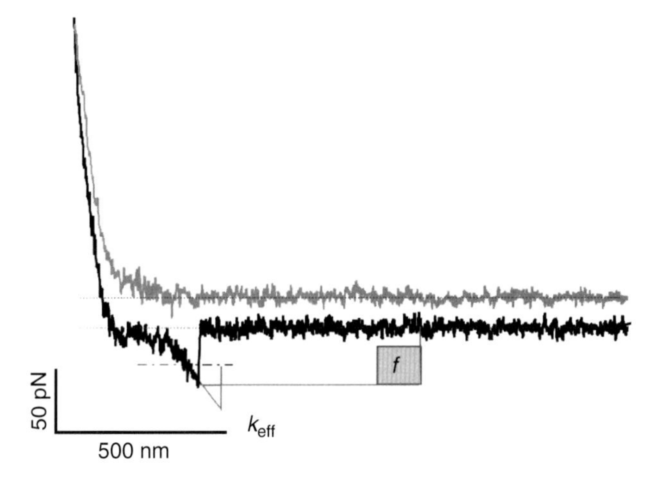

Figure 12.9 Representative F–D curve recorded in AFM–SCFS experiments. To achieve single-molecule binding, short contact time and low contact force are chosen. From the F–D curve, the rupture force (*f*) and the spring constant of the linker k_{eff} can be extracted (see Section 12.10).

the probability that this rupture event corresponds to the unbinding of a single receptor–ligand bond is ~86%. At binding frequencies below 10%, more than 95% of all rupture events correspond to single bond rupture (Figure 12.9). In practice, contact conditions that produce adhesion frequencies of ~10% are a feasible compromise between the need to ensure single-bond rupture and to maintain experimental throughput. Suitable interaction conditions have to be optimized for each cell–substrate pair. During single-molecule experiments, the interaction frequency has to be constantly monitored and, if necessary, adjusted by changing contact force or time. Because the desired low binding frequency requires a large number of force cycles, it is useful to implement an automatic force-curve-saving procedure. Commercial AFM instrumentations allow fully automated force cycle control, and single-molecule measurements only require short pulling distances; several hundred force curves per cell can be easily recorded.

12.10
Dynamic Force Spectroscopy

Interactions between CAMs and their ligands are governed by weak noncovalent interactions, such as van der Waals, electrostatic, hydrophobic–hydrophilic, and hydrogen bond interactions. The dissociation kinetics of these bonds is determined by the landscape of energy barriers, or free energy profile along an energetically favorable unbinding pathway [41]. In general, the energy landscape is described by a minimum (bound state), which is separated by one or several energy barriers from the unbound state (Figure 12.10). The energy barrier height influences the kinetics of bond dissociation since the bond lifetime τ increases exponentially with the energy barrier height

$$\tau = \tau_0 \exp\left(\frac{E_b}{k_B T}\right) \tag{12.4}$$

where τ is the bond lifetime ($\tau = 1/k_{off}$), τ_0 is the reciprocal of the natural frequency of oscillations, E_b is the height of the energy barrier, k_B the Boltzmann constant, and T the absolute temperature [40].

In contrast to equilibrium methods in which bond dissociation is measured under free diffusion and thermal activation, SCFS probes tethered bonds rupturing under an applied external force. An externally applied force, distorts the energy landscape along the unbinding pathway, reducing the activation energy required to break the bond and the lifetime of the bond (Figure 12.11) [50].

By probing bond rupture forces over a range of loading rates, a DFS can be generated, from which parameters describing the energy landscape of the bond, such as the number of energy barriers and their heights and widths, can be extracted. Thus, force spectroscopy provides a detailed view of the inner workings of a receptor–ligand bond and gives information about the unbinding mechanisms. In DFS experiments, force measurements with single-molecule sensitivity are

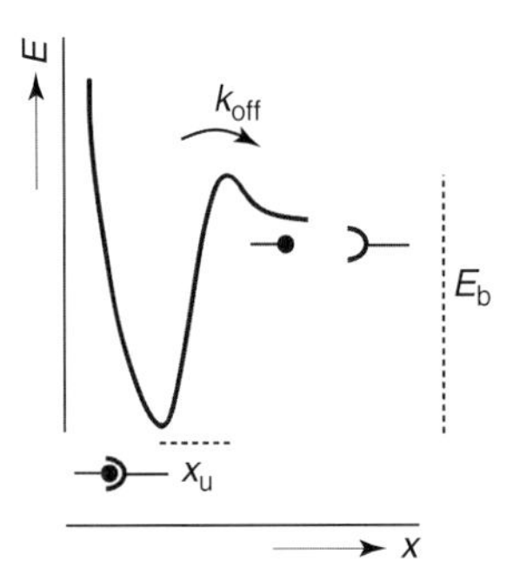

Figure 12.10 Schematic representation of receptor–ligand bond dissociation. A potential barrier (height E_b, width x_u) has to be overcome to dissociate a ligand–receptor bond. For simplicity, a model with a single barrier is shown. The energy landscapes of biomolecular bonds are expected to be more complex, since many interaction sites may be involved. (Figure modified from Ref. [49].)

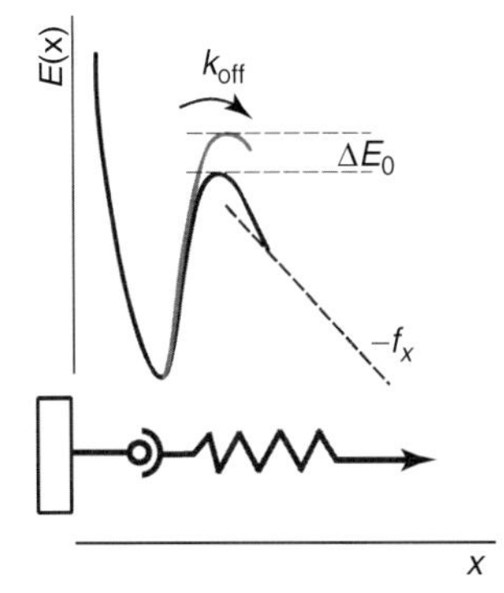

Figure 12.11 Influence of an external force on a single energy barrier separating the bond and unbound states. Along the reaction coordinate x, an external force adds a mechanical potential $-f_x$ that tilts the energy landscape and lowers the dissociation barrier. (Figure modified from Ref. [41].)

performed with different force loading rates, which is achieved by varying the cantilever pulling speed. A total of 100 F–D curves per loading rate value can be easily collected with a single cell because only minimal contact is made between cell and surface. From the obtained force curves (Figure 12.9), the rupture force f and the effective spring constant k_{eff} of the cantilever-cell-substrate bond system is extracted. The rupture force f corresponds to the force difference between the peak and the baseline, while the effective spring constant k_{eff} is obtained by the slope of a straight line fit to the part of the F–D curve that precedes the point of bond rupture. This slope approximates the effective spring constant k_{eff} of the cantilever-cell-substrate bond system and, when multiplied with the cantilever retraction speed ν, yields the effective bond loading rate (r_{eff})

$$r_{eff} = \nu k_{eff} \tag{12.5}$$

During bond rupture, bonds will exist in different thermal states, leading to a distribution of rupture forces (Figure 12.12a) for each loading rate value. Here, bond strength is interpreted as the mean of the rupture force distribution (f_m). Alternative approaches, such as taking either the most probable rupture force [22] or the median of the rupture force distribution, yield similar values because of the symmetry of the force distribution. The mean rupture force f_m is then plotted against the natural logarithm of r_{eff} (Figure 12.12b).

From the resulting dynamic force spectrum, different bond parameters can be extracted, such as the width of the potential barrier x_u and the unstressed dissociation constant k_{off} (or bond life time $1/k_{off}$). According to the Bell–Evans theory of kinetic bond rupture [50–52], the adhesion force acting on a single receptor–ligand bond increases linearly with the natural logarithm of the loading rate r_{eff} applied to the bond

$$f_m = \left(\frac{k_B T}{x_u}\right) \ln(r_{eff}) + \left(\frac{k_B T}{x_u}\right) \ln\left(\frac{x_u}{k_B T k_{off}}\right) \tag{12.6}$$

K_{off} and x_u can be obtained by fitting the data points (f_m versus r_{eff}) with Eq. (12.6) or by fitting a line to the data points (f_m versus $\ln(r_{eff})$) of the force spectrum (Figure 12.12a)

$$f_m = a \ln(r_f) + b \text{ with } a = \frac{k_B T}{x_u} \text{ and } b = a \ln\left(\frac{1}{a k_{off}}\right) \tag{12.7}$$

The slope a permits calculation of x_u by

$$x_u = \frac{k_B T}{a} \tag{12.8}$$

The y-intercept, b, and the slope, a, are used to extract k_{off}

$$k_{off} = \frac{1}{a} \exp\left(-\frac{b}{a}\right) \tag{12.9}$$

The Bell–Evans model has been used extensively to characterize binding interactions between adhesion molecules *in vitro* and in living cells [9, 37, 53, 54].

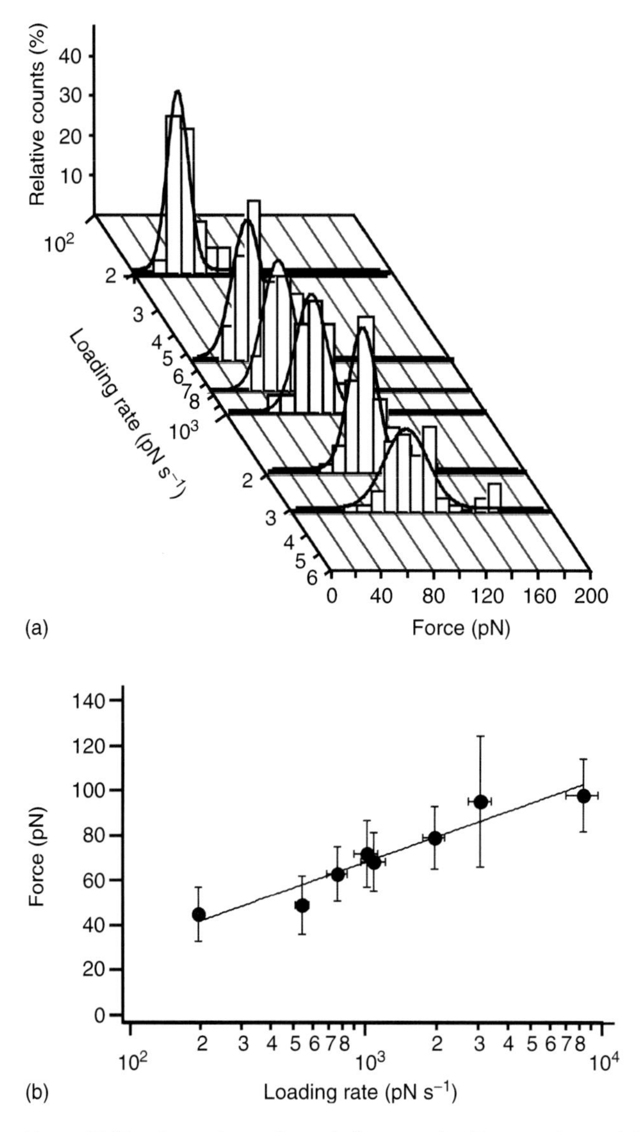

Figure 12.12 Generating a dynamic force spectrum. (a) Distribution of rupture forces measured at different loading rates (n > 50 adhesion events). Histograms containing well-defined single peaks indicate single-molecule unbinding events. (b) Dependence of rupture force (mean ± SD) on loading rate (mean ± SD). The mean rupture forces were fitted with a line ($r^2 = 0.9$), and from the obtained fit parameters (slope, y-intercept), the bond parameters k_{off} (unstressed dissociation rate), and x_u (barrier width) can be calculated. (Modified from Ref. [16].)

However, it is not applicable to all types of bonds. For instance, the so-called "catch bonds" display quick unbinding in the low loading rate regime but increased lifetimes at high force loading [55]. Catch bonds have been reported for a number of receptor–ligand pairs [55–58], including for P-selectin/PSGL. The catch bond behavior of P-selectin may indicate an adaptation to the physiological environment of leukocytes expressing this molecule. While leukocytes move freely in the blood stream below a certain shear threshold, at increased shear, they start rolling along blood vessels because of transient binding events. Thus, force measurements performed at the single-molecule scale can reveal molecular mechanisms that are leveled out in ensemble experiments and unravel subtle variations in the unbinding mechanism of different bond types.

Because bond rupture forces depend on the speed of force application, one must consider the effective loading rate acting on the bond at the moment of bond rupture together with the bond strength value when comparing measurements performed using different cellular systems. Furthermore, dissociation rates determined under free diffusion conditions are often orders of magnitude lower than the off rates determined when the same receptor–ligand pair is ruptured under force application [50], as anchoring the binding partners affects their free enthalpy and increases their dissociation rate [59]. Therefore, results obtained under equilibrium and nonequilibrium conditions have to be interpreted carefully to describe the functional role of the receptor under physiological conditions: depending on the forces a cell experiences in its native environment, the lifetime of bonds may differ greatly between experimental and physiological conditions.

12.11 Measuring Cell–Cell Adhesion

The methods described in the previous section can be easily adapted to investigate cell–cell adhesion. A number of cell–cell adhesion receptors have been analyzed by SCFS, both on the single-molecule and on the cell level. These include E- and N-cadherins [20, 22, 24, 25, 60, 61]; E-, L-, and P-selectin-mediated interactions [26–28]; as well as integrin-mediated cell–cell contacts [20, 22, 61, 62]. AFM–SCFS is especially useful when bulk adhesion assays are not practicable because the number of cells available for adhesion measurements is limited. For instance, during embryonic development, the specific adhesive properties of different cell lineages are thought to play an important role in driving tissue formation but collectively probing adhesion of the cells is difficult because of their relatively small number in the developing embryo. Using different markers, a limited number of lineage-specific cells can be isolated at defined developmental stages and homo- or heterotypic adhesion of cell pairs can be measured [63]. Cell–cell experiments have also provided insight into the abnormal interaction of cancer cells with tissues. For instance, it could be shown that myeloid progenitor cells expressing the fusion gene BCR-ABL, the hallmark of chronic myeloid leukemia, show increased adhesion to bone marrow stromal cells because of increased integrin beta1 expression [64].

12.12 Conclusions and Outlook

AFM-based SCFS provides a powerful setup to study cell adhesion from the single-molecule level to overall adhesion in the same experimental setup [65]. However, so far, cell adhesion has been mainly used to study cell adhesion to homogeneously coated substrates. In contrast, the native environment of cells is a flexible scaffold of structured macromolecules. Transferring experimental results from SCFS adhesion measurements onto physiologically relevant systems will benefit from using structured cell adhesion substrates with tunable stiffness. Through recent advances in surface functionalization techniques, such as microcontact printing, laser lithography, or self-assembling nanoarrays, cell adhesion substrates with reproducible micro- and nanostructures have now becoming available and are tested in SCFS experiments [30]. Determining the impact of the structure of the presented substrate on cell adhesion may lead to a better understanding of the molecular and geometric cues that drive adhesion formation. Techniques producing array of individual ligands with nanometer precision have recently been used in SCFS and provided fascinating insight into the importance of intermolecular spacing of adhesion receptors for cell adhesion strength [15]. Likewise, microcontact printing can be used to fabricate heterofunctional adhesion substrates carrying multiple ECM components in close proximity. In this way, directly comparative cell adhesion measurements using the same cell can be performed. In a fortunate coincidence, the very same AFM instrument, used in its imaging capacity, can be used to inspect the surface of such advanced adhesion substrates [16].

A further challenge is to perform SCFS measurements while imaging cells using advanced light microscopic techniques. For instance, cells expressing fluorescently labeled adhesion molecules can be imaged by confocal laser scanning (CLS) or total internal reflection fluorescence (TIRF) microscopy to monitor the establishment of macroscopic adhesion clusters. Subsequently, the rupture of these adhesion complexes can be monitored visually during cell detachment and correlated with the force step information contained in the force curves [30].

A current drawback of SCFS compared to other adhesion techniques, such as washing assays, is still the relatively high cost of a suitable AFM setup. Furthermore, because only one cell can be analyzed at a time, SCFS is time consuming compared to bulk adhesion assays and involves a relatively long experimental procedure to obtain statistically relevant data. Finally, the limited strength of the cantilever–cell coupling and thermal drift problems confine SCFS measurements to short contact times of up to 20 min. However, in our minds, these drawbacks are offset by the unique ability to quantitate receptor-mediated cell adhesion under physiological conditions over a unique force range starting from the single-molecule level to overall adhesion in the same experimental setup.

References

1. Kemler, R. (1992) Classical cadherins. *Semin. Cell Biol.*, **3**, 149–155.
2. Humphries, M.J. (2000) in *Extracellular Matrix Protocols* (eds C.H. Streuli and M.E. Grant), Humana Press, pp. 279–285.
3. Klebe, R.J. (1974) Isolation of a collagen-dependent cell attachment factor. *Nature*, **250**, 248–251.
4. Dobrowsky, T.M., Panorchan, P., Konstantopoulos, K., and Wirtz, D. (2008) Chapter 15: live-cell single-molecule force spectroscopy. *Methods Cell Biol.*, **89**, 411–432.
5. Evans, E., Ritchie, K., and Merkel, R. (1995) Sensitive force technique to probe molecular adhesion and structural linkages at biological interfaces. *Biophys. J.*, **68**, 2580–2587.
6. Neuman, K.C. and Block, S.M. (2004) Optical trapping. *Rev. Sci. Instrum.*, **75**, 2787–2809.
7. Kollmannsberger, P. and Fabry, B. (2007) High-force magnetic tweezers with force feedback for biological applications. *Rev. Sci. Instrum.*, **78**, 114301.
8. Benoit, M. (2002) Cell adhesion measured by force spectroscopy on living cells. *Methods Cell Biol.*, **68**, 91–114.
9. Helenius, J., Heisenberg, C.P., Gaub, H.E., and Muller, D.J. (2008) Single-cell force spectroscopy. *J. Cell Sci.*, **121**, 1785–1791.
10. Friedrichs, J., Helenius, J., and Muller, D.J. (2010) Quantifying cellular adhesion to extracellular matrix components by single-cell force spectroscopy. *Nat. Protoc.* **5**, 1353–1361.
11. Benoit, M., Gabriel, D., Gerisch, G., and Gaub, H.E. (2000) Discrete interactions in cell adhesion measured by single-molecule force spectroscopy. *Nat. Cell Biol.*, **2**, 313–317.
12. Puech, P.H., Poole, K., Knebel, D., and Muller, D.J. (2006) A new technical approach to quantify cell-cell adhesion forces by AFM. *Ultramicroscopy*, **106**, 637–644.
13. Gallant, N.D. and Garcia, A.J. (2006) Model of integrin-mediated cell adhesion strengthening. *J. Biomech*, **40**(6), 1301–1309.
14. Lotz, M.M., Burdsal, C.A., Erickson, H.P., and McClay, D.R. (1989) Cell adhesion to fibronectin and tenascin: quantitative measurements of initial binding and subsequent strengthening response. *J. Cell Biol.*, **109**, 1795–1805.
15. Selhuber-Unkel, C., Lopez-Garcia, M., Kessler, H., and Spatz, J.P. (2008) Cooperativity in adhesion cluster formation during initial cell adhesion. *Biophys. J.*, **95**, 5424–5431.
16. Taubenberger, A., Cisneros, D.A., Friedrichs, J., Puech, P.H., Muller, D.J., and Franz, C.M. (2007) Revealing early steps of alpha2beta1 integrin-mediated adhesion to collagen type I by using single-cell force spectroscopy. *Mol. Biol. Cell*, **18**, 1634–1644.
17. Li, F., Redick, S.D., Erickson, H.P., and Moy, V.T. (2003) Force measurements of the alpha5beta1 integrin-fibronectin interaction. *Biophys. J.*, **84**, 1252–1262.
18. Sun, Z., Martinez-Lemus, L.A., Trache, A., Trzeciakowski, J.P., Davis, G.E., Pohl, U., and Meininger, G.A. (2005b) Mechanical properties of the interaction between fibronectin and alpha5beta1-integrin on vascular smooth muscle cells studied using atomic force microscopy. *Am. J. Physiol. Heart Circ. Physiol.*, **289**, H2526–H2535.
19. Trache, A., Trzeciakowski, J.P., Gardiner, L., Sun, Z., Muthuchamy, M., Guo, M., Yuan, S.Y., and Meininger, G.A. (2005) Histamine effects on endothelial cell fibronectin interaction studied by atomic force microscopy. *Biophys. J.*, **89**, 2888–2898.
20. Thie, M., Rospel, R., Dettmann, W., Benoit, M., Ludwig, M., Gaub, H.E., and Denker, H.W. (1998) Interactions between trophoblast and uterine epithelium: monitoring of adhesive forces. *Hum. Reprod.*, **13**, 3211–3219.
21. Zhang, X., Craig, S.E., Kirby, H., Humphries, M.J., and Moy, V.T. (2004) Molecular basis for the dynamic strength of the integrin alpha4beta1/VCAM-1 interaction. *Biophys. J.*, **87**, 3470–3478.
22. Zhang, X., Wojcikiewicz, E., and Moy, V.T. (2002) Force spectroscopy

of the leukocyte function-associated antigen-1/intercellular adhesion molecule-1 interaction. *Biophys. J.*, **83**, 2270–2279.

23. Wojcikiewicz, E.P., Abdulreda, M.H., Zhang, X., and Moy, V.T. (2006) Force spectroscopy of LFA-1 and its ligands, ICAM-1 and ICAM-2. *Biomacromolecules*, **7**, 3188–3195.
24. Baumgartner, W., Hinterdorfer, P., Ness, W., Raab, A., Vestweber, D., Schindler, H., and Drenckhahn, D. (2000) Cadherin interaction probed by atomic force microscopy. *Proc. Natl. Acad. Sci. U.S.A.*, **97**, 4005–4010.
25. Panorchan, P., Thompson, M.S., Davis, K.J., Tseng, Y., Konstantopoulos, K., and Wirtz, D. (2006) Single-molecule analysis of cadherin-mediated cell-cell adhesion. *J. Cell Sci.*, **119**, 66–74.
26. Fritz, J., Katopodis, A.G., Kolbinger, F., and Anselmetti, D. (1998) Force-mediated kinetics of single P-selectin/ligand complexes observed by atomic force microscopy. *Proc. Natl. Acad. Sci. U.S.A.*, **95**, 12283–12288.
27. Hanley, W., McCarty, O., Jadhav, S., Tseng, Y., Wirtz, D., and Konstantopoulos, K. (2003) Single molecule characterization of P-selectin/ligand binding. *J. Biol. Chem.*, **278**, 10556–10561.
28. Hanley, W.D., Wirtz, D., and Konstantopoulos, K. (2004) Distinct kinetic and mechanical properties govern selectin-leukocyte interactions. *J. Cell Sci.*, **117**, 2503–2511.
29. Tulla, M., Helenius, J., Jokinen, J., Taubenberger, A., Muller, D.J., and Heino, J. (2008) TPA primes alpha2beta1 integrins for cell adhesion. *FEBS Lett.*, **582**, 3520–3524.
30. Selhuber-Unkel, C., Erdmann, T., Lopez-Garcia, M., Kessler, H., Schwarz, U.S., and Spatz, J.P. (2010) Cell adhesion strength is controlled by intermolecular spacing of adhesion receptors. *Biophys. J.* **98**, 543–551.
31. Wojcikiewicz, E.P., Zhang, X., and Moy, V.T. (2004) Force and compliance measurements on living cells using atomic force microscopy (AFM). *Biol. Proced. Online*, **6**, 1–9.
32. Hoffmann, S.C., Wabnitz, G.H., Samstag, Y., Moldenhauer, G., and Ludwig, T. (2011) Functional analysis of bispecific antibody (EpCAMxCD3) mediated T-lymphocyte and cancer-cell interaction by single cell force spectroscopy. *Int. J. Cancer*, **128**(9), 2096–2104.
33. Hutter, J.L. and Bechhoefer, J. (1993) Calibration of atomic-force microscope tips. *Rev. Sci. Instrum.*, **64**, 1868–1873.
34. Weder, G., Blondiaux, N., Giazzon, M., Matthey, N., Klein, M., Pugin, R., Heinzelmann, H., and Liley, M. (2010) Use of force spectroscopy to investigate the adhesion of living adherent cells. *Langmuir*, **26**, 8180–8186.
35. Friedrichs, J., Helenius, J., and Muller, D.J. (2010) Stimulated single-cell force spectroscopy to quantify cell adhesion receptor crosstalk. *Proteomics*, **10**, 1455–1462.
36. Lehenkari, P.P. and Horton, M.A. (1999) Single integrin molecule adhesion forces in intact cells measured by atomic force microscopy. *Biochem. Biophys. Res. Commun.*, **259**, 645–650.
37. Alcaraz, J., Buscemi, L., Grabulosa, M., Trepat, X., Fabry, B., Farre, R., and Navajas, D. (2003) Microrheology of human lung epithelial cells measured by atomic force microscopy. *Biophys. J.*, **84**, 2071–2079.
38. Janovjak, H., Struckmeier, J., and Muller, D.J. (2005) Hydrodynamic effects in fast AFM single-molecule force measurements. *Eur. Biophys. J.*, **34**, 91–96.
39. Franz, C.M., Taubenberger, A., Puech, P.H., and Muller, D.J. (2007) Studying integrin-mediated cell adhesion at the single-molecule level using AFM force spectroscopy. *Sci. STKE*, **2007**, pl5.
40. Evans, E. and Ritchie, K. (1997) Dynamic strength of molecular adhesion bonds. *Biophys. J.*, **72**, 1541–1555.
41. Evans, E.A. and Calderwood, D.A. (2007) Forces and bond dynamics in cell adhesion. *Science*, **316**, 1148–1153.
42. Sun, M., Graham, J.S., Hegedus, B., Marga, F., Zhang, Y., Forgacs, G., and Grandbois, M. (2005a) Multiple membrane tethers probed by atomic force microscopy. *Biophys. J.*, **89**, 4320–4329.

43. Hochmuth, F.M., Shao, J.Y., Dai, J., and Sheetz, M.P. (1996) Deformation and flow of membrane into tethers extracted from neuronal growth cones. *Biophys. J.*, **70**, 358–369.
44. Krieg, M., Helenius, J., Heisenberg, C.P., and Muller, D.J. (2008b) A bond for a lifetime: employing membrane nanotubes from living cells to determine receptor-ligand kinetics. *Angew. Chem. Int. Ed. Engl.*, **47**, 9775–9777.
45. Marcus, W.D., McEver, R.P., and Zhu, C. (2004) Forces required to initiate membrane tether extrusion from cell surface depend on cell type but not on the surface molecule. *Mech. Chem. Biosyst.*, **1**, 245–251.
46. Jokinen, J., Dadu, E., Nykvist, P., Kapyla, J., White, D.J., Ivaska, J., Vehvilainen, P., Reunanen, H., Larjava, H., Hakkinen, L., and Heino, J. (2004) Integrin-mediated cell adhesion to type I collagen fibrils. *J. Biol. Chem.*, **279**, 31956–31963.
47. Chesla, S.E., Selvaraj, P., and Zhu, C. (1998) Measuring two-dimensional receptor-ligand binding kinetics by micropipette. *Biophys. J.*, **75**, 1553–1572.
48. Tees, D.F., Waugh, R.E., and Hammer, D.A. (2001) A microcantilever device to assess the effect of force on the lifetime of selectin-carbohydrate bonds. *Biophys. J.*, **80**, 668–682.
49. Evans, E. (1999) Energy landscapes of biomolecular adhesion and receptor anchoring at interfaces explored with dynamic force spectroscopy. *Faraday Discuss.*, **111**, 1–16.
50. Evans, E. (2001) Probing the relation between force – lifetime – and chemistry in single molecular bonds. *Annu. Rev. Biophys. Biomol. Struct.*, **30**, 105–128.
51. Bell, G.I. (1978) Models for the specific adhesion of cells to cells. *Science*, **200**, 618–627.
52. Merkel, R., Nassoy, P., Leung, A., Ritchie, K., and Evans, E. (1999) Energy landscapes of receptor-ligand bonds explored with dynamic force spectroscopy. *Nature*, **397**, 50–53.
53. Chang, M.I., Panorchan, P., Dobrowsky, T.M., Tseng, Y., and Wirtz, D. (2005) Single-molecule analysis of human immunodeficiency virus type 1 gp120-receptor interactions in living cells. *J. Virol.*, **79**, 14748–14755.
54. Konstantopoulos, K., Hanley, W.D., and Wirtz, D. (2003) Receptor-ligand binding: 'catch' bonds finally caught. *Curr. Biol.*, **13**, R611–R613.
55. Marshall, B.T., Long, M., Piper, J.W., Yago, T., McEver, R.P., and Zhu, C. (2003) Direct observation of catch bonds involving cell-adhesion molecules. *Nature*, **423**, 190–193.
56. Barsegov, V. and Thirumalai, D. (2005) Dynamics of unbinding of cell adhesion molecules: transition from catch to slip bonds. *Proc. Natl. Acad. Sci. U.S.A.*, **102**, 1835–1839.
57. Evans, E., Leung, A., Heinrich, V., and Zhu, C. (2004) Mechanical switching and coupling between two dissociation pathways in a P-selectin adhesion bond. *Proc. Natl. Acad. Sci. U.S.A.*, **101**, 11281–11286.
58. Thomas, W. (2008) Catch bonds in adhesion. *Annu. Rev. Biomed. Eng.*, **10**, 39–57.
59. Nguyen-Duong, M., Koch, K.W., and Merkel, R. (2003) Surface anchoring reduces the lifetime of single specific bonds. *Europhys. Lett.*, **61**, 845.
60. du Roure, O., Buguin, A., Feracci, H., and Silberzan, P. (2006) Homophilic interactions between cadherin fragments at the single molecule level: an AFM study. *Langmuir*, **22**, 4680–4684.
61. Zhang, X., Wojcikiewicz, E.P., and Moy, V.T. (2006) Dynamic adhesion of T lymphocytes to endothelial cells revealed by atomic force microscopy. *Exp. Biol. Med. (Maywood)*, **231**, 1306–1312.
62. Alon, R., Feigelson, S.W., Manevich, E., Rose, D.M., Schmitz, J., Overby, D.R., Winter, E., Grabovsky, V., Shinder, V., Matthews, B.D. *et al.* (2005) Alpha4beta1-dependent adhesion strengthening under mechanical strain is regulated by paxillin association with the alpha4-cytoplasmic domain. *J. Cell Biol.*, **171**, 1073–1084.
63. Krieg, M., Arboleda-Estudillo, Y., Puech, P.H., Kafer, J., Graner, F., Muller, D.J., and Heisenberg, C.P. (2008a) Tensile forces govern germ-layer organization in zebrafish. *Nat. Cell Biol.*, **10**, 429–436.

64. Fierro, F.A., Taubenberger, A., Puech, P.H., Ehninger, G., Bornhauser, M., Muller, D.J., and Illmer, T. (2008) BCR/ABL expression of myeloid progenitors increases beta1-integrin mediated adhesion to stromal cells. *J. Mol. Biol.*, **377**, 1082–1093.

65. Franz, C. and Puech, P.-H. (2008) *Cellular and Molecular Bioengineering*, Springer, New York, pp. 289–300.

13
Nanosurgical Manipulation of Living Cells with the AFM

Atsushi Ikai, Rehana Afrin, Takahiro Watanabe-Nakayama, and Shin-ichi Machida

13.1
Introduction: Mechanical Manipulation of Living Cells

Atomic force microscopy is now widely used to measure forces to disrupt covalent or noncovalent bonds in biological and nonbiological systems, including separation of the double helical structure of DNA [1], forced stretching of helical polypeptides [2, 3], unfolding of globular proteins [4–6], unbinding of antigen–antibody [7, 8] and other ligand–receptor complexes [9–13], and recognition of membrane proteins by force [14], or in a more general term, artificial manipulations at the molecular and cellular levels [15–17].

Among contemporary uses of the atomic force microscope (AFM), the direct manipulation of biological materials at the molecular and cellular levels utilizes one of the most unique features of AFM, that is, its capability of directly contacting atomic/molecular-sized specimens with a probe of corresponding dimensions [18, 19]. The size of the AFM probe at its very tip is most fitted to touch, push, and pull biological macromolecules for probing their physical, that is, mechanical properties at the single-molecule level [20, 21]. This unique feature of the AFM has been exploited in combination with the use of chemically or physically modified probes for the purpose of innovative biological operations. In this chapter, we describe how we and other researchers have tried to manipulate biological specimens, especially living cells, using this feature of AFM. The manipulation modes introduced in this chapter are limited to what may be called *nanosurgery* or *nano-operation* of living cells. Examples are still not abundant and most of them described in this chapter are those reported from the authors' group on cultured fibroblast cells. Other types of molecular manipulation are found in Chapters 11 and 12 of this book.

13.2
Basic Mechanical Properties of Proteins and Cells

The basic knowledge of biological materials such as proteins, nucleic acids, carbohydrates, lipid membranes, and live cells is essentially required for their

Atomic Force Microscopy in Liquid: Biological Applications, First Edition.
Edited by Arturo M. Baró and Ronald G. Reifenberger.

mechanical manipulations. Using the pulling capability of the AFM on a single polymer molecule, it has been shown that protein molecules have locally different mechanical strength. It has also been revealed that the folded conformation of proteins can be unraveled by a tensile force of less than a few hundreds of piconewtons, often <100 pN [5, 6, 22]. Double stranded B-DNA (ds-DNA) was shown to be reversibly stretched into a new conformation called *S-DNA* by a tensile force of about 80 pN [23]. Interbase hydrogen bonding in ds-DNA was shown to be disrupted by a shear force of few tens of piconewtons [1]. In the following sections, several different examples of mechanical manipulations of biological materials, in particular, living cells are discussed.

13.3 Hole Formation on the Cell Membrane

If a hole can be created and resealed on the cell surface without compromising the viability of the cell, it could be used as a window for imaging intracellular structures, as a passage for transporting bioactive molecules into and out of the cell, and for many other nanotechnological operations. Two different methods have so far been tested to make a hole on the cell surface: first, by forcefully pushing an AFM tip through the membrane [17, 18, 24] and second, by locally dissolving phospholipid molecules by applying a glass bead that was coated with phospholipase A_2 (PLA_2) on the cell membrane [24]. Here, we describe the second method of hole creation, while examples of the first method are given in Sections 13.4 and 13.5.

Afrin *et al.* [24] modified a glass bead with PLA_2 through covalent cross-linkers and attached it to an AFM cantilever. The modified bead was positioned above a target live cell under an optical microscope equipped with an AFM. The cells were labeled with the lipophilic fluorescent dye, 3,3′-dioctadecyloxacarbocyanine perchlorate (DiO), for fluorescence visualization of the cell membrane. The bead was brought into contact with the cell surface and a further force of 5–10 nN was applied for 1–5 min. The process of hole creation was monitored with an optical microscope equipped with a charge coupled device (CCD) video camera and AFM.

In Figure 13.1a–c, phase contrast micrographs of the target cell, respectively, before, during, and after the contact with a phospholipase A_2 (PLA_2)-modified bead are given.

A circular structure having a darker rim was created in the same area where the bead was pressed. The fluorescence image in Figure 13.1d clearly showed a discolored region corresponding to the removal of the cell membrane on the apical side of the cell due to the action of PLA_2. The authors commented that adhesion of the apical cell membrane to the bead followed by its partial peeling may be an additional factor in hole creation.

When the AFM image of the hole area of the cell was taken after fixing of the cell, as given in Figure 13.1e, the color scale in height image and a cross section along the dotted line demonstrated the presence of a hole with a depth of about 150 nm and a width of 5–7 μm (Figure 13.1e,f). A few filamentous bundles, probably

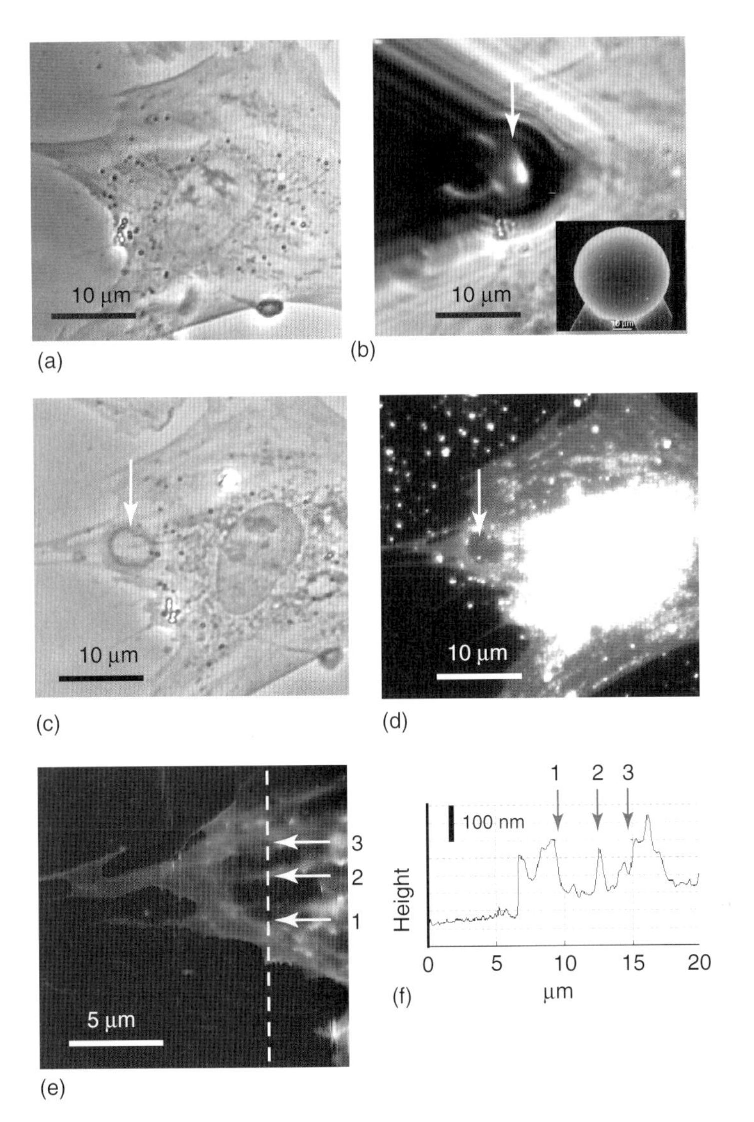

Figure 13.1 An example of hole creation on the cell surface using a PLA_2-modified poly bead as observed by phase contrast microscopy. (a) Before, (b) during, and (c) after application of the bead at the position indicated by a white arrow in b, c and d. In (c), the location of a newly created hole surrounded by a dark rim is also indicated by an arrow. (d) Fluorescence imaging of the same cell indicating the less fluorescent area corresponding to the hole. (e) Imaging of the hole area with the intermittent contact mode AFM after light fixation with glutaraldehyde, which presents the hole area with a darker hue. (f) A height cross section along the dotted line in the image in (e) showing an average depth of 150 nm. The positions of three arrows in (f) correspond to those in (e). High profiles in the hole presumably correspond to the actin fibers. (Reproduced from Ref. [24], with permission.)

actin fibers, were also imaged as a continuous structure from the undigested area. Similar attempts with beads modified with serum albumin did not create holes.

The cell membrane can be removed not only with lipid-digesting enzymes but also with the nonionic detergent, Triton X-100, presumably with little damages to the intracellular structures such as the cytoskeleton. The Triton-treated cells are often called *semi-intact cells* because parts of their structures retain the capacity to respond to the addition of ATP. For example, their stress fibers (SFs) regain the prestressed state that is once destroyed by the addition of the detergent. A good example of using semi-intact cells for imaging intracellular structures under semi-native conditions was carried out by Berdyyeva *et al.* [25] to directly visualize cytoskeletal fibers in human foreskin epithelial cells. Subcellular structures and molecules not anchored to the SFs were washed away from the cell. The authors especially emphasized the measurement of the volume of SFs.

13.4 Extraction of mRNA from Living Cells

Applying the first method of hole creation on the cell surface as mentioned above, Uehara *et al.* [26–29] demonstrated a successful harvest of intracellular mRNA localized to specific regions in the cytoplasm. The commonly used fluorescence *in situ* hybridization (FISH) method is a powerful and useful method for detecting the localization of mRNAs and proteins, but it does not allow time-lapse analysis of mRNA expression and/or localization in a single living cell because cells have to be fixed for fluorescence labeling.

Uehara *et al.* inserted an AFM tip into a living cell to collect mRNAs after their physical adsorption to the tip and subjected the adsorbed mRNA to RT-PCR by placing the tip in a test tube containing RT-PCR solution kit. The process was followed by nested PCR and quantitative PCR methods to amplify specific mRNAs. By applying this method, they successfully performed a quantitative measurement of β-actin mRNA at different loci within individual living cells.

Since β-actin mRNA production is known as the *primary gene response* induced by the addition of fetal bovine serum (FBS) [30], Uehara *et al.* examined β-actin mRNA distribution in individual cells under two different conditions, A (inactive in the absence of calf serum) and B (activated in the presence of calf serum). The results obtained for two cells under each condition are shown in Figure 13.2.

In the vicinity of the nucleus (distance from nuclear membrane: 0–18 μm), no clear difference in the number of positive results between conditions A and B was found. On the other hand, in the peripheral region far from the nucleus (distance from nuclear membrane: more than 18 μm), a clear difference in the number of positive results between conditions A and B was observed. These results agreed with previous experiments of *in situ* hybridization that indicated β-actin mRNA was localized to the leading edge of the cell after FBS stimulus.

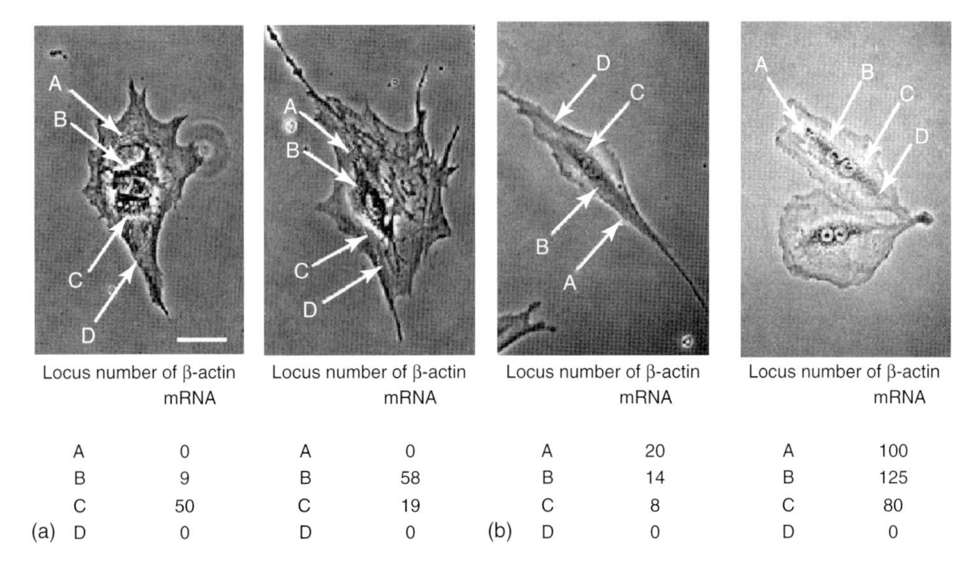

Figure 13.2 Single cell analysis of mRNA. (a) When cells were inactive in serum-depleted medium, β-actin mRNAs were detected predominantly near the nucleus at the positions B and C. (b) When cells were actively moving in serum-supplemented medium, β-actin mRNAs were detected not only near the nucleus but also at A, that is, far from the nucleus and in the direction of cell movement. Scale bar, 50 μm. (Reproduced from Ref. [26], with permission.)

Uehara *et al.* [27] compared their results of the single cellular mRNA analysis based on AFM with those of the conventional FISH method and found a good correlation between them.

13.5
DNA Delivery and Gene Expression

An AFM cantilever can be used to insert plasmid DNA into a living cell [16, 17, 24, 30–32]. Nakamura and his colleagues [16, 17] used a specially fabricated cantilever for easy insertion into living cells and demonstrated insertion and expression of plasmid DNA coding for fluorescent proteins. The AFM probe called *nanoneedle* was fabricated by etching AFM tips with a focused ion beam (FIB) into a sharpened needle between 200 and 800 nm in diameter. They evaluated the proper diameter of a needle required for insertion into human cells without causing cell death and achieved a highly efficient gene expression method for human cells using a nanoneedle mounted on an AFM. The tip of the nanoneedle is shown in Figure 13.3a.

A fluorescein-isothiocyanate (FITC)-modified needle was inserted into a breast-cancer-derived MCF-7 cell, and the decrease in force was monitored. The positions of the vertical-sectional image (Figure 13.3c) and cross-sectional image (Figure 13.3d) are indicated by white lines in the stack of confocal images in

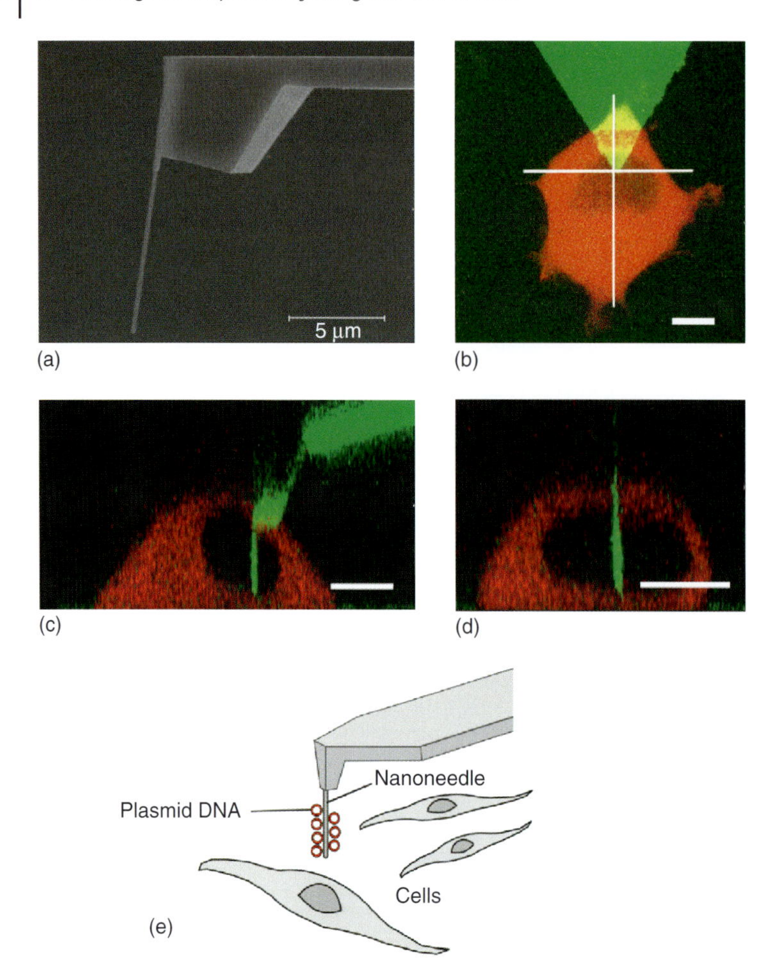

Figure 13.3 DNA insertion into a living cell using a nanoneedle. (a) An SEM image of the nanoneedle. (b) Stack of confocal slices of MCF-7 cells expressing DsRed2-nuclear exporting signal (NES) excited at 543 nm and a fluorescein isothiocyanate (FITC)-labeled nanoneedle excited at 488 nm when the nanoneedle was inserted. (c) Cross-sectional image of green and red emissions processed from the confocal slices. (d) Vertical-sectional image of green and red emissions processed from the confocal slices. (e) Schematic representation of a single cell manipulation using AFM and a nanoneedle. Scale bars for all images indicate 5 μm. (Reproduced from Ref. [31], with permission.)

Figure 13.3b. The vertical- and cross-sectional images indicated that the nanoneedle was accurately inserted into the nucleus. In addition, the shape of the plasma membrane and nucleus was undistorted. The schematic diagram in Figure 13.3e presents their single cell manipulation technology and cell surgery system using a nanoneedle. By applying the nanoneedle technology of their own development, Nakamura and his colleagues succeeded in accurate insertion of plasmid DNA into living cells, with minimum invasiveness.

Cuerrier *et al.* [32] and Afrin *et al.* [24] used conventional AFM tips to insert plasmid DNA into living cells and demonstrated its expression in similar manners. The method and the result from Afrin *et al.* are discussed below.

A AFM cantilever modified with plasmid DNA was positioned onto the cell surface, and insertion into the cell was performed under a high loading force by stepper motor commands. After 1 min of the insertion, the AFM cantilever was pulled out from the cell. After the insertion procedure, the cells were rinsed three times with a serum-free Dulbecco's modified Eagle medium (DMEM)/F12 medium and then incubated at 37 °C for 24 h in 5% CO_2. DNA release from an AFM tip was confirmed by comparing the fluorescence intensity of the DNA stained with 4′,6-diamidino-2-phenylindole (DAPI, a fluorescent stain that binds strongly to DNA) on the tip before and after insertion into the cell. A separate reference experiment was also performed by first spreading DNA solution into a petri dish where cells were cultured and then a bare AFM tip was used to puncture holes on the cells.

Following overnight incubation, successful DNA transfections were observed in the cells that were manipulated by a DNA-coated AFM tip for insertion but not in cells having holes created with a bare tip in a DNA-containing solution.

A single DNA-modified cantilever could be used to transform multiple, up to six, cells, as presented in Figure 13.4 (A–G) and (A'–G') where eight cells were sequentially punctured with the same tip modified with DNA at the beginning, resulting in six consecutive cells expressing green fluorescent protein (GFP) but not in the seventh and eighth cells.

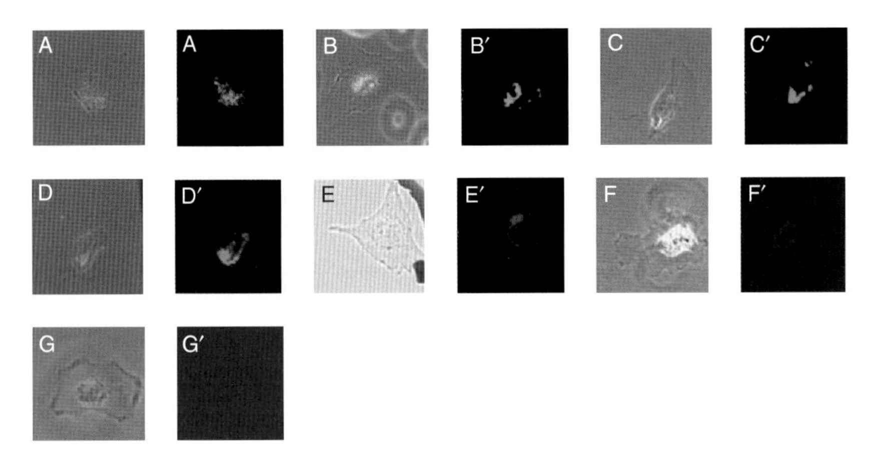

Figure 13.4 Observation of successive DNA transfection (A–G: phase contrast images and A′–G′: fluorescence images) using a single gold-coated AFM tip. Green fluorescence image signifies expression of GFP. (Reproduced from Ref. [24], with permission.)

13.6
Mechanical Manipulation of Intracellular Stress Fibers

13.6.1
AFM Used as a Lateral Force Microscope

In recent years, the mechanical properties of cytoskeletal structures, in particular, those of actin stress fibers (SFs), have been under vigorous scrutiny. SFs are composed of tens and hundreds of actin filaments bundled in a side-by-side fashion and, at the same time, transversely cross-linked non-covalent interactions with mainly α-actinin through and other minor proteins. Association with non-muscle-myosin II filaments renders a motile function to SFs so that they acquire a contractile capacity driven by the hydrolysis of ATP [33]. According to the intracellular location and the mode of attachment to the cell membrane, SFs are classified into ventral, transverse, and dorsal fibers. Owing to the contractile property in the presence of Mg^{++}-ATP, those SFs with their both ends anchored to a solid substrate through the transmembrane structure called the *focal adhesion* are considered to be under a prestressed and therefore prestrained condition. Prestressed SFs are known to change their size, spanning direction, and chemical compositions according to the external chemical and mechanical signals [34]. The prestressed condition is considered to be appropriate to sense and react to such stimuli with high precision. It is, therefore, of special interest to understand the precise behavior of the SFs against externally and internally mediated stimuli from a biochemical as well as mechanical point of view. In this chapter, we are primarily concerned with their response to mechanical stimuli and therefore in their mechanical properties *in vivo* and *in vitro*. To understand their fundamental mechanical property and the response mechanism, work that covers the measurement of the whole cell to single fiber response to externally applied mechanical stimuli is required.

On the whole cell level, much work has been done on the rearrangement of SFs according to the unidirectional cyclic stretch of the cell body placed on various types of reversibly stretchable substrates [35]. The ventral SFs respond to the stretching stress in such a way that they rearrange their spanning direction approximately normal to the direction of stretch force application. When placed under a shear flow, SFs align themselves parallel to the flow direction. In both cases, SFs seem to rearrange themselves in less stressful directions.

To understand the physical basis of such "biological" behavior of the SFs, we need to have every piece of information on the physical nature of fibers themselves and their interactions with other intracellular proteins and structures. Precise physical characteristics of the SFs can be obtained from the application of well-established engineering methods in materials science to single SFs under controllable conditions. For example, Deguchi *et al.* [36] measured the mechanical properties of isolated SFs by using a glass rod as a force transducer. They determined the Young modulus and breakdown force using an isolated single SF. They obtained a clue on the degree of prestrain of SFs *in vivo* by comparing their length measured *in vivo* and *in vitro*.

Kumar *et al.* [34] employed the femtosecond laser ablation method to cut a hole in the middle of an SF or dissect it without disturbing other cellular structures such as the cell membrane. They also collected some suggestive observations for the prestressed condition of SFs and the presence of interfiber supporting interactions. Silberberg *et al.* [37] recently reported on induced moves of subcellular organelles by a gentle push of the cell with an AFM probe at distant sites and suggested the presence of matrix connectivity between cytoskeleton and other subcellular structures.

For a more direct examination of the SF network inside a living cell, Hakari *et al.* [38] applied a lateral force to individual SFs inside a cell and recorded the force versus displacement relations.

Although in most common uses of the AFM in its force mode, a vertical force, either tensile or compressive, is applied to the experimental specimen, the method of scanning an AFM probe laterally to obtain the frictional force between the probe and the sample surface has also been developed since the early days of AFM application [39–43]. Hakari *et al.* exploited this type of AFM application to probe the mechanics of SFs inside a living cell.

In their work, an AFM probe tip was first pushed into the cell and then force was laterally applied to a selected fiber, that is, the fiber was pushed sideways along the substrate surface to which the cell was adhered. Since the cell was transfected with β-actin gene fused with GFP, the displacement of the fiber was visualized on a fluorescence microscope with simultaneous recording of torsional deflection of the AFM cantilever by the quadrant photodetector of AFM.

In lateral force application and its measurement, one has to determine the torsional force constant of cantilever together with its torsional or lateral sensitivity.

To obtain the lateral sensitivity of a cantilever (ratio between the torsional deflection angle against the photodetector output), a method of trapping the probe tip in a small hole of an Anopore filter (inorganic aluminum oxide membrane filter) having a pore diameter of $\sim$0.1 μm was used. The AFM tip was laterally scanned over the filter, and the horizontal output from the quadrant photodetector was defined as the lateral sensitivity of the cantilever [41, 42].

The torsion angle of the cantilever was calculated from $\tan\varphi = \mathrm{d}L/h$, where $\mathrm{d}L$ and$h$ are, respectively, the lateral displacement and the height of the probe. $\mathrm{d}L$ was determined by multiplying the scanning speed of the cantilever (micrometer per second) and the capture time of the probe in the Anopore hole.

Conversion of torsional deflection, θ, to the lateral force, F_{L}, was done by using Sader's equation [44]. First, the torque force, T, at the tip of the probe when the cantilever is tilted by an angle of θ is

$$T = k_{\varphi}\theta \tag{13.1}$$

where k_{ϕ} is the lateral force constant and given by Green *et al.* [44] as follows:

$$k_{\varphi} = k_{z}\frac{2L^{2}}{3\,(1+\nu)} \tag{13.2}$$

where k_{z}, L, and ν are, respectively, the vertical force constant, length of the cantilever, and the Poisson ratio of the cantilever material. Once the torque is obtained, the lateral force F_L is related to T as $T = F_L h$, where h is the height of the probe tip. The equation is valid for cantilevers made from isotropic material and with a high aspect ratio. Since noncrystalline silicon nitride is considered to be isotropic and the aspect ratio of our cantilever is 10, Eq. (13.2) was used for the estimation of lateral force constant.

13.6.2 Force Curves and Fluorescence Images under Lateral Force Application

Application of lateral force to the ventral SFs residing at the base of a cultured cell was performed on the principle stated above [38]. First, an AFM probe was vertically inserted into a live cell in a serum-free medium by forcefully breaking the cell membrane with the tip of the probe. The probe was set beside a fluorescently visualized SF with GFP and then scanned laterally in an approximately normal direction to the long axis of the SF. In the following section, two types of fiber response to the lateral force are explained.

13.6.2.1 Case 1

In this case, the AFM probe tip was scanned over more than six SFs and along its way cut at least four fibers with a force ranging from 200 to 600 nN. Since the break force in this case should depend not only on the stiffness of the fibers but also on the shape of the probe, it is not mechanical constant. What seemed more interesting was the shape of SFs under lateral force and the clear evidence of attachment of SFs to membranous structures, which were also dimly fluorescent because of the GFP-tagged β-actin molecules. In Figure 13.5, the nonlinear displacement of one of the manipulated SFs (fiber 4) and the image of the supporting structure between SFs are shown as reproduced from [38].

When several SFs were laterally displaced and cut with an AFM tip, discolored areas around the SFs with the loss of GFP-β-actin fluorescence appeared. This observation led Hakari *et al.* to the conclusion that these SFs were associated with the ventral cell membrane so that when severed and recoiled, they pulled away together with a part of the cell membrane, creating a large dark hole in the cell membrane.

13.6.2.2 Case 2

In Case 2 in Figure 13.6a, the displacements of the two major SFs, fibers 2 and 3 were found to be nonlinear again, that is, the magnitude of displacement from the original state is not proportional to the contour distance from the loading point. The displacement feature of the SFs was reproduced in the manual tracing of fibers in Figure 13.6b (dotted lines, before application of force; solid lines, at the height of maximum displacement of fiber 3 before it was cut).

The observed nonlinear displacement of the fiber 3 was fitted to the theoretical curve for a thin plate under in-plane point loading [45]. The difference between the solid and dotted lines in Figure 13.6b was obtained and plotted against the distance from the point of force loading in Figure 13.6c. The experimentally obtained curve

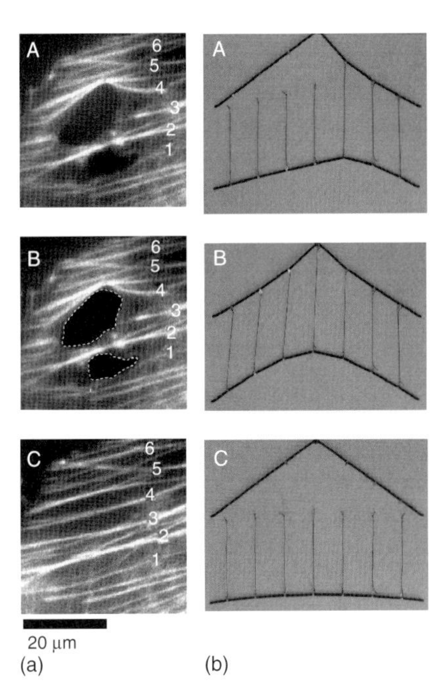

Figure 13.5 Displacement of SFs in a live cell under the application of lateral force by the AFM probe. (a) In A, fiber 4 is under maximum displacement due to the AFM probe stationed at the inflection site approximately in the middle of the fiber. B: the dark hole created by the rupture of dimly fluorescent membranous sheet by the scanning probe is emphasized. C: the same area before the application of the lateral force. (Reproduced from Ref. [38], with permission.) (b) A model explaining the displacement of the upper thick fiber. A: the upper thick horizontal string is pulled upward with support (vertical thin strings) from the lower thick fiber on the right half but without support on its left half. B: the upper string is pulled in full support from the lower one. C: no supports between the two horizontal strings.

was then compared with theoretical curves calculated according to the thin plate theory that is, strictly speaking, applicable for a small displacement of an isotropic and elastic thin plate.

The theoretical displacement of the 2D plate as given in [46] is briefly reproduced here. Using a polar coordinate, the radial and angular displacements at position (r, θ) of an infinitely large 2D plate when a concentrated in-plane load is applied at $(0, 0)$ in the direction of x-axis are given as follows:

$$u_r = \frac{(1+\nu)P_x}{4\pi E}[(1+\nu) - (3-\nu)\ln r]\cos\theta \tag{13.3}$$

$$u_\theta = \frac{(1+\nu)P_x}{4\pi E}[(3-\nu)\ln r]\sin\theta \tag{13.4}$$

where $u_r, u_\theta, P_x, E,$ andν are, radial and angular displacements, the magnitude of applied load in x-direction per unit thickness of the plate, the Young modulus and Poisson ratio of the plate material, respectively.

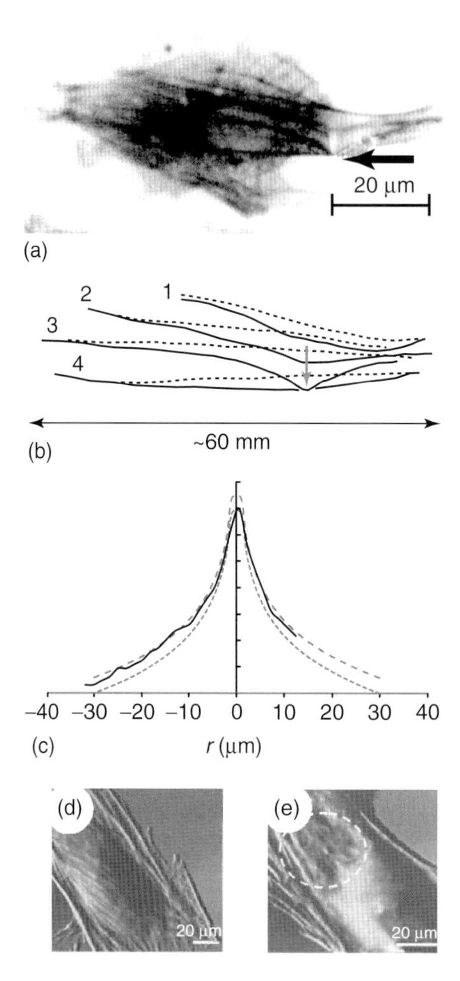

Figure 13.6 Analysis of nonlinear displacement of SFs in a living cell according to the result of thin plate theory. (a) Fluorescent image of displaced SFs. The AFM probe is positioned at the arrowed site. (b) Traces of four major SFs before the application of force (dotted line) and at the maximum displacement (solid line) of fiber 3. The arrow indicates the position of force loading. (c) The absolute value of the maximum displacement of fiber 3 in (b) is plotted against the distance, r, from the point of force loading (solid line). Two other lines are theoretical curves based on Eq. (13.5), with $r_0 =$ 30 μm (dotted line) and $r_0 = 40$ μm (dashed line). The theoretical curves should diverge at $r = 0$, but here, they are rounded. The ordinate is of arbitrary value proportional to the displacement of fiber 3. (d) AFM tapping-mode image of a live cell and (e) a similar image over a hole of approximate diameter of 5 μm as produced by the application of phospholipase-A_2-modified AFM probe. Filamentous structures laterally connecting wide SFs are observed in the hole area as circled in (e). (Reproduced from Ref. [38] for a–c, and from Ref. [24] for d and e, with permission.)

Since for fiber 3, θ is equal to $\pi/2$ and $3\pi/2$ for regions $r > 0$ and $r < 0$, respectively, the magnitude of u_x is equal to u_θ with $|\sin\theta| = 1$. By setting r_0 as an arbitrary chosen reference radius as explained for the Flamant problem in [45], the relative magnitude of $u_x(r)$ for a fiber initially sitting on the r_0-axis ($x = 0$) is obtained as,

$$u_x = \frac{(1+\nu)\,P_x}{4\pi E}\,(3-\nu)\ln\left(\frac{r_0}{r}\right) \tag{13.5}$$

For $r < r_0$, $u_x(r)$ is a positive and logarithmically decreasing function of r.

As shown in Figure 13.6c, the theoretical curve with $r_0 = 40\,\mu\text{m}$ was found to be a good fit to the experimental result. A trial calculation of the Young modulus of the sample based on the prelogarithmic factor in Eq. (13.5) assuming $\nu = 0.5$, $P_x \sim 1\,\mu\text{N}/\mu\text{m}$, $u_x \sim 1\,\mu\text{m}$ gave E in the order of 10^5 Pa, which is not an unreasonable value for explaining the cellular stiffness [47].

The nonlinear displacement of some of the SFs suggested the presence of matrix material that links the movement of neighboring SFs in the cell. Such interconnecting structures between SFs may correspond to the filamentous materials visualized inside the cell (Figure 13.6d,e) by the AFM probe inserted into the cell through a hole created, as discussed above and in [24].

Another possibility for the nonlinear displacement of an SF exists when it is progressively hardened toward its two ends and both ends are rigidly supported. This possibility should not be ignored before its disproval.

13.7 Cellular Adaptation to Local Stresses

Mechanosensory systems in the cell and associated cytoskeletal proteins are one of the main targets in biophysics since they have a crucial role in energy conversion and signal transduction [48]. To investigate the biophysical properties of these proteins and the mechanics of the cell, it is required to quantitatively manipulate single molecules or single cells with pico- to micronewton order and to detect the precise length of their extension or motion in response to the applied force. The most commonly used techniques are optical tweezers, magnetic tweezers, and AFM [49].

The capacity of conventional optical and electromagnetic tweezers covers the force range from 0.1 to 100 pN and from 0.01 pN to 10 nN, respectively [49]. These two methods are also widely used in single cell observations in combination with fluorescence microscopy to display how and where mechanical stress is transmitted and spread in the cell [50, 51], as well as to analyze the cell's mechanical property. Cell culturing on a silicone rubber substrate is also used in molecular cell biological studies to apply a stretching force to cells for the purpose of causing a large deformation of the cell [52, 53].

AFM allows the measurement of mechanical properties of various materials from single molecules to single cells in a larger force range as well as the ability to capture sample topography [19]. Watanabe-Nakayama *et al.* [54] studied the time-dependent process of physical and biological responses of the cell against localized application of cyclic pressing and pulling force under a spherical bead attached to the AFM cantilever. The bead was coated with the matrix protein, fibronectin, to facilitate the binding of the bead to the cell surface through integrin-dependent focal adhesion formation. They pulled the glass bead after waiting for the formation of a tight connection between the bead and the cell surface and then found an initial relaxation of the force followed by a slow increase in the tensile force because of the reinforcement of the newly formed connection between the bead and the cellular cytoskeleton. Since the probe was fixed on the cantilever, however, the throughput of measurements was low. To overcome this problem, Watanabe-Nakayama *et al.* [55] introduced a new cantilever (created by FIB technology) with a semicircular cut at the edge that gave a scooping function to the cantilever (Figure 13.7).

Using cantilevers with a scooping function, they analyzed the extracellular matrix (ECM) adhesion force of a cell and the cell extension when the probe was detached from the cell surface. The experimental setup is depicted in Figure 13.7a.

In rat fibroblasts spread on a human-fibronectin-coated glass surface, β-actin, α_V-Integrin, and paxillin were marked with different fluorescent proteins. α_V-Integrin is an ECM receptor protein [56], and paxillin is a scaffold protein in the focal adhesion, having binding sites with β-Integrin, focal adhesion kinase, and vinculin [57]. Therefore, paxillin has been thought to relate to a key factor in the mechanosensory system [58].

Carboxylated functionalized polystyrene microspheres were covalently modified with human recombinant fibronectin type III domain 7–10 (hrFNIII7–10) and allowed to adhere on the cell surface in advance before the manipulation experiment. Cells were then plated on a petri dish. As shown in Figure 13.7b, actin SFs and focal adhesion proteins were accumulated around the bead. The bead was then scooped up with the newly fabricated cantilever and completely separated from the cell. The images of a typical separated bead with some of the fluorescently visible focal adhesion proteins and the derived force–distance curves are shown in Figure 13.7c,d, respectively.

The images revealed that fluorescently labeled focal adhesion proteins and actin cytoskeleton were successfully recovered on the bead surface. This is the first clear demonstration of recovering focal adhesion proteins, using an AFM cantilever.

13.8 Application of Carbon Nanotube Needles

Carbon nanotubes are an attractive choice for the manipulation of individual cells and possibly subcellular structures because of their small diameter, extremely high aspect ratio, robust mechanical property, and the relative easiness of *in situ* growth on various microscopic probes. Other types of carbon-based probes aimed at nanomanipulation at the cellular and molecular levels have also been developed as introduced below.

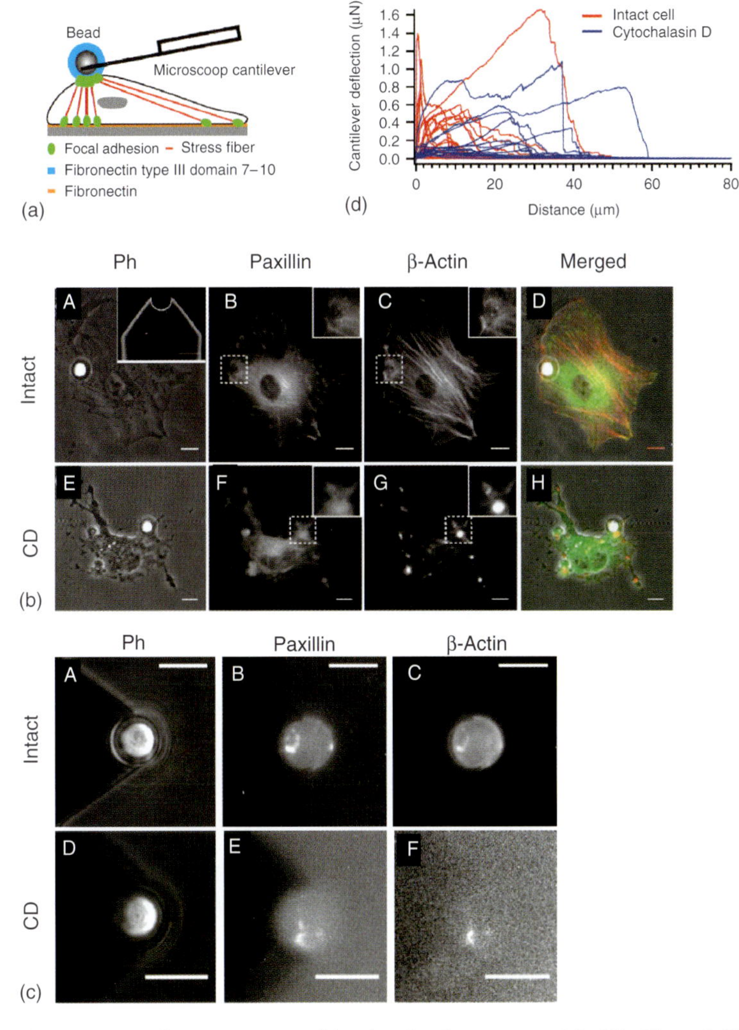

Figure 13.7 Picking-up process of beads adhered to fibroblast cells. (a) Schematic view of manipulation of a hrFNIII7–10-coated bead on a cell. (b) The micrographs of rat fibroblasts expressing AG1-paxillin and KO1-β-actin (E–H) with or (A–D) without cytochalasin D (CD) Ph here and in c means phase contrast microscopic images and the inset in A is an electron microscopic image of a scoop tip. The cells had the adhered hrFNIII7–10-coated beads. (A, E), (B, F), and (C, G) are phase contrast, AG1-paxillin fluorescence, and KO1-β-actin fluorescence images, respectively. (D, H): merged images from (A–C) and (E–G), respectively. The insets in (B), (C), (F), and (G) are magnified images around the bead shown as a dashed box in each image. (c) The micrographs of the beads after being picked up from the cells expressing AG1-paxillin and KO1-β-actin (D–F) with or (A–C) without CD. (A, D), (B, E), and (C, F) are phase contrast, AG1-paxillin fluorescence, and KO1-β-actin fluorescence images, respectively. Bar represents 10 μm. (d) Force versus bead–sample separation curves derived from bead-picking-up experiments for intact cells (red) and CD-treated cells (blue). (Reproduced from Ref. [55], with permission.)

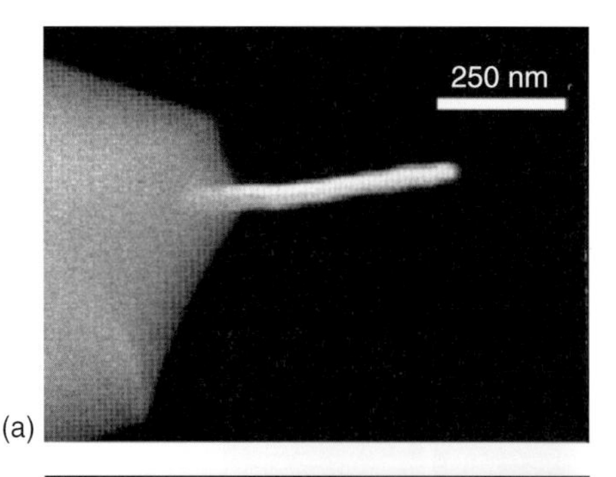

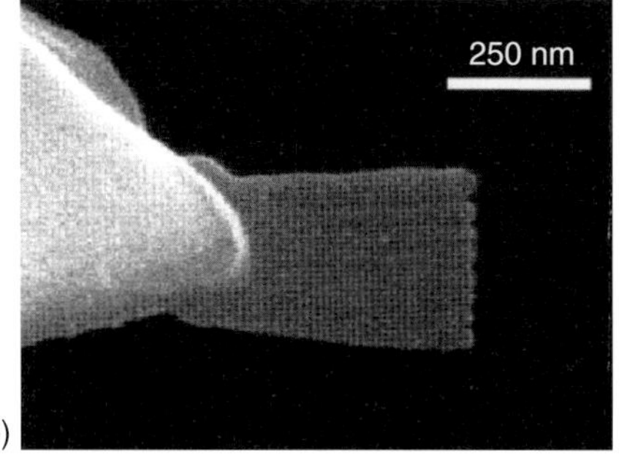

Figure 13.8 (a) Top view (as grown) and (b) side views of a nanoscalpel grown on the apex of an AFM tip. The scalpel blade was grown on the edge of a silicon AFM cantilever. In the growth process using EBID, carbon was deposited when hydrocarbon contaminants present in the SEM chamber vacuum were decomposed by secondary electrons (SEs) produced under electron beam irradiation. (Reproduced from Ref. [60], with permission.)

Singhal *et al.* [59] connected a glass capillary and carbon nanotube by a flow technique and sealed the gap between them with either conducting or nonconducting epoxy glue. By inserting the newly formed apparatus called as *cellular endoscope,* they explored the intracellular space injecting fluorescent dyes and particle.

Beard *et al.* describes fabrication of a tool they named *nanoscalpel* that has the shape of a flat blade or spatula as grown on the apex of a standard AFM probe using an electron-beam-induced deposition (EBID) process (Figure 13.8a,b).

The blade was used to cut lithographic lines on relatively hard inorganic and soft biological samples [60]. Vakarelski *et al.* [61] studied the efficiency of penetrating the cell membrane with the fortified carbon nanotube tips.

13.9
Use of Fabricated AFM Probes with a Hooking Function

Another type of AFM cantilever having a hooking function was introduced by Machida *et al.* [62] to facilitate direct manipulation of fibrous materials of micrometer sizes. Advanced TEC™ series cantilevers were chosen and processed

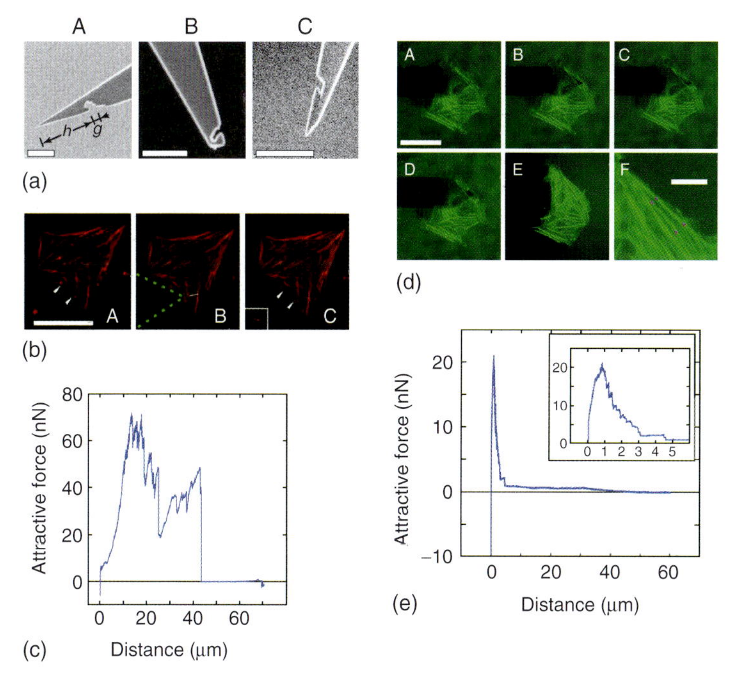

Figure 13.9 Scanning ion microscopy (SIM) images of the fabricated nanohooks on AFM cantilevers. (a) The gap size (*g*) and bait point height (*h*) are (A) 0.6 and 2.2 μm, (B) 0.09 and 0.3 μm, and (C) 0.1 and 0.8 μm, respectively. Each scale bar indicates 1 μm. (b) Fluorescence microscopy images of a semi-intact cell expressing red fluorescent protein (RFP) on actin (scale bar = 50 μm). (A) The initial situation of the sample. (B) A single SF was captured by a modified AFM probe hook. The green dotted line and the white arrow indicate, respectively, the cantilever position and the holding point. (C) The SF was severed and the captured SF fragment is shown in the inset. The bar in (A) represents 50 μm for (A–C). (c) Force curve obtained while pulling the SF in this procedure. (d) Fluorescence microscopy images of a living cell expressing green fluorescent protein (GFP) on actin (scale bar = 50 μm). (A) The initial situation of the sample. (B) Two SFs were snagged using a modified AFM probe hook. (C) Two SFs were severed at the hold point. (D) The severed SFs were instantly shrunken. (E) The cell was cured in 2 h after this operation. (F) The magnified image of the repaired part in (E) (scale bar = 10 μm). The recovered SFs are indicated by solid magenta triangles. (e) Force curve obtained while pulling the SFs using the probe hook in this procedure. The inset shows an enlarged view of the attractive force peak. (Reproduced from Ref. [62], with permission.)

under FIB because their shapes were suited for the manipulation of samples after implementation of a hooking function. The tips of the AFM cantilevers were processed into shapes with a potential hooking function, as shown in Figure 13.9a (A–C).

Machida *et al.* used these cantilevers to pick up individual SFs from inside semi-intact (membrane disrupted with Triton X-100) and intact cells as discussed below.

13.9.1 Result for a Semi-Intact Cell

The SF in the bottom left-hand corner of the cell was captured using a hook-shaped probe and pulled upward (Figure 13.9b (A–C)). In this case, the fiber was severed at its two ends around focal adhesions and a fiber fragment was captured by the probe hook. The force curve given in Figure 13.9c recorded during this measurement had two large attractive force peaks at extension distances of around 15 and 40 μm. The first large peak corresponded to severing one of the two ends of the fiber fragment and the second large peak to severing of the other end. The fracture strength of the fiber was ~70 nN in the first case and ~50 nN in the second case. The two large peaks were composed of several subpeaks. It means that the fiber was cut gradually in several discreet steps [63]. The steps were considered to be derived from the snapping action of smaller bundles of the actin filaments forming the SF. In the first peak, we observed at least five subpeaks; in the second, at least three.

13.9.2 Result for a Living Cell

Figure 13.9d (A–D) shows the process of manipulating a living fibroblast cell, and Figure 13.9d (E, F) shows the once broken parts of the SF that were repaired. In Figure 13.9e is given the corresponding force measurement result. Here, two SFs were seen to be pulled in one action and both SFs were severed at the hooking point (Figure 13.9e) after an extension of ~1 μm. The SFs were observed to quickly recoil and shrink after being severed, suggesting that the SFs were under a high tension (prestressed). In this case, the probe hook did not capture the fragmented fibers. There was one sharp peak in the force curve followed by a gradual decrease to the zero force level. The steepness of the curve in the initial part confirmed that the fibers were in prestressed state as mentioned above and as also suggested by Ingber [64]. Comparing with the results of previous studies reporting 377 ± 214 nN for the fracture strength of isolated SFs [36], Machida *et al.* gave smaller values of fracture strength. The fabricated hooks could have worked as cutting blades. Interestingly, the damaged part of the cell was reported to be regenerated in 2 h after the manipulation.

13.10 Membrane Protein Extraction

The intracellular space of mammalian red blood cells (RBCs) surrounded by the lipid bilayer membrane is largely filled with hemoglobin, the oxygen carrier, and other proteins [65, 66]. The volume of the cell is about 70% of the maximum volume of a sphere having the same surface area [66]. On the basis of such uniform and unstressed conditions, the cell with the well-known biconcave form can return to its original shape after deformation during its tortuous passages through the capillaries with diameters less than that of the cell itself [67]. This remarkable resilience is believed to from the mechanical properties of the composite cell membrane of a lipid bilayer and the spectrin-based cytoskeletal structure [68–70].

The highly convoluted cytoskeleton of RBCs made mainly of spectrin tetramers and short F-actins is believed to allow a large extent of stretching when freed of lipid bilayer [67], but the true mechanical properties of the spectrin network, however, has not yet been studied. Recently, Afrin *et al.* [70] reported tensile experiments of spectrin–actin network after exposing it to the solvent by delipidation treatment of RBCs with PLA_2.

Of the major intrinsic membrane proteins on the RBC surface, Band 3 is known to be linked to spectrin through the linker protein ankyrin [71]. Glycophorin A, another major RBC membrane protein, is known to be largely free of such association with the cytoskeletal structure [72], although partial association of glycophorin A with Band 3 has been proposed [73, 74]. Afrin *et al.* pulled RBC membrane proteins using AFM probes modified with lectins with specific binding capacities to the carbohydrate moieties of Band 3 or glycophorin A [75] and showed that the spectrin-based network had a slackness that allowed its easy extension up to a few micrometers with a force <50 pN. Force curves obtained on intact RBCs with an AFM probe coated with anti-F-actin antibody are given in Figure 13.10.

The force curves given in Figure 13.10a are characterized with a long extension of plateau force of $\sim$40 pN and extending up to 2–3 μm, which was followed by a gradual increase in the tensile force of up to 60–70 pN before the final rupture of the linkage. The plateau force of about 40 pN was attributed to the extension of lipid tether from the cell surface, and the final increase of the force beyond the plateau level was assumed to result from the extension and breakdown of cytoskeletal

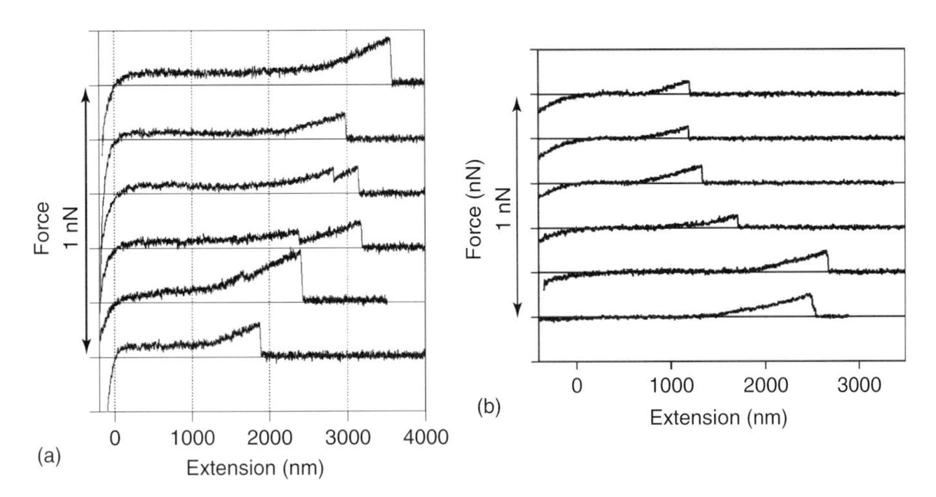

Figure 13.10 Force curves obtained on (a) intact and (b) delipidated RBCs using Con-A-modified AFM tip in the former and anti-F-actin-antibody-modified tips in the latter case. (Reproduced from Ref. [75], with permission.)

structure, which was extended together with the lipid tether up to the limit of the tensile strength of the noncovalent linkage extending from Con A on the probe to the cytoskeletal structure.

The extensibility of the cytoskeletal structure assumed above was proved in the results of pulling experiments of delipidated RBCs using AFM probes coated with the anti-F-actin antibody as given in Figure 13.10b where the results of pulling experiments obtained with an AFM probe modified with anti-F-actin antibody are given.

The force curves in Figure 13.10b are characterized with an almost flat, low force extension of up to 3 μm, which was terminated by a steep increase in the tensile force of up to a few tens of piconewtons. The magnitude of the final rupture force was 70 ± 30 pN, and the mean of the largest extension length was 2 μm. The magnitude of the force in the flat part of the force curve was <10 pN.

13.11 Future Prospects

At this moment, AFM is the most suited instrument for direct manipulation of biological materials and living cells. Manipulation on biological materials as large and complex as the living cell can be considered as cellular level nano surgery since it aims at not only the measurement of their physical properties but also changing their biological functions. Manipulation with AFM provides parallel information of the force level being applied to the cell and *in situ* images of operation when AFM is combined with optical microscopes of high resolution. The fluorescence microscope with its unique capability of distinguishing biomolecules and structures according to the difference in labeling dyes having distinguishable fluorescence properties is a powerful example. It is important to obtain mechanical information associated with the direct manipulation of the biological samples because we still have very little knowledge about their relative, if not absolute, strength that would be the basis of interpretation of the result of manipulation. Mechanical manipulation of living cell will be established as a useful method in cell biology in the near future.

Acknowledgments

This work was supported by Grants-in-Aid for Exploratory Research for RA (19651058) and for Scientific Research (S) (#15101004) and Creative Scientific Research (19GS0418) to Atsushi Ikai from the Japan Society for the Promotion of Science (JSPS).

References

1. Lee, G.U., Chrisey, L.A., and Colton, R.J. (1994) Direct measurement of the forces between complementary strands of DNA. *Science*, **266**, 771–773.

2. Idiris, A., Alam, M.T., and Ikai, A. (2000) Spring mechanics of alpha-helical polypeptide. *Protein Eng.*, **13**, 763–770.
3. Afrin, R., Takahashi, I., Shiga, K., and Ikai, A. (2009) Tensile mechanics of alanine-based helical polypeptide: force spectroscopy versus computer simulations. *Biophys. J.*, **96**, 1105–1114.
4. Mitsui, K., Hara, M., and Ikai, A. (1996) Mechanical unfolding of alpha2-macroglobulin molecules with atomic force microscope. *FEBS Lett.*, **385**, 29–33.
5. Alam, M.T., Yamada, T., Carlsson, U., and Ikai, A. (2002) The importance of being knotted: effects of the C-terminal knot structure on enzymatic and mechanical properties of bovine carbonic anhydrase II. *FEBS Lett.*, **519**, 35–40.
6. Rief, M., Gautel, M., Oesterhelt, F., Fernandez, J.M., and Gaub, H.E. (1997) Reversible unfolding of individual titin immunoglobulin domains by AFM. *Science*, **276**, 1109–1112.
7. Dammer, U., Hegner, M., Anselmetti, D., Wagner, P., Dreier, M., Huber, W., and Guntherodt, H.J. (1996) Specific antigen/antibody interactions measured by force microscopy. *Biophys. J.*, **70**, 2437–2441.
8. Hinterdorfer, P., Baumgartner, W., Gruber, H.J., Schilcher, K., and Schindler, H. (1996) Detection and localization of individual antibody-antigen recognition events by atomic force microscopy. *Proc. Natl. Acad. Sci. U.S.A.*, **93**, 3477–3481.
9. Yan, C., Yersin, A., Afrin, R., Sekiguchi, H., and Ikai, A. (2009) Single molecular dynamic interactions between glycophorin A and lectin as probed by atomic force microscopy. *Biophys. Chem.*, **144**, 72–77.
10. Florin, E.L., Moy, V.T., and Gaub, H.E. (1994) Adhesion forces between individual ligand-receptor pairs. *Science*, **264**, 415–417.
11. Baumgartner, W., Hinterdorfer, P., Ness, W., Raab, A., Vestweber, D., Schindler, H., and Drenckhahn, D. (2000) Cadherin interaction probed by atomic force microscopy. *Proc. Natl. Acad. Sci. U.S.A.*, **97**, 4005–4010.
12. Dufrene, Y.F. and Hinterdorfer, P. (2008) Recent progress in AFM molecular recognition studies. *Pflugers Arch.*, **456**, 237–245.
13. Yersin, A., Osada, T., and Ikai, A. (2008) Exploring transferrin-receptor interactions at the single-molecule level. *Biophys. J.*, **94**, 230–240.
14. Hinterdorfer, P. and Dufrene, Y.F. (2006) Detection and localization of single molecular recognition events using atomic force microscopy. *Nat. Methods*, **3**, 347–355.
15. Xu, X.M. and Ikai, A. (1998) Retrieval and amplification of single-copy genomic DNA from a nanometer region of chromosomes: a new and potential application of atomic force microscopy in genomic research. *Biochem. Biophys. Res. Commun.*, **248**, 744–748.
16. Obataya, I., Nakamura, C., Han, S., Nakamura, N., and Miyake, J. (2005) Nanoscale operation of a living cell using an atomic force microscope with a nanoneedle. *Nano Lett.*, **5**, 27–30.
17. Obataya, I., Nakamura, C., Han, S., Nakamura, N., and Miyake, J. (2005) Mechanical sensing of the penetration of various nanoneedles into a living cell using atomic force microscopy. *Biosens. Bioelectron.*, **20**, 1652–1655.
18. Butt, H.-J., Cappella, B., and Kappl, M. (2005) Force measurements with the atomic force microscope: technique, interpretation and applications. *Surf. Sci. Rep.*, **59**, 1–152.
19. Binnig, G., Quate, C.F., and Gerber, C. (1986) Atomic force microscope. *Phys. Rev. Lett.*, **56**, 930–933.
20. Ikai, A. (2008) *Nano-Biomechanics*, Elsevier, Amsterdam.
21. Dufrene, Y.F. (2008) AFM for nanoscale microbe analysis. *Analyst*, **133**, 297–301.
22. Carrion-Vazquez, M., Oberhauser, A.F., Fisher, T.E, Marszalek, P.E., Li, H., and Fernandez, J.M. (2000) Mechanical design of proteins studied by single-molecule force spectroscopy and protein engineering. *Prog. Biophys. Mol. Biol.*, **74**, 63–91.
23. Bustamante, C., Smith, S.B., Liphardt, J., and Smith, D. (2000) Single-molecule

studies of DNA mechanics. *Curr. Opin. Struct. Biol.*, **10**, 279–285.

24. Afrin, R., Zohora, U.S., Uehara, H., Watanabe-Nakayama, T., and Ikai, A. (2009) Atomic force microscopy for cellular level manipulation: imaging intracellular structures and DNA delivery through a membrane hole. *J. Mol. Recognit.*, **22**, 363–372.
25. Berdyyeva, T., Woodworth, C.D., and Sokolov, I. (2005) Visualization of cytoskeletal elements by the atomic force microscope. *Ultramicroscopy*, **102**, 189–198.
26. Uehara, H., Osada, T., and Ikai, A. (2004) Quantitative measurement of mRNA at different loci within an individual living cell. *Ultramicroscopy*, **100**, 197–201.
27. Uehara, H., Kunitomi, Y., Ikai, A., and Osada, T. (2007) mRNA detection of individual cells with the single cell nanoprobe method compared with in situ hybridization. *J. Nanobiotechnol.*, **5**, 7.
28. Uehara, H., Ikai, A., and Osada, T. (2009) Detection of mRNA in single living cells using AFM nanoprobes. *Methods Mol. Biol.*, **544**, 599–608.
29. Osada, T., Uehara, H., Kim, H., and Ikai, A. (2003) mRNA analysis of single living cells. *J. Nanobiotechnol.*, **1**, 2.
30. Herschman, H.R. (1991) Primary response genes induced by growth factors and tumor promoters. *Annu. Rev. Biochem.*, **60**, 281–319.
31. Han, S.W., Nakamura, C., Kotobuki, N., Obataya, I., Ohgushi, H., Nagamune, T., and Miyake, J. (2008) High-efficiency DNA injection into a single human mesenchymal stem cell using a nanoneedle and atomic force microscopy. *Nanomedicine*, **4**, 215–225.
32. Cuerrier, C.M., Lebel, R., and Grandbois, M. (2007) Single cell transfection using plasmid decorated AFM probes. *Biochem. Biophys. Res. Commun.*, **355**, 632–636.
33. Pellegrin, S. and Mellor, H. (2007) Actin stress fibres. *J. Cell Sci.*, **120**, 3491–3499.
34. Kumar, S., Maxwell, I.Z., Heisterkamp, A., Polte, T.R., Lele, T.P., Salanga, M., Mazur, E., and Ingber, D.E. (2006) Viscoelastic retraction of single living stress fibers and its impact on cell shape, cytoskeletal organization, and extracellular matrix mechanics. *Biophys. J.*, **90**, 3762–3773.
35. Takemasa, T., Sugimoto, K., and Yamashita, K. (1997) Amplitude-dependent stress fiber reorientation in early response to cyclic strain. *Exp. Cell Res.*, **230**, 407–410.
36. Deguchi, S., Ohashi, T., and Sato, M. (2006) Tensile properties of single stress fibers isolated from cultured vascular smooth muscle cells. *J. Biomech.*, **39**, 2603–2610.
37. Silberberg, Y.R., Pelling, A.E., Yakubov, G.E., Crum, W.R., Hawkes, D.J., and Horton, M.A. (2008) Mitochondrial displacements in response to nanomechanical forces. *J. Mol. Recognit.*, **21**, 30–36.
38. Hakari, T., Sekiguchi, H., Osada, T., Kishimoto, K., Afrin, R., Machida, S., Nakayama, T., and Ikai, A. (2011) Displacement of intracellular stress fibers under an externally applied lateral force by AFM. *Cytoskeleton* (in revision).
39. Clear, S.C. and Nealey, P.F. (2001) Lateral force microscopy study of the frictional behavior of self-assembled monolayers of octadecyltrichlorosilane on silicon/silicon dioxide immersed in n-alcohols. *Langmuir*, **17**, 720–732.
40. Munz, M. (2010) Force calibration in lateral force microscopy: a review of the experimental methods. *J. Phys. D: Appl. Phys.*, **43**, 063001.
41. Quintanilla, M.A. and Goddard, D.T. (2008) A calibration method for lateral forces for use with colloidal probe force microscopy cantilevers. *Rev. Sci. Instrum.*, **79**, 023701.
42. Ling, X., Butt, H.-J., and Kappl, M. (2007) Quantitative measurement of friction between single microspheres by friction force microscopy. *Langmuir*, **23**, 8392–8399.
43. Lievonen, J. and Ahlskog, M. (2009) Lateral force microscopy of multiwalled carbon nanotubes. *Ultramicroscopy*, **109**, 825–829.

44. Green, P., Jason, H.L., Cleveland, P., Proksch, R., and Sader, J.E. (2004) Normal and torsional spring constants of atomic force microscope cantilevers. *Rev. Sci. Instrum.*, **75**, 1988–1996.
45. Timoshenko, S.P. and Goodier, J.N. (1970) *Theory of Elasticity*, McGraw Hill Higher Education.
46. Nakahara, I., Tsuchida, E., Tsuji, T., and Shibuya, J. (2001) *Danseigaku Handbook (Handbook of Elasticity)*, Asakura Publisher, Tokyo.
47. Rotsch, C. and Radmacher, M. (2000) Drug-induced changes of cytoskeletal structure and mechanics in fibroblasts: an atomic force microscopy study. *Biophys. J.*, **78**, 520–535.
48. van den Heuvel, M.G. and Dekker, C. (2007) Motor proteins at work for nanotechnology. *Science*, **317**, 333–336.
49. Neuman, K.C. and Nagy, A. (2008) Single-molecule force spectroscopy: optical tweezers, magnetic tweezers and atomic force microscopy. *Nat. Methods*, **5**, 491–505.
50. Wang, Y., Botvinick, E.L., Zhao, Y., Berns, M.W., Usami, S., Tsien, R.Y., and Chien, S. (2005) Visualizing the mechanical activation of Src. *Nature*, **434**, 1040–1045.
51. Na, S., Collin, O., Chowdhury, F., Tay, B., Ouyang, M., Wang, Y., and Wang, N. (2008) Rapid signal transduction in living cells is a unique feature of mechanotransduction. *Proc. Natl. Acad. Sci. U.S.A.*, **105**, 6626–6631.
52. Naruse, K., Yamada, T., and Sokabe, M. (1998) Involvement of SA channels in orienting response of cultured endothelial cells to cyclic stretch. *Am. J. Physiol.*, **274**, H1532–H1538.
53. Yoshigi, M., Clark, E.B., and Yost, H.J. (2003) Quantification of stretch-induced cytoskeletal remodeling in vascular endothelial cells by image processing. *Cytometry A*, **55**, 109–118.
54. Watanabe-Nakayama, T.S., Machida, I., Harada, H., Sekiguchi, R., Afrin, and Ikai, A. (2011) Direct detection of cellular adaptation to local cyclic stretching at the single cell level by atomic force microscopy. *Biophys. J.*, **100**, 564–572.
55. Watanabe-Nakayama, T., Machida, S., Afrin, R., and Ikai, A. (2010) Micro-scoop for manipulation of micro objects: use of fabricated cantilever with atomic force microscope. *Small*, **6**, 2853–2857.
56. Humphries, J.D., Byron, A., and Humphries, M.J. (2006) Integrin ligands at a glance. *J. Cell Sci.*, **119**, 3901–3903.
57. Deakin, N.O. and Turner, C.E. (2008) Paxillin comes of age. *J. Cell Sci.*, **121**, 2435–2444.
58. Afrin, R., Arakawa, H., Osada, T., and Ikai, A. (2003) Extraction of membrane proteins from a living cell surface using the atomic force microscope and covalent crosslinkers. *Cell Biochem. Biophys.*, **39**, 101–117.
59. Singhal, R., Orynbayeva, Z., Kalyana Sundaram, R.V., Niu, J.J., Bhattacharyya, S., Vitol, E.A., Schrlau, M.G., Papazoglou, E.S., Friedman, G., and Gogotsi, Y. (2011) Multifunctional carbon-nanotube cellular endoscopes. *Nat. Nanotechnol.*, **6**, 57–64.
60. Beard, J.D., Burbridge, D.J., Moskalenko, A.V., Dudko, O., Yarova, P.L., Smirnov, S.V., and Gordeev, S.N. (2009) An atomic force microscope nanoscalpel for nanolithography and biological applications. *Nanotechnology*, **20**, 445302.
61. Armini, S., Vakarelski, I.U., Whelan, C.M., Maex, K., and Higashitani, K. (2007) Nanoscale indentation of polymer and composite polymer-silica core-shell submicrometer particles by atomic force microscopy. *Langmuir*, **23**, 2007–2014.
62. Machida, S., Watanabe-Nakayama, T., Harada, I., Afrin, R., Nakayama, T., and Ikai, A. (2010) Direct manipulation of intracellular structure using a hook-shaped AFM probe. *Nanotechnology*, **21**, 385102.
63. Fantner, G.E., Oroudjev, E., Schitter, G., Golde, L.S., Thurner, P., Finch, M.M., Turner, P., Gutsmann, T., Morse, D.E., Hansma, H., and Hansma, P.K. (2006) Sacrificial bonds and hidden length: unraveling molecular mesostructures in tough materials. *Biophys. J.*, **90**, 1411–1418.
64. Ingber, D.E. (2008) Tensegrity-based mechanosensing from macro to micro. *Prog. Biophys. Mol. Biol.*, **97**, 163–179.

65. Yawata, Y. (2003) *Cell Membrane*, Wiley-VCH GmBH & Co. KGaA, Weinheim.
66. Fung, Y.C. (1998) *Biomechanics: Motion, Flow, Stress, and Growth*, Springer-Verlag, Berlin.
67. Boal, D. (2002) *Mechanics of the Cell*, Cambridge University Press, Cambridge.
68. Evans, E.A. (1989) Structure and deformation properties of red blood cells: concepts and quantitative methods. *Methods Enzymol.*, **173**, 3–35.
69. Ethier, C.R. and Simmons, C.A. (2007) *Introductory Biomechanics: From Cells to Organisms*, Cambridge University Press, Cambridge.
70. Afrin, R., Nakaji, M., Sekiguchi, H., Lee, D., Kishimoto, K., and Ikai, A. (2010) Tensile property of delipidated red blood cell cytoskeleton. *Ultramicroscopy*, submitted.
71. Bennett, V. (1982) The molecular basis for membrane – cytoskeleton association in human erythrocytes. *J. Cell Biochem.*, **18**, 49–65.
72. Chasis, J.A., Reid, M.E., Jensen, R.H., and Mohandas, N. (1988) Signal transduction by glycophorin A: role of extracellular and cytoplasmic domains in a modulatable process. *J. Cell Biol.*, **107**, 1351–1357.
73. Nigg, E.A. and Cherry, R.J. (1980) Anchorage of a band 3 population at the erythrocyte cytoplasmic membrane surface: protein rotational diffusion measurements. *Proc. Natl. Acad. Sci. U.S.A.*, **77**, 4702–4706.
74. Nigg, E.A., Bron, C., Girardet, M., and Cherry, R.J. (1980) Band 3-glycophorin A association in erythrocyte membrane demonstrated by combining protein diffusion measurements with antibody-induced cross-linking. *Biochemistry*, **19**, 1887–1893.
75. Afrin, R. and Ikai, A. (2006) Force profiles of protein pulling with or without cytoskeletal links studied by AFM. *Biochem. Biophys. Res. Commun.*, **348**, 238–244.

Index

Atomic Force Microscopy in Liquid: Biological Applications, First Edition.
Edited by Arturo M. Baró and Ronald G. Reifenberger.

e

f

k

l

m